An Introduction to

Statistical Problem Solving in Geography

Fourth Edition

Arthur J. Lembo, Jr.
Salisbury University

J. Chapman McGrew, Jr.
late of Salisbury University

Graphics by John Patrick Soderstrom

WAVELAND PRESS, INC.
Long Grove, Illinois

For information about this book, contact:
Waveland Press, Inc.
4180 IL Route 83, Suite 101
Long Grove, IL 60047-9580
(847) 634-0081
info@waveland.com
www.waveland.com

10-digit ISBN: 1-4786-4946-1
13-digit ISBN: 978-1-4786-4946-5

Printed in the United States of America

7 6 5 4 3 2 1

Contents

Preface

Time marches on, and having published the third edition of *An Introduction to Statistical Problem Solving in Geography* over 7 years ago, it was time to introduce a fourth edition with much needed updates. Sadly, due to the untimely passing of Chapman McGrew, I took this new work on myself. In doing so, I wanted to make sure that all new chapters and updates to the examples would honor Chapman McGrew's vision for creating a readable text in quantitative geography that was rigorous, yet understandable for undergraduates. For this, I followed the original premise that Chapman and I had many years ago:

> What is the most effective way to introduce statistical methods and techniques to undergraduate and beginning graduate geography students?
>
> How can we best demonstrate the usefulness and practical benefits of statistics to geography students in a way that holds their interest and generates enthusiasm?
>
> Can students who believe they are weak in math and statistics achieve a strong fundamental competency in the use of statistical methods?

The positive feedback from the third edition showed that these questions can all be answered successfully if the statistical procedures are presented in the context of solving real-world geography problems—problems that already interest geography students. As much as possible, the focus of the book continues to center on actual geographic problems and issues. Previous editions of this text already pointed strongly in this direction, but in consultation with many users of the book it is clear that "geographic problem-solving" is the best guiding principle and contextual framework through which to present statistical techniques. With feedback from faculty members and students using the previous edition, an emphasis was placed on fixing mistakes that existed in the third edition, updating examples with more recent data, and also introducing more content in the form of three new chapters.

As a result, in this fourth edition you will see continued emphasis placed on the development of descriptive statements and statistical hypotheses that relate directly to contemporary problems in geography.

Chapter 1 illustrates this "geography-first" approach by discussing several contemporary spatial patterns, all presented in the context of the scientific research process. Right from the beginning, students are exposed to critical geographic issues including the distribution of obesity levels across the United States, the global variation of life expectancy levels, the timing of the last spring frost across the southeastern United States, and the recent changes in population patterns for both states and counties in the U.S. These issues continue to appear as geographic variables to illustrate new statistical procedures as they are introduced throughout the text. Additionally, brief conversations about the use of quantitative geography and the use of statistics in response to the COVID-19 pandemic are introduced to illustrate the usefulness of quantitative analysis on what is likely the most impactful event of our lives.

With attention centered on real-world geographic problem-solving using statistics, the text presents more than 60 map examples throughout the following pages. It is worthwhile comparing this text to others that deal with statistics in geography, where maps are virtually nonexistent. Chapman believed that geography students will learn how to use statistics more effectively when presented with real, contemporary problems and maps.

Generally speaking, what must a geography student learn to become skilled in applying statistics to solve real-world problems? While something of an oversimplification, two essential skills should be mastered. First, when presented with a geographic situation or problem to solve, *you must be able to select the correct statistical technique* (or set of techniques) that allow you to approach that situation or problem in the most effective and productive way. Second, when you have results from a statistical analysis (the computer output, for example), *you must fully and properly interpret that output to reach the correct conclusions.* Simply stated, *you must fully understand how to interpret the results of statistical analysis and use that understanding to recommend appropriate geographic policies and plans.*

What began as a fresh coat of paint on what Chapman and I considered a good book almost a decade ago, ended with the third edition being a complete rewrite over a 3-year period. In this fourth edition, the general chapter-by-chapter organization remains relatively unchanged, with updates to most examples to include recent data. In addition to the existing sections (Part 5, 6) devoted to inferential spatial statistics and multivariate techniques of regression analysis, three new chapters are added to include multifactor ANOVA, logistic regression, and spatial estimation. These new chapters are likely beyond the level of an introductory course in quantitative geography, but will serve useful for more advanced class offerings.

In addition to maintaining the successful general organization of the previous editions, the fourth edition keeps several features that characterized the earlier editions. The book continues to stress the importance of written narratives that clearly explain each statistical technique in ways that undergraduate and beginning graduate students in geography can understand. This is accomplished without compromising or oversimplifying the statistical integrity of the material. This new edition also continues to enthusiastically advocate for the exploratory, investigative approach. Therefore, in many parts of the text you will see the calculation of descriptive summaries and graphics used as intermediate steps in a multistage geographic research process.

In some cases, inferential statistics are used as exploratory and descriptive tools to reinforce the basic concept of learning as much as possible about a situation in geography. More specifically, the magnitudes of multiple test statistic values can be compared to gain geographic insights are demonstrated, even though the statistical test is not being used to make formal probabilistic statements.

Also, in several instances, a slightly stronger parametric inferential test (such as ANOVA) is "paired together" with a slightly weaker nonparametric test (such as Kruskal-Wallis). This "pairing" allows a researcher to gather as much useful information as possible about a geographic problem or situation from appropriate statistical techniques. For example, in some cases it may be uncertain if the data fully meet all of the required assumptions needed to run a stronger parametric test. If both a parametric and corresponding nonparametric test are run, one can examine the differences in test statistic results and better evaluate the comparative advantages and disadvantages each test offers.

It is likely that the reader will use a statistical software package to do most of the analysis, rather than making extensive calculations manually. Nevertheless, when presenting many of the statistical techniques, the knowledge and understanding gained through showing the basic steps in the calculation procedure will help the reader better understand the goals and objectives of the technique.

Clearly, too much information is included to cover everything in a single undergraduate course in statistics and geography. I defer to an instructor's best judgment as to the proper pace to follow when using the text in a course. Nonetheless, I encourage instructors to strive for a positive student experience in the hope that an appreciation for the role of statistics in geography will encourage the student to become a lifetime learner in the discipline. One possible strategy for instructors wishing to reduce the volume of material presented in a single course is to consider using certain chapters in more advanced courses. For instance, the inferential spatial statistics in Part V (Chapters 13–15) might be introduced in an undergraduate GIS course. Similarly, the multiple regression and cluster analysis material in Chapter 18 and Part VII (Chapters 20–22) could be presented to students in an advanced statistical applications course in geography.

Acknowledgments

This edition is dedicated to J. Chapman McGrew, Jr. As I collaborated with Chapman over a decade ago to rewrite the third edition, I was inspired by his excitement and passion for teaching quantitative geography to undergraduate students. Chapman was thrilled to see the third edition go to press, but sadly passed away unexpectedly before the book was actually used in courses. I know that he would have thoroughly enjoyed interacting with faculty and students who would use the book, and would eagerly look for ways to improve it. While Chapman always treated me as an equal partner in our endeavor, it was clear to me that I had a lot to learn from him about how to communicate complex technical material to undergraduate students who might be intimidated by the mathematical concepts in a book on quantitative geography. As I take on the responsibility of delivering this fourth edition, my hope is that this book continues the tradition of excellent undergraduate instruction that Chapman began many decades ago.

Many people have helped with the development of the fourth edition of this book, especially as I added three additional chapters on the more advanced techniques.

In this context, certain individuals deserve special recognition. Larence Becker of the Salisbury University Department of Psychology is a kindred spirit with a commitment and skill for teaching statistics to undergraduate students. Larence was instrumental in helping me better communicate the complex ideas of multifactor ANOVA and logistic regression, while Barbara Wainwright of the Salisbury University Department of Mathematics was a great resource on the mathematical concepts of both chapters. When taking on the very complicated chapter of spatial estimation, I turned to my former PhD students Antoni Magri and Steven Lyon for guidance, allowing them to critique my work for a change. I was not disappointed, as their knowledge of inverse distance weighting and kriging far exceeds my own. What a privilege it was to have these former students contribute to my own knowledge on the subject. Brent Skeeter of the Salisbury University Department of Geography and Geosciences gave great feedback on each chapter, providing advice on ways to better communicate the material to undergraduate readers.

The personnel at Waveland Press have been incredibly supportive throughout the entire process. Don Rosso, an Editor and the Production Manager at Waveland, has always been available if we had a question or issue that needed to be addressed, and he has constantly maintained an enthusiastic and positive attitude about our project. Perhaps as important as anything else, he encouraged me to keep improving the manuscript, even if it meant lengthening the amount of time involved. Producing the finest product has always been his stated goal. Dakota West has done an extraordinary job in editing the manuscript and preparing it for distribution. I thank them for their efforts.

I also thank my wife, Karen, and my three children—Emily, Arthur, and Katie—for their support as I made the transition from the private sector to academia. I am honored to be partners with them as we've spent the last 25 years as a family making a difference in the lives of college students. Finally, I would be remiss to overlook Chapman's wife, Kathy McGrew, for her support over the decades as Chapman created one of the finest undergraduate textbooks on quantitative geography. Thank you, Chapman, for being such an inspiration to my career, and a great example of what a quality teacher looks like.

Arthur J. Lembo, Jr.

PART 1

Basic Statistical Concepts in Geography

1

Introduction

The Context of Statistical Techniques

Geography is an integrative spatial science that attempts to explain and predict the spatial distribution and variation of human activity and physical features on earth's surface. Geographers study how and why things differ from place to place, as well as how spatial patterns change through time. Of particular interest to geographers are the relationships between human activities and the environment and the connections between people and places. People have been interested in such geographic concerns for thousands of years. Early Greek writers such as Eratosthenes and Strabo emphasized the earth's physical structure, its human activity patterns, and the relationships between them.

These traditions remain central to the discipline of geography today. Contemporary geography continues to be an exciting discipline that attempts to solve a variety of problems and issues from spatial and ecological perspectives. The spatial perspective focuses on patterns and processes on the earth's surface, and the ecological perspective focuses on the complex web of relationships between living and nonliving elements on the earth's surface (Geography Education Standards Project, 1994).

The geographer starts by asking *where* questions. Where are things located on the earth's surface? Where are features distributed on the physical or cultural landscape? What spatial patterns are observable, and how do phenomena vary from location to location? Historically, geographers focused on trying to answer these questions. In fact, the popular image of the discipline of geography probably remains almost exclusively focused on the location of places.

The recent dramatic increase in the use of Geographic Information Science (GIS) technology and the internet make the assembly of relevant data for traditional geographic analysis even easier. Not only are GIS software packages easier than ever to use, but an almost endless supply of data are now collected and distributed by stakeholders worldwide. Nevertheless, although GIS and the internet provide additional new sets of data and technologies, it is still the responsibility of geographers to evaluate this spatial information properly and to make intelligent decisions. Unfortunately, unprecedented access to geographic data without good quantitative training can often lead to suboptimal (or even poor) decision-making. The accomplished GIS technician therefore needs to properly understand statistical problem solving in geography.

Clearly, professional geographers are no longer content to limit themselves to spatial or locational description. After a spatial pattern is described and the *where* questions are adequately answered, attention shifts to *why* questions. Why does a particular spatial pattern exist? Why does a locational pattern vary in a specific observable way? What spatial or ecological processes have affected a pattern? Why are these processes operating? As such questions are answered, or as speculation occurs about why a spatial pattern has a particular distribution, we gain a better understanding of the processes that create the pattern. Sometimes we find different variables related spatially, thus providing insights into underlying spatial processes. In other instances, geographers try to determine if circumstances differ between locations or regions and seek to understand why such differences exist.

In recent decades, geographers are increasingly concerned with the practical applications and policy implications of this spatial information. Frequently, we ask *what to do* questions that involve the development of spatial policies and plans. More geographers now want to be active participants in both public and private decision-making. Some questions geographers might explore include: What type of policy might best achieve more equal access for urban residents to city services and facilities? What should be done to recommend a balance of wetlands protection and economic development in a fragile environment?

Geography is now a problem-solving discipline, and we are concerned with applying spatial knowledge and understanding to the problems facing the world today. Noted geographer Risa Palm once stated, "[G]eography involves the study of major problems facing humankind such as environmental degradation, unequal distribution of resources and international conflicts. It prepares one to be a good citizen and educated human being" (Association of American Geographers, n.d.).

1.1 The Role of Statistics in Geography

The term **statistics** is generally defined as the collection, classification, presentation, and analysis of numerical data. Statistical techniques and procedures are applied in all fields of academic research. In fact, whenever data are collected and summarized or whenever numerical information is analyzed or scientific research conducted, statistics are needed for sound analysis and interpretation of results.

Beyond the widespread application of statistics in academic research, use of statistics is also seen in many aspects of everyday life. Just consider for a moment this sampling of common applications of statistics:

1. In sports, statistics are frequently cited when measuring both individual and team performance. The success of a team (and the job security of the coach!) is often gauged by the team's winning percentage. Baseball fans are well-versed in such individual player statistics as batting average, earned run average, and slugging percentage. It is not uncommon to see a well-performing pitcher removed during a critical time in the game (to the casual baseball fan, this removal may seem inexplicable) because a left-handed or right-handed batter is coming to the plate. The manager has very likely based his decision to remove the pitcher on a statistical analysis of the pitcher's performance in past similar situations. Goalies in hockey may be benched if their "goals against average" gets too high. Quarterbacks in professional football are evaluated by their quarterback rating—a multifaceted descriptive statistic with a number of components that few fans can understand.
2. Statistical procedures are used routinely in political polling and opinion gathering. When statistical sampling is applied properly, certain characteristics about a statistical population can be inferred based solely on information obtained from the sample. Political opinion polling is commonplace before any major election and exit polling of a sample of voters is used to predict the election results as quickly as possible.
3. Statistics are widely used in market analysis. Businesses constantly scrutinize all aspects of consumer behavior and details of consumer purchasing patterns because a fuller understanding of consumer actions and spending often translates into greater profit. Major business decisions, such as where to advertise and which markets are best to introduce a new product, depend on sample statistics. Many of us are familiar with the ratings systems used to estimate the relative popularity of various television programs. These decisions, involving literally millions of dollars of television advertising and variable advertising rates, rely on well-designed statistical samples of the viewing population.
4. Most of us use a wide variety of statistical measures and procedures in making personal financial decisions. Whether purchasing a home, buying auto or life insurance, trying to develop a workable budget, or setting up an investment or savings plan, an understanding of different monetary terms and statistical measures is indispensable. Knowledgeable reading of stock trends and financial data requires an understanding of various statistical measures. Two common examples are price-earnings ratio (the closing price of stock divided by the company's earnings per share for the latest twelve-month period) and yield (current annual dividend rate divided by the closing price of a stock, expressed as a percentage).
5. Weather is an everyday practical concern for many of us. We want to know which coat to wear in the morning or whether we will need an umbrella. Perhaps a winter storm system is approaching and we want to know the likelihood of getting measurable snow as the system moves through. A farmer may want to know the probability of getting sufficient precipitation over the next two weeks as an aid to scheduling spring planting. Weather forecasting models based on statistics attempt to provide answers to such questions, often using probability estimates from similar historical weather situations.

Geographers use statistics in numerous ways. Statistical analysis benefits geographic investigation by helping answer the *where, why,* and *what-to-do* questions posed in the introductory discussion. Among many general applications, the use of statistics allows the geographer to:

- describe and summarize spatial data,
- make generalizations concerning complex spatial patterns,
- estimate the likelihood or probability of outcomes for an event at a given location,

- use limited geographic data (sample) to make inferences about a larger set of geographic data (population),
- determine if the magnitude or frequency of some phenomenon differs from one location to another, and confirm that an actual spatial pattern matches some expected pattern.

Here are several typical examples:

1. Medical geographers attempt to identify patterns across the landscape to determine if a phenomenon is clustered, random, or dispersed. For instance, a health organization might look for cancer clusters to see whether a pattern of cancer occurrence is concentrated in a specific part of an area. Similarly, law enforcement officials often try to identify patterns of crime within a city, and ecologists try to determine if the distribution of diseased trees in a forest follows a specific pattern.
2. Economic geographers might want to determine if there is a negative correlation between monthly apartment rental prices and the distance from a large college or university. That is, the further the distance from the school, the lower the monthly rental price. This relationship identifies a variation of a basic principle in economic geography called *distance decay.*
3. Natural resource planners may want to confirm that the size of fish in one lake is significantly different than the size of fish in another lake, assuming an experimental growth hormone was introduced in one of the two lakes. Similarly, an agronomist may want to see if different fertilizer treatments on various experimental plots produce different crop yields.
4. Planners may want to determine if average home sale prices in one neighborhood are higher than average home sale prices in another neighborhood. When determining recipients for a community grant, an urban planner might want to know if a particular community has a larger percentage of children under the age of 18 than the state percentage.

The use of statistics must be placed within the context of a general scientific research process. Virtually all geographers recognize the overall importance of statistics in research. Some may view the application of statistical methods as an essential element of any scientific geographic research, whereas others view statistical methods as one of many approaches that geographers can apply. Whatever one's particular perspective, however, the general methodological shift or "revolution" in the geographer's view of the world occurred in the late 1950s and early 1960s. During that time, our discipline began to move from qualitative descriptions of the spatial distribution and variation of human and physical features to quantitative analyses of the same features. The application of quantitative methods (including a variety of statistical techniques) now serves as a fundamental methodological approach or paradigm for geographic research.

The geographic research process and the roles of statistics in that process are summarized in a multiple-step organizational framework (figure 1.1). Very generally, the left column of the figure lists the series of steps that lead to the formulation of inferential hypotheses. The right column diagrams additional possible steps involved in scientific research after an inferential hypothesis is stated. Clearly, formulating hypotheses is central to the research process. In many different ways statistical techniques are often involved in geographic research both before and after hypotheses are generated.

While the sequence of tasks outlined in Figure 1.1 is typical of the process many geographers follow when conducting research, it is not the only mode of geographic research. Moreover, this framework should not be viewed as a rigid series of steps, but rather as a general, flexible guide used in geographic research.

The research process generally begins when we identify a worthwhile geographic problem to investigate. To recognize a productive research problem, the geographer must have background knowledge and experience in the area studied. There is simply no substitute for having a strong background in the appropriate branch of the discipline.

A **hypothesis** is an unproven or unsubstantiated general statement concerning the problem under investigation. Sometimes you have identified a possible problem or research area, but are not yet ready to formulate a hypothesis. Perhaps more information is needed, or preliminary questions related to the problem require an answer. In these cases, one or more **descriptive statements** can be formulated. Geographers often create such statements by collecting location-based data and presenting summaries of this information using graphical procedures, descriptive statistics, and maps. During the descriptive phase of the investigation, statistical analysis can provide quantitative summaries or numerical descriptions of the data, allowing the geographer to explore the data through these methods to identify useful hypotheses. The information gathered allows us to draw conclusions about some research questions, develop a model of the spatial situation, or possibly proceed further and generate **inferential hypotheses** about a population based on the collection of sample data. It is important to note that we must always have sample data if we are to make inferences about a population. In other cases, we may have sufficient background knowledge or information from previous research (such as a review of the literature) to readily develop inferential hypotheses. Often, hypotheses are formulated using a **model**, which is a simplified replication of the real world. A well-known model in geography is the spatial interaction model that predicts the amount of movement expected between two places as a function of their populations and the distance that separates them.

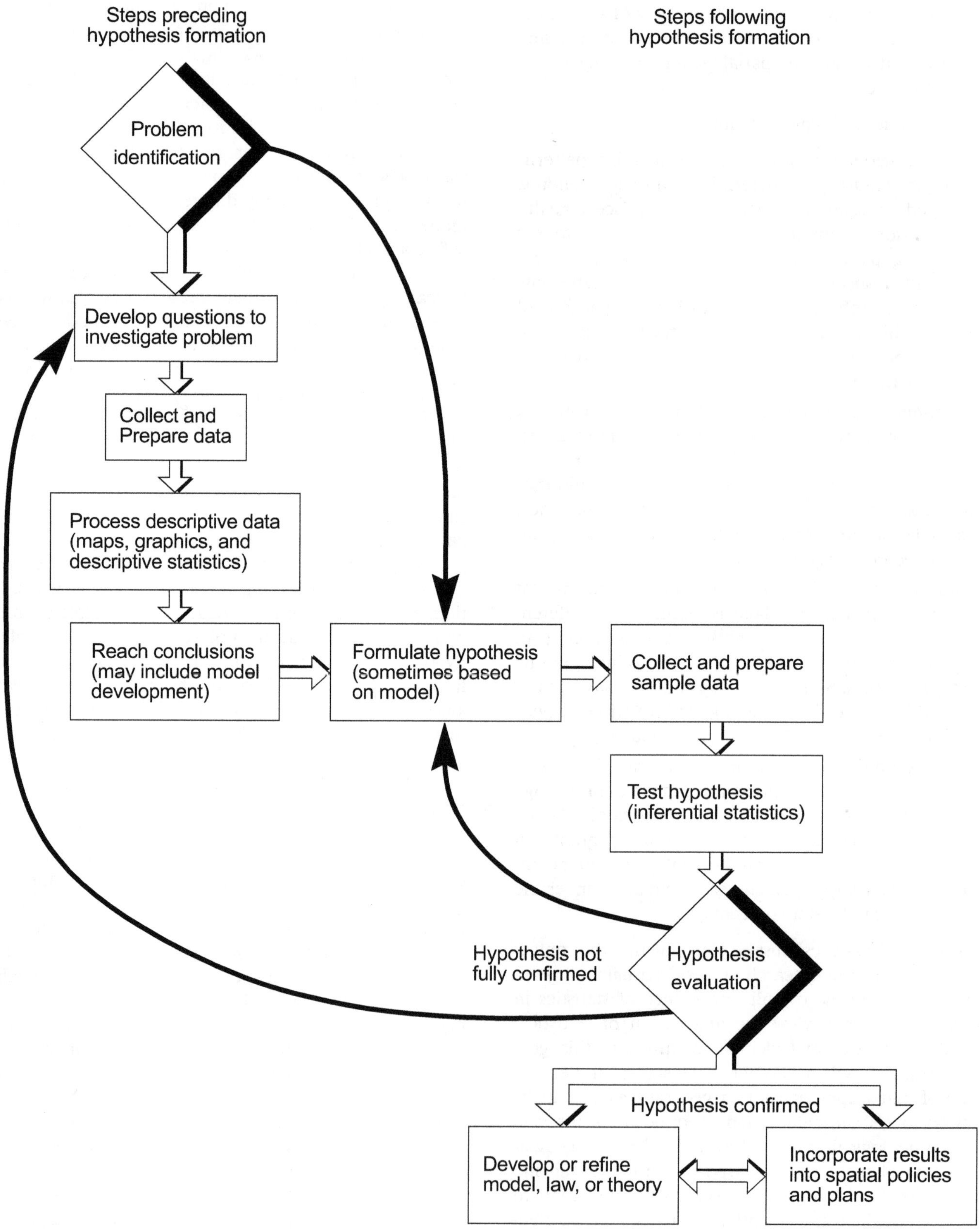

FIGURE 1.1
The Role of Statistics in the Geographic Research Process

Clearly, different research scenarios are possible. In some cases, it may not be realistic or practical to go any further than the collection and descriptive processing of a set of data. For example, we would be limited to descriptive statistics and exploratory techniques if it were impractical or impossible to collect sample data for the problem variable being examined. In other cases, when it is realistic and practical to collect sample data or analyze a set of sample data that has already been collected by someone else, it may be appropriate to formulate inferential hypotheses and the research process then continues down the right column of Figure 1.1. We will examine these possibilities in more detail later.

At this point in the research process, several options are possible:

1. Incorporate the research findings into actual or recommended spatial policies and plans.
2. Further refine the results into a spatial model predicting what is likely to occur under various scenarios. If a hypothesis is repeatedly verified as correct under a variety of circumstances (perhaps at various locations or times) it may gradually take on the stature of a **law**. If various laws are combined, they constitute a **theory**.
3. If a hypothesis is tested and found partially or completely incorrect, we may need to return to an earlier step in the geographic research process. With only a partially validated hypothesis, it might be best for us to conduct further exploratory and descriptive analysis—collecting additional information or refining the original hypothesis. If a hypothesis is proven totally wrong, we might need to return to the initial question development step.

Descriptive statistics provide a concise numerical or quantitative summary of the characteristics of a variable or data set. Descriptive statistics describe—usually with a single number—some important aspect of the data, such as the "center" or the amount of spread or dispersion. For most geographic problems, using such descriptive summary measures is superior to working directly with a large group of values. Descriptive statistics allow us to work efficiently and communicate effectively.

Replacing a set of numbers with a summary measure necessarily involves the loss of information. Various descriptive statistics are available, each with different advantages and limitations. It is important to select the descriptive measure whose characteristics seem appropriate to the geographic problem being analyzed. In many geographic situations, the objective is to minimize the loss of relevant information when moving from individual observations to descriptive measures.

The purpose of **inferential statistics** is to create and test hypotheses about a statistical population based on information obtained from a sample of that population. In the context of inferential statistics, a **statistical population** is the total set of information or data under investigation in a geographic study. A **sample** is a clearly identified subset of the observations in a statistical population. If we are to make proper inferences about the population based on the collection of sample data, then the sampled subset must be representative of the entire statistical population from which it is drawn.

The most basic element in statistics is **data** or numerical information. We often use groups of data, referred to as a **data set** and presented in tabular format (Table 1.1). A data set consists of observations, variables, and data values. The elements or phenomena under study for which information (data) are obtained or assigned are often referred to as **observations**. Observations are sometimes called *individuals* or *cases*. Geographers use many types of observations in their research. Some are spatial locations such as cities or states, and others are nonspatial items such as people or households. A property or characteristic of each observation that can be measured, classified, or counted is called a **variable** because its values vary among the set of observations. The resulting measurement, code, or count of a variable for each observation is a **data value**.

The data set in Table 1.1 is typical—several variables are chosen from a set of observations. Variables such as population in millions and COVID-19 deaths per 1 million people (as of January 2022) are presented in the columns of the table and observations (Canada and the United Kingdom, for example) are displayed in

TABLE 1.1

A Typical Set of Geographic Data

Country	Population in Millions (2020)	Population Per Square Kilometer	Infant Mortality*	COVID-19 Deaths per 1 M (as of January 2022)
Canada	38.2	3	4.5	862
Chile	19.2	25	6.6	2,086
Guatemala	18.2	182	21.3	980
Kenya	54.9	69	37.1	105
United Kingdom	68.2	229	4.3	2,295
Vietnam	98.1	297	17.3	382
United States	332.9	35	5.8	2,633

* Annual number of deaths of infants under age one per thousand live births.
Source: Population Reference Bureau (PRB), World Population Data Sheet, 2020; Johns Hopkins University.

TABLE 1.2

Number of Interprovincial Migrants, Canada, 2016 - 2020

	Province of Destination												
Province of Origin	**AB**	**BC**	**MB**	**NB**	**NF**	**NT**	**NS**	**NU**	**ON**	**PE**	**QB**	**SK**	**YT**
Alberta	—	99,124	9,540	7,520	6,005	2,053	11,142	399	66,794	1,900	11,203	20,996	1,416
British Columbia	68,388	—	6,785	3,343	1,250	943	7,056	278	53,902	1,180	10,692	7,643	1,921
Manitoba	16,461	15,695	—	871	531	295	1,644	216	23,199	141	2,418	5,987	193
New Brunswick	6,081	2,689	735	—	1,079	192	8,050	105	12,464	1,254	5,776	627	85
Newfoundland and Labrador	8,885	2,323	470	1,741	—	276	4,588	257	9,201	526	988	469	78
Northwest Territories	3,010	1,471	217	160	248	—	480	130	1,311	79	274	323	436
Nova Scotia	9,314	5,118	995	6,869	2,354	429	—	245	19,381	1,837	2,912	796	181
Nunavut	331	252	207	189	228	257	412	—	1,353	51	393	118	76
Ontario	54,614	65,869	12,992	14,973	7,880	1,342	25,378	1,037	—	6,578	49,927	9,226	1,238
Prince Edward Island	1,494	1,185	120	1,313	298	39	2,057	37	4,762	—	500	114	27
Quebec	10,133	13,618	1,663	5,617	710	387	3,185	373	71,670	559	—	1,069	368
Saskatchewan	35,110	17,262	5,925	846	387	343	1,329	118	19,112	246	1,722	—	220
Yukon Territory	1,056	1,922	82	105	56	141	196	30	654	28	163	164	—
Total In-migration	214,877	226,528	39,731	43,547	21,026	6,697	65,517	3,225	283,803	14,379	86,968	47,532	6,239
Total Out-migration	238,092	163,381	67,651	39,137	29,802	8,139	50,431	3,867	251,054	11,946	109,352	82,620	4,597
Net Migration	−23,215	63,147	−27,920	4,410	−8,776	−1,442	15,086	−642	32,749	2,433	−22,384	−35,088	1,642

Source: Statistics Canada

the rows of the table. A value or unit of data inside the table represents the magnitude of a single variable for a particular observation. Examples of data values are Canada's 2020 population of 38 million, Kenya's infant mortality rate of 37.1, and Chili's COVID-19 deaths of 2,086 people per 1 million.

An alternative type of geographic **data matrix or array** is shown in Table 1.2. This data set depicts multiple data values for a single variable—the number of Canadian interprovincial migrants from 2016 to 2020. We are often interested in the description and analysis of such spatial interaction (origin-destination) data. It is interesting to note the second largest data value is the number of people moving from British Columbia to Alberta (with 68,388 migrants). This certainly confirms the recent employment opportunities in energy development and shale oil mining. Maybe not too surprisingly, less than 30 people moved from Yukon to Prince Edward Island (or vice versa). The two provinces with the largest positive net migration values are British Columbia and Ontario, and the province with the largest negative net migration was Saskatchewan, while Alberta also saw a large decline. Many who previously migrated to Alberta and Saskatchewan during the oil boom in the early 2000s, left as the energy prices fell in 2014.

1.2 Examples of Statistical Problem Solving in Geography

Four different illustrations of the geographic research process are now introduced. In two of these examples the data of the problem variable encompass an entire population (all country-level data globally or all state-level data in the United States). Because sample data are not utilized in these examples, the statistical analysis later in the text is limited to the development of descriptive statements and summaries. These circumstances apply to the exploration of the global pattern of life expectancy from country to country and to the examination of recent population growth rates for states and counties in the United States.

In the other two example problems, suitable sample data are available making it possible later in the text to create inferential hypotheses and conduct inferential tests as well as explore descriptive statements. Samples are available with obesity rate patterns across the United States and the timing of the last spring frost in a portion of the Southeastern US.

All four of these examples have interesting location-based patterns which are the result of complex spatial processes that are not completely understood. The upcoming figures and graphic displays provide some introductory information about *where* a topic or variable of interest is distributed and how the magnitude of the values seems to vary from place to place. This is often an excellent starting point for developing geographic summary statements and hypotheses. However, the figures do not explain *why* these particular spatial patterns exist—this is where descriptive statements and inferential hypotheses enter the geographic research process. Many of the descriptive statements and hypotheses are explored in later chapters as we consider each of the example problems in greater detail.

Example: Obesity Levels for States of the United States

Obesity is a growing concern in the United States and other developed nations. According to the Centers for Disease Control and Prevention (CDC), American society has become increasingly "obesogenic," characterized by environments that promote greater consumption of less healthy food and discourage physical activity. Obesity is identified as a leading cause of coronary heart disease, certain types of cancer, and type 2 diabetes. Many other health risks (including high total cholesterol, liver and gallbladder disease, and even reproductive and mental health conditions) are also associated with obesity. As concerned citizens, we should recognize that the aforementioned health problems have far reaching societal consequences, including rising healthcare costs and lower worker productivity. The CDC estimates that the overall annual medical costs related to obesity could be as high as $206 billion.

In response to this rapidly growing epidemic, many individuals and organizations are attempting to take prescriptive action. The CDC's Division of Nutrition, Physical Activity, and Obesity (DNPAO) provides funding to states, with the expressed goal to: "Create *policy* [italics added] and environmental changes . . . to address obesity and other chronic diseases through a variety of nutrition and physical activity strategies."

Other policymakers are attempting to address the problem through legislation that is considered overly intrusive by some—banning children's meals at fast food restaurants, placing a surcharge or "fat tax" on certain foods and beverages, or requiring restaurant menus to provide nutritional information on the food they serve. White House initiatives attempted to confront the problem of childhood obesity with programs sponsored by former First Lady Michelle Obama, with the hope that they would increase the amount of exercise children received and better educate families about healthy food choices.

Over the last few decades, the CDC's Behavioral Risk Factor Surveillance System (BRFSS) collected obesity-related data through monthly telephone interviews. The CDC report that "obesity rates for all population groups—regardless of age, sex, race, ethnicity, socioeconomic status, educational level, or geographic region—have increased markedly."

Since 1995, the BRFSS has gathered fully comparable data from all states, and the results show a rapid and frightening trend toward obesity. Obesity is defined using the Body Mass Index (BMI), a measure of an adult's weight in relation to his or her height, specifically by dividing weight in pounds (lbs) by height in inches (in) squared and multiplying by a conversion factor of 703.

Example: Weight = 150, Height = 5′ 5″ (65″)

Calculation: $[150 / (65)^2] \times 703 = 24.96$

Of course, a comparable index is available in the metric system, using kilograms and meters (or centimeters). For adults 20 years old and older, BMI is interpreted similarly for all ages and for both men and women. CDC has created standard weight status categories associated with BMI ranges:

obese: BMI $\geq$ 30.0

overweight: BMI 25.0 to 29.9

neither overweight nor obese: BMI $\leq$ 24.9

Using these three standard weight status categories, we can display the rapid nationwide trend toward obesity. The prevalence and trend data from 1995 to 2020 are shown in both numeric (Table 1.3) and graphic form (Figure 1.2) with **bar** and **pie charts**. A bar chart displays the distribution of a variable whose observations are allocated to categories, while a pie chart shows the entire data set divided into pie-sliced shaped portions of the whole. To emphasize the rapid change over this 25-year period, only the 1995 and 2020 bar and pie charts are shown in Figure 1.2.

Looking at the nationwide obesity trends certainly illustrates the rapid changes in obesity levels in the United States. Geographers, however, are specifically interested in analyzing the geographic patterns and trends of obesity in the United States, not just nationwide trends. By looking at a map showing the percent of adults who are obese from state to state (Figure 1.3), we can determine where the highest and lowest levels of obesity exist, and then further explore why the pattern has the distribution it does.

Only Colorado has a prevalence of adult obesity less than 25% (2020 data). Thirty-six states have obesity levels of at least 30% and eleven of those states have 36% or more of their adult population classified as obese. Looking at Figure 1.3, it appears that the states with the highest obesity levels are located mostly in the mid-South (especially in the Mississippi River Valley) and in Appalachia. The very highest obesity rate (39.7%) is in Mississippi. Conversely, states with the lowest levels of obesity seem to form two distinct clusters—one in the western and Rocky Mountain states and another in the northeastern states, from New Jersey and New York north through much of New England.

How can we now proceed through the scientific research process? More specifically, in what ways can we use statistics to help answer various where and why questions associated with this spatial distribution of obesity levels? It certainly seems that there is a distinct regional patterning associated with obesity, but how can it best be confirmed and how can we use statistics to help approach the question of why the obesity level pattern varies regionally in the way that it does?

The general geographic research procedure (Figure 1.1) recommends starting with the following sequence of steps: develop questions to investigate regarding the obesity problem, collect and prepare data to answer these questions, and

TABLE 1.3

Percent of Total U.S. Population by Weight Status Category: 1995 to 2020

Year	Neither Overweight or Obese	Overweight	Obese
1995	47.9	35.5	15.9
2000	42.9	36.7	20.1
2005	38.5	36.7	24.4
2010	35.3	36.2	27.6
2015	35.4	35.7	28.9
2020	33.3	34.8	31.9

Source: Centers for Disease Control (CDC)

(continued)

then process these data through the use of maps, graphic summaries, and descriptive statistics. The following is an example of a straightforward descriptive question and descriptive statement that seems worthy of investigation:

- Do southern and Appalachian states have higher obesity levels than other states outside these regions? Visually, it appears that the logical answer to this question is yes. We can rephrase this question as a descriptive statement: "Southern and Appalachian states have higher obesity levels than states outside these regions."

Using similar logic, we can create other examples of descriptive statements:

- "States with higher obesity levels have a higher percentage of population that is black (African American)." This statement is correct to the degree that there is a higher percentage of black population in southern states compared to the rest of the country.
- "States with lower income (or education) levels have higher obesity levels." This descriptive statement will be confirmed to the extent that there is a state-level association between low income or education and high obesity rate.
- "States in which a lower percentage of the population is physically active will have higher obesity levels."
- "States in which a higher percentage of the population eats a 'healthy and nutritious' diet will have lower obesity rates."
- "States in which a higher percentage of the population has type 2 diabetes will have higher obesity levels."
- "States having higher obesity levels in 2010 will have larger increases in the percentage of their population that is obese from 2010–2020 than will states having lower obesity levels in 2010." If this descriptive statement is correct, it would suggest that the level of "obesity inequality" is increasing and that the "obesity gap" between high and low obesity rate states is widening, not narrowing.

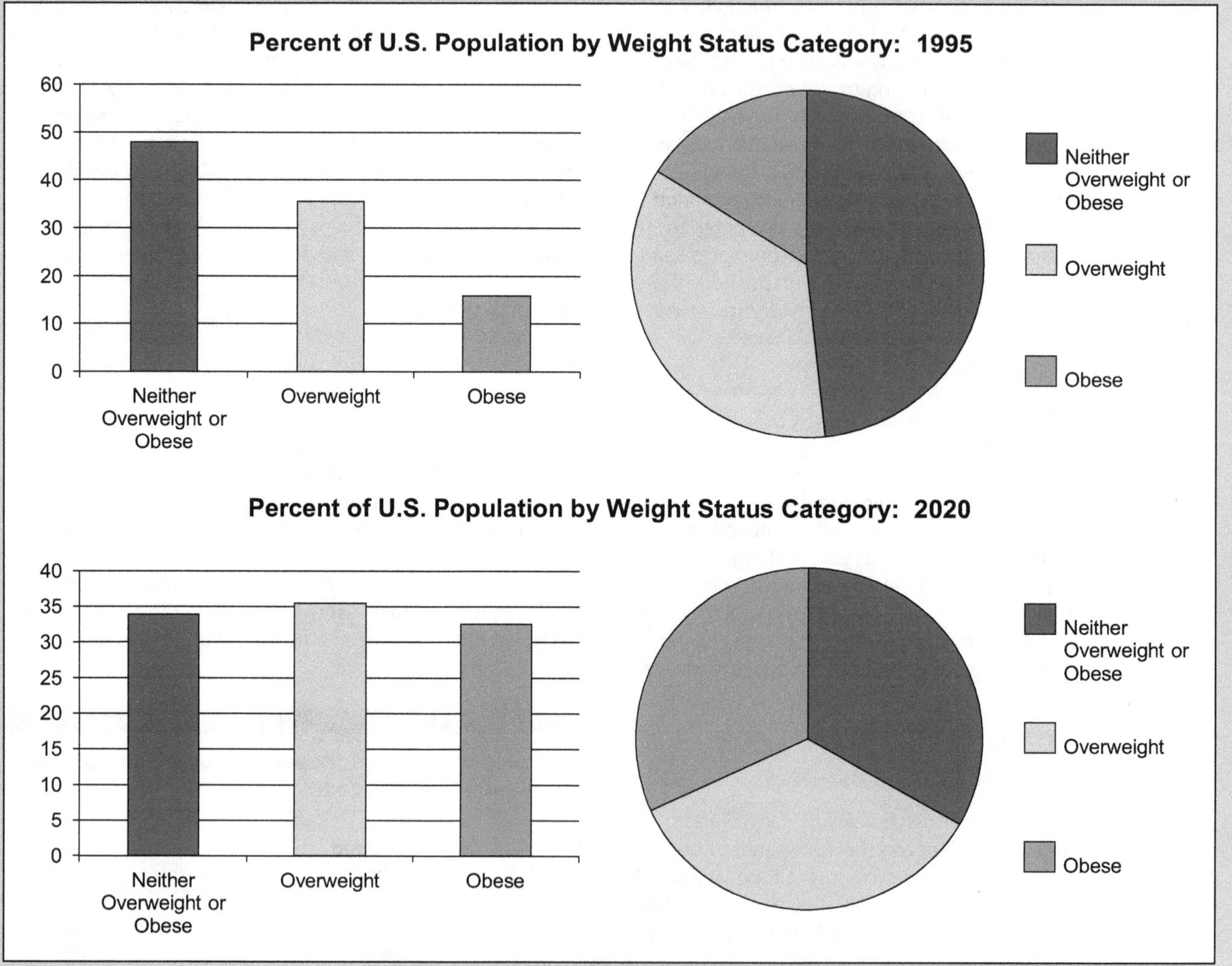

FIGURE 1.2
Bar and Pie Charts of Total U.S. Population by Weight Status Category: 1995 and 2020
Source: Centers for Disease Control and Prevention (CDC)

Descriptive questions and statements such as these can often be definitively answered without too much difficulty, simply because data are already available and are summarized at the state level. This may allow us to make "conclusive" descriptive summary statements, and perhaps develop geographic models to predict obesity levels based on some explanatory factor or variable, such as being physically active or eating a healthy and nutritious diet.

To proceed further in the geographic research process and formulate inferential hypotheses, it is necessary to collect, prepare, and analyze *sample* data. Only then can we use inferential statistics to their ultimate or maximum potential. Keep in mind that the following are just a few representative examples from many inferential hypotheses that one could create using obesity along with some of these other variables. In many cases, it is most appropriate to create inferential hypotheses that are applicable at a disaggregated level, using smaller areas or spatial units such as counties, or using individual observations. Using disaggregated data allows us to take proper samples for subsequent inferential projection to entire populations.

Here are a few examples showing how the use of disaggregated levels might work with obesity patterns and some of the variables taken from the descriptive statement list:

- Take two representative samples of individuals—one sample from the southern and Appalachian states and another from states outside this region. Based on these sample data, we can formulate an inferential hypothesis, such as: "The percent of population that is obese in the southern and Appalachian states is higher than the percentage of population that is obese in other states outside this region." Notice that this first inferential hypothesis is worded in a similar fashion as the first descriptive statement, but the inferential hypothesis is fundamentally different because it is evaluated based on sample data. In this case, we can use an inferential statistical test to determine whether we should reject or not reject (accept) the hypothesis and estimate the probability or likelihood that the correct conclusion was reached.
- Select a random sample of individuals from two states one state having a relatively high percentage of the population that is black (say Alabama) and another state having a relatively low percentage of the population that is black (say South Dakota). Based on this sample data, we can create an inferential hypothesis: "The percentage of Alabama adult population that is obese is higher than the percentage of South Dakota adult population that is obese."
- Select a random sample of counties from each of the four major census regions (Northeast, South, Midwest, and West) and record the county obesity rate for multiple dates (1995, 2002, 2009, and 2020 for example). An inferential hypothesis suitable for testing this sample data is: "There

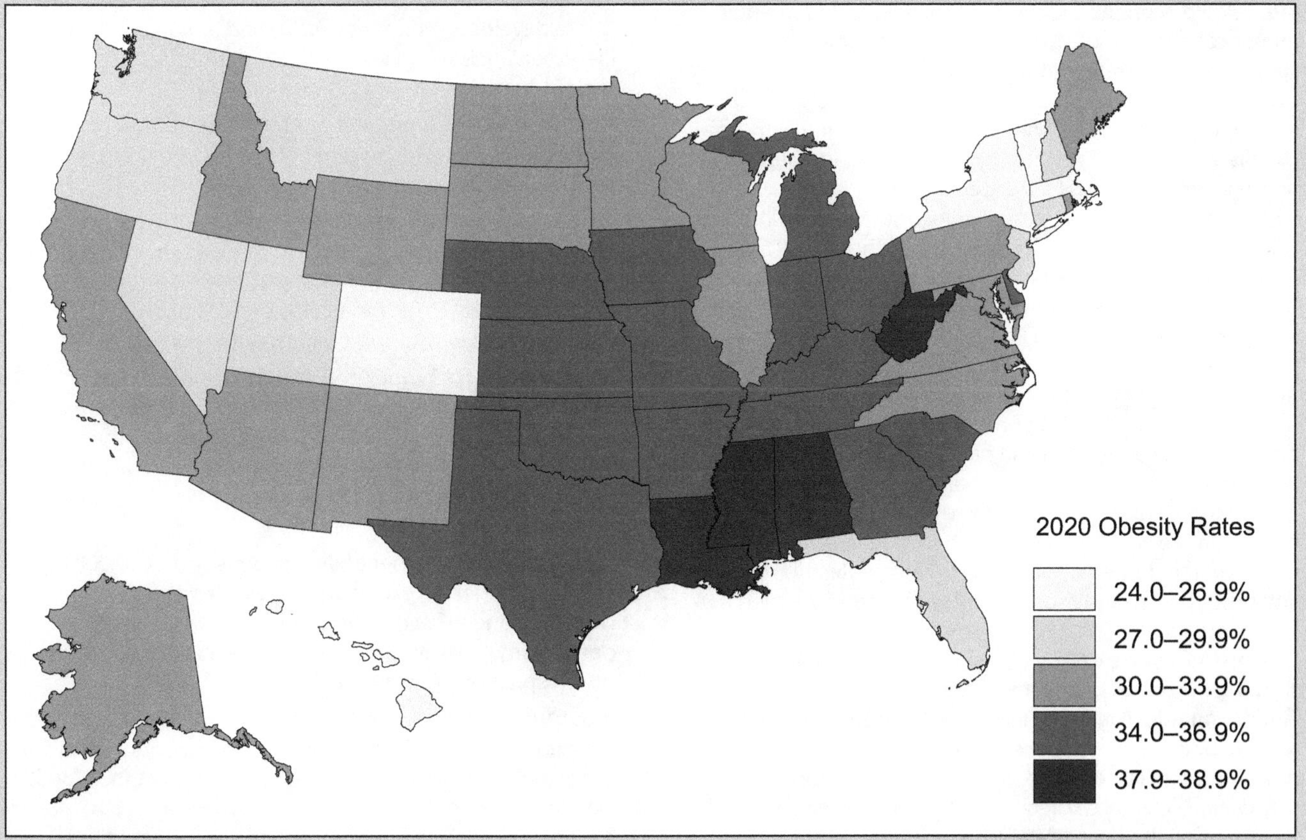

FIGURE 1.3
Percent of Adults Classified as Obese by State, 2020
Source: Centers for Disease Control and Prevention (CDC)

(continued)

are significant differences in obesity rates between the four census regions."

- Select a random sample of individuals from the nationwide population. For each person sampled, indicate whether the person is "physically active" or "physically inactive" (of course, these terms must be defined in a consistent and useful manner) and also whether the person is classified as obese or not. An inferential hypothesis based on this sample data is: "The number of physically active people that are actually obese (observed to be obese) is significantly lower than the number of physically active people that are expected to be obese if no relationship exists between obesity and the level of physical activity."

Of course, unless we actually analyze these descriptive statements and inferential hypotheses, it is difficult or impossible to determine whether any of them are correct or incorrect. Suppose, however, we discover through descriptive analysis that states where a higher percentage of the population eats a "healthy and nutritious diet" generally have lower obesity rates. Further suppose we determine through inferential analysis that a sample of the population from a state having an "healthy and nutritious diet" has a significantly lower obesity rate than a sample of the population from a state having an "unhealthy and less nutritious" diet. These findings would send a clear signal that policies improving diet and nutrition levels should help lower obesity levels in places that currently have poor dietary and nutritional habits. In other words, it now becomes reasonable for us to incorporate spatial policies and plans in those places needing help to lower their obesity levels. Stated in yet another way, we can now prescribe *what to do* at any particular location, and thereby help policymakers in a meaningful way. Without this analysis, policymakers are simply basing decisions on "gut reactions" or "educated guesses." Sadly, many policy decisions are made in exactly this way—without strong scientific rationale. (Several of the descriptive statements and inferential hypotheses just discussed are explored and analyzed in later chapters using various statistical techniques).

Example: Life Expectancy at the Country and World Region Levels

Life expectancy at birth is one of the most frequently used variables to measure the overall health of a country. This is not surprising, as length of life is the cumulative result of so many interrelated economic, social, political, and environmental influences. From cradle to grave, many favorable circumstances must operate for the typical or average citizen of a country to have a long life expectancy. Conversely, it seems logical that many individuals in countries plagued with famine, disease, war, and minimal health services and personnel will have relatively short life expectancies.

Numerous organizations (including the World Health Organization [WHO] of the United Nations and the United States Central Intelligence Agency) collect life expectancy data and monitor changes over time. In addition, many private organizations such as the Kellogg Foundation and the Bill and Melinda Gates Foundation spend billions of dollars in humanitarian relief to alleviate human suffering in countries suffering from low life expectancy. Countless studies have shown the overall spatial distribution and pattern of life expectancy worldwide, confirming exactly where life expectancies are lowest and then trying to understand why they are lowest in those places. On the basis of this information, one hopes to determine what to do, perhaps developing more effective medical and nutritional health care delivery systems and devising better financial policies that help raise life expectancies.

Once again, an excellent place for us to start is with a map showing the overall pattern of life expectancy at birth for the world's countries (Figure 1.4 shows the 2019 pattern). Life expectancy at birth is defined as the average number of years a live newborn infant can expect to live under current mortality levels (Population Reference Bureau—PRB). The life expectancy pattern seems to show considerable regional differences and spatial variation from country to country. Some countries have very low average life expectancies, only in the mid to high-fifties. Other countries have average life expectancies in the low eighties.

Following the general geographic research process, and using the life expectancy map as a starting point, we can generate a series of basic descriptive statements:

- "More developed counties (MDCs) and regions have longer life expectancies than less developed countries (LDCs) and regions."
- "Countries on the continent of Africa have shorter life expectancies than countries on other continents." This observation seems to hold particularly for sub-Saharan Africa.
- "Countries with higher incidence rates of tuberculosis (per 100,000 people per year) have shorter life expectancies." Of course, we could substitute many other cause-specific or infectious diseases for tuberculosis (cholera, HIV/AIDS, malaria, measles, mumps, poliomyelitis, yellow fever, etc.).
- "Countries with a higher percentage of the population using improved drinking water sources have longer life expectancies." Here again, we could substitute other risk factors for good drinking water (improved sanitation, level of alcohol consumption, prevalence of smoking a tobacco product, etc.).
- "Countries with a higher physician density (per 10,000 people) have longer life expectancies." Other measures of health workforce and infrastructure could be substituted for physician density (environmental and public health worker density, hospital bed density, etc.).
- "Countries with lower adult literacy rates have shorter life expectancies." We could substitute other demographic and social measures indirectly related to life expectancy for literacy rate (percent population living in urban areas, total fertility rate per woman, net percentage primary school enrollment rate, gross national income per capita measured in Purchasing Power Parity [PPP] in international dollars, etc.).

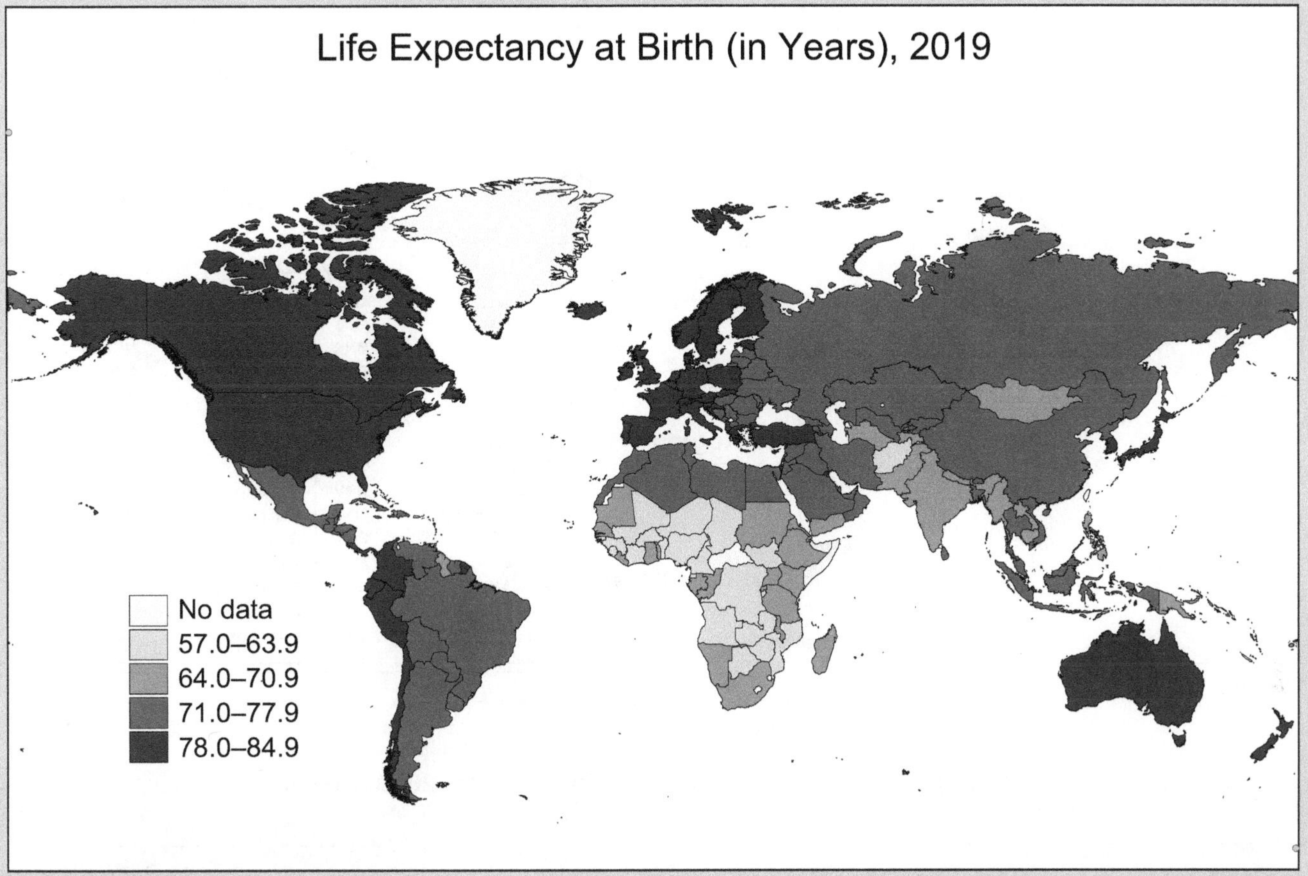

FIGURE 1.4
Life Expectancy at Birth (in Years), 2019
Source: World Health Organization (WHO), World Health Statistics, 2020

- "Countries with a higher per capita total expenditure on health (expressed in PPP—international dollars) have longer life expectancies." In economics, PPP determines how much money is needed to purchase the same goods and services in two countries, which is then used to calculate an implicit foreign exchange rate. Using that PPP rate, an amount of money therefore has the same purchasing power in different countries. The actual nature of this relationship, with countries identified by world region, is shown in Figure 1.5. In this figure, the assignment of countries to world region follows the system used by the Population Reference Bureau and the source of data is *World Health Statistics 2020*, published annually by the WHO.

Figure 1.5 is an example of a **scatterplot**, a graphic technique that shows the relationship between two quantitative variables. On a typical scatterplot, every observation has a value on both variables and is plotted on a case-by-case basis. The scatterplot shown in Figure 1.5 indicates there may be a rather intriguing and complex relationship between these two variables. The appearance of three distinct stages as per capita health expenditures increase may suggest a "life expectancy transition" to population geographers.

1. Countries in the "Initial Stage" (Stage 1) are spending little money on health. Per capita total expenditure on health is generally less than $300 annually (PPP—international dollars) and average life expectancies are quite low, ranging in years from the low 50s to the upper 60s. The countries with the very lowest life expectancies (below 60) are virtually all in sub-Saharan Africa. Somewhat better are several of the poorest Asian and Latin American nations (Bhutan, Haiti, Laos, Nepal, Turkmenistan) with life expectancies in the low- to mid-60s but still with annual per capita total health expenditures below $300. Somewhat surprisingly, perhaps, there is quite a range of life expectancy values among these countries, even though the amount of money spent on health care per person is about the same.
2. Countries in the "Transition Stage" (Stage 2) are experiencing concurrent or simultaneous changes in both life expectancy and per capita total expenditure on health. All of the countries spending between $300 and $1,500 per person annually seem to have life expectancies ranging from the low 70s to the upper 70s. Several world regions and subregions are well represented in this group, including countries from North Africa, Latin American and the Caribbean, Asia, and Eastern Europe and the Russian Federation. Although a lot of "scatter" appears among the countries in this stage, a rough general relationship seems to exist. Countries with lower total expenditures on health per capita (in the $300 to $700 range) generally have life expectan-

(continued)

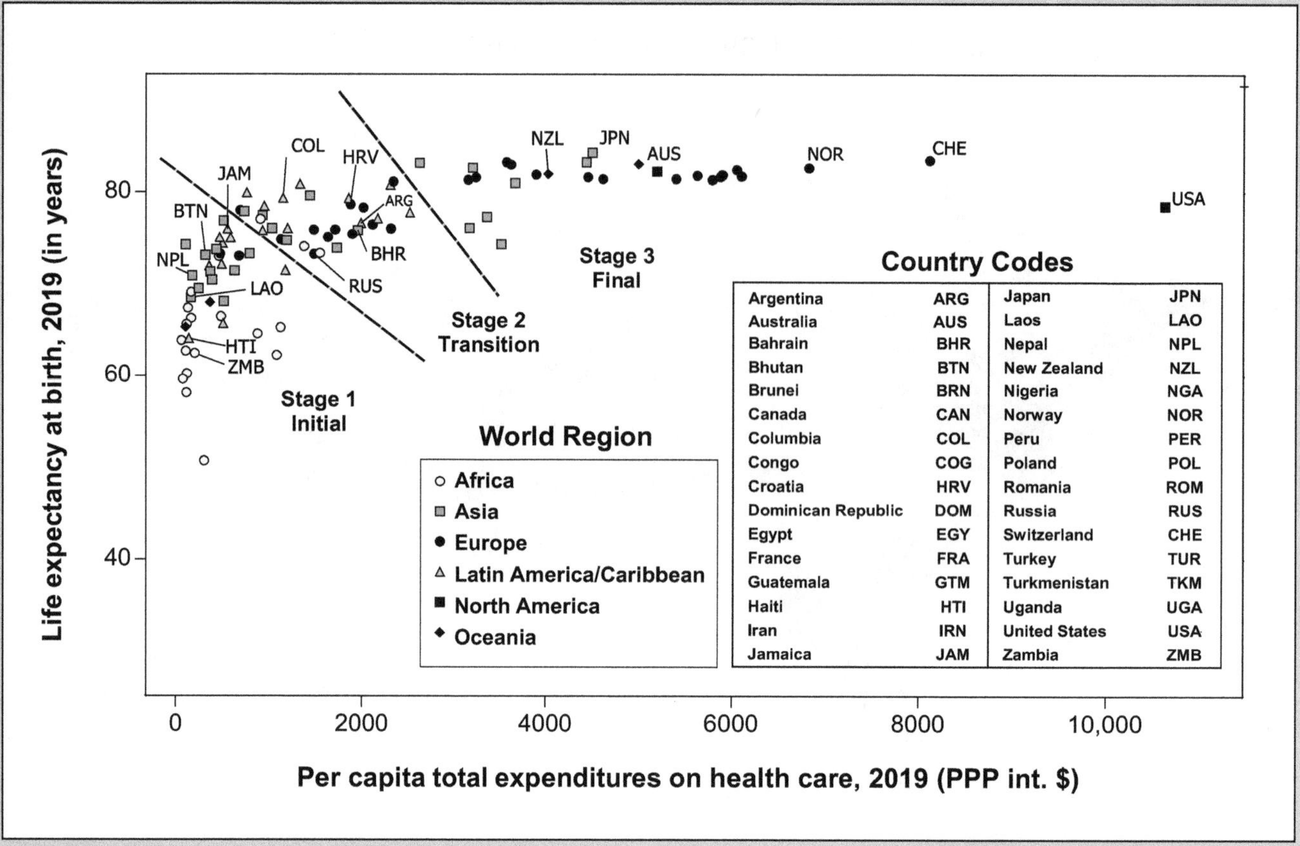

FIGURE 1.5

Scatterplot Showing Relationship Between Life Expectancy at Birth and Total Health Care Expenditures Per Capita, Selected Countries, 2019 (Note: Regional Assignment of Countries Based on Population Reference Bureau)

Source: World Health Organization (WHO), World Health Statistics, 2020

cies in the low 70s (e.g., Bolivia—life expectancy 72, health expenditures per capita $496; Moldova—LE 73, $479). Countries with somewhat higher expenditures on health per capita (in the $1,100 to $2,500 range) mostly have slightly longer life expectancies in the mid- to upper 70s (e.g., Colombia—LE 79, $1,155; Romania—LE 75, $1,555).

3. Countries in the "Final Stage" (Stage 3) are all classified as "more developed" by the United Nations. This category includes all of Western Europe, the United States, Canada, Japan, Australia, and New Zealand. In these countries, life expectancies are in the very high 70s (78 or 79) or very low 80s (80 to 83). However, there seems to be little or no relationship between life expectancy and total healthcare expenditures per capita. For countries that spend more than $2,500 per capita, life expectancies differ by only a very small amount. Beyond this $2,500 threshold, it does not seem to matter how much money is spent on health care. By far, the largest total expenditure on health care per capita is in the United States, with $10,623, yet life expectancy is only 78 years. Contrast US performance with Japan (the country with the world's longest life expectancy at 84 years, but a total healthcare expenditure of only $4,503) or any other more developed country (e.g., Australia—LE 83, $5,004; Canada—LE 82, $5,199; or France—LE 82, $5,250).

As a curious geographer, a number of interesting inferential hypotheses could be created to examine the precise nature of this relationship between life expectancy and per capita total health care expenditures. To what extent are there differing cultural or social values between countries with regard to spending a lot of additional money to increase life expectancy only very slightly? If samples of elderly seniors were followed in different MDCs, would the United States healthcare delivery system spend more money per person to support very expensive end-of-life care? By contrast, do other MDCs have more triage-based decision-making policies that discourage or not allow certain expensive end-of-life medical procedures? Is there a spatial (regional) pattern for countries having lower life expectancies than predicted given their total health expenditure levels?

To proceed further with the general research process for the life expectancy problem will require the availability of comparable sample data from country to country at the global level. Unfortunately, there does not seem to be directly comparable, sample-based life expectancy figures for many of the very poor less developed counties. Certainly there are life expectancy estimates for every country, and excellent descriptive statements can be made about the spatial pattern of these estimates, however, the actual "in-the-field" sampling procedures used to get these estimates vary significantly from country to country. For example, sample sizes and sample proportions will almost certainly vary from country to country. Infield data collection procedures will necessarily be quite different from one country to another, especially in poorer LDCs. A multiplicity of economic, political, and environmental circumstances will likely make it literally impossible to collect comparable life expectancy sample data for all of the world's countries. We may also have no idea whatsoever about the amount of error introduced in samples of life expectancy estimates from some countries. Even if background information regarding sample characteristics exists, reports or documents may not provide this information (this limitation seems to be the case in the WHO annual World Health Statistics reports). All of these problems and issues make the use of inferential test statistics a risky proposition with uncertain results and questionable conclusions. Therefore, no sample data dealing with life expectancy are analyzed in this text.

However, *if* appropriate sample data were available, it would certainly be possible for us to create any number of inferential hypotheses:

- Take two representative samples of individuals—one sample from MDCs and another sample from LDCs. Based on these sample data, test the inferential hypothesis: "The life expectancy of those individuals sampled from MDCs is longer than the life expectancy of those individuals sampled from LDCs." Analysis of this hypothesis leads to a statistical conclusion that the life expectancy for the entire population of all MDCs is longer than the life expectancy for the entire population of all LDCs.
- Select a random sample of individuals from two countries one sample from an MDC (e.g., Norway) and another sample from an LDC (e.g., Egypt). Based on these sample data, the following inferential hypothesis is reasonable: "The average life expectancy of those individuals sampled from Norway is longer than the average life expectancy of those individuals sampled from Egypt." If a significantly longer life expectancy is confirmed for Norway than Egypt, using an inferential statistical test on the samples from these two countries, then it is reasonable to infer that life expectancy is longer for the entire population of Norway than for the entire population of Egypt.
- Select a random sample of countries whose total annual expenditure on health per capita is high (say $2,500 or more per capita) and another random sample of countries whose annual total expenditure on health per capita is low (say $500 or less). Based on these two samples, a possible inferential hypothesis is: "Countries with a high total annual expenditure on health per capita have longer life expectancies than countries with a low total annual expenditure on health per capita."

Example: Timing of the Last Spring Frost in the Southeast United States

The length of the growing season in any agricultural area is of critical importance in determining the spatial patterns of different agricultural crops. Systematic changes and natural variations from year to year in growing season length will have profound effects on the management of agricultural systems. Especially crucial is the timing of the last spring frost, for that date strongly influences the timing of all subsequent management activities, from planting to harvesting. In moderate climates with distinct seasonal changes (such as the Southeastern United States), a later-than-normal spring frost can have numerous adverse socioeconomic impacts on both vegetable crops and citrus.

Given the potential impact of the last spring freezes on the agricultural industry across the Southeast United States, it is not surprising that a great deal of research is conducted over a wide variety of crops to determine the spatial patterns and temporal trends of the growing season in this area. One recent study is unique because it examines the characteristics of severe late spring frost events in new, innovative ways. This study (along with more recent updates) collected late spring frost data from a sample of weather stations distributed throughout a large portion of the Southeast US from 1950 to 2010. The author has graciously allowed the use of some of these sample data, which provides us with another practical example of the geographic research process.

The spatial pattern of average last spring frost dates is shown (Figure 1.6) for a selected sample of 76 weather stations distributed across a large portion of the Southeastern United States. From this map pattern a set of research hypotheses can be proposed. Since the last spring frost data represent a sample of weather stations in this region, these hypotheses are inferential in nature. This will allow statistical prediction and estimation of last spring frost dates for any location within this study region, not just the original sample of weather stations.

At the general regional level, the following inferential hypotheses are worth considering:

- "Weather stations with more northerly latitudes (further from the equator) have average last spring frost dates later in the spring than weather stations with less northerly latitudes (closer to the equator)."
- "Weather stations with higher elevations above sea level have average last spring frost dates later in the spring than weather stations at lower elevations."
- "Weather stations further from the coast (the Atlantic Ocean and Gulf of Mexico) have average last spring frost dates later in the spring than weather stations closer to the coast."

These three inferential hypotheses suggest that latitude, elevation, and distance to the coast are all potential influences on the timing of the last spring frost date. In Chapter 6 we return to this topic and look at the effectiveness of a regional probability map showing the likelihood of a late spring frost occurring on or after April 1. Then, in the final part of the book (Chapters 16 and 17), we examine the spatial relationships between these three "explanatory variables" and last spring frost and explore the effectiveness of some predictive models.

(continued)

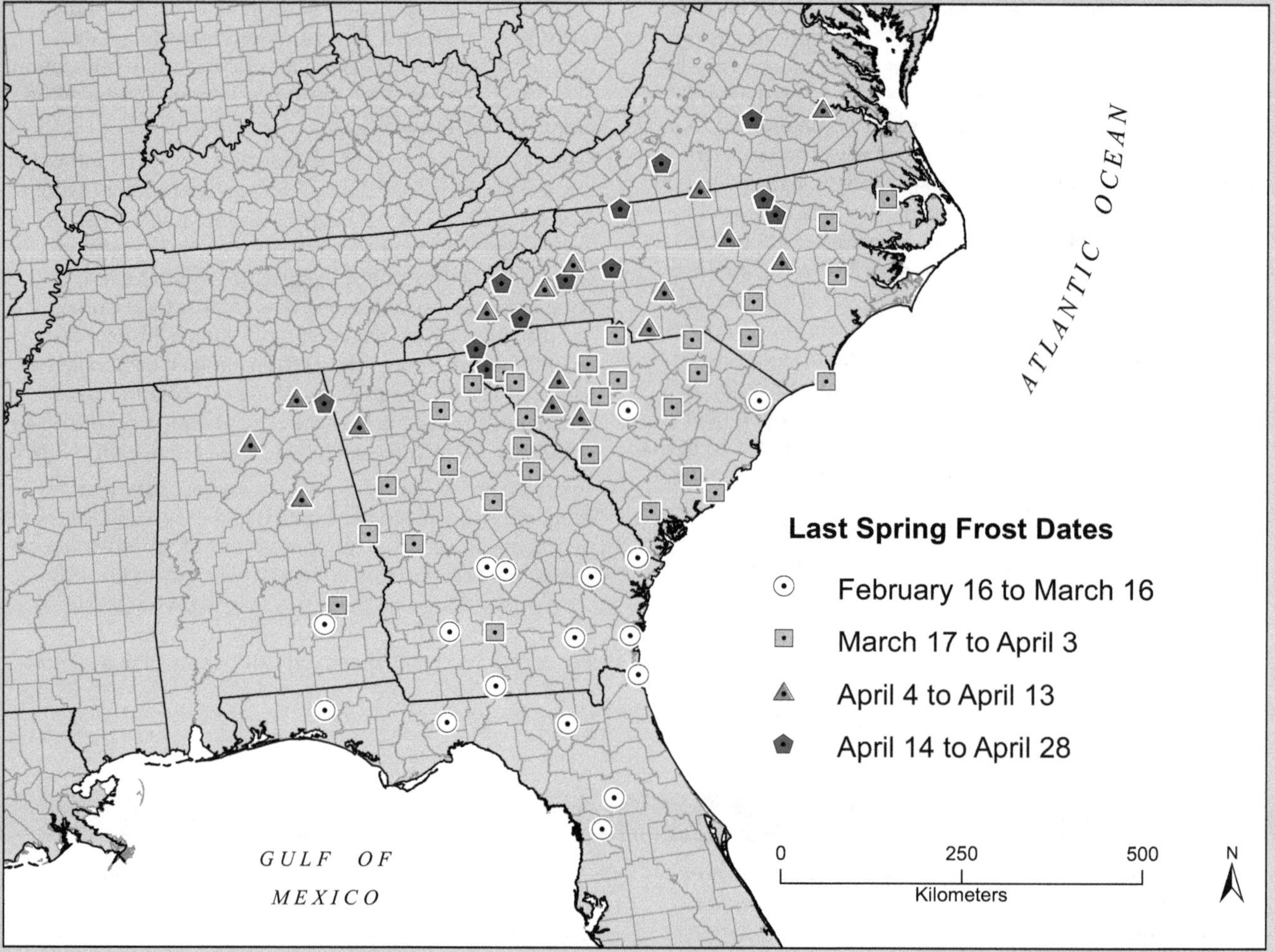

FIGURE 1.6
Average Date of Last Spring Frost, Selected Weather Stations in Southeast United States, 1950 to 2010
Source: Parnell, 2013

Example: Patterns of Population Change in the Contiguous United States

Suppose we are interested in analyzing state-level growth trends in the United States over the last few decades (Figure 1.7). The fastest growing states over a recent 30-year period (1990–2020) appear to be generally located in the South and West, with consistency from decade to decade. Conversely, many of the states which have experienced slow growth (less than 13%) or even population loss (as is the case with a few states during the period from 2000 to 2010) are located in the Northeast and Midwest. However, further examination of the maps reveals exceptions to this rule. For example, during the 1980s Wyoming lost population while New Hampshire witnessed relatively high growth (above 15%). In addition, the growth rates for many states vary quite a bit from decade to decade, indicating possible changes in the pattern of economic growth and employment opportunities or changes in people's preferences for different places at different times.

Given these maps of state-level population growth, we might ask why these spatial patterns exist. Why do these growth rate patterns vary in the ways that they do? What factors can we suggest to help explain the nature of these spatial distributions? What spatial processes might have been operating and what policies (if any) might alleviate any problems caused by excessive population growth or excessive population loss?

As with previous geographic examples in this chapter, a set of descriptive statements is developed:

- "Southern and western states with warmer climates have higher population growth rates than northeastern and midwestern states having colder climates." Many people feel that climatic amenities such as a warmer winter and associated environmental and recreational considerations influence the spatial pattern of population change in the United

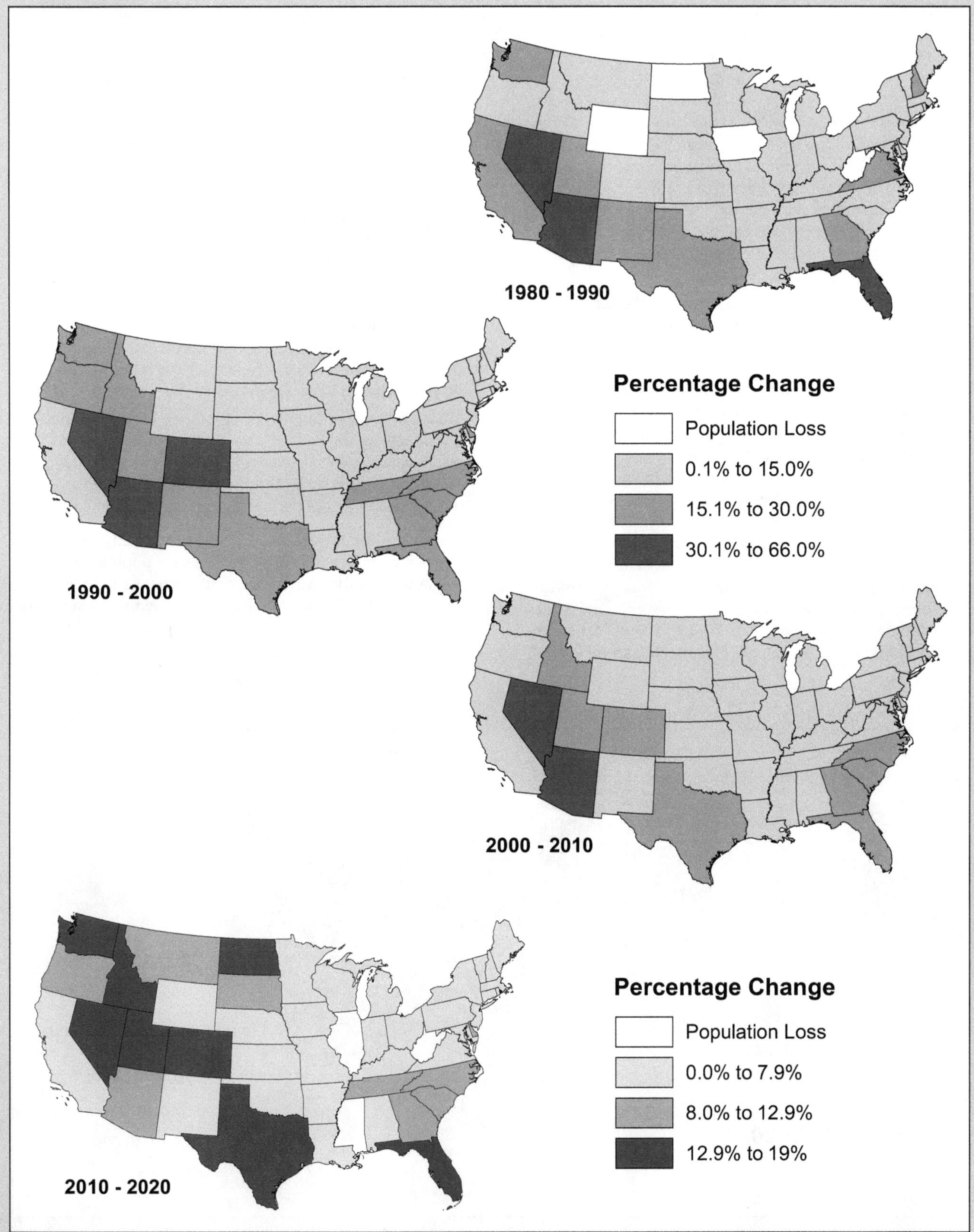

FIGURE 1.7
Percent Population Change by State, 1980 to 2020
Source: United States Bureau of the Census

(continued)

States. Some states in the South and West—including Florida, Texas, Arizona, and California—are considered to be in the "sunbelt" because of their warm and sunny climates. Conversely, the northeastern and midwestern states that are either losing population or experiencing slow growth are sometimes classified as being in the "Snowbelt" or "Frost Belt." The implication is that people are fleeing the cold winters and snow for year-round warm weather and outdoor recreation opportunities. We have a problem, however: how are we going to operationally define "warmer climate" for a state? After all, a state is an area, not a point like a weather station, and some states have wildly variable climates within their borders.

- "States with healthier and stronger economies will have higher population growth rates than states whose economies are weaker and less vibrant." Economic factors undoubtedly influence growth in a number of ways. A key task, of course, will be to define what is meant by "healthier" or "stronger" economy, so that it can be measured and analyzed. In surveys examining why people change residence, respondents frequently cite job opportunities and related economic reasons. Over recent decades, many more new jobs were created in the southern and western states than in the Northeast and Midwest. The region containing the states of Pennsylvania, Ohio, Michigan, Indiana, and Illinois is sometimes called the "Rust Belt" because of its traditional economic reliance on heavy industry and manufacturing sectors of the US economy that have been suffering a relative decline for several decades. There is some recent compelling evidence, however, that this spatial pattern is changing again.
- "States attracting larger numbers of immigrants will have higher population growth rates than states that attract few immigrants." Many migrants from Asia and Latin America have recently settled in California, Florida, and Texas (among other places), contributing to higher growth rates in these states. However, many immigrants have also settled in Illinois and New York—states that have experienced little recent change in population size.
- "States with smaller populations and lower population densities have higher growth rates than states with larger populations and higher population densities." Geographers studying recent migration trends sometimes suggest that low-density residential areas such as rural regions and small towns are increasingly attractive residential destinations for many Americans. To the extent that this trend toward nonmetropolitan growth or decentralization of the population is true, states with smaller populations and lower population densities should have higher growth rates.
- "States with ocean-oriented, water-based recreation and economic activity have higher population growth rates than states lacking these amenities." To the degree this is the case, rapidly growing coastal (peripheral) communities should contrast with stagnant or declining interior (heartland) locations.

A decidedly different set of descriptive statements might emerge if we look at a county-level map of population change, such as the map of percentage change from 2010–2020 (Figure 1.8). The spatial pattern of growth now seems to be much more precise and complicated than at the state level, and this allows us to speculate specifically about what spatial processes might be operating that influence growth patterns.

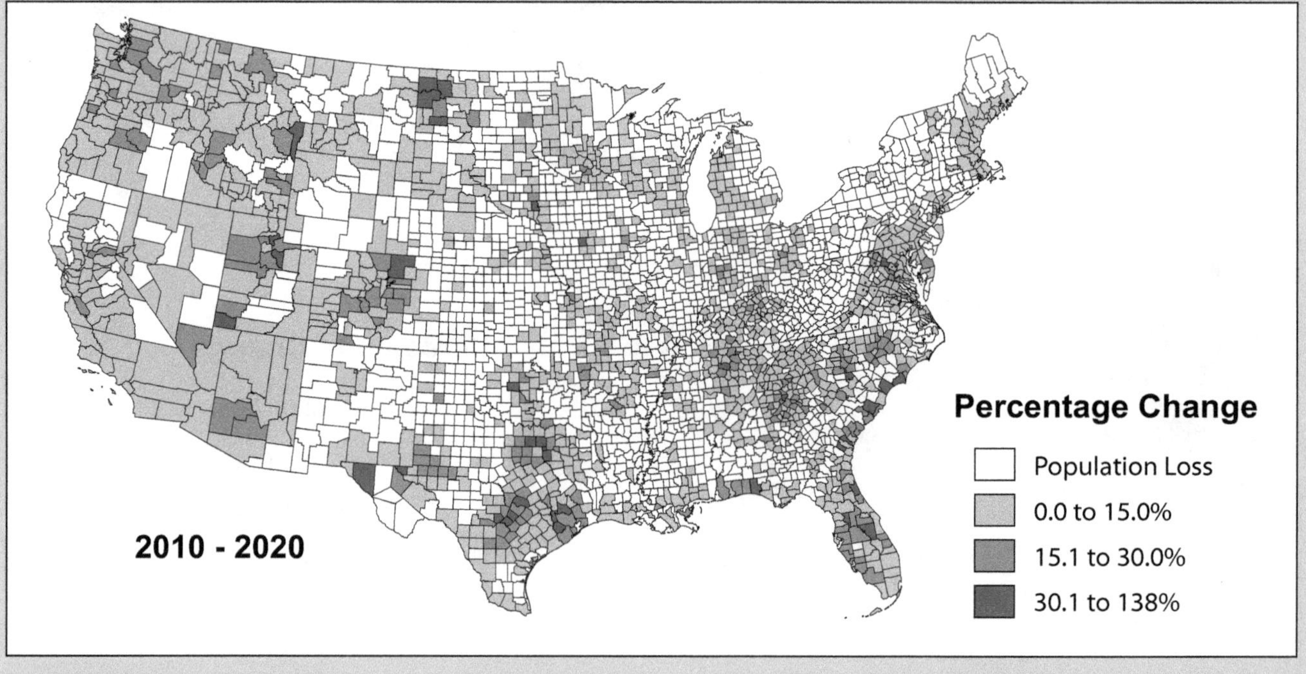

FIGURE 1.8
Percent Population Change by County, 2010 to 2020
Source: United States Bureau of the Census

For example, visually contrast the growth rate patterns of New Mexico and Colorado. The map indicates that the many counties in New Mexico gained population at about the same rate from 2010–2020, with relatively little difference from county to county. In Colorado, however, you can see considerable variation in growth rates during the decade from county to county. What spatial processes related to growth are operating to cause only small differences in growth rate among New Mexico counties but much larger differences among Colorado counties? Perhaps you can create a descriptive statement suggesting some possible reason or reasons. Finally, those who have used the Third Edition of this textbook that showed population change from 2000–2010 will notice a dramatic increase in the number of counties that have lost population from 2010–2020. Also, the maximum increase in population has grown from 109% in 2010 to 138% in 2020. It is very possible that a few counties in the US are seeing significant growth in the last decade, at the expense of many smaller, rural counties losing population.

As stated when introducing these four geographic examples at the beginning of this section, we are going to explore many of the descriptive statements and inferential hypotheses in later chapters. Each of these topics and problem settings (obesity, life expectancy, last spring frost, and population change) deserves further analysis, and they all lend themselves to various types of statistical procedures as we continue.

COVID-19 Conversation: The Context of Statistics During the COVID-19 Pandemic

Perhaps no other event in our lifetimes has had a greater impact than the COVID-19 pandemic. Coronavirus disease (COVID-19) is an infectious disease caused by the SARS-CoV-2 virus. It is part of the coronavirus family, which includes common viruses that cause a variety of diseases from head or chest colds to more severe diseases like acute respiratory syndrome (SARS) and Middle East respiratory syndrome (MERS). The impact of COVID-19 on the world was staggering. With over 670 million cases, and 6.8 million deaths (as of March 2023), COVID-19 shut down economies, governments, and led to international debates regarding policies to mitigate the impact it had on the world.

The World Heath Organization (WHO) first learned of COVID-19 in December 2019, following a report of a cluster of cases of "viral pneumonia" in Wuhan, People's Republic of China. And while the worldwide medical establishment joined together to understand the virus, spatial scientists were also early contributors to understanding the spread of the virus. In fact, by January 2020, the now-famous Johns Hopkins COVID-19 dashboard was already monitoring the spread of COVID-19 (Figure 1.9). Within months, the geospatial community created dashboards for individual states and counties to better understand the impact of COVID-19 in local geographic areas. Clearly, understanding the spread of

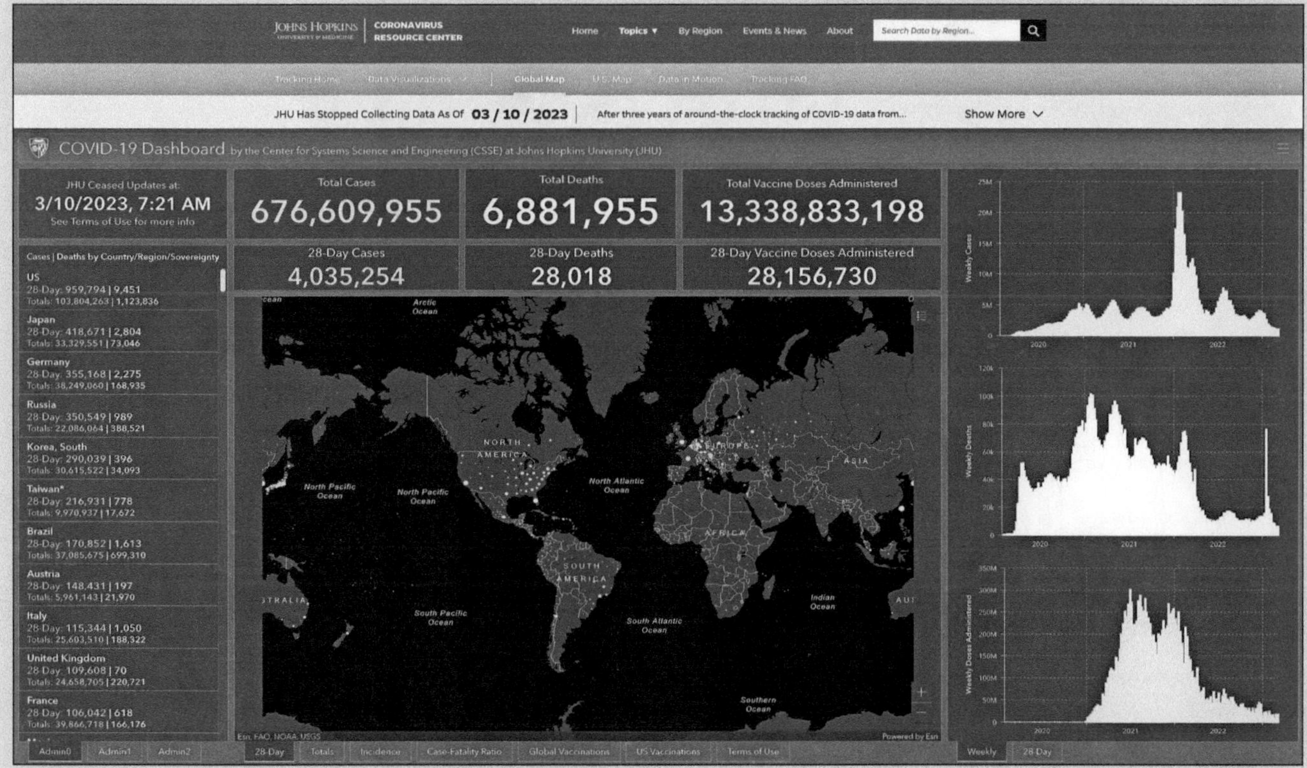

Figure 1.9
Data on the Johns Hopkins COVID-19 Dashboard
Source: Used with permission from Johns Hopkins University and Medicine

(continued)

infectious disease and developing sound policy is aided by properly analyzing the disease over time and space. Since we first discovered the existence of COVID-19, geographers and spatial scientists have played an important role in understanding its spread, communicating its impact on our lives, and shaping policy decisions for mitigating its impact.

While we are still in the midst of battling the COVID-19 pandemic, the situation has become much less severe, with hopes that it will soon change to a less deadly endemic. Yet, there are important lessons we can learn that relate to the use of statistics in understanding when analyzing any world-altering crisis.

Figure 1.10 shows the county-level deaths for COVID-19 in 2020 and 2021 per 100,000 people.

While the local news often showed raw numbers of deaths, causing largely populated states to stand out, normalizing the data to show deaths per 100,000 people provides us with a more refined view of the distribution. While many factors contributed to COVID-19 deaths, Figure 1.10 does appear to exhibit a pattern of where large portions of those deaths occurred, and may potentially inform us regarding hypothesis formulation. We can even combine these observations with the other data sets we explored, for example:

- Counties located in states that had lower obesity rates appear to have less deaths per 100,000. Could we quantitatively find a relationship between the obesity level in a county and the rate of COVID-19 deaths?
- Southern and Midwestern counties appear to have higher death rates than the Northeast and West. We can explore if there is a statistical difference in death rates by census region.
- Urban areas appear to have higher death rates than rural areas. Perhaps there is a statistical difference among metropolitan and nonmetropolitan areas as it relates to COVID-19 death rates.

One thing we may discover is that while our eyes help us develop interesting hypotheses, they may also deceive us in assessing whether a true statistical relationship exists. By actually collecting data, we can perform statistical tests to determine whether a hypothesis is valid. Throughout this book, we will explore the use of statistics as it relates to the patterns and data for COVID-19. Sometimes we will have robust data to perform good quantitative analysis, and other times we will lack the data to arrive at a proper conclusion. In either case, we can learn a great deal about the benefits and limitations of statistically analyzing data on a fast-moving event.

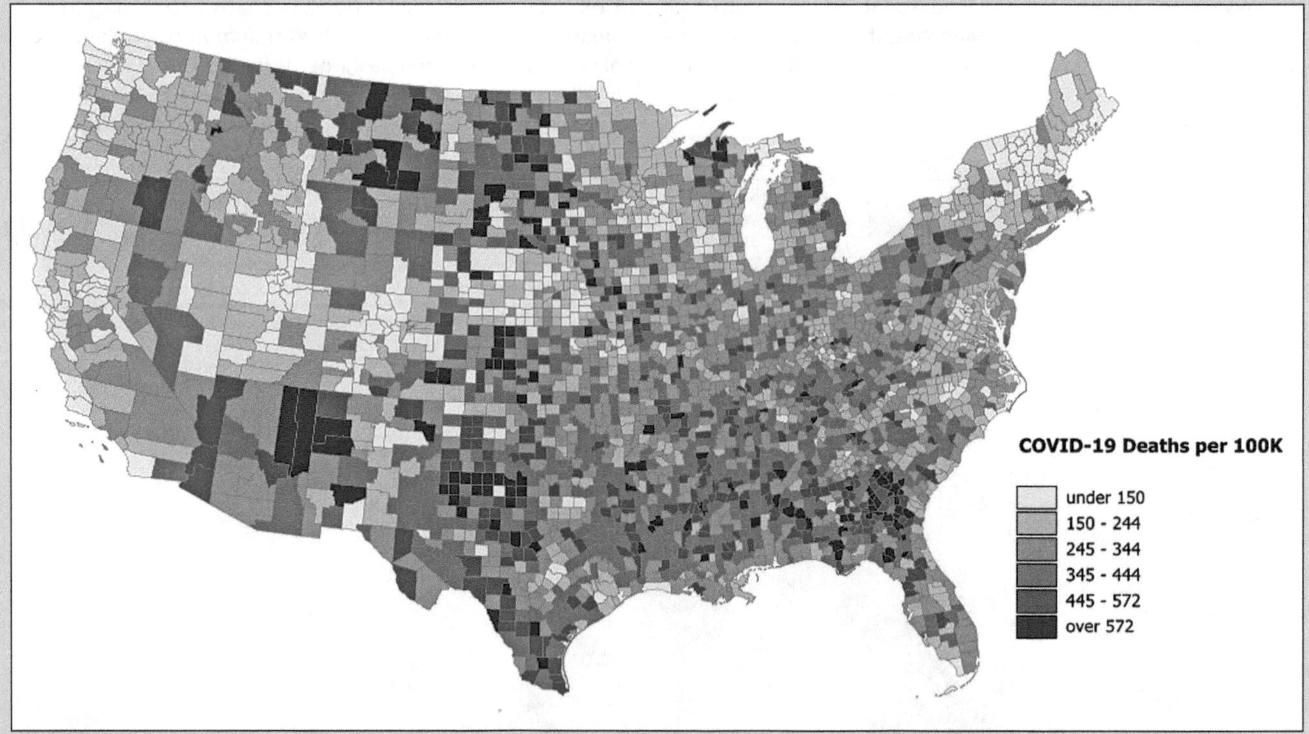

Figure 1.10
County-level deaths for COVID-19 in 2020 and 2021 per 100,000 People
Source: National Center for Health Statistics, AH County of Occurrence COVID-19 Death Counts

Key Terms

bar chart, *9*
data, data set, data value, *7*
data matrix (array), *8*
descriptive statement, *5*
descriptive statistic, *7*
geography, *3*
hypothesis, *5*
inferential hypothesis, *5*
inferential statistic, *7*
law, *7*
model, *5*
observation, *7*
pie chart, *9*
sample, *7*
scatterplot, *13*
statistical population, *7*
statistics, *4*
theory, *7*
variable, *7*

Major Goals and Objectives

If you have mastered the material in this chapter, you should now be able to:

1. Identify the types of questions geographers ask.
2. Understand the general importance and uses of statistics.
3. List the potential applications of statistics in geography.
4. Understand the role of statistics in the geographic research process.
5. Explain the role of statistics both before and after hypothesis formulation.
6. Formulate possible descriptive statements and inferential hypotheses when presented with a spatial pattern or locational data set.
7. Distinguish between questions, hypotheses, laws, theories, and models in geography.
8. Explain the typical organization or relationship of observations, variables, and data values in geographic data sets.
9. Explain the basic difference between descriptive statistics and inferential statistics.
10. Distinguish between a statistical population and a sample.
11. Describe or summarize a given set of data graphically with a bar chart, pie chart, and scatterplot.

References and Additional Reading

Abler, R. F., J. S. Adams, and P. R. Gould. *Spatial Organization: The Geographer's View of the World.* Englewood Cliffs, NJ: Prentice-Hall, 1971.

Amedeo, D., and R. G. Golledge. *An Introduction to Scientific Reasoning in Geography.* New York: John Wiley and Sons, 1975.

American Statistical Association. "What is Statistics?" Accessed September 15, 2022, via www.amstat.org/education/what-is-statistics-

Association of American Geographers. *Geography: Today's Career for Tomorrow.* (informational brochure, discontinued). Washington, DC: AAG, no date.

Association of American Geographers. *Practicing Geography: Careers for Enhancing Society and the Environment.* Upper Saddle River, NJ: Prentice Hall, 2012.

Burton, I. "The Quantitative Revolution and Theoretical Geography." *The Canadian Geographer* 7 (1963): 151–62.

Cobb, G. "Reconsidering Statistics Education: A National Science Foundation Conference." *Journal of Statistics Education* 1 (1993).

Geography Education Standards Project. *Geography for Life: National Geography Standards 1994.* Washington, DC: National Geographic Research and Exploration, 1994. (2nd ed. now available, 2012). www.ncge.org/geography-for-life

Haring, L. L., and J. F. Lounsbury. *Introduction to Scientific Geographic Research.* 4th ed. Dubuque, IA: Wm. C. Brown, 1992.

Hubbard, R. "Assessment and the Process of Learning Statistics." *Journal of Statistics Education* 5 (1997). Accessed September 15, 2022, via www.tandfonline.com/doi/full/10.1080/10691898.1997.11910522

Kuhn, T. S. *The Structure of Scientific Revolutions.* 3rd ed. Chicago, IL: University of Chicago Press, 1996.

Montello, D. R., and P. C. Sutton. *An Introduction to Scientific Research Methods in Geography and Environmental Studies.* 2nd ed. Thousand Oaks, CA: Sage Publications, 2012.

Parnell, D. B. *A Climatology of Frost Extremes Across the Southeast United States, 1950–2009.* Unpublished manuscript, Salisbury, MD: Salisbury University, 2013.

If you are interested in exploring further the geographic examples mentioned in this chapter, the following are good places to start. For obesity-related data and information, see the Centers for Disease Control and Prevention (CDC; www.cdc.gov). For life expectancy data and information regarding many other health-related variables worldwide, see the World Health Organization (WHO), especially the annual *World Health Statistics* publications (www.who.int/en and www.who.int/data/gho/publications/world-health-statistics). A wide variety of climatological data (including last spring frost information) is available at the National Climatic Data Center (NCDC); www.ncdc.noaa.gov). The sample data set we use for last spring frost analysis in the Southeast US is provided by Darren Parnell of the Department of Geography and Geosciences at Salisbury University, Maryland. For information concerning population growth and change patterns, see the United States Census Bureau (www.census.gov). Specifically see the series of Census Briefs dealing with population-related issues (www.census.gov/prod/cen2010). Another excellent source for demographic data is the Population Reference Bureau (PRB; www.prb.org). Especially useful is their DataFinder, from which you can browse a variety of both US and International topics (www.prb.org/usdata).

2

Geographic Data

Characteristics and Preparation

Before performing statistical processing and analysis we must first understand a number of characteristics of spatial data. One needs to understand how variables are organized as well as how data are arranged. In this chapter, basic concepts are introduced that provide the background information needed to characterize data before statistical techniques are applied.

Questions about data and variables arise early in the scientific research process. When identifying an appropriate geographic research problem and formulating meaningful hypotheses, you must make decisions about the sources of available data, the method of collecting data, and the variables to include in the analysis. In Section 2.1, these dimensions of geographic decision-making are discussed.

Several measurement issues are considered before conducting any statistical analysis. Variables may be organized and displayed in various ways, and different levels of measurement are used depending on the geographic problem. The characteristics of nominal, ordinal, and interval-ratio measurement scales are reviewed in Section 2.2. In Section 2.3 we discuss potential measurement errors and address the issues of precision, accuracy, validity, and reliability.

Basic methods of data classification are reviewed in Section 2.4, emphasizing the goals and purposes of classification, including its importance in geographic research. The two fundamental classification strategies of subdivision and agglomeration are defined and several specific operational methods or rules of classification are applied to a set of spatial data.

2.1 Selected Dimensions of Geographic Data

In the scientific research process, questions about data arise almost immediately. When trying to identify an appropriate geographic research problem or formulate a hypothesis, questions like these usually emerge: What sources of data are available? Which method(s) of data collection should be used? What type of data will be collected and then analyzed statistically? The various aspects of a geographic problem must be carefully considered to ensure that when data are collected and analyzed, you can answer research questions effectively and reach meaningful conclusions. The dimensions of geographic data discussed here include sources of information, methods of data collection, and selected characteristics of data that distinguish geographic research problems.

A simple distinction is made between primary and secondary data sources. **Primary data** are acquired directly from the original source. If you are collecting primary data, then you are probably obtaining this information "in the field." Primary data collection is often quite time-consuming and generally involves making decisions about sample design to acquire a representative set of data.

Secondary (or archival) data are generally collected by some organization or government agency and can be used by the geographer. Because the data are already collected and are probably organized in an accessible, convenient form—such as a written report, DVD, database file, or download from the internet—secondary sources are generally less expensive and time-consuming to use than primary sources of data. In addition, many of the problems associated with sampling and survey design may not be experienced with archival sources. Secondary sources are often very comprehensive, including a census or total enumeration from a very large population. Duplicating these efforts with primary data collection is virtually impossible for the researcher.

Potential difficulties can also occur with secondary data sources, such as the data being improperly collected, organized, or summarized. Errors may have occurred in the editing and collating of data, especially if the information was obtained from a number of different original sources. Information from secondary sources is not always measured properly, resulting in other potential problems. Finally, an old axiom states that data immediately becomes out-of-date the moment it is collected.

Several basic procedures are used to collect geographic data. If primary data are necessary for the study, sampling will almost certainly be part of the data collection process and may require the design of survey questionnaires. The options for primary data collection include direct observation, field measurement (especially in physical geography research), mail questionnaires, personal interviews, and telephone interviews.

To select the appropriate method of data collection, you must evaluate the nature of the research problem very carefully. Even if a suitable method is selected, problems in survey design are common. When using a survey to collect data, each question must be properly worded, all possible responses to a question must be considered in advance, and the sequence of questions must be determined. In fieldwork, logistic problems often occur, or special arrangements are necessary. Preliminary site reconnaissance may not reveal all the difficulties. A more detailed discussion of data collection methods is found in Chapter 7, "Basic Elements of Sampling."

Other dimensions or characteristics of data help distinguish geographic research problems. Some studies are considered **explicitly spatial** because the locations or placement of the observations or units of data are themselves directly analyzed. For example, a geographer responsible for selecting potentially profitable locations for a new retail store might calculate the "center of gravity" or average location of certain "target" households in the area. In another application, a biogeographer might analyze the spatial pattern of a sample of diseased trees in a national forest to determine whether these trees are randomly distributed throughout the area or clustered in certain places. In both of these examples, the data are spatially explicit because the locations of the observations or units of data are analyzed directly. An important set of spatial statistics is used to investigate these problems. Later in the book, a variety of descriptive spatial statistics are discussed (Chapter 4), and in Chapters 13–15 issues such as autocorrelation are introduced and several inferential spatial statistics are applied to point and area patterns to test for randomness and measure key aspects of spatial distributions.

Other geographic studies are **implicitly spatial**. An implicitly spatial situation exists when the observations or units of data represent locations or places, but the locations themselves are not analyzed directly. For example, a geomorphologist might wish to determine if significant differences occur in alluvial fan development in two different basins in the basin-and-range region of the American Southwest. A random sample of alluvial fans from each of the basins could be taken and relevant aspects of alluvial fan development compared. In a suburban neighborhood, a geographer may be investigating the relationship between the assessed valuation of homes and their ages. In both of these examples, the observations (alluvial fans in a basin or homes in a suburban neighborhood) obviously have locations on the earth's surface, *but the locational pattern itself is not under scrutiny.* Looking ahead, a two-sample difference test might be appropriate in the alluvial fan study (Chapter 10) while a correlation analysis may be needed to study the suburban housing question (Chapter 16).

Another important dimension of geographic research is the contrast between **individual-level** and **spatially aggregated** data sets. In some geographic problems each data value represents an individual element or unit of the phenomenon under study. In other problems, each "value" entered into the statistical analysis is a summary or spatial aggregation of individual units of information for a particular place or area. The best way to see this distinction is by discussing an example. Suppose a population geographer is researching current fertility patterns in Nigeria. One approach would be to collect a set of individual-level data, perhaps through personal interviews of a random sample of Nigerian women. Another possible approach would be to obtain birth rate estimates from officials in each of Nigeria's administrative divisions (21 states and 1 territory) and use these 22 spatially aggregated values as data units to estimate the nationwide fertility pattern.

Using spatially aggregated data raises special issues. Geographers must always be extremely cautious when trying to transfer results or apply conclusions "down" from larger areas to smaller areas or from smaller areas to individuals. If conclusions are derived from the analysis of data spatially aggregated for large areas, it may not be valid to reach conclusions about smaller areas or individuals. In the study of Nigerian fertility patterns, for example, valid spatially aggregated conclusions can be drawn about the degree of acceptance or rejection of family-planning programs in each of these states by using birth rate estimates from the administrative divisions. However, taking these aggregate conclusions down to the level of the individual family will almost certainly result in deductive errors. Even in the Nigerian state with the lowest birth rate, a number of families will not be practicing any form of birth control. This invalid transfer of conclusions from spatially aggregated analysis to smaller areas or to the individual level is known as the **ecological fallacy**.

Conversely, taking individual-level data and aggregating it to larger spatial units is generally not a problem. In fact, Nigerian officials probably collected data from individuals and selected villages, then aggregated that information to obtain state-level estimates. Some of the effects of level of spatial aggregation on descriptive statistics are discussed in Chapter 3.

Variables in a data set are characterized as either discrete or continuous. A **discrete variable** has some restriction placed on the values the variable can assume. A **continuous variable** has an infinitely large number of possible values along some interval of a real number line. In general, discrete data are the result of counting or tabulating the number of items, and potential values are limited to whole integers. Continuous data are the result of measurement, and values can be expressed as decimals. Examples of discrete variables include the number of households in a county with cell phones, the number of immigrants currently living in a city, the number of survey respondents in favor of a local bond issue, and the number of active volcanoes in a country. Examples of continuous variables include inches of precipitation at a weather station collected over a period of time, total area under irrigation in a country, distance traveled by a family on its annual vacation, and average wind speed at the summit of a mountain.

The "rounding-off" of data values must not be confused with the distinction between discrete and continuous data. For example, the elevation of a mountain is almost always expressed to the nearest foot or meter. Representation of elevation as a whole number may give the impression that this variable is discrete. However, since elevation can be measured more precisely than the nearest foot or meter, it is considered a continuous variable.

When discussing probability distributions, the distinction between discrete and continuous data is important. Geographic problems with discrete variables often require the application of different probability distributions than problems with continuous variables. Several practical geographic examples of both discrete and continuous distributions are presented in Chapters 5 and 6.

Variables in a set of data are either quantitative or qualitative. If a variable is **quantitative**, the observations or responses are expressed numerically—that is, units of data are assigned numerical values. On the other hand, if a variable is **qualitative**, each observation or response is assigned to one of two or more categories. Suppose an agricultural geographer asked 80 farmers to identify their primary cash crop and received the following responses: 43 farmers said corn, 28 said wheat, and 9 said barley. It might be tempting to conclude that these are quantitative data, since 43, 28, and 9 are clearly numerical values. However, the *variable responses* (corn, wheat, barley) are nonnumeric. The values of 43, 28, and 9 are not the raw data, but rather *frequency counts* of observations assigned to the nonnumerical categories, making this is an example of qualitative (or **categorical**) data. Other examples of qualitative variables are type of land use, gender (male or female), political party affiliation, religious preference, and climate type. Special types of statistical tests are discussed in Chapter 12 to handle qualitative (categorical) data.

2.2 Levels of Measurement

Just as the organization of variables determines how data can be analyzed statistically, so too does the related issue of variable measurement. That is, the level of measurement of units of data is considered when selecting an appropriate statistical technique to solve a geographic problem. Several different levels of measurement are described (Table 2.1), and different statistical procedures are appropriate to each.

Nominal Scale

The simplest scale of measurement for variables is the assignment of each value or unit of data to one of at least two qualitative classes or categories. In **nominal scale** classification of variables, each category is given some name or title, but no assumptions are made about any relationships between categories—only that they are different. Values are "different" if they are assigned to different categories or "similar" if assigned to the same category. Thus, problems using variables placed on a nominal scale are considered categorical (qualitative).

An urban planner could assign each parcel of land in a city to one of several nominal land use categories (residential, commercial-retail, industrial, recreation-open space, etc.) without inferring that residential land use is "greater than" recreation-open space or "less than" industrial. In fact, the only necessary conditions for a proper nominal scale classification of variables are that the categories are **exhaustive** (every value or unit of data can be assigned to a category) and **mutually exclusive** (it is not possible to assign a value to more than one category because the categories do not overlap). Oftentimes, cartographers utilize the category of "other" to ensure the mutually exclusive category.

Geographers create nominal variables in many ways: individuals are classified by religious affiliation (Baptist, Catholic, Methodist, Presbyterian, etc.) or political party (Democrat, Republican, Independent); cities are classified by primary economic function (manufacturing,

TABLE 2.1

Summary: Levels of Measurement

Level of Measurement	Brief Description
Nominal	Each value or unit of data assigned to one of at least two categories or qualitative classes; no assumptions made about relationships between categories—only that they are "different."
Ordinal	Values themselves are placed in some rank order.
Weakly ordered	Each value or unit of data assigned to a category, and the categories are then rank ordered.
Strongly ordered	Each value or unit of data is given a particular position in a rank order sequence; that is, each value is assigned its own particular rank.
Interval	Each value or unit of data placed on a measurement scale, and the interval between any two units of data on this scale can be measured; origin or zero starting point is assigned arbitrarily (that is, origin does not have a "natural" or "real" meaning).
Ratio	Each value or unit of data placed on a measurement scale, and the interval between any two units of data on this scale can be measured; origin or zero starting point is "natural" or nonarbitrary, making it possible to determine the ratio between values.

retail, mining, transportation, tourism, etc.); counties are organized by primary type of home heating fuel (fuel oil, utility gas, coal, wood, electricity, etc.); and countries are distinguished by predominant language family (Indo-European, Afro-Asiatic, Sino-Tibetan, Ural-Altaic, etc.).

If a variable has only two categories, a special subset of nominal classification is used. This sort of dichotomous (binary) assignment is used when no greater degree of qualification is necessary or possible. For example, each person could be assigned a "1" if they attend a private elementary school and a "0" if they do not. Although this type of information is clearly limited, statistical analysis on the frequency counts of the number of values or individuals in the categories can be conducted. Many geographic variables have only "yes-no" or "presence-absence" data available.

Variables that are organized nominally limit the types of numerical analysis that can be applied. Nevertheless, appropriate statistical techniques exist and are presented in Chapter 12.

Ordinal Scale

The next higher level of measurement involves placement of values in rank order to create an **ordinal scale** variable. The relationship between observations takes on a form of "greater than" and "less than." With data in rank order, more quantitative distinctions are possible than with nominal (qualitative) scale variables.

Geographers can easily identify examples of ordinal scale variables. An important distinction needs to be made, however, between a strongly-ordered ordinal variable and a weakly-ordered ordinal variable. When each value or unit of data is given a particular position in a rank order sequence, the variable is considered **strongly-ordered**. The city ranking schemes that occasionally appear in newspapers or magazines, such as the "10 best places to live" or "50 best American cities," are typical examples of this sort of survey. A popular publication is the *Places-Rated Almanac*, which provides a ranking of US cities based on the aggregate compilation of many variables. Since each city is assigned its own particular rank, these "preference rankings" are examples of strongly-ordered ordinal variables. Other examples of strongly-ordered variables include the ranking of countries by gross national product per capita or the ranking of states in terms of dollars spent per resident on higher education. Realize, however, that even strongly-ordered data is limited in the assumptions that can be made. For example, because a city is ranked second in a list of 100 best places to live does not imply that the city is twice as good as a city ranked fourth.

By contrast, in a **weakly-ordered** variable, the values are placed in categories, and the categories themselves are rank ordered. Suppose one is constructing a choropleth map showing the percentage of population change in each county of the United States from 2015 to 2020. To depict the population change cartographically, suppose six ordinal categories are selected (15% increase or more, 12% to 14.9% increase, and so on), and each of the more than 3,000 counties nationwide is assigned to one of these six categories. When frequency counts of counties are made for each category, the variable is weakly- rather than strongly-ordered. It is "weak" or "incomplete" in the sense that two counties assigned to the same category on the map cannot be distinguished, even though in reality the counties almost certainly have different population change values.

As with nominally scaled variables, specific statistical tests are designed to deal with both strongly- and weakly-ordered ordinal variables. Some of these techniques will be discussed in Chapters 10, 11, and 12.

Interval and Ratio Scales

With variables measured on either an **interval or ratio scale**, you can determine the magnitude of the difference between values. That is, the length of an interval between any two units of data can be measured on the scale. This means that not only is the relative position of each value known (a value is above or below another), but also known is the magnitude of difference of each observation from all other observations on the measurement scale.

Interval and ratio measurement scales are distinguished by the way in which the origin or zero starting point is determined. With interval scale measurement the origin or zero starting point is assigned arbitrarily. The Fahrenheit and Celsius scales used in the measurement of temperature are two widely known interval scales. The placement of the zero degree point on both of these measurement scales is arbitrary. With the Fahrenheit scale, the zero-degree mark is the lowest temperature attained with a mixture of ice, water, and common salt, whereas with the Celsius scale the zero-degree mark corresponds to the melting point of ice.

In ratio scale measurement, by contrast, a natural or nonarbitrary zero is used, making it possible to determine the ratio between values. If Montreal, Canada, receives 40 inches of annual precipitation and Chihuahua, Mexico, receives only 10 inches, the ratio between these two measures is easily calculated (40/10 = 4). Furthermore, it is correct to conclude that Montreal received four times as much precipitation as Chihuahua. Zero inches of rainfall represent a natural or nonarbitrary zero.

It should be noted that "ratio-type" statements cannot be made with interval-scale variables. For example, since zero is an arbitrary value on the Fahrenheit scale, 60 degrees Fahrenheit cannot be considered twice as warm as 30 degrees. Many other variables of interest to geographers are measured on a ratio scale, including distance, area, and such demographic and socioeconomic variables as infant mortality rate and median family income.

Observations from the same variable can be expressed at different measurement scales, depending on how they are measured, organized, and displayed. For example, a state resource planner interested in the type of energy used in homes across the state could use data collected and measured in several different ways. Data could be collected at the individual household level and organized nominally by the primary type of energy used (number of homes in the state whose principal energy source is coal, utility gas, fuel oil, etc.). Alternatively, data might be available as county-level summaries. Counties in the state could then be arranged in a strongly-ordered sequence by the percent of households in each county using natural gas (with the county having the highest percentage of households using natural gas assigned rank one, and so on). A choropleth map of the county values could display a weakly ordered graphic representation of the percentage of households using natural gas. As yet another alternative, the number of households per county using natural gas could be organized and displayed on a ratio scale.

2.3 Measurement Concepts

As proficiency improves when working with descriptive and inferential statistics, it becomes tempting to believe that the analysis is truly error free and that the geographic research problem is solved. Results from statistical analysis often seem very exact and definitive. If the data are analyzed on the computer, the results are nicely displayed and the same outcome is obtained if the data are resubmitted. It may seem appropriate to accept the answers as error-free automatically. However, an error-free result cannot be guaranteed just because a set of data is submitted correctly into some statistical software package. In fact, several interrelated sources of measurement error can operate separately or in combination to produce problems for the geographer.

Precision

Precision refers to the level of exactness associated with measurement. Precision is often associated with the calibration of a measuring instrument, such as a rain gauge. As a frontal system moves through an area, suppose that the amount of rainfall is recorded by two standard rain gauges having different calibration systems. On the coarsely calibrated gauge, the amount of rainfall might be estimated as somewhere between 1.2 and 1.3 inches. However, the more finely calibrated gauge provides a more precise estimate between 1.26 and 1.27 inches.

In many geographic problems, the issue of **spurious precision** must be considered. The computer (or calculator) output will often provide statistical outcomes with six or more decimal places, even when the input data are in integer form. Reporting seemingly precise statistics based on less precise input is a misleading but relatively commonplace occurrence. Unless confidence in such a level of measurement precision is warranted, it should be avoided.

Accuracy

The concept of **accuracy** refers to the extent of system-wide bias in the measurement process. It is quite possible for measurement to be very precise, yet inaccurate. Return to the rain gauge example for a moment. Suppose another more finely calibrated (more precise) rain gauge is used, but the gauge was not calibrated properly. The person reading the gauge estimates the amount of rainfall to be 1.19 inches rather than the actual rainfall total of about 1.26 or 1.27 inches, resulting in an inaccurate reading. Unfortunately, discovering systematic bias in a measurement instrument is often quite difficult.

To understand the relationship between precision and accuracy, consider this "target analogy" showing

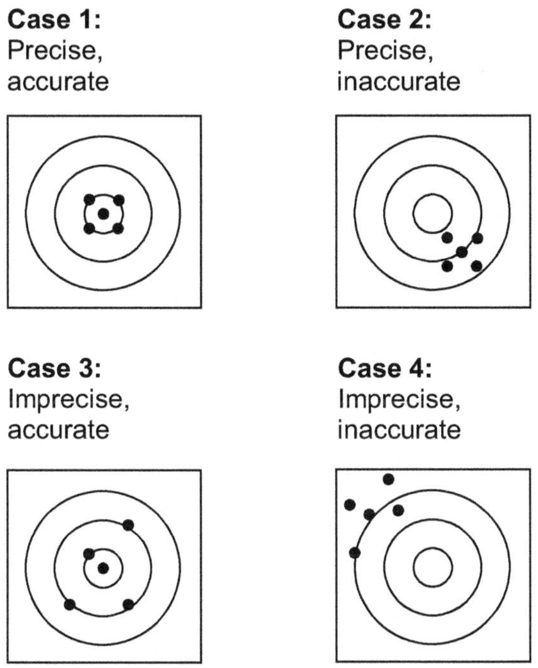

FIGURE 2.1
The Measurement Concepts of Precision and Accuracy: The Target Analogy

results from successive firings of a gun at a target (Figure 2.1). In Case 1, the five bullet holes are closely clustered (precise) and centered on the middle of the target (accurate), making this the best of the four alternatives presented. Cases 2 and 3 are both flawed—the inaccuracy of the bullet holes in Case 2 and the imprecision of the holes in Case 3 result in different types of errors. Case 4 appears to have the most severe problems, with an inaccurate systematic bias toward the upper left corner of the target as well as considerable scatter of bullet holes that also make the results imprecise. Clearly, you must take care to distinguish between the interrelated concepts of precision and accuracy when working with any data.

Validity

In many geographic problems, the spatial distribution or locational pattern being analyzed is the result of complex processes. In such situations it is understandably difficult to express the "true" or "appropriate" meaning of that concept through the measurement of any simple variable or set of variables. **Validity** addresses measurement issues related to the nature, meaning, or definition of a concept or variable. The discipline of geography is saturated with multifaceted, complex variables, such as "level of physical inactivity," "environmental quality," "economic well-being," "quality of education," "nutritious diet," and "quality of life." To express the true meaning of such concepts is often not possible, so we find it necessary to create **operational definitions** that can serve as indirect or surrogate measures. The question then becomes whether the operational definition is valid. For example, a geographer studying the spatial pattern of "quality of education" in a metropolitan area might evaluate elementary schools on the basis of "average student score on the California Achievement Test (CAT)" and evaluate high schools by "percentage of graduates who subsequently go to college." Clearly, the concept "quality of education" involves much more than what is reflected by these operational definitions, and their validity must be questioned. In this case, it should be asked whether "average CAT score" is a fully valid, somewhat valid, or invalid measure of "quality of education" received.

Admittedly, the degree of validity in a geographic problem may prove difficult (even impossible) to determine. Consequently this question is often ignored and problems with validity are sometimes assumed away as being inconsequential. A good geographic study involving complex variables will discuss the degree of validity of any operational definitions used in the analysis.

Reliability

A final measurement concept of concern in many geographic problems is **reliability**. When data are collected over time or when changes in spatial patterns are analyzed over time, you should question the consistency and stability of the data. For example, if repeated or replicate samples of water are taken from the same set of locations over time, a consistent and well-defined method of sampling is necessary or the results of subsequent statistical analysis may be unreliable. Suppose you want to examine the spatial pattern of poverty across the United States—does it vary more today than it did 50 or 100 years ago? A reliable and consistent measure of poverty is needed to answer this question. It could be that poverty was defined differently 50 or 100 years ago.

Reliability problems often occur when using international data. Fully comparable and totally consistent methods of collecting data rarely exist from country to country. A developing nation has fewer resources (e.g., personnel, money) than one that is more developed, and sources of measurement error inevitably affect the data collection process. International comparative statistics are often unreliable. Even within the same country or region, locational variations in data collection and processing methods can render data unreliable. Data collection procedures can change drastically from one time period to the next, again resulting in unreliable data. Some of these issues were discussed in the life expectancy example in Chapter 1.

One way to assess the degree of reliability of a measurement instrument is to compare at least two applications of the data collection method used at different times. This "test-retest" procedure is one way to evaluate the reliability of IQ test scores, medical diagnoses, and SAT scores, for example. Reliability checks need to be used more frequently in behavioral geography, particularly when analyzing the results from a survey or questionnaire that contains individual attitudes and opinions, which may be uncertain and subject to frequent change.

2.4 Basic Classification Methods

Although we have already introduced categories and classification, we have not examined the methods describing *how* to classify data into categories and reasons *why* one method of classification is better than another. In this section, the purposes and importance of classification in geographic research are explained and basic classification methods are reviewed.

Geographers regularly face the problem of deciding how to classify or group spatial data. Classification is used for several important reasons. Classification schemes organize, simplify, and generalize large amounts of information into effective or meaningful categories, bringing relative order and simplicity to complexity. As a result, communication is enhanced, detailed spatial information is better understood, and complex spatial patterns are represented more clearly. Maps created with properly classified data result in more effective graphic communication. Classification is also an integral part of the scientific research process, helping in the formation of hypotheses and guiding further investigation.

In classification, values are organized according to their degree of similarity. That is, similar values are generally placed in the same category and dissimilar values are generally placed in different categories. The result is a grouping of data that seems to minimize the amount of fluctuation or dispersion of values within the same category and maximizes the dispersion of values between different categories. Whatever the specific method of classification, however, the resulting categories should be both mutually exclusive and exhaustive.

Many specific classification methods are available, and each method approaches these general goals from a slightly different perspective. No matter what specific method of classification is used, some information is invariably lost when large amounts of information are simplified and generalized. Information is also lost if individual-level values are spatially aggregated and only the aggregated data are available. Similarly, information is lost if values are classified and only the classified data are available. In fact, spatially aggregated data are simply individual-level data classified by location.

At the most fundamental level, classification uses one of two *conceptual strategies.* The first of these is **subdivision** (sometimes called **logical subdivision**). At the start of the subdivision process, all units of data in a population are grouped together. Then, through a series of steps or iterations, individual values are allocated to an appropriate subdivision using carefully defined criteria. This strategy "works down" by disaggregating all values into logically subdivided classes. Most practical geographic examples of subdivision are hierarchical, with multiple levels of subdivision, depending on the problem or situation. A clear and consistent set of rules is always needed to assign values to the proper category at each stage of the subdivision procedure. The characteristics or values associated with each category are defined before the classification procedure begins.

The subdivision strategy of classification is illustrated with two examples. The United States Geological Survey leads a partnership of Federal agencies that has created a Multi-Resolution Land Characteristics (MRLC) consortium. One of the many products of this Consortium is the National Land Cover Database (NLCD) classification system, which serves as the definitive Landsat-based, 30-meter resolution, land cover database for the nation. The Level I and Level II classes are shown in Table 2.2. This classifica-

TABLE 2.2

A Portion of the National Land Cover Database System

1. Water		2. Developed		3. Barren		4. Forested Upland		5. Shrubland	
11	Open Water	21	Low Intensity Residential	31	Bare Rock/Sand/Clay	41	Deciduous Forest	51	Shrubland
12	Perennial Ice/Snow	22	High Intensity Residential	32	Quarries/Strip Mines/Gravel Pits	42	Evergreen Forest		
		23	Commercial/Industrial/Transportation	33	Transitional	43	Mixed Forest		

6. Non-Natural Woody		7. Herbaceous Upland Natural/Semi-Natural Vegetation		8. Herbaceous Planted/Cultivated		9. Wetlands	
61	Orchards/Vineyards/Other	71	Grasslands/Herbaceous	81	Pasture/Hay	91	Woody Wetlands
				82	Row Crops	92	Emergent Herbaceous Wetlands
				83	Small Grains		
				84	Fallow		
				85	Urban/Recreational Grasses		

Source: United States Geological Survey (USGS), USGS Land Cover Institute

TABLE 2.3

A Portion of the North American Industry Classification System (NAICS)

2 Digit		3 Digit		4 Digit		5 Digit	
11	Agriculture, Forestry, Fishing and Hunting						
21	Mining, Quarrying, and Oil and Gas Extraction						
22	Utilities						
23	Construction						
31–33	Manufacturing	311	Food Mfg.				
		312	Beverage and Tobacco Product Mfg.	3121	Beverage Mfg.	31211	Soft Drink and Ice Mfg.
						31212	Breweries
						31213	Wineries
						31214	Distilleries
				3122	Tobacco Mfg.		
		313	Textile Mills				
		314	Textile Product Mills				
		315	Apparel Mfg.				
42	Wholesale Trade						
44–45	Retail Trade						
48–49	Transportation and Warehousing						
51	Information						
52	Finance and Insurance						
53	Real Estate and Rental and Leasing						
54	Professional, Scientific, and Technical Services						
55	Management of Companies and Enterprises						
56	Administrative and Support and Waste Management and Remediation Services						
61	Educational Services						
62	Health Care and Social Assistance						
71	Arts, Entertainment, and Recreation						
72	Accommodation and Food Services						
81	Other Services (except Public Administration)						
92	Public Administration						

Source: United States Bureau of the Census

tion system supports a wide variety of applications that seek to assess ecosystem status and health, understand the spatial patterns of biodiversity, predict effects of climate change, and develop land management policy.

The North American Industry Classification System (NAICS) can also be used to illustrate subdivision. Applied throughout the United States, Canada and Mexico, this hierarchical system classifies establishments (businesses) by what they produce or supply and allows for comparability of data among the three trading partners of the North American Free Trade Association (NAFTA). In the United States, data on business activities are summarized and reported by NAICS groupings every five years in the *Economic Census*. The NAICS classification subdivides all production or service activity according to numerical code, ranging from a two-digit general economic sector grouping at the top level of the hierarchy to more specific six-digit breakdown of activity at the lowest level. A small portion of NAICS is shown in Table 2.3.

The second conceptual classification strategy is **agglomeration**. With this general approach, each observation in a population or data set is separate and distinct from others at the start of the classification process. The agglomeration procedure then "works up" by allocating values into classes according to well-defined grouping criteria. Agglomeration is accomplished when similar values are combined into the same category and dissimilar values are placed in different categories. The agglomeration method of grouping is the conceptual opposite of subdivision. This classification strategy is very important in the geographic research process and data for many geographic problems are summarized numerically or graphically using agglomeration.

Beyond these general conceptual strategies of logical subdivision and agglomeration, a variety of specific *operational procedures or rules* are applied in practical classification. These operational methods are not necessarily "pure" logical subdivision or agglomeration, but may contain elements of both. The remainder of this section focuses on the first four of the six simple operational methods or rules of classification that can be applied in geography (Table 2.4). The standard deviation breaks and Jenks natural breaks methods will be discussed in the next chapter.

1. **Equal intervals based on range**. The **range** is simply the difference in magnitude between the largest and smallest values in an interval-ratio set of data. To determine **class breaks** (the values that separate one class from another), the range is divided into the desired number of equal-width class intervals. The procedure is easy to use and results in class intervals of equal width, which may be an advantage for some applications. However, because all class breaks are derived from the two units of data with the most extreme values, results are sometimes misleading. To maintain the equal width of each class, the class break

numbers are not usually rounded-off. The number of values in each category may also vary considerably. This may be an advantage or a disadvantage, depending on the purposes of your classification and the goals of the analysis.

2. **Equal intervals not based on range**. This method of classification also designates class breaks to create equal interval classes, but the exact range is not used to select the class breaks. Instead, a convenient or practical interval-width is selected arbitrarily, based on rounded-off class break values. Units of data are then assigned to the categories. This method of classification is preferred for constructing a frequency distribution, histogram, or ogive to represent the data graphically (as we will see in the next chapter). Many institutions and government agencies use this method to map complex spatial patterns. The convenient class break values generally result in maps that are easy to understand and interpret. The number of values in each category could vary widely, which may be either an advantage or disadvantage, depending on your classification goals.
3. **Quantile breaks**. This method approaches classification from a somewhat different perspective. The total number of values is divided as equally as possible into the desired number of classes. Two frequently used alternatives are the division of data into quartiles (four categories) or quintiles (five categories). The allocation of an equal number of values to each category is often an advantage in choropleth (area pattern) mapping, particularly if an approximately equal area on the map is desired for each category. However, the possible disadvantages with quantile breaks classification should be evaluated before deciding to use this method. Class breaks are frequently not convenient rounded-off values and class interval widths are almost never equal for different categories. If a large number of observations are clustered relatively close together—a frequent occurrence with geographic data—these similarly sized values are likely to be "split unnaturally" by a class break to keep an equal number of values in each category.
4. **Natural breaks**. Yet another perspective is the natural breaks method of classification. There are many methods of natural breaks, but the most elementary is known as the **single-linkage** approach. The logic is to identify natural breaks in the data and separate values into different classes based on these breaks. The process is done iteratively, with the largest gap or separation between adjacent values on a number line selected as the first class break location. The next largest gap is selected second, and so on until the desired number of classes has been created. With this classification process, similar values are kept together in the same category, dissimilar values are separated into different categories, and gaps in the data are incorporated directly in the grouping procedure. This method will highlight extreme values, often placing unusually large positive or negative observations into their own unique categories. Depending on your research problem, highlighting extreme values

TABLE 2.4

Brief Description of Selected Single-Variable Classification Methods

Classification Method	Brief Description
Equal intervals based on range	Class breaks determined by dividing range (difference between the lowest- and highest-valued units of data) into desired number of equal-width class intervals.
Equal intervals not based on range	Based on convenience or practical considerations; rounded-off class breaks and class interval widths arbitrarily selected.
Quantile breaks	Equally divide the total number of values into the desired number of classes; two commonly used divisions are quartiles (4 categories) and quintiles (5 categories).
Natural breaks (Single-Linkage Method)	Place units of data in rank order, identify "natural breaks" or separations between adjacent ranked values, and locate class breaks in the largest of these natural breaks. Iterative process if single-linkage version used, with largest natural break selected as first class break location, next largest natural break selected second, and so on until desired number of classes created.
Standard Deviation Breaks *	After calculating the mean and standard deviation of all the observations, class breaks are determined by dividing the data at rounded-off standard deviation intervals (*z*-values) such as 1.0 or 0.5. This method is often used when a data set is normally distributed.
Natural Breaks (Jenks Method) *	Class breaks are determined by "natural" groupings inherent in the data. This is an iterative process that attempts to minimize the variation of observations within classes and maximize the variation between classes.

* Discussion of this classification method is delayed until Chapter 3, in which the terms *mean*, *standard deviation* and *variance* are formally defined.

may (or may not) be a primary goal of the classification. Another common consequence of natural breaks is the clustering of large numbers of values into one or two categories. Again, you must decide whether this is an advantage or disadvantage.

To illustrate how each method works, 2020 state-level adult obesity rates are classified. Even a simple list of 51 ranked obesity rates (Washington, DC, is included) is too detailed and cumbersome for direct study and interpretation (Table 2.5). This is particularly true if one goal is to illustrate the spatial pattern of obesity in a choropleth map, with each state allocated to one of several classes.

Selecting the *ideal number of classes* is very important in choropleth mapping. If data are allocated into too few categories, key details in the spatial pattern are likely to be lost. On the other hand, if too many categories are used, the map reader becomes overwhelmed with detail and could miss certain important generalizations of the spatial pattern. Your decision on the number of classes to use always involves the trade-off between effective generalization and communication of sufficient detail. Five classes are used in this obesity example; many cartographers consider this number reasonable for choropleth mapping of geographic patterns.

Each simple method of classification is applied to the percent of adults defined as obese for each area. For consistency of discussion and interpretation, exactly five classes and four class breaks are used with each method. The results are shown on a series of choropleth maps (Figure 2.2, Cases 1–5). On each choropleth map the class intervals and class breaks derived from application of each classification scheme are shown in the legend.

Application of the various classification methods results in dramatically different obesity pattern maps. It is important to realize that such startling visual differences on choropleth maps are expected from one method of classification to another, even when all the maps are created from the same data set. These sharp visual contrasts do not mean that any of the classification methods are inaccurate or biased. Rather, this situation illustrates the distinctiveness of the goals in classification. It seems that very different impressions regarding the "reality" of the situation emerge with application of different classification methods.

Perhaps the greatest visual contrast exists between the maps of quantile breaks and natural breaks (Figure 2.2, Cases 3 and 5). The objective in using quantile breaks is to allocate an equal number of values to each category. Applying quantiles, Mississippi (the state with the highest percent of adults considered obese; 39.7%) is grouped with many other states. The overall impression from the quantile map is that high obesity rates can be found in many parts of the South, Midwest, and Appalachia, and there is no particular visual focus on Mississippi.

TABLE 2.5

Ranked Obesity Levels by State, 2020 (in percent)

State	Percent Obese	State	Percent Obese
Colorado	24.2	Virginia	32.2
Washington, DC	24.3	Wisconsin	32.3
Massachusetts	24.4	Illinois	32.4
Hawaii	24.5	North Dakota	33.1
New York	26.3	South Dakota	33.2
Vermont	26.3	North Carolina	33.6
New Jersey	27.7	Missouri	34.0
Washington	28.0	Nebraska	34.0
Oregon	28.1	Georgia	34.3
Florida	28.4	Michigan	35.2
Montana	28.5	Kansas	35.3
Utah	28.6	Ohio	35.5
Nevada	28.7	Tennessee	35.6
Connecticut	29.2	Texas	35.8
New Hampshire	29.9	South Carolina	36.2
Rhode Island	30.1	Oklahoma	36.4
California	30.3	Arkansas	36.4
Minnesota	30.7	Delaware	36.5
Wyoming	30.7	Iowa	36.5
New Mexico	30.9	Kentucky	36.6
Arizona	30.9	Indiana	36.8
Maryland	31.0	Louisiana	38.1
Maine	31.0	Alabama	39.1
Idaho	31.1	West Virginia	39.1
Pennsylvania	31.5	Mississippi	39.7
Alaska	31.9		

* Obesity is defined as a body mass index (BMI) of 30 or greater. BMI is calculated from a person's weight and height and provides a reasonable indicator of body fatness and weight categories.

Source: Centers for Disease Control and Prevention (CDC)

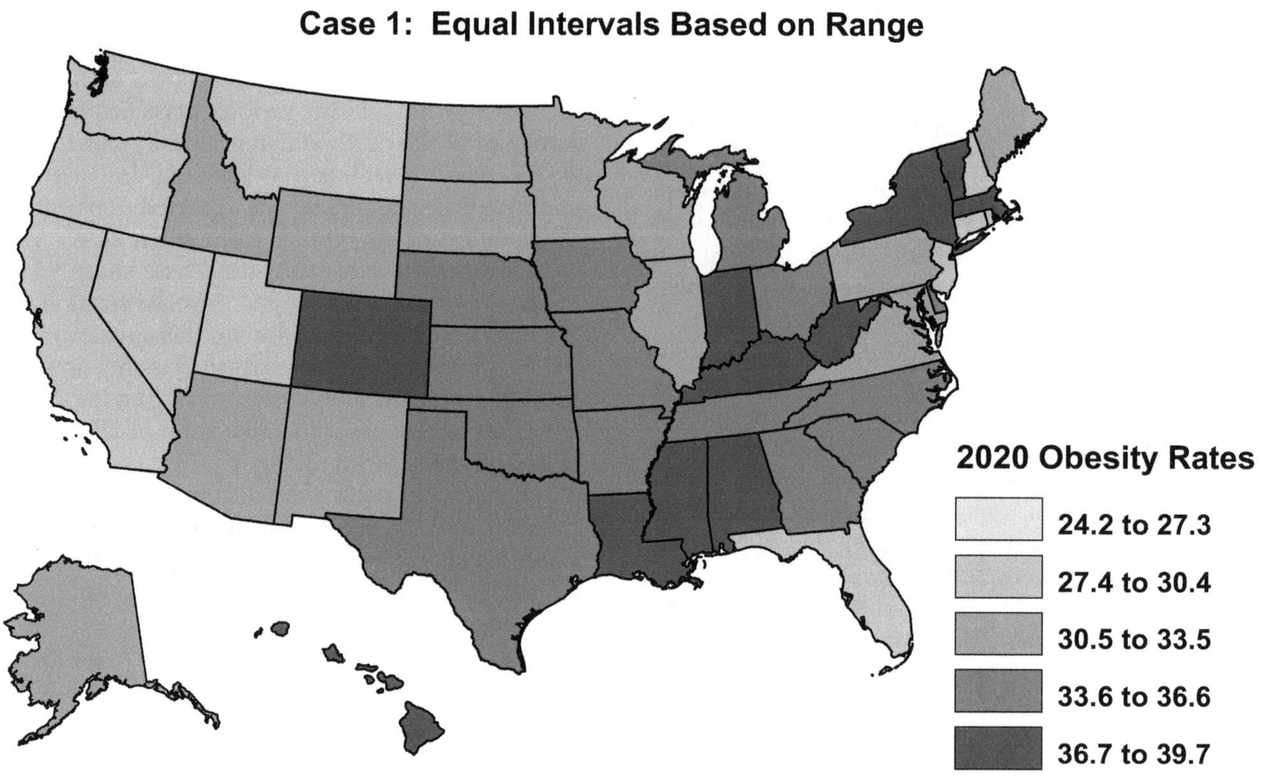

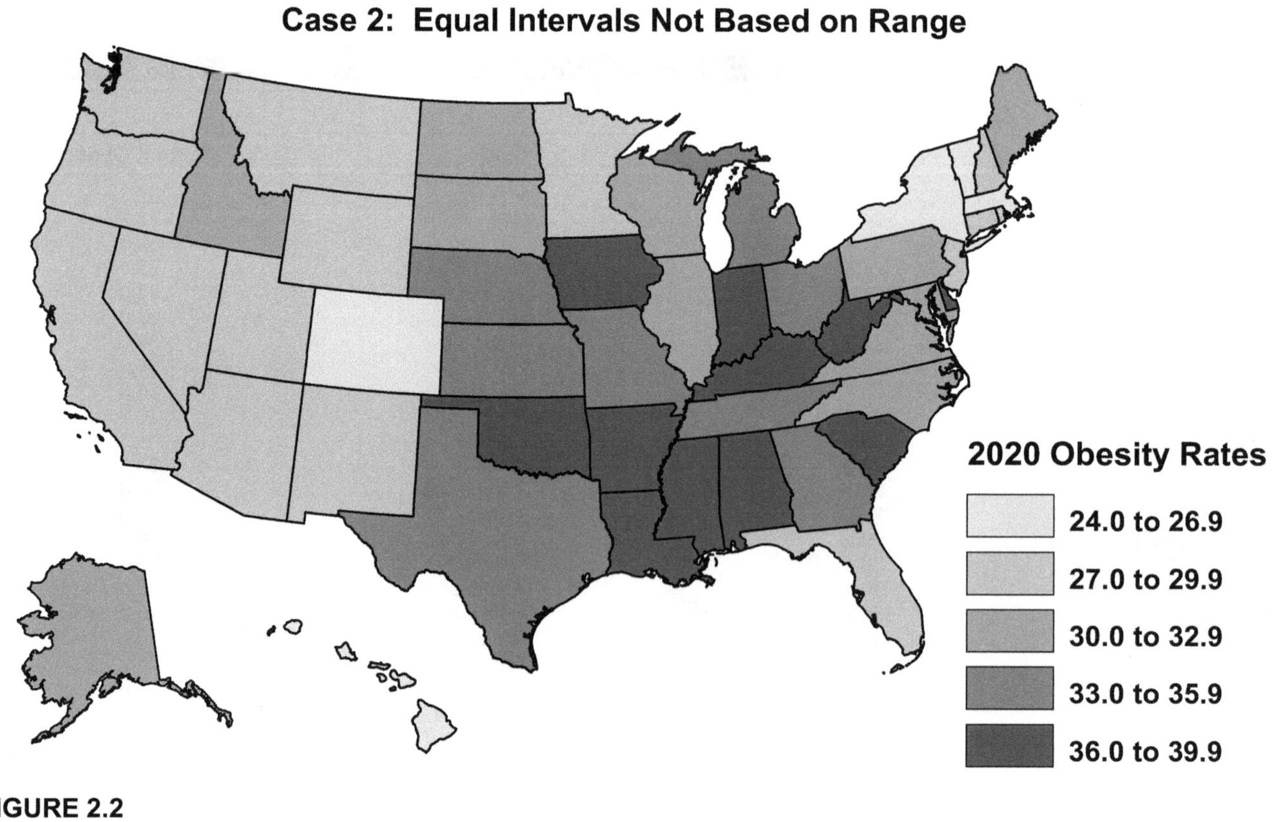

FIGURE 2.2

Percent of Adults Considered Obese, by State, 2020: Several Classification Methods

Source: Centers for Disease Control and Prevention (CDC), 2020

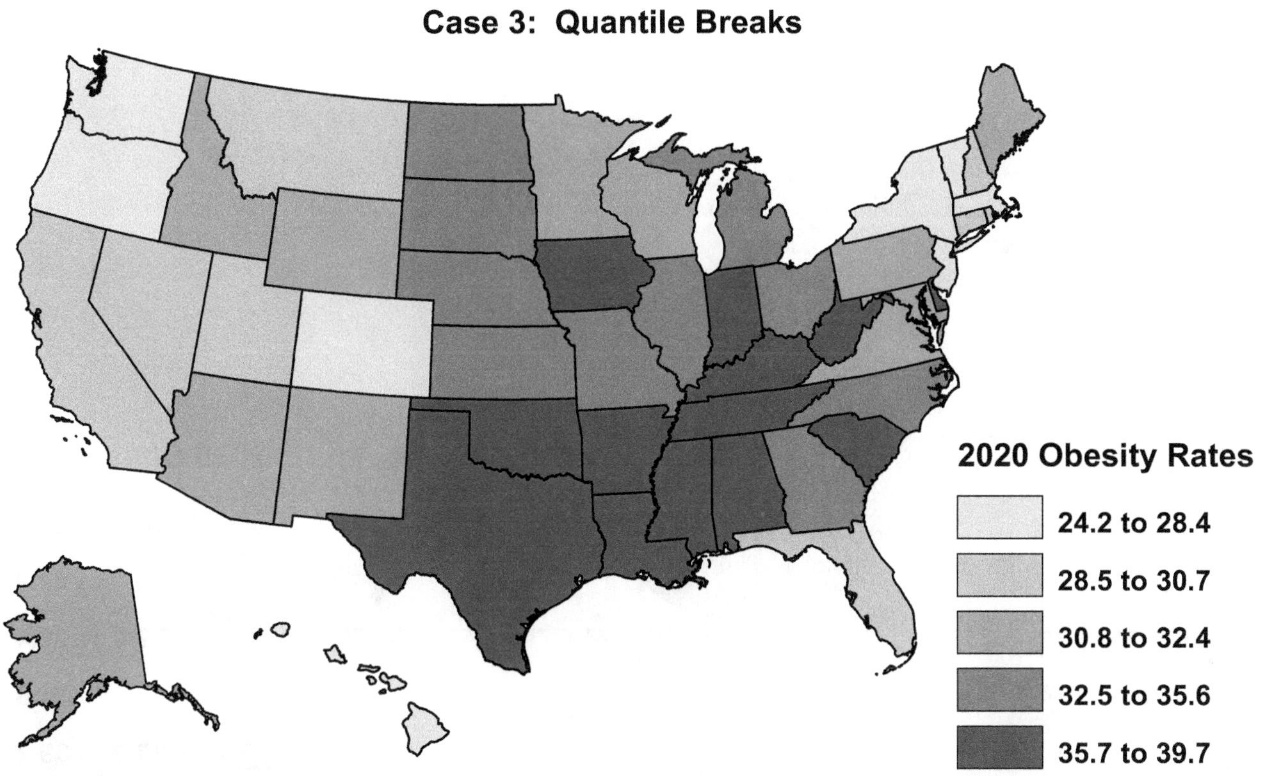

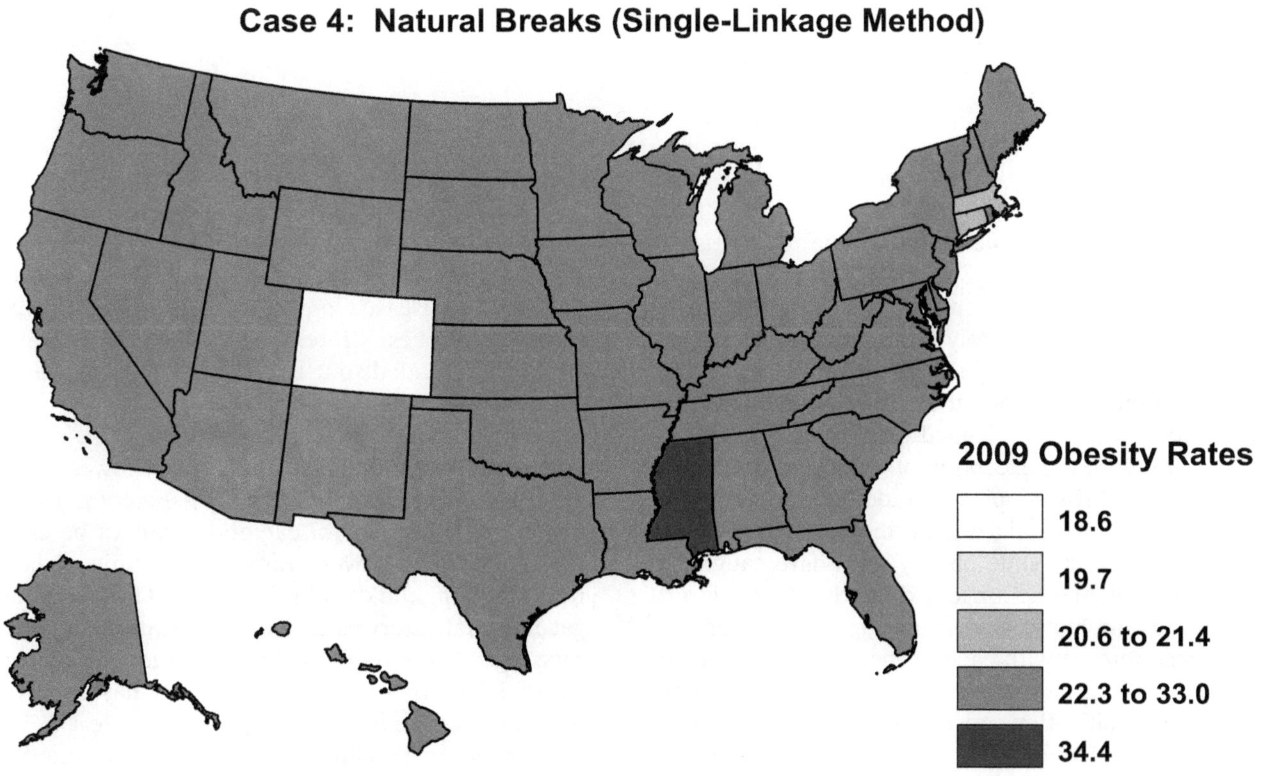

FIGURE 2.2 (continued)
Percent of Adults Considered Obese, by State, 2020 and 2009: Several Classification Methods
Source: Centers for Disease Control and Prevention (CDC), 2020

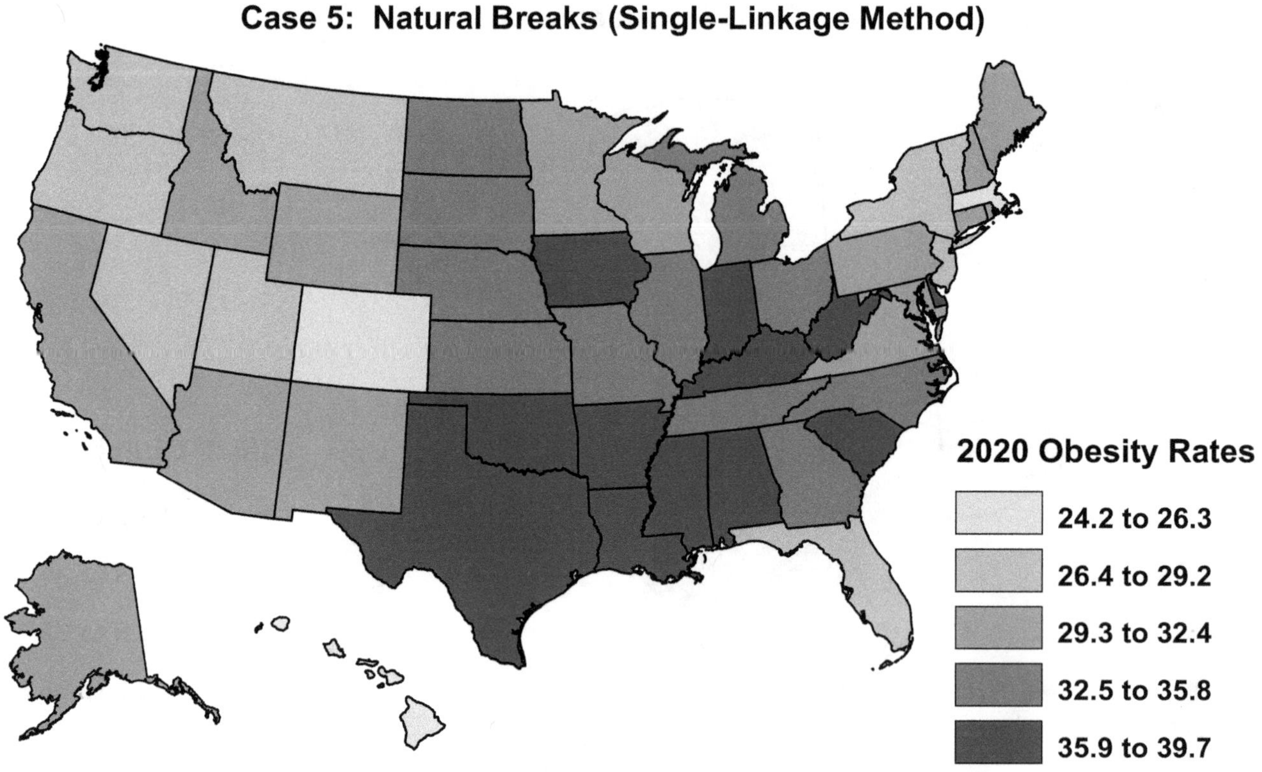

FIGURE 2.2 (continued)
Percent of Adults Considered Obese, by State, 2020: Single-Linkage Method
Source: Centers for Disease Control and Prevention (CDC), 2020

By contrast, the goal in natural breaks is to separate those places that are very different in value from all other places. To illustrate this, we also show obesity rates from 2009 in addition to the 2020 rates, as the recent rise in obesity rates in the United States in 2020 has exhibited less extreme differences among the states. Exploring the 2009 mapped result is sharply different, as Mississippi's high obesity rate is highlighted graphically. In fact, the dominant impression from the natural breaks map in 2009 is that most of the United States has comparable obesity rates, with the exception of a few extreme observations such as Mississippi, Colorado, Connecticut, and Massachusetts. Rather alarmingly, the 2020 results seem to indicate that other state obesity rates have caught up with Mississippi, and no longer exhibit this separation.

In this example problem, minor operational difficulties are encountered with each of the simple methods of classification. In addition, elements of subjectivity seem to enter the classification process no matter which procedure we use, and trade-offs or compromises are sometimes necessary. One limitation is that the obesity data are expressed in tenths of a percent. Of course, this is a reasonably precise degree of measurement, but it causes application problems with some methods of classification. As a result of this precision level, some ties exist in the data set. For example, two states are reported with a 36.5% adult obesity rate. Obviously these two states do not all have an *exact* rate of 36.5%, so they are not really tied, but the full reality of the situation is not known. The Centers for Disease Control recognizes that more precise estimates of obesity rates (say, to the nearest hundredth) could be provided, but they would create meaningless, spurious precision.

The first method (Figure 2.2, Case 1) is equal intervals based on range. The range from lowest obesity in Colorado to highest obesity in Mississippi is 15.5%, and five *exactly equal interval* categories cannot be created at this precision level over a range of 15.8 units of data without modifying the data in some way. The closest one can get to equal intervals is the classification shown in the map legend, where the first four categories have an interval of 3.1 units, and the last category has an interval of 3.0 units (see the legend in Figure 2.2, Case 1).

The second method (Case 2) is equal intervals not based on range. Again, without creating spurious precision it is not possible to create five exactly equal interval categories where the lower bound of the first category is 24.2% and the upper bound of the fifth category is

39.7%. The solution shown on the map uses a class interval of 3.0% obesity, with the first category "anchored" on 24.0% and the last category on 40.0%. While these class breaks are convenient, a disconcerting feature is that no state has an obesity rate as low as 24.0% or as high as 40.0%. If the class interval width is changed from 3.0% or the number of categories is changed to something other than five, a variation of this same problem will still exist. Nevertheless, this is a fully valid and objective application of equal intervals not based on range, working within the constraints of the data.

The third method (Case 3) is quantile breaks. Tied observations create a minor problem here as well. It is not possible to allocate *exactly* ten states to each of the five categories without splitting multiple states that are tied. Also, the inclusion of Washington, DC, makes such a perfect division impossible. A valid operational rule would be to minimize the disparities between groups, creating a classification that is as close as possible to having an equal number of observations in each category. One such allocation is shown on the map, where the maximum number of areas in any category is 11 and the minimum number is 10 (Figure 2.2, Case 3).

The final method (Case 5) is the natural breaks, single-linkage method. Here, the lack of precision once again generates a problem. The single largest natural break in the data set is 1.8% obesity units, separating Hawaii (24.5%) from New York (26.3%). If the classification process were to stop at this juncture, there would be no problem, but the resultant map would only have two categories—one category of size 4 containing places with obesity rates of 24.5% or less, and another category of size 46 containing places of 26.3% or more. A two-category choropleth map will reveal almost none of the actual spatial pattern and is clearly not the best strategy. Also, a condition in this problem was to create classification schemes having exactly five categories, so this objective would also not be attained.

Proceeding iteratively using the single-linkage approach, the next largest natural break in the data set is 1.4 units, separating Vermont (26.3%) from New Jersey (27.7%). The next largest natural break is 1.0 units—the separation of Louisiana (38.1%) from Alabama (39.1%). After this iteration, exactly five distinct categories are created, as shown in Figure 2.2, Case 5.

A much better way of showing this entire step-by-step process of natural breaks single linkage is graphically with a **dendrogram** (Figure 2.3). Many statistical software packages are now capable of creating nice-looking dendrograms. The first subdivision occurs when Colorado, Massachusetts, DC, and Hawaii separate from all the others at a distance (difference) of 1.8 variable units. This is clearly seen at the top-left of Figure 2.3. The second subdivision occurs when New York and Vermont separate from all others at a distance of 1.4 units, and so on. The horizontal line added to the dendrogram at a distance below 0.9 but above 0.8 shows the grouping of areas if the process is stopped with exactly five categories.

Despite the impressive dendrogram, the resulting map is not very satisfying geographically, and the process creates some other problems as well. First of all, a few extreme values in the data set (often called **outliers**) are identified, but the overall map contains very little information about the nationwide obesity pattern. Furthermore, the physical area of one of these outliers (Washington, DC) is so small it is not really visible on the map and both Connecticut and Massachusetts are just barely noticeable.

What should you conclude about these disparities among classification methods? Depending on the method used, outcomes can be quite different, even though the same data set is used and the same number of classes created. The visually distinctive choropleth maps in Figure 2.2 illustrate the substantial effects of various classification decisions on spatial patterning and the resultant conclusions that can be drawn about these spatial patterns. The logical conclusion is for you to recognize that any observed spatial pattern (map) is a function of the specific classification method applied and that using a different method of classification will likely result in a visually distinctive map.

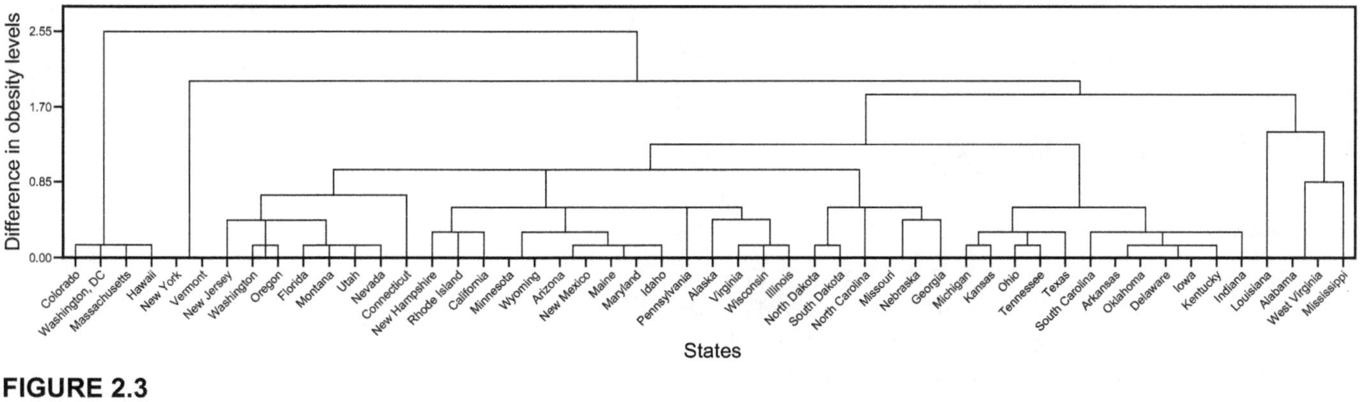

FIGURE 2.3
Dendrogram: State Obesity Rates, 2020

In addition to the four simple classification methods presented in this chapter, a number of other classification methods will be introduced later in the text. As footnoted in Table 2.4, the standard deviation breaks and Jenks natural breaks methods are discussed in Chapter 3. These methods are not discussed here, because the statistics of *mean, standard deviation,* and *variance* have not yet been formally defined. Classification techniques will appear again toward the end of the text (Chapter 17), when the multivariate classification method of cluster analysis is applied to aspects of the obesity and life expectancy examples introduced in the first chapter.

Key Terms

classification strategies:
- subdivision, *28*
- agglomeration, *29*

discrete and continuous variables, *24*
dendrogram, *35*
ecological fallacy, *24*
exhaustive and mutually exclusive categories, *24*
explicitly spatial and implicitly spatial data, *23*
individual-level and spatially aggregated data, *23*
measurement concepts:
- precision, *26*
- accuracy, *26*
- validity, *27*
- reliability, *27*

measurement scales:
- nominal, *24*
- ordinal, *25*
- interval, *26*
- ratio, *26*

methods of classification (selected):
- equal intervals based on range, *29*
- equal intervals not based on range, *30*
- quantile breaks, *30*
- natural breaks (single linkage), *30*
- operational definition, *27*
- outlier, *35*
- primary and secondary (archival) data, *22, 23*
- quantitative and qualitative (categorical) variables, *24*
- spurious precision, *26*

strongly ordered and weakly ordered ordinal variables, *25*

Major Goals and Objectives

If you have mastered the material in this chapter, you should now be able to:

1. Categorize geographic variables and data sets on a variety of dimensions (e.g., primary or secondary [archival], explicitly or implicitly spatial, individual-level or spatially aggregated, discrete or continuous, categorical [qualitative] or quantitative variables).
2. Identify the measurement scale (nominal, ordinal, interval, or ratio) of a variable and understand the characteristics of each measurement scale.
3. Describe the measurement concepts of precision, accuracy, validity, and reliability. When presented with data from any particular geographic problem or situation, evaluate the variable in terms of these potential sources of measurement error.
4. Explain the general logic associated with each of the two conceptual classification strategies (subdivision and agglomeration).
5. Understand the specific procedures, as well as both the advantages and disadvantages associated with each of the operational classification methods (equal intervals based on range, equal intervals not based on range, quantile breaks, and natural breaks—single linkage). When given a set of geographic data, select an appropriate classification method.
6 Describe or summarize a given set of data graphically with a dendrogram.

References and Additional Reading

Blalock, H. M., Jr. *Conceptualization and Measurement in the Social Sciences.* Beverly Hills, CA: Sage Publications, 1982.

Brewer, C. A. *Designing Better Maps: A Guide for GIS Users.* 2nd ed. Redlands, CA: ESRI Press, 2016.

Dramowicz, E., and K. Dramowicz. "Choropleth Mapping with Exploratory Data Analysis." *Directions Magazine,* 2004. Article 718. www.directionsmag.com

Jenks, G. F., and R. C. Coulson. "Class Intervals for Statistical Maps." *International Yearbook of Cartography* 3 (1963):119–34.

Monmonier, M. S. *How to Lie with Maps.* 3rd ed. Chicago: University of Chicago Press, 2018.

Robinson, A. H., J. L. Morrison, P. C. Muehrcke, A. J. Kimerling, and S. C. Guptill. *Elements of Cartography.* 6th ed. New York: John Wiley and Sons, Inc., 1995.

Sokal, R. R. "Numerical Taxonomy." *Scientific American* 215, No. 6 (1966):106–16.

Stauffer, C. L. "Making Maps: The Untold Story." *Population Today* 30 (2002): 3, 6.

Part 2

Descriptive Problem Solving in Geography

3

Descriptive Statistics and Graphics

A basic distinction between descriptive and inferential statistics is made in Chapter 1. Recall that the overall goal of descriptive statistics is to provide a concise and easily understood summary of the characteristics of a particular data set. For most geographic problems, such quantitative or numerical summary measures are clearly superior to working with unsummarized raw data. With these easily understood and widely used descriptive statistics, geographers can effectively communicate the characteristics of the data.

Some of the advantages of summarizing spatial information were demonstrated in the discussion of basic classification methods (Section 2.4). In this chapter, using basic descriptive statistics and related graphics, we show how numerical or quantitative summary measures and complementary graphics can "describe" a data set.

We can summarize a data set in several different ways:

- **Measures of central tendency**—numbers that represent the center or typical value of a frequency distribution, such as mode, median, and mean (Section 3.1).
- **Measures of dispersion**—numbers that depict the amount of spread or variability in a data set, such as range, interquartile range, standard deviation, variance, and coefficient of variation (Section 3.2).
- **Measures of shape or relative position**—numbers that further describe the nature or shape of a frequency distribution, such as skewness, which indicates the amount of symmetry of a distribution, or kurtosis, which describes the degree of flatness or peakedness in a distribution (Section 3.3).

Choosing the proper descriptive statistic and related graphics display for a particular geographic problem depends partly on the level of measurement—whether the data are nominal, ordinal, or interval-ratio. In addition, different calculation procedures are applied if the data are grouped (weighted) or ungrouped (unweighted).

You must be cautious when applying descriptive statistics to spatial or locational data. The way in which a geographic problem is structured can affect the resulting descriptive statistics. Issues related to the structure of statistical problems and the resultant impact on the magnitude of descriptive statistics are discussed in Section 3.4, including: (1) the effects of external boundary line delineation and study area location on the values of descriptive measures; (2) the effects of altering internal subarea boundaries within the same overall study area—a component of the so-called modifiable areal units problem (MAUP) known as the grouping or zoning problem; and (3) the impact on descriptive statistics when using different levels of spatial resolution or spatial acuity—another component of the MAUP known as the scale or aggregation problem.

3.1 Measures of Central Tendency

The central or typical value of a data set is described numerically in several different ways. Each measure of central tendency has advantages and disadvantages, and the logic underlying the calculation procedure for each measure is different. To select the most appropriate measure in a particular geographic situation requires an

TABLE 3.1

Annual Precipitation for Washington, DC: A Ranked 40-Year Record (in inches)

26.87	35.20	39.86	45.62
26.94	35.38	40.21	46.02
28.28	35.96	40.54	47.73
29.48	36.02	41.11	47.90
31.56	36.65	41.34	48.02
32.78	36.83	41.44	50.50
33.07	36.99	41.46	51.17
33.62	38.15	41.94	51.97
34.98	39.34	43.30	54.29
35.09	39.62	43.53	57.54

Source: National Climatic Data Center (NCDC)

understanding of each measure and its characteristics. The discussion in this section is limited to three widely used measures of central tendency: mode, median, and mean. Annual precipitation data from Washington, DC, collected over a 40-year period are used to present the calculation procedures and related graphics (Table 3.1).

Mode

The **mode** is simply the value that occurs most frequently in a set of ungrouped data. When nominal data are used, the mode would be the category containing the largest number of observations. With ordinal or interval-ratio data grouped into classes, the category with the largest number of observations is defined as the **modal class**. The midpoint of the modal class interval is the **crude mode**. A mode can be calculated for data at all levels of measurement. For nominal data, however, the mode is the only available descriptive measure of central tendency.

While the mode is sometimes a useful measure of central tendency, it often does not provide a practical result. For example, the mode would not be an appropriate measure for large data sets having few or no tied observations. In the Washington, DC, annual precipitation data set, no value occurs more than once over the 40-year sample period, making the mode ineffective. In fact, a large number of tied values is not likely to occur in most geographic situations where data are interval or ratio scale. However, if the precipitation data are grouped (Table 3.2) the modal class is 35–39.99 (with 12 values) and the crude mode of 37.5 is the midpoint of this modal class interval.

This is an appropriate time to introduce some graphics that are often associated with grouped data: (1) the **histogram** (relative frequency histogram) and related **frequency polygon**, and (2) ogive (cumulative frequency distribution). Graphic summaries of grouped data are constructed using several different formats, but they all share some common characteristics. Usually the frequency of values is shown on the vertical (Y) axis and the range of data is presented on the horizontal (X) axis. Information may be shown as absolute frequencies (actual frequency counts) or as relative frequencies (percentages or probabilities).

In a histogram, the frequency of values is shown as a series of vertical bars, one for each class of values. For the Washington, DC, annual precipitation, both Table 3.2 and Figure 3.1 are constructed using the equal intervals not based on range classification technique. With this procedure, the class breaks occur at convenient, rounded-off positions, and widths of class intervals are uniform. Since the precipitation data for Washington, DC, are continuous and extend from a low of 26.87 inches to a high of 57.54 inches, the histogram is constructed using a series of five-inch intervals from 25 inches to 60 inches.

A frequency polygon is similar to a histogram, except the vertical position of each class is shown as a point rather than a bar. If the values have been categorized into groups or classes, the single point for displaying the frequency is placed at the midpoint of the class interval. The points are then connected as straight lines to produce the frequency polygon. In Figure 3.1, the frequency polygon is superimposed on the histogram.

Instead of displaying the absolute frequency count on the vertical axis, a histogram or frequency polygon is easily converted to relative frequency. For example, by dividing the individual frequency values by the sum of all frequencies for the data set, the diagrams would display the frequency percentages or proportions for each class. This change would not affect the general shape of the graphic, only the scale of values along the vertical axis.

TABLE 3.2

Classification of Washington, DC, Precipitation Data Using Equal Intervals Not Based on Range Method

Category	Interval	Absolute Frequency	Cumulative Absolute Frequency	Relative Frequency	Cumulative Relative Frequency
1	25–29.99	4	4	0.100	0.100
2	30–34.99	5	9	0.125	0.225
3	35–39.99	12	21	0.300	0.525
4	40–44.99	9	30	0.225	0.750
5	45–49.99	5	35	0.125	0.875
6	50–54.99	4	39	0.100	0.975
7	55–59.99	1	40	0.025	1.000
TOTAL		40		1.000	

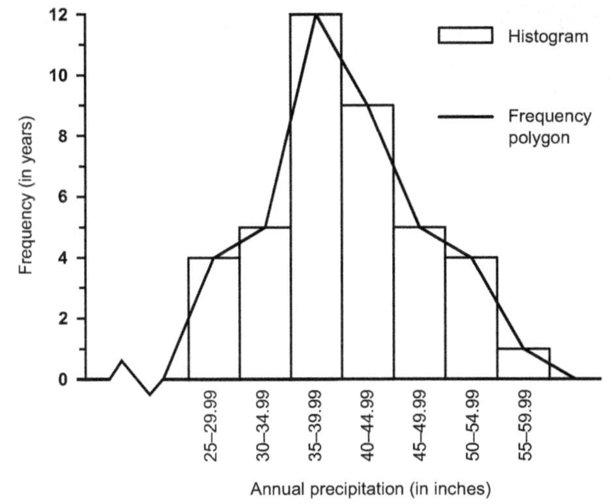

FIGURE 3.1
Histogram and Frequency Polygon for Washington, DC, 40-Year Annual Precipitation Data
Source: National Climatic Data Center (NCDC)

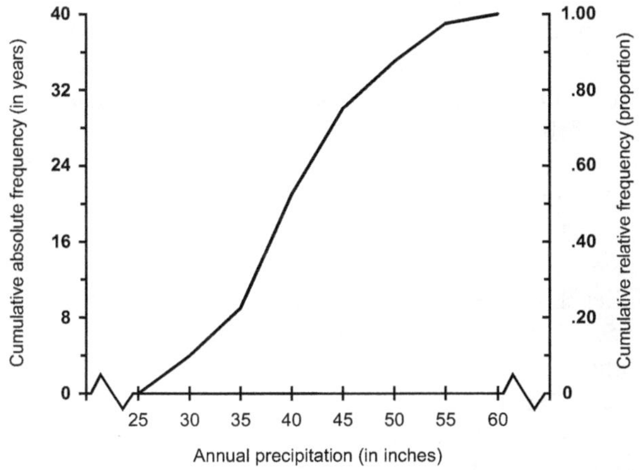

FIGURE 3.2
Cumulative Frequency Polygon (Ogive) for Washington, DC, 40-Year Annual Precipitation Data

Another useful method for displaying data in a relative frequency format is a cumulative frequency diagram, or **ogive** (Figure 3.2). This graphic aggregates frequencies from class to class and displays cumulative frequencies at each position. By starting at the lowest value and cumulating higher values, this technique is equivalent to presenting the number of values that are "less than or equal to" each value or class along the horizontal axis. The result often produces an ogive having a logistic-curve shape. In addition to displaying cumulative absolute frequencies, the data is also shown as cumulative relative frequencies (or cumulative proportions)—see the right vertical axis label on Figure 3.2.

Median

The **median** is the middle value from a set of ranked observations and is therefore the value with an equal number of data units both above it and below it. With an odd number of observations, the middle value is unique and defines the median. With an even number of observations, the median is defined as the midpoint of the values of the two "middle ranks." The Washington, DC, data has 40 values, so the two middle values are rank 20 (39.62 inches of precipitation) and rank 21 (39.86 inches), and the median is their midpoint (39.74 inches). The median is therefore not affected by extreme high or low values.

Mean

The **mean** (also called the arithmetic mean or the average) is the most widely used measure of central tendency. It is usually the most appropriate measure when using interval or ratio data. The arithmetic mean ($\overline{X}$) is the sum of a set of values divided by the number of observations in the set. In standard statistic notation, the mean is defined as follows:

$$\overline{X} = \frac{\sum_{i=1}^{n} X_i}{n} = \frac{X_1 + X_2 + \ldots X_n}{n} \quad \textbf{(3.1)}$$

where $\overline{X}$ = mean of variable X

X_i = value of observation i

Σ = summation symbol (upper case sigma)

n = number of observations

It is generally understood that summation is over all n observations, so the symbols above and below sigma are usually omitted:

$$\overline{X} = \frac{\sum X_i}{n} \quad \textbf{(3.2)}$$

The calculation of mean annual precipitation for the Washington, DC, example data is shown in Table 3.3.

In many geographic problems, a sample mean must be differentiated from a population mean. Using conventional notation, lower-case n refers to sample size and upper-case N to population size. Population characteristics are customarily defined using Greek letters, but sample measures are not. The formula for a population mean (μ = the Greek letter mu) with N values is:

$$\mu = \frac{\sum X_i}{N} \quad \textbf{(3.3)}$$

Because it is possible to draw many samples of size n from a large population of size N, a different-magnitude

TABLE 3.3

Worktable for Calculating Arithmetic Mean of Washington, DC, Precipitation Data

Observation i	Precipitation X_i
1	26.87
2	26.94
3	28.28
...	...
38	51.97
39	54.29
40	57.54
TOTAL	1598.3

$$\bar{X} = \frac{\Sigma X}{n} = \frac{26.87 + 26.94 + \ldots + 57.54}{40}$$

$$\frac{1598.30}{40} = 39.96$$

TABLE 3.4

Worktable for Calculating Weighted Mean of Washington, DC, Precipitation Data

Class interval j	Class midpoint X_j	Class frequency f_j	$X_j f_j$
25–29.99	27.5	4	110.0
30–34.99	32.5	5	162.5
35–39.99	37.5	12	450.0
40–44.99	42.5	9	382.5
45–49.99	47.5	5	237.5
50–54.99	52.5	4	210.0
55–59.99	57.5	1	57.5
TOTAL		40	1610.0

$$\bar{X}_w = \frac{\Sigma X_j F_j}{n} = \frac{1610.0}{40} = 40.25$$

sample mean may result from each sample drawn. The population mean is a fixed value, however. In Chapter 8 the distinction between samples and populations is explored further when applying sample information to estimate population characteristics.

The mean can also be calculated for grouped data. In some practical situations, computing a grouped or weighted mean is the only viable option. Suppose, for example, the only information available is in summary form, like the histogram and frequency polygon for the Washington, DC, annual precipitation data shown in Figure 3.1. Perhaps the source data from which the histogram has been constructed (like the National Climatic Data Center data in Table 3.1) is not available. In other practical situations, it may be most effective first to classify a set of data, then to use the classified (grouped) information to calculate the weighted mean. Generally, if a graphic representation such as a histogram is part of the analysis, the conventionally used method of classification is equal intervals not based on range, which is how the data in Table 3.2 and Figure 3.1 are classified.

Whatever the specific circumstance, a **weighted mean** is calculated from only the class intervals and class frequencies presented in Table 3.2 and Figure 3.1, using this formula:

$$\bar{X}_w = \frac{\sum_{j=1}^{k} X_j f_j}{n} \quad \textbf{(3.4)}$$

where $\bar{X}_w$ = weighted mean

X_j = midpoint of class interval j

f_j = frequency of class interval j

k = number of class intervals

n = total number of values = $\sum_{j=1}^{k} f_j$

To permit this calculation, two related assumptions are made about the distribution of values within each category or class interval: (1) without any information to the contrary, one assumes the data are evenly distributed within each class interval; therefore, (2) the best summary representation of the values in each interval is the **class midpoint**. Quite simply, the class midpoint is located exactly midway between the extreme values that identify the class interval. So, for example, the class midpoint of the interval from 25 to 29.99 is 27.5, which also happens to be the mean of those two values (Table 3.4).

A quick comparison of the unweighted and weighted means for the Washington, DC, precipitation data reveals that the unweighted mean is only slightly smaller than the weighted mean (the unweighted mean is 39.95, whereas the weighted mean is 40.25). In general, weighted and unweighted means should differ only slightly, depending on the overall effect of using class midpoints as representative summary values for each category. For example, if the majority of original values in most of the categories happen to fall below the class midpoints, the weighted mean will be higher than the corresponding unweighted mean.

Selecting the Proper Measure of Central Tendency

Deciding which measure of central tendency to use depends both on the geographic application and certain key characteristics of the data set. On the surface, the mean seems to have certain advantages. It is the most widely applied and generally understood measure of central tendency. In addition, it is always affected to some extent by a change or modification of any data set value. This is not always true of either the mode or the median. The mean is also an element in many tests of statistical inference when estimates of population parameters are made from sample statistics. When working with interval-ratio data, the mean is often the measure of choice.

Selection of the "best" measure of central tendency may depend on other characteristics of the distribution of the data, as the following situations illustrate:

- If a distribution is **unimodal** (having a single distinct mode) and **symmetric** (evenly balanced around a single distinct mode or central vertical line), then all three "centers" are likely to be similarly located and no advantage accrues to any statistic (Figure 3.3, Case 1).
- If a frequency distribution is not completely symmetric about a central line, it is said to be asymmetric or skewed, and the various measures of central tendency are positioned at different places along the distribution. Additional discussion and a formal definition of skew are provided in Section 3.3. With slight positive skew (Figure 3.3, Case 2), the median is pulled slightly to the right toward the positive tail of the distribution. The location of the mean is affected even more significantly, positioned even further out the positive tail than the median. With large negative skew (Figure 3.3, Case 3), the magnitude of displacement of the median and, especially the mean, is even more dramatic.
- If a frequency distribution is **bimodal** (having two distinct modes) or **multimodal** (with more than two modes), the mean may be located on the distribution at a place that would not be considered "typical." In Figure 3.4, neither the mean nor median depicts useful central locations; providing a separate value for mode A and mode B is probably more informative.
- If a frequency distribution contains one or more extreme or atypical values **(outliers)** the mean is heavily influenced by these values and its effectiveness as a measure of centrality may be reduced. The existence of an extreme value or values is actually a form of skew.

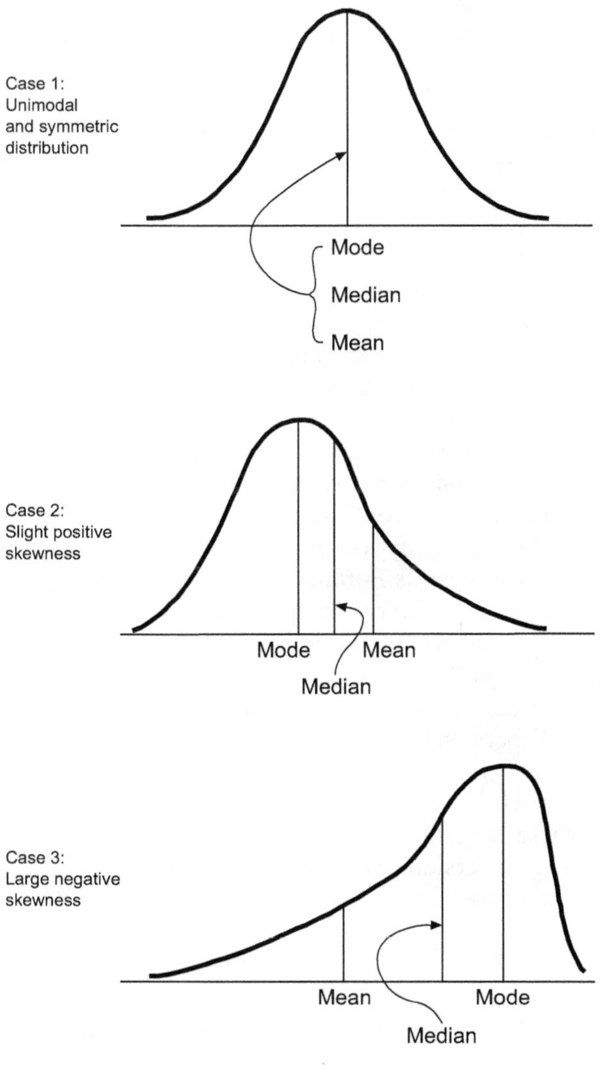

FIGURE 3.3
Measures of Central Tendency Placed on Symmetric and Asymmetric (Skewed) Frequency Distributions

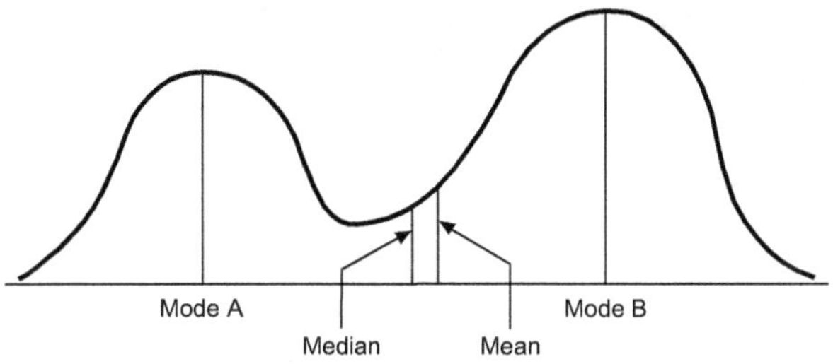

FIGURE 3.4
Measures of Central Tendency Placed on Bimodal Frequency Distribution

A simple example using the incomes of seven households illustrates the sensitivity of the mean to a single outlier (Table 3.5). The mean is ineffective as a measure of central tendency in this example due to the income of a single atypical wealthy household. The mode is too low an income to represent a useful "central" value from this frequency distribution. The median provides the most suitable measure in this situation. In fact, with variables such as income that often contain a large amount of skewness, the median is generally the descriptive statistic of choice.

TABLE 3.5

Sensitivity of the Mean to a Single Extreme Value (Outlier)

Household Income	Descriptive Statistics		
\$31,000	Sum	=	\$659,000
\$31,000			
\$32,000	Mode	=	\$31,000
\$36,000			
\$37,500	Median	=	\$36,000
\$42,500			
\$449,000	Mean	=	\$659,000 / 7
		=	\$94,142.86

3.2 Measures of Dispersion and Variability

Simple Measures of Dispersion

The three measures of central tendency—mode, median, and mean—specify the center of a distribution, but provide no indication of the amount of spread or variability in a set of values. Dispersion can be calculated in several different ways. The level of measurement and nature of the frequency distribution determine the most appropriate dispersion statistic.

The simplest measure of variability is the range, already defined in the previous chapter as the difference between the largest and smallest values in an interval-ratio set of data. Because it is derived solely from the two most extreme or atypical values and ignores all other values, the range is often a misleading measure. The range measured for the 40 annual precipitation values in the Washington, DC, example is (57.54 – 26.87) = 30.67 inches.

If data are grouped, the range is defined as the difference between the upper value in the highest-numbered class interval and the lower value in the lowest-numbered class interval. With the grouped precipitation data, these values are (59.99 – 25) = 34.99 inches. As with the ungrouped range, the grouped range can be a deceptive measure of dispersion. Given only the range, the degree of clustering or dispersion of the values between these two extremes is unknown.

To obtain more information, you can examine certain intervals, portions, or percentiles within a frequency distribution. If data are divided into equal portions or percentiles (**quantiles**) then the range of values within any quantile can be calculated and graphed. Although any logical subdivision is possible, data are often classified into quartiles (fourths), quintiles (fifths), or deciles (tenths). The median, which is the 50th percentile, may also be one of these subdivisions.

When data are divided into four portions or quartiles, the **interquartile range** (IQR) is defined as the difference between the 25th percentile value (the lower quartile) and the 75th percentile value (the upper quartile). The interquartile range thus encompasses the "middle half" of the data. In the Washington, DC, precipitation example, the interquartile range is (43.53 – 35.20) = 8.33 inches.

A **boxplot** (or box-and-whiskers plot) is a very useful graphic extension of the interquartile range concept that is particularly effective when comparing groups and displaying outliers. Many different statistics are shown on a boxplot, and you can gain additional insights by comparing multiple boxplots on the same diagram. For each group, the typical boxplot will display the median line, the interquartile range box (encompassing the middle 50% of the data), an **upper whisker** (extending to the maximum data point within 1.5 box-heights from the top of the IQR box), a **lower whisker** (extending to the minimum data point within 1.5 box-heights from the bottom of the IQR box), and outliers (all observations beyond the upper or lower whisker). Figure 3.5 is a generalized boxplot diagram with these statistics labeled.

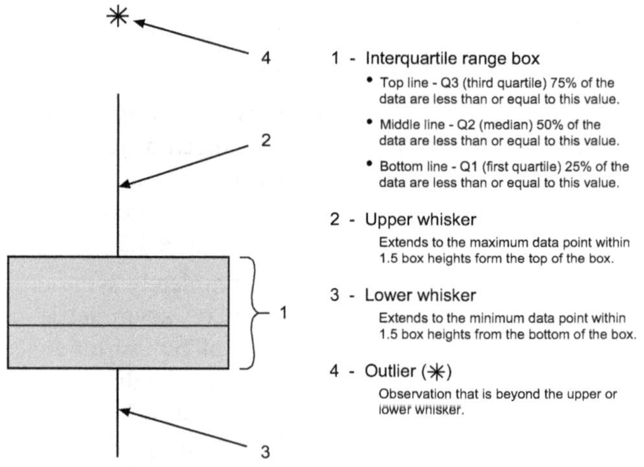

FIGURE 3.5
Generalized Diagram of a Boxplot

The value of comparing multiple boxplots on the same diagram becomes clear when we look at 40 years of annual precipitation data from Buffalo, NY, St. Louis, MO, and San Diego, CA. (Figure 3.6). The median line values for Buffalo and St. Louis appear almost identical (in fact, the medians are 39.7 inches and 37.78 inches, respectively), while the median line value for San Diego is much less (9.29 inches). You can also see from a quick visual examination of the boxplots that the interquartile range box for San Diego is nearly as large as that for Buffalo, even though Buffalo's annual precipitation is about four times that of San Diego. The **individual point distribution** (or **individual value plot**) for each of the three cities also seems to confirm that San Diego has almost as much dispersion or variation in annual precipitation from year to year as Buffalo and St. Louis. We will look at additional descriptive statistics and measures for the annual precipitation data of these three cities as we continue through this chapter.

Continuing our discussion of dispersion measures—other dispersion statistics are based on the individual deviations of each observation from the mean value. The difference between each value and the mean is calculated and these deviations are used as the building blocks to measure dispersion. An individual deviation (d_i) is calculated as follows:

$$d_i = \left(X_i - \bar{X}\right) \qquad \textbf{(3.5)}$$

where d_i = deviation of value i from the mean
X_i = value of observation i
$\bar{X}$ = mean of variable X

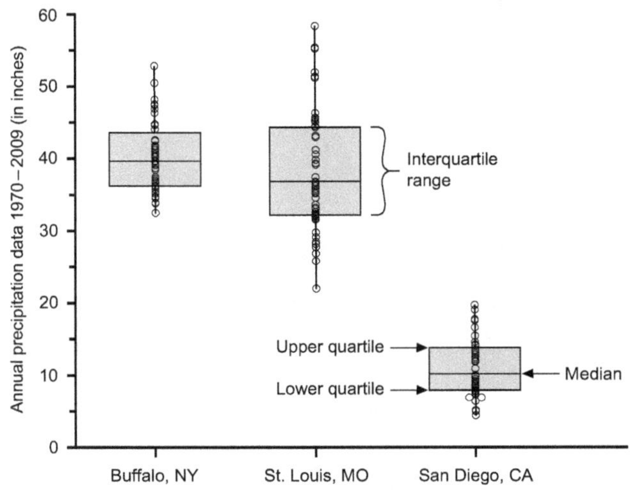

FIGURE 3.6
Comparative Boxplots of Annual Precipitation Data for Three Cities
Source: National Climatic Data Center (NCDC)

The **average deviation** or **mean deviation** is based on the mean of the set of individual deviations. That is, the mean deviation (m) for a sample of values is

$$m = \frac{\Sigma|X_i - \bar{X}|}{n} \quad \textbf{(3.6)}$$

where $|X_i - \bar{X}|$ equals the absolute value of the difference between X_i and $\bar{X}$.

Some units of data are larger than the mean, making the difference $(X_i - \bar{X})$ a positive number. Conversely, values below the mean make $(X_i - \bar{X})$ a negative number. However, the sum of the deviations about a mean is always zero. To avoid the problem of negative differences offsetting positive differences, the absolute value of each individual deviation is calculated. With absolute values, negative deviations are converted to positive ones. For example: $|-2| = |2| = 2$.

Another important property associated with the mean is its "least squares" character. Not only is the sum of deviations (that is, unsquared deviations) of all observations about a mean always zero, but the sum of *squared* deviations of all observations about a mean is the minimum sum possible. That is, the sum of squared deviations about a mean is less than the sum of squared deviations about any other number. This important attribute is called the **least squares property of the mean**, and can be summarized:

$$\bar{X}: \min \Sigma(X_i = \bar{X})^2 \quad \textbf{(3.7)}$$

Standard Deviation and Variance

The least squares property of the mean carries over into the most common measure of variability or dispersion—**standard deviation**. Given a sample of values, the standard deviation (s) is defined as:

$$s = \sqrt{\frac{\Sigma(X_i - \bar{X})^2}{n-1}} \quad \textbf{(3.8)}$$

In the numerator of Equation 3.8, the deviation of each value from the mean is squared and all of the squared deviations are summed, thus incorporating the least squares property into this measure of dispersion. Note also that the squaring process removes the problem of negative deviations offsetting positive deviations in a more effective way than absolute values. This least squares sum is then divided by (n −1) to reduce sampling bias. Finally, to reverse the effect of squaring, the square root of this quotient is taken.

For a population, the standard deviation formula is written:

$$\sigma = \sqrt{\frac{\Sigma(X_i - \mu)^2}{N}} \quad \textbf{(3.9)}$$

where σ = lower case Greek sigma

Note the different denominators of the two standard deviation formulas. The denominator of a sample is $n - 1$, but the denominator becomes N when dealing with a population. If the standard deviation is being calculated for a large sample, ($n > 30$), the difference between n and ($n - 1$) is small, and the difference between s and σ will also be small. However, with a small sample ($n < 30$), the true population standard deviation (σ) is slightly underestimated if division is by n. When dividing by ($n - 1$), s becomes larger, correcting the underestimation problem. The incorporation of this sample size correction provides a better estimate of the true population standard deviation for smaller samples.

The **variance** of a data set, defined as the square of the standard deviation (in the original units) provides a measure of the average squared deviation of a set of values around the mean. The magnitude of the variance statistic is generally a very large number and is sometimes difficult to interpret. Nevertheless, it is an extremely important descriptive measure that has many applications in inferential statistics. The formulas for sample variance (s^2) and population variance (σ^2) are as follows:

$$s^2 = \frac{\Sigma(X_i - \bar{X})^2}{n-1} \quad \textbf{(3.10)}$$

$$\sigma^2 = \frac{\Sigma(X_i - \mu)^2}{N} \quad \textbf{(3.11)}$$

Both standard deviation and variance are fundamental building blocks in statistical analysis. When discussing sampling procedures and estimation in Chapter 8, these descriptive measures are used as the theoretical basis for making probabilistic statements. Confidence intervals are placed around estimates derived from samples to predict population parameters. These two statistics are also integral components in inferential hypothesis testing. For example, variance is used in analysis of variance (ANOVA), a statistical technique that determines whether multiple (more than two) samples differ significantly from one another (Chapter 11).

A final issue regarding standard deviation and variance requires clarification—the procedural question of whether to use a more convenient computational formula or a more cumbersome and time-consuming definitional formula. Formulas based on the definition of standard deviation are valuable because they show how variability is a direct function of individual deviations $\left(X_i - \bar{X}\right)$ or $(X_i - \mu)$. However, if it is necessary to calculate standard deviation manually without computer assistance, more efficient computational formulas are available. These alternative formulas, based on algebraic manipulations of the definitional formulas, are summarized in Table 3.6. Since variance is the square of the standard deviation, it is calculated by simply removing the square root symbol from each formula in Table 3.4.

TABLE 3.6

Definitional and Computational Formulas for Standard Deviation of Sample and Population

	Definitional Formula	Computational Formula
Sample (s)	$\sqrt{\dfrac{\Sigma(X_i - \bar{X})^2}{n-1}}$	$\sqrt{\dfrac{\Sigma X_i^2 - n\bar{X}^2}{n-1}}$
Population (σ)	$\sqrt{\dfrac{\Sigma(X_i - \mu)^2}{N}}$	$\sqrt{\dfrac{\Sigma X_i^2}{N} - \mu^2}$

TABLE 3.7

Worktable for Calculating Sample Standard Deviation of Washington, DC, Precipitation Data

Observation i	X_i	$X_i - \bar{X}$	$(X_i - \bar{X})^2$	X_i^2
1	26.87	−13.09	171.28	722.00
2	26.94	−13.02	169.46	725.76
3	28.28	−11.68	136.36	799.76
...	...	...	...	...
38	51.97	12.01	144.30	2700.88
39	54.29	14.33	205.42	2947.40
40	57.54	17.58	309.14	3310.85
TOTAL	1598.30		2142.36	66,006.44

$n = 40$ $\qquad n - 1 = 39$ $\qquad \bar{X} = \dfrac{\Sigma X_i}{n} = \dfrac{1598.30}{40} = 39.96$

Using the definitional formula for sample standard deviation:

$$s = \sqrt{\frac{\Sigma(X_i - \bar{X})^2}{n-1}} = \sqrt{\frac{2142.36}{39}} = 7.41$$

Using the computational formula for sample standard deviation:

$$s = \sqrt{\frac{\Sigma X_i^2 - n\bar{X}^2}{n-1}} = \sqrt{\frac{66{,}006.44 - 40(39.96)^2}{39}} = 7.40$$

The calculation procedures for both the definitional and computational formulas of sample standard deviation are demonstrated in Table 3.7 using the Washington, DC, precipitation data. The statistical parity of the definitional and computational formulas is easily seen.

For calculating standard deviation and variance from grouped data, the same assumptions are made as when calculating a weighted mean. That is, the values are assumed to be evenly distributed within each class interval, making the midpoint the best representation of the middle of that class interval. As with the weighted average, the weighted standard deviation (s_w) and variance (s_w^2) formulas use these class midpoints (X_j) and class frequency counts (f_j). Although a definitional formula for weighted standard deviation exists, Table 3.8 uses the simpler computational formula for calculating this index with the Washington, DC, precipitation data.

The weighted standard deviation ($s_w = 7.59$) is similar in magnitude to the unweighted standard deviation ($s = 7.40$). A slight difference between the two descriptive statistics is expected given the contrasting assumptions that underlie the calculation procedures.

Now that you have been introduced to the basic descriptive measures of dispersion—standard deviation and variance—it is an appropriate time to explain the standard deviation breaks and Jenks natural breaks methods of classification that were cited in Table 2.4. Once again using the 2020 state-level obesity rate data set, we can now compare the patterns generated by these two additional classification methods with the other four classification methods shown in Figure 2.2.

The **standard deviation breaks** method determines class break values after calculating the mean and standard deviation of the data set. Rounded-off standard deviation intervals such as 1.0 or 0.5 are utilized to divide observations into categories. This method of classification is most effective when a data set is normally distributed. Note: we will discuss the normal distribution in Chapter 6.

If we want an even number of categories, the mean itself is used as a class break value, and the resultant map will clearly show which areas have values above and below the mean. On the other hand, if we want an odd number of categories, the "middle category" would be centered on the mean. All five of the choropleth maps showing state-level obesity rates (Figure 2.2) used exactly five categories, so we will again use a five-category classification with the standard deviation breaks method, to allow consistent comparisons of classification methods. The result is shown in Figure 3.7.

The **Jenks method of natural breaks** determines class breaks by finding "natural" groupings inherent in the overall distribution of data values. Each observation is assigned to a class based on its location relative to all other observations in the data set. Using an iterative (step-by-step) algorithm or procedure, the Jenks method identifies class breaks that minimize the variation in values within classes and maximize the variation in values between classes. The computational specifics of this iterative algorithm are too lengthy and complex to show in an introductory textbook, but many statistical software and GIS packages include the Jenks method among their set of univariate (single variable) classification options. The result of applying the Jenks method of natural breaks to the 2020 obesity data is shown in Figure 3.8.

TABLE 3.8

Worktable for Calculating Weighted Standard Deviation of Washington, DC, Precipitation Data

Class interval j	Class midpoint X_j	Class frequency f_j	X_jf_j	$(X_j)^2f_j$
25–29.99	27.5	4	110.0	3025.00
30–34.99	32.5	5	162.5	5281.25
35–39.99	37.5	12	450.0	16875.00
40–44.99	42.5	9	382.5	16256.25
45–49.99	47.5	5	237.5	11281.25
50–54.99	52.5	4	210.0	11025.00
55–59.99	57.5	1	57.5	3306.25
TOTAL		40	1610.0	67050.00

Using the computational formula for sample standard deviation:

$$s_w = \sqrt{\frac{\Sigma(X_j)^2 f_j - \left((\Sigma X_j f_j)^2/n\right)}{n-1}}$$

$$= \sqrt{\frac{67050 - ((1610)^2/40)}{39}} = 7.59$$

Coefficient of Variation

In geographic research, it is extremely valuable to directly compare the amount of variability in different spatial patterns to see which has the greatest spatial variation. In other problems it may be useful to compare variability in some phenomenon as it changes over time. For example, a climatologist might want to compare the variability in annual rainfall at different meteorological stations to learn which locations have the greatest variation from year to year. An economic development planner might be interested in comparing household income in several different counties to learn which region has the greatest internal variation and income inequality. The general issue of income inequality itself is now a key political and social issue. As a geographer, you might want to look at various aspects of the distribution and variation of income. Asking these investigative, comparative questions allows you to explore practical problems in a highly productive manner, often leading to better policy decisions.

Using standard deviation or variance to compare locations or regions directly is often inappropriate because they are both **absolute measures**. That is, their values depend on the size or magnitude of the units from which they are calculated. A data set with large numbers

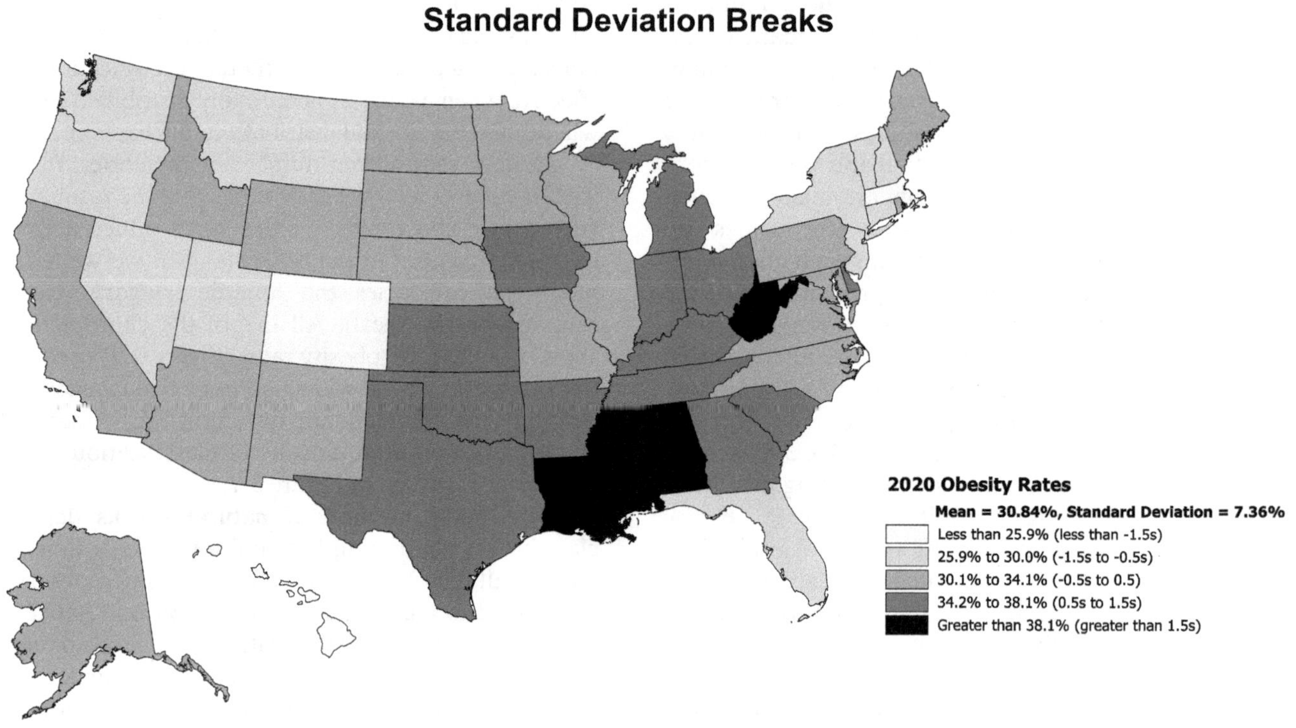

FIGURE 3.7
Percent of Adults Classified as Obese, by State, 2020: Standard Deviation Breaks
Source: Centers for Disease Control and Prevention (CDC), 2021

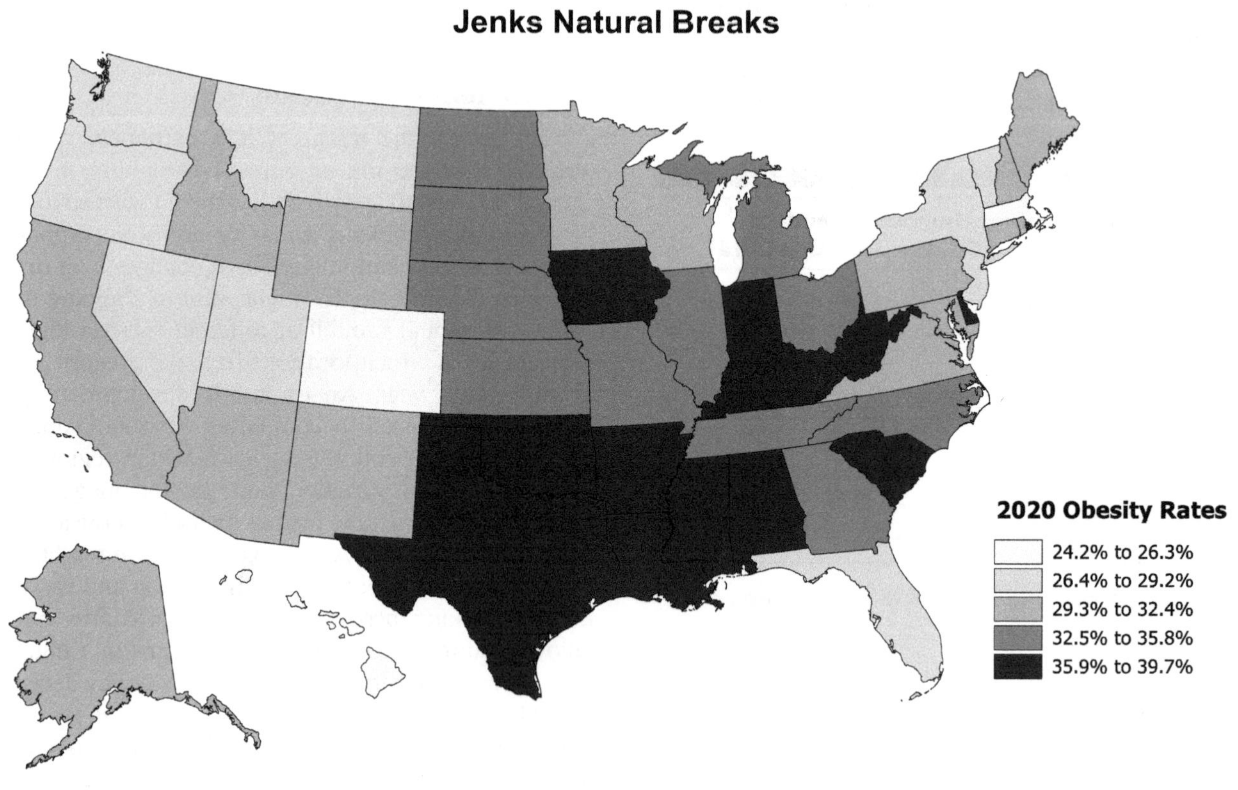

FIGURE 3.8
Percent of Adults Classified as Obese, by State, 2020: Jenks Natural Breaks
Source: Centers for Disease Control and Prevention (CDC), 2021

(e.g., in the millions) is described with a large average, standard deviation, and variance. Conversely, analysis of a data set containing small numbers results in small absolute descriptive measures. Clearly, direct comparison of averages, standard deviations, and variances is limited across data sets with different magnitudes.

To resolve this problem and compare the spatial variation of two or more geographic patterns, a **relative measure** of dispersion called the **coefficient of variation** or **coefficient of variability** (CV) is used. The CV is simply the standard deviation expressed relative to the magnitude of the mean:

$$CV = \frac{s}{\overline{X}} \quad \text{or} \quad CV = \frac{s}{\overline{X}}(100) \qquad \textbf{(3.12)}$$

The coefficient of variation is such a simple method with such great potential, it is surprising that it is so underutilized by geographers and statisticians. The CV is usually expressed as a proportion or percentage of the mean and may be used with either sample or population data. Dividing the standard deviation by the mean removes the influence of the magnitude of the data and allows direct comparison of relative variability in different data sets.

Only ratio-scale data should be used to calculate the coefficient of variation. Data measured at the interval scale are not appropriate because the interval metric has an arbitrary zero. As a result, numeric manipulations involving multiplication or division are meaningless. Given that the coefficient of variation is a ratio of the standard deviation to the mean, the coefficient of variation value has no meaning when calculated from interval scale data.

Using 40 years of annual precipitation data from three weather stations—Buffalo, NY; St. Louis, MO; and San Diego, CA (Table 3.9)—sharp contrasts between basic descriptive statistics are evident. The absolute descriptive statistics for Buffalo and St. Louis are fairly similar, but still have some notable differences. The mean annual average precipitation values are very close (Buffalo with 40.38 inches and St. Louis with 39.26 inches). However, St. Louis has a notably larger absolute amount of variability in precipitation from year to year (with a standard deviation of 8.12) than Buffalo (whose standard deviation is only 4.90). Situational factors related to climate characteristics accounts for the different standard deviation levels of these two cities. St. Louis is found in the middle of the continent (subject to changeable continental influences and extremes as well as being far from any large body of water), whereas Buffalo is close to a consistent source of moisture with prevailing winds coming off Lake Erie. By the way, you should notice that these absolute descriptive statistics confirm the graphic evidence shown in the boxplots of these three cities (Figure 3.6).

TABLE 3.9

Descriptive Statistics Using 40 Years of Annual Precipitation Data for Three Cities (In Inches)

City	Mean	Standard deviation (absolute)	Coefficient of variation (relative)
Buffalo, NY	40.38	4.90	12.14
St. Louis, MO	39.26	8.12	20.68
San Diego, CA	10.23	3.98	38.91

Of the three cities, however, semiarid San Diego has by far the greatest degree of relative variability in precipitation from year to year. Although the standard deviation for San Diego is only 3.98 inches (the smallest of the three cities), that magnitude of variability is *relatively large* when compared to San Diego's rather low average annual precipitation of 10.23 inches. The much higher coefficient of variation for San Diego clearly reveals that the amount of precipitation occurring there from year to year fluctuates much more dramatically than in Buffalo or St. Louis. Those of you interested in climatology would start to explain this phenomenon by citing the semiarid climate and the variable air and water currents that affect rainfall patterns in southern California, especially this far to the south. Therefore, one expects some degree of consistency in rainfall in Buffalo and St. Louis, while San Diego experiences greater fluctuations in rainfall in proportion to the mean.

Example: Maps of Annual Precipitation Statistics by US Climate Division

The comparative annual precipitation data from Buffalo, St. Louis, and San Diego may have whetted your appetite for more descriptive statistics and climatology. What if this brief three-city example is expanded to encompass all of the conterminous United States (excluding the climatic outliers of Alaska and Hawaii)? Can you fashion a couple of descriptive statements related to the spatial distribution and variation of precipitation across the country? Where in the US do you think the absolute measures (mean, standard deviation) of annual precipitation are the highest or the lowest? It might be a bit more difficult to hypothesize about where the relative measure (coefficient of variation) is highest or lowest. Fortunately, we can investigate these questions and reach definitive conclusions because plentiful data are available from the National Climatic Data Center (NCDC).

Located in Asheville, North Carolina, NCDC collects "normals" (means) and standard deviations of precipitation and other key climatic indicators (temperature, heating degree days, cooling degree days) for thousands of weather stations across the United States. Since 1931, NCDC has consolidated this voluminous amount of weather station data into climate division summaries. The conterminous US has 344 climate divisions, plus climate divisions in Alaska and Hawaii. According to the NCDC website, "climatic divisions are regions within each

(continued)

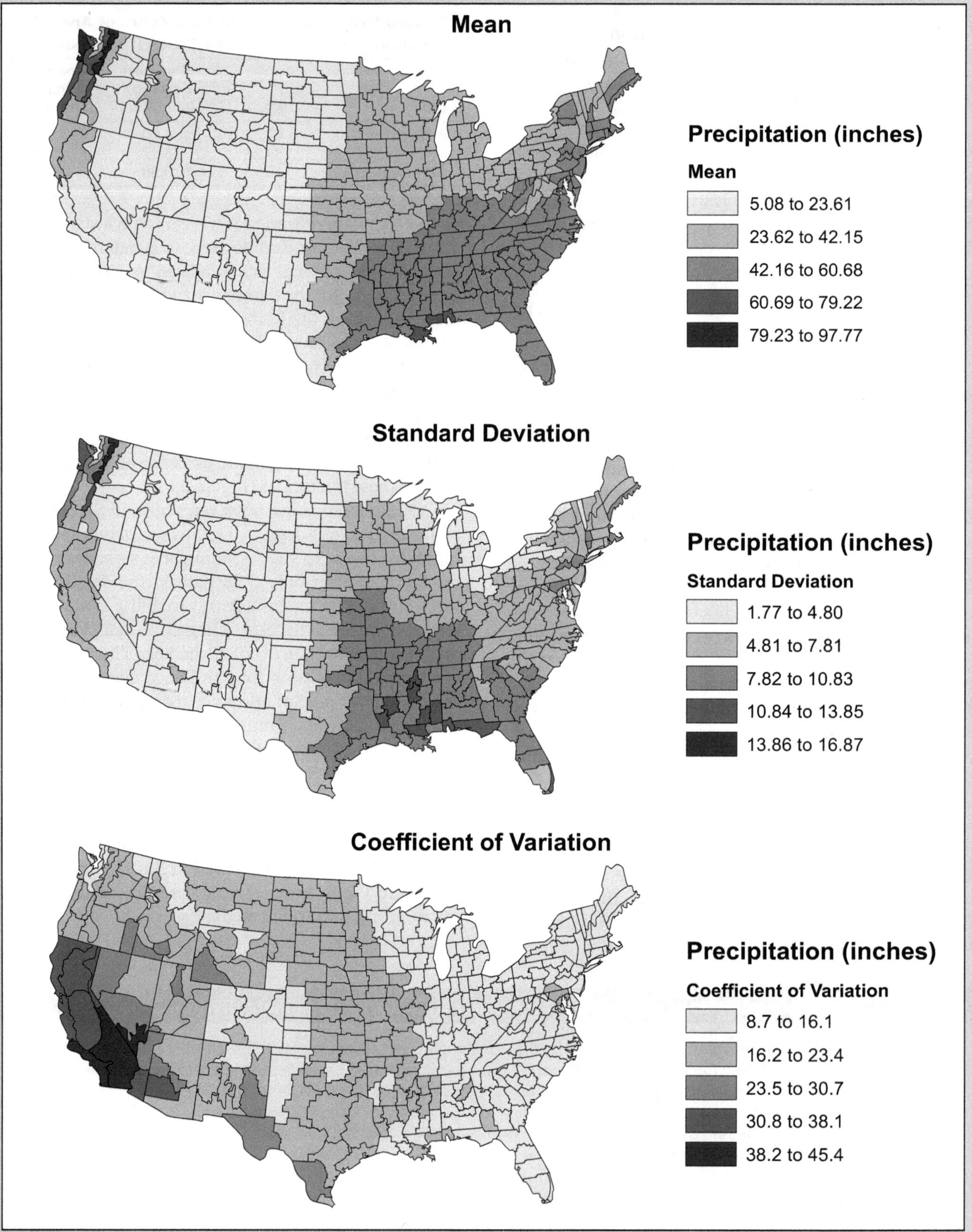

FIGURE 3.9

Maps of Annual Precipitation Statistics by US Climate Division, 1971–2000

Source: National Climatic Data Center (NCDC)

state that have been determined to be reasonably climatically homogeneous."[1] These divisions were established to encourage research in hydrology, agriculture, biogeography and other disciplines that sometimes require data averaged over an area of a state rather than just at selected points (weather stations).

NCDC derived the climate division summaries by using information from all weather stations that consistently report precipitation within a climate division. The number of reporting stations within a climate division varies from year to year, but this variation was ignored by the agency in the computation of the mean and standard deviation values.

The top portion of Figure 3.9 shows mean annual precipitation for each of the 344 climate divisions in the conterminous United States. Mean annual precipitation ranges from 5.08 inches in the very arid southwestern corner of Arizona to 97.77 inches on the northwestern coast of Washington. Relatively low mean annual precipitation values are found throughout the western third of the United States, with climate divisions in several states (Arizona, California, Nevada, Utah, and Wyoming) averaging less than 10 inches per year in precipitation. The highest mean annual precipitation values (above 60 inches per year) are concentrated in the Pacific Northwest, either along the coast of Washington and Oregon or in the Cascade Mountains of those two states. A second concentration of high mean annual precipitation values is found in climate divisions along the Gulf of Mexico coastline—from Louisiana, through Mississippi and Alabama, to Florida.

In general, we should expect climate divisions having the lowest mean annual precipitation to also have the lowest standard deviation values from year to year, and this seems to be the case, looking at the middle portion of Figure 3.9. Many of the same climate divisions in Arizona, Idaho, Nevada, Utah, and Wyoming with low average annual precipitation also have low standard deviation values from year to year. Conversely, the climate divisions having the greatest annual precipitation also generally have the greatest absolute variation in precipitation from year to year (see the areas in the Pacific Northwest and along the Gulf of Mexico).

NCDC has produced excellent multicolor maps of annual precipitation by climate division for both mean and standard deviation. With multi-color mapping (as opposed to shades of gray), spatial patterns have the potential to be shown more precisely, and NCDC personnel are effectively able to use a larger number of map categories or classes. In fact, the NCDC map of annual precipitation normals by climate division contains fifteen categories and the corresponding standard deviation map has fourteen categories. In addition, NCDC cartographers are able to use composite classification methods, with class intervals that vary by the magnitude of annual precipitation, thereby allowing them to highlight extremes to some extent while maintaining equal interval categories throughout the middle portions of the classification scheme.

By contrast, the maps shown in our text are limited to five categories (using shades of gray), and we use equal intervals based on range throughout our classification scheme. It is certainly a worthwhile exercise to compare the NCDC maps online with the maps shown in Figure 3.9.

The bottom portion of Figure 3.9 diagrams the spatial pattern of the coefficient of variation. This map illustrates the amount of relative variation in annual precipitation from year to year for each climate division. Recall that the coefficient of variation is simply the standard deviation divided by the mean. A large CV for a climate division would therefore indicate that annual precipitation fluctuates a great deal from year to year relative to the average annual precipitation, while a small CV would indicate very little fluctuation in precipitation from year to year relative to the average annual precipitation.

Being a relative descriptive measure, the coefficient of variation pattern differs markedly from the absolute descriptive measures (mean and standard deviation) and provides a greatly expanded array of geographic insights regarding the spatial patterns of annual precipitation across the United States. Of the 344 climate divisions in the conterminous US, only 11 of them have CV values greater than 30. This region of highly variable and uncertain annual precipitation from year to year is located almost entirely in the state of California, with adjacent climate divisions in southwestern Arizona and southern Nevada. In fact, the entire state of California has highly variable precipitation from year to year—all 7 of California's climate divisions have CV values greater than 30. This is clearly not just a function of low average annual precipitation in arid locations resulting in large precipitation fluctuations from year to year. The climate division along the northern California coast has a substantial average annual precipitation of 41.25 inches, but a very high CV value of 31.3.

What climatological processes are operating to cause such variability in precipitation from year to year throughout California but almost nowhere else in the United States? Current climatological thinking suggests the answer is to be found by associating Pacific Coast precipitation levels with El Niño and La Niña cycles. The term "El Niño" refers to the dramatic warming of sea-surface temperatures in the eastern and central tropical Pacific Ocean. Historic monitoring of these warming events suggests they occur in an irregular cycle, generally ranging from 2 to 7 years and lasting from as little as 6 months to as long as 4 years. During an El Niño event, increased precipitation is expected in California and adjacent portions of the southwest. This is due to more southerly zonal storm tracks.

The "cool" event counterpoint of El Niño is La Niña. During La Niña events, the polar and Pacific jet streams shift northward and increased precipitation is diverted into the Pacific Northwest due to more northerly zonal storm tracks.

Recent studies have shown that heavy precipitation years in southern California (especially in Climate Division 6, which represents the southern California coast from Santa Barbara south to San Diego) are strongly associated with moderate-to-strong El Niño events, while years with below-normal precipitation are generally associated with moderate-to-strong La Niña occurrences. These relationships seem to be strongest in the San Diego area.

Another phenomenon worth noting in the coefficient of variation portion of Figure 3.9 is the general west-to-east decline in CV values across the US. What descriptive hypotheses would you propose to account for this decline in precipitation variation as we move eastward?

[1] "US Climate Normals 1971–2000, Products," National Centers for Environmental Information (NCEI)

3.3 Measures of Shape or Relative Position

Skewness and Kurtosis

In addition to the coefficient of variation, two other relative measures—skewness and kurtosis—further describe the nature or character of a frequency distribution. As already mentioned, **skew** measures the degree of symmetry in a frequency distribution by determining the extent to which the values are evenly or unevenly distributed on either side of the mean. **Kurtosis** measures the flatness or peakedness of a data set along with the size of the tails. Like the coefficient of variation, geographers underutilize these indices, yet they can sometimes provide important descriptive insights about a frequency distribution and offer considerable potential in spatial research.

Introducing the concept of moments of a distribution about the mean provides a more complete understanding of skewness and kurtosis. The first moment is the sum of individual deviations about the mean and must equal zero:

$$\text{First moment} = \Sigma d_i = \Sigma\left(X_i - \bar{X}\right)^1 = 0 \qquad \textbf{(3.13)}$$

The second moment of a frequency distribution is the numerator in the expression that defines variance:

$$\text{Second moment} = \Sigma\left(X_i - \bar{X}\right)^2 \qquad \textbf{(3.14)}$$

Skewness involves use of the third moment of a frequency distribution. One commonly used measure of relative skewness contains the third standardized moment in the numerator:

$$\text{Skewness} = \frac{\Sigma\left(X_i - \bar{X}\right)^3}{ns^3} \qquad \textbf{(3.15)}$$

The denominator of this expression contains the cubed standard deviation, which effectively standardizes the third moment. This allows you to compare the amount of relative skewness in different frequency distributions directly.

If a frequency distribution is roughly symmetric, with a roughly equal number of values on either side of the mean, the distribution has little or no skew. If a value in a distribution is greater than the mean, its cubed deviation is positive. However, if a value is less than the mean, it will produce a negative cubed deviation. In a symmetric distribution, these positive and negative cubed deviations counterbalance each other and the sum (the third moment) is close to zero. In a distribution with a longer tail to the left, large negative cubed deviations cause the sum of all deviations (the third moment) to be negative. In this case, the resultant distribution is said to be **negatively skewed**. In a distribution with a longer tail to the right, large positive cubed deviations will dominate the sum, resulting in a **positively skewed** distribution.

Another measure of skewness, known as Pearson's coefficient, is based on a comparison of the mean and median:

$$\text{Pearson's Skewness} = \frac{3\left(\bar{X} - \text{median}\right)}{s} \qquad \textbf{(3.16)}$$

With a unimodal and symmetric distribution, the mean and median have similar values, and the skewness coefficient is close to zero (Figure 3.3, Case 1). When the mean is greater than the median (as in Case 2), positive skew generally results, and when the mean is less than the median (as in Case 3), the skew measure is generally negative. Division by the standard deviation provides a standardization of the Pearson's skewness values, allowing direct comparisons.

Kurtosis measures the fourth standardized moment of a frequency distribution using the following formula:

$$\text{Kurtosis} = \frac{\Sigma\left(X_i - \bar{X}\right)^4}{ns^4} \qquad \textbf{(3.17)}$$

If a large proportion of all values are clustered in one part of the distribution, it will have a "pointed" or "peaked" appearance, a high level of kurtosis, and be considered **leptokurtic** (Figure 3.10), having a somewhat *heavier* tail. In a data set having low kurtosis (a "flat" or **platykurtic** distribution), values are dispersed more evenly over many different portions of the distribution and exhibit a flatter tail. Some distributions are considered **mesokurtic**, having a "moderate" or "bell-shaped" appearance, which is neither very peaked nor very flat.

The interpretation of kurtosis is enhanced by comparing the peakedness of a distribution to that of a normal probability distribution. Although the importance of the normal curve is discussed in more detail in Chapter 6, it is worth mentioning here because kurtosis formulas assign characteristic values to a normal distribution. In Equation 3.17, a leptokurtic distribution has a kurtosis greater

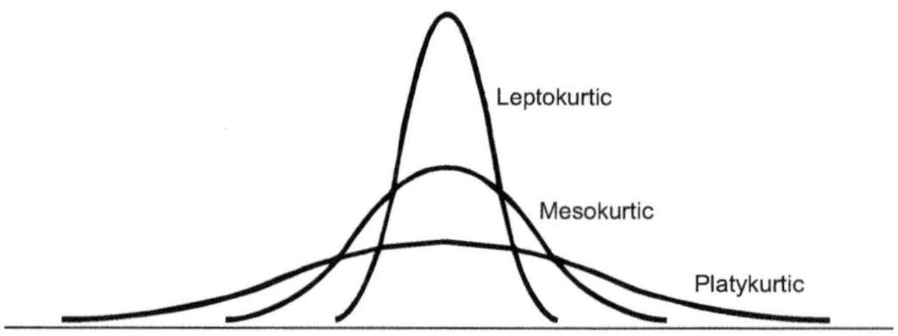

FIGURE 3.10
Different Levels of Kurtosis

than 3.0, a normal distribution is mesokurtic, with a kurtosis value of 3, and a platykurtic distribution has a value less than 3. Some computer software packages use Equation 3.18, subtracting 3 from the quotient, so that a normal distribution has zero kurtosis. With this alternative, platykurtic distributions are negative, mesokurtic distributions are close to zero, and leptokurtic distributions are positive.

$$\text{Kurtosis} = \frac{\Sigma(X_i - \bar{X})^4}{ns^4} - 3 \qquad \textbf{(3.18)}$$

Exploratory Comparison of Skewness and Kurtosis Values

Can the examination of skew and kurtosis values provide any useful insights about the comparative nature of spatial patterns? Geographers sometimes want to compare descriptive statistics for a problem variable at different locations during a single time period. They may ask such questions as: "During this period of time, which locations had the greatest relative variability? . . . the greatest amount of skew? . . . the most leptokurtic distribution?" We can then continue the investigation by asking *why* these observed spatial patterns of skew or kurtosis occurred. By exploring data sets more thoroughly through the application of descriptive statistics, geographers will gain a better understanding of the problem variable.

You must proceed with caution, however, as recent research shows both skew and kurtosis statistics appear very dependent on sample size. While skew and kurtosis provide fully valid descriptive measures of a data set, they may provide very imprecise population parameter or probability model estimates. In fact, in many cases even sample sizes as large as several hundred fail to give good estimates of the true probability model skewness and kurtosis. We will discuss techniques for estimating population parameters from sample statistics in later chapters.

Given these limitations, we can still make some valid descriptive statements regarding the 40 years of annual precipitation data from Buffalo, St. Louis, and San Diego. In Table 3.10, intercity comparisons about precipitation distributions are made for skew and kurtosis. All three cities show slightly positive skew. A quick glance at their boxplots (Figure 3.6) suggests a slight amount of skew in each case, because the upper quartile portion of the interquartile range is larger than the lower quartile portion. In addition, there seems to be more dispersion of individual values plotted above the interquartile range compared to the dispersion of individual values plotted below the interquartile range. These observations indicate a slightly higher chance of these cities getting an annual precipitation well above their averages than getting an annual precipitation well below their averages. You might speculate a bit about whether these facts concerning year-to-year differences in annual precipitation have any practical economic implications, such as future irrigation needs or flood management protection strategies.

The kurtosis values are all very close to zero, and don't seem to reveal anything meaningful about the flatness or peakedness of the distributions. We might gain some geographic insight, however, if we extended the data collection to hundreds of cities across the United States. Would any nonrandom kurtosis pattern be revealed?

TABLE 3.10

Additional Descriptive Statistics Using 40 Years of Annual Precipitation Data for Three Cities

City	Mean	Standard deviation	Skewness	Kurtosis
Buffalo, NY	40.38	4.90	0.74	0.15
St. Louis, MO	39.26	8.12	0.47	−0.25
San Diego, CA	10.23	3.98	0.59	−0.25

3.4 Selected Issues: Spatial Data and Descriptive Statistics

Geographers need to recognize several potential complications associated with the analysis of spatial or location-based data. Aspects of these problems and issues are also discussed to some extent later in the text, (for example, in Chapter 16 we examine how correlations can be dramatically affected by the spatial structure of a data set). However, we concentrate here solely on how spatial data can affect the value and magnitude of various descriptive statistics. Problems addressed in this section include how each of the following can affect the summary statistics you calculate from a set of data: (1) alteration of the external boundary of a study area; (2) modification of internal (subarea) boundaries; and (3) change in the level of spatial resolution by using a different scale or level of aggregation.

As an introductory text, our discussion of these complications is deliberately limited to a few basic comments and rather simplistic examples. If you are interested in following up on these ideas, many intermediate and advanced references are available—a few of these are identified at the end of this chapter. Sadly, these issues are often not discussed sufficiently outside the discipline. However, a basic understanding of the nature of these problems is essential for conducting geographic research.

Impact of External Boundary Delineation

The location and spatial extent of the overall study area boundary (external boundary) is likely to have a significant impact on the magnitude of any summary descriptive statistics that you calculate. Let's look at a simple example.

Suppose you are assisting local planners who are interested in analyzing the number and spatial pattern of persons below the poverty level in the Wicomico County, Maryland area (Wicomico County is the home county of Salisbury University). There are several alternatives for the external study area boundary, including: (1) all of Wicomico County; (2) the Salisbury/Wicomico Metropolitan Planning Organization (S/WMPO) area boundary; and (3) the incorporated area of Salisbury (Figure 3.11).

Suppose you have detailed locational information regarding the number and distribution of people below the poverty level (perhaps down to the census tract area), so any of these three external study area boundaries is feasible. It is generally known that dense clusters of poverty are located near the Salisbury city center, in multiple directions. If you analyze only those census tracts within the incorporated area of Salisbury, you suspect the average number (and percentage) of people below the poverty level per census tract will be higher than if you analyze all census tracts in the S/WMPO area or all census tracts in the county.

Other absolute measures whose values are partially a function of the mean, such as standard deviation and variance, will also be affected by the external study area boundary selected. These effects may be complicated and rather hard to predict. For example, is absolute variability in poverty levels greatest at the city, metro planning area, or county level? The answer to this question is not obvious, but the standard deviation and variance values will likely differ depending on the study area boundary selected. It is clear, therefore, that absolute descriptive statistics should be evaluated primarily in relation to a particular external study area boundary.

Modifiable Areal Unit Problem: Modification of Internal Subarea Boundaries

Descriptive statistics can also be influenced significantly by using alternative subdivision or regionalization schemes within the same overall study area. This is an important component of the Modifiable Areal Unit Problem (MAUP) often referred to as the **grouping or zoning problem**. In many geographic problems, the external study area may be fixed, but the internal subarea boundaries may be drawn in many different ways, resulting in different descriptive statistical summaries when the subarea data is analyzed.

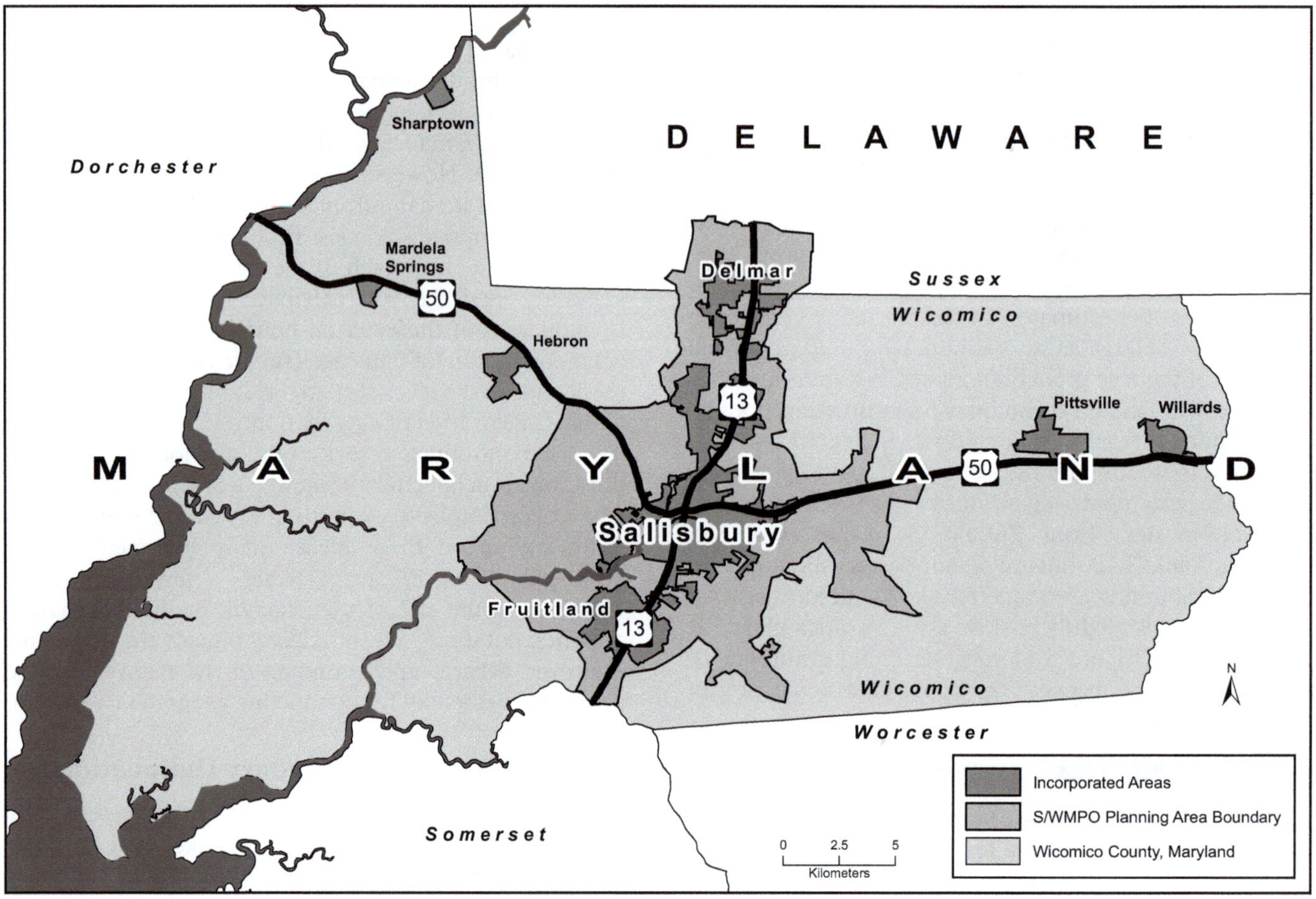

FIGURE 3.11

Alternative External Study Area Boundaries: Wicomico County, MD and Salisbury/Wicomico Metropolitan Planning Organization (S/WMPO) Area Boundary

Source: Salisbury/Wicomico Metropolitan Planning Organization, Incorporated Areas and Boundary Map (2000)

The MAUP grouping or zoning problem is illustrated with a very simple example. Suppose the shaded area in Figure 3.12 represents the location of a high concentration of a particular demographic group. In both Case 1 and Case 2 the external boundary (study area) is the same, the location of the demographic group within that external study area boundary is the same, and the number of internal subareas in the overall study area is the same (6 subareas in each case). However, the positioning of the internal subarea boundaries seems to create the perception that the group is more "integrated" (Case 1) or more "segregated" (Case 2) within the study area. When basic descriptive statistics (mean, percentage, standard deviation, etc.) are calculated using these alternative subdivision schemes, dramatically different results are likely to occur. Again, caution is advised when interpreting descriptive statistics. It is likely that summary measures will have fully valid comparative interpretations only if calculated over the same area and subarea configuration.

A practical example further illustrates the difficulties that the grouping or zoning component of the MAUP can uncover. Consider an actual experimental farm in upstate New York. Agronomists took 72 soil samples and

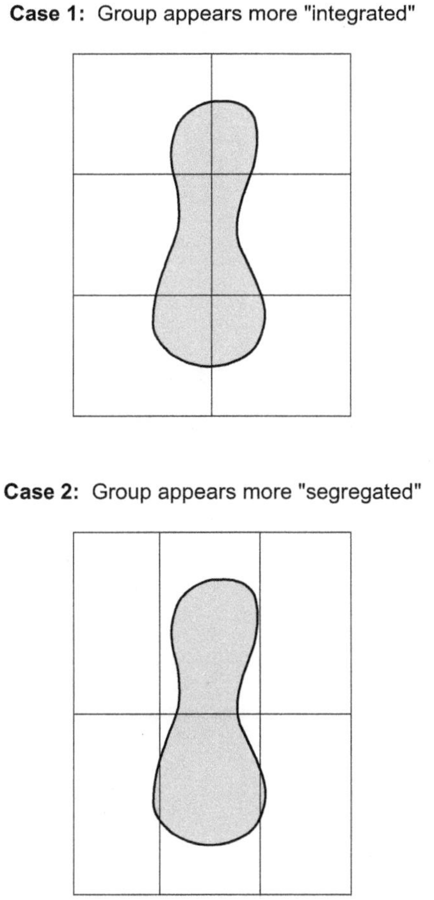

FIGURE 3.12
Perceived Affect of Alternative Internal Subarea Boundaries

analyzed them in a lab to determine numerous chemical concentrations for each sample. In Figure 3.13 we show the chemical concentrations of phosphorous (P) and potassium (K) in parts per million (ppm) at each sample point in the farm field. Phosphorus is a macronutrient essential for crop production as it stimulates early plant growth, gives plants a good and vigorous start, and allows the crop to mature more quickly. Potassium is associated with the movement of water, nutrients, and carbohydrates in plant tissue. If potassium is lacking, growth is stunted and yields are reduced. Therefore, an agronomist may be interested in understanding the relationship between potassium and phosphorous throughout a field to better enact crop management strategies.

The phosphorous level appears *above* each of the 72 sample point symbols, while the potassium level appears *below* each sample point symbol. The overall average phosphorous level (across all 72 sample points) is 5.6 with a variance of 21.77 and the overall average potassium level (again, across all 72 sample points) is 69 with a variance of 1056. Each sample point symbol indicates whether the concentration values are above or below the field-wide averages (see the legend in Figure 3.13 for details). A glance at Figure 3.13 illustrates that both the potassium and phosphorous values generally increase from north to south. Noticeable differences in pattern are apparent, however. The "above average potassium zone" includes all sample points in the southeast corner of the field, while the "above average phosphorous zone" does not. Other slight differences can be noted where the "above average" zones do not coincide.

Agricultural studies often segment a study area (field) into different subareas or regions, allowing them to collect fewer samples by associating the sample averages with each subarea or region. The grouping or zoning component of MAUP describes the general potential problem that different regionalization or internal subdivision schemes may yield different statistical summaries. Given that the underlying geographic patterns of potassium and phosphorous are generally similar, will different regionalization schemes result in significantly different summary statistics?

For this example, we created three different regionalization patterns, and constructed scatterplots showing the relationships between phosphorous and potassium. For each example there were six groups, and each group contained 12 sample points.

The regionalization schemes were chosen to extract different geographic patterns, to the extent that they exist at all. In Figure 3.14, Case 1 is designed to pick up east/west variation, Case 2 is designed to pick up north/south variation, and Case 3 is a compromise, attempting to capture both east/west and north/south variability. The Figure shows all three regionalization schemes and the mean and variance for both potassium and phosphorous in each scheme.

While the means in each regionalization scheme remain the same (simply because we are computing the mean of the means), the variances differ. The variance relates to the actual regionalization or internal subarea boundary scheme, thus presenting a real interpretive challenge. If we look at a series of scatterplots of the values for potassium and phosphorous overall and in each of the three cases, the results can be interpreted very differently. Figure 3.15 shows each scatterplot along with a basic interpretation.

When the regionalization schemes (internal subarea boundaries) are modified, we notice there are no differences in averages; however, there are significant differences in variances. The greater problem introduced with the different areal units is that one can obtain diametrically opposed interpretations of the relationship between potassium and phosphorous. Oddly enough, even when using the same set of underlying sample data, we can identify a possible negative relationship between potassium and phosphorous in Case 1, but in Case 2 we might conclude there is a positive relationship. And remember, in each instance, we are using exactly the same spatial pattern of data! Only the regionalization scheme has changed. It would appear that Case 3 most closely matches the overall data, probably because this regionalization scheme sought to accommodate the variation in both north/south and east/west directions.

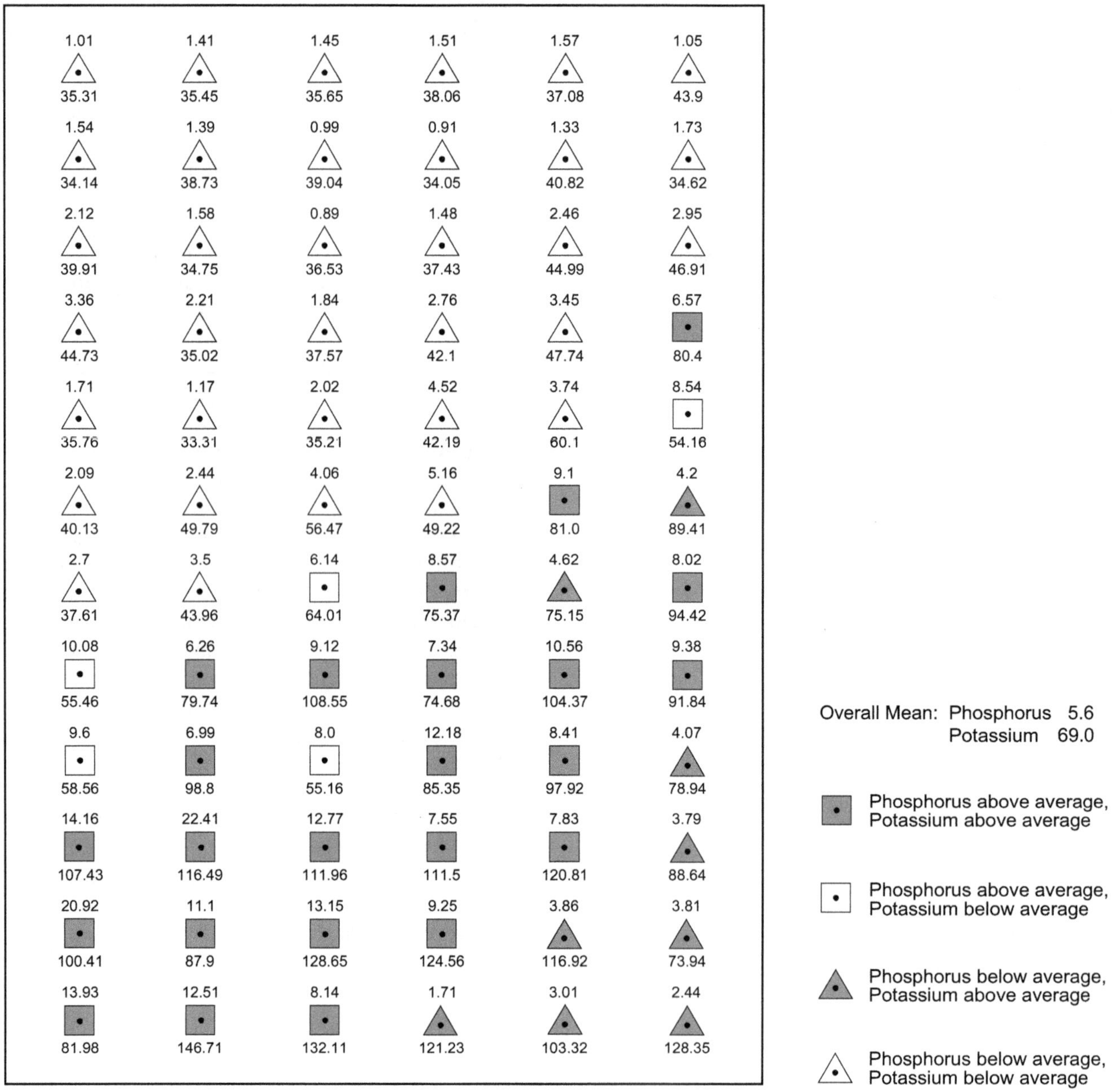

FIGURE 3.13

Farm Field Showing the Spatial Patterns of Phosphorus Levels (above) and Potassium Levels (below) for 72 Sample Points

Source: Lembo, A., Lew, M., Laba, M., Baveyye, P. 2006

So what should a geographer conclude? This real-world situation certainly illustrates the potential magnitude of problems associated with the grouping and zoning component of MAUP, and clearly demonstrates why it is considered one of the more vexing problems in geography. This example also demonstrates that understanding the nature of the underlying data may actually help the geographer determine the most appropriate regionalization (internal subarea boundary) scheme to use. This is an important point. The use of GIS tools to overlay and aggregate data may lead to embarrassing results if the geographer does not understand the data being analyzed.

Modifiable Areal Unit Problem: Change in Scale or Level of Spatial Aggregation

When analyzing data, you can often choose from many different levels of spatial aggregation or scale. Many socioeconomic variables, for example, are often available at the block, census tract, enumeration district, election district, county, planning region, and state levels. When the same data are aggregated at different spatial levels or scales, the resultant descriptive statistics will vary, sometimes in a systematic, predictable fashion and sometimes in an uncertain way.

Population size is a basic variable, readily available at virtually any level of aggregation. In this simple example, we present population figures for two dates (2010 and 2020) at three different levels of aggregation (state, census division [CD], and census region [CR]). Figure 3.16 shows all three scales—50 states aggregated to 9 census divisions, then 9 census divisions subsequently aggregated to 4 census regions. Many of the basic descriptive statistics (both absolute and relative) introduced in this chapter are calculated for both dates and all three scales or levels of aggregation (Table 3.11).

As you might expect, the mean and standard deviation values for each date increase in magnitude as the scale increases in level of aggregation. For example, in 2010 the average state population was 6,053,830 with a standard deviation of 6,823,980. When these 50 state populations were aggregated to nine census divisions, the average CD population was 34,305,600 with a standard deviation of 16,093,910. Further aggregating the nine CDs to four census regions resulted in an average CR population of 77,186,380 with standard deviation of 25,867,930. Clearly, by aggregating data into larger subareas, the magnitude of the absolute descriptive measures increases.

In contrast to the absolute measures, the columns of Table 3.11 showing relative measures are more challeng-

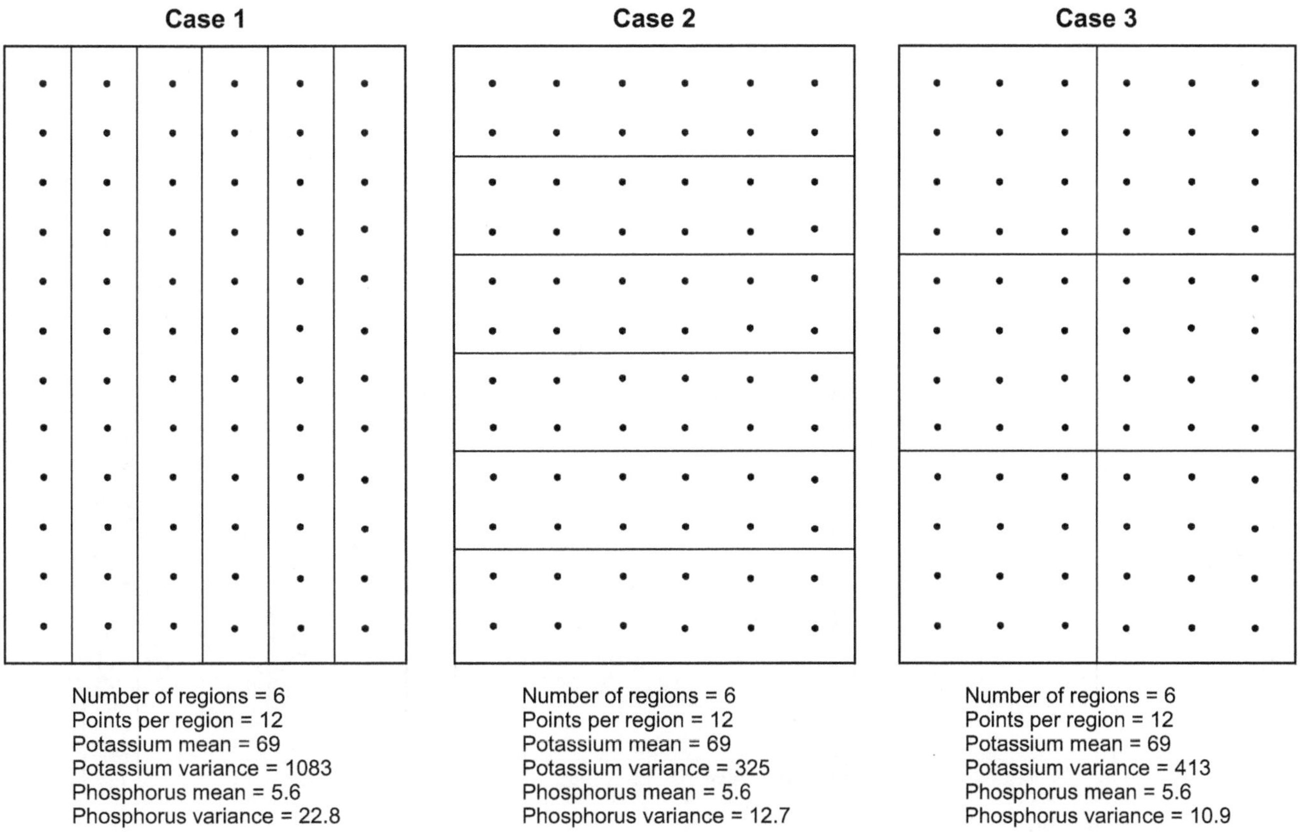

FIGURE 3.14

Three Regionalization Schemes (Different Internal Subarea Boundaries) for the Same Field (Same External Study Area Boundary)

Source: Lembo, A., Lew, M., Laba, M., Baveyye, P. 2006

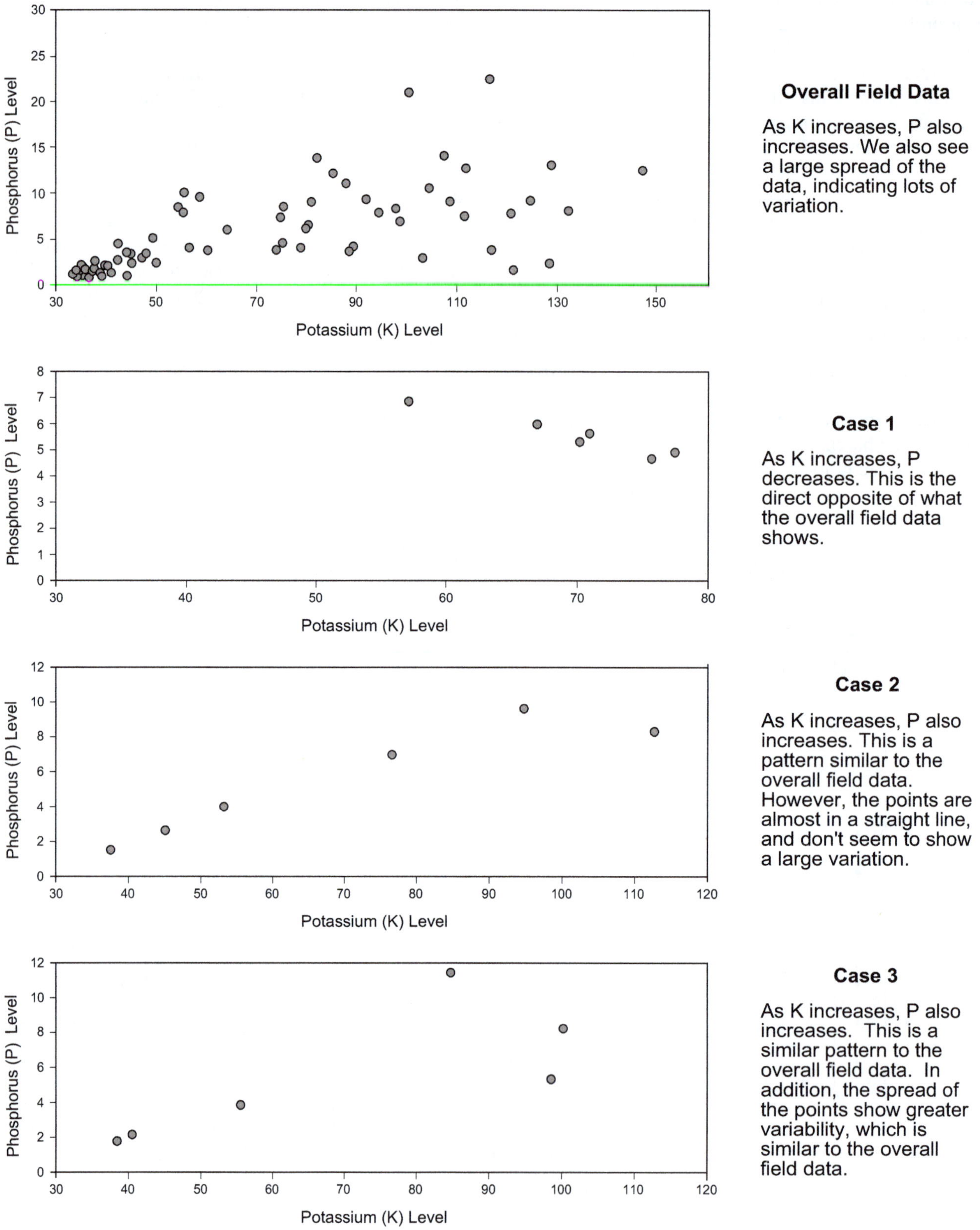

FIGURE 3.15

Scatterplots Showing Various Relationships Between Potassium and Phosphorus: Overall Field Summary and Three Different Internal Subarea Boundary Schemes

Source: Lembo, A., Lew, M., Laba, M., Baveyye, P. 2006

ing to interpret geographically. For both 2010 and 2020, notice that the magnitude of the coefficient of variation decreases as the level of spatial aggregation increases. Why would this occur? At the state level, population figures vary greatly, from very large states like California and New York to very small states like Alaska and Vermont. Therefore the standard deviation of these 50 state populations is large *relative to* the average of those populations. As we shift to the census division scale, and especially to the highly aggregated census region scale, the population averages become much larger, but the standard deviation increases at a much slower pace, making the ratio of standard deviation to mean (that is, the coefficient of variation) a smaller number.

The interpretive value of skew and kurtosis seems very limited in this example. This is not surprising, since these measures of shape or relative position generally become useful only when the sample size is quite large. However, the skew values for the State aggregation moves dramatically from 2.65 to –1.50. The 2020 negative skew

TABLE 3.11

Descriptive Statistics of Total Population by Date and Level of Spatial Aggregation

	Absolute measures (in 1000s)		Relative measures		
2010	**Mean**	**Standard deviation**	**Coefficient of variation**	**Skewness**	**Kurtosis**
State	6,053.83	6,823.98	112.72	2.65	8.82
Census Division	34,305.60	16,093.91	46.91	0.23	–1.45
Census Region	77,186.38	25,867.93	33.51	1.56	2.81
2020	**Mean**	**Standard deviation**	**Coefficient of variation**	**Skewness**	**Kurtosis**
State	6,499.01	7,408.02	113.99	–1.5	8.49
Census Division	36,827.70	17,488.94	47.49	0.32	–1.09
Census Region	82,862.32	29,745.58	35.90	1.51	2.54

Source: United States Bureau of the Census

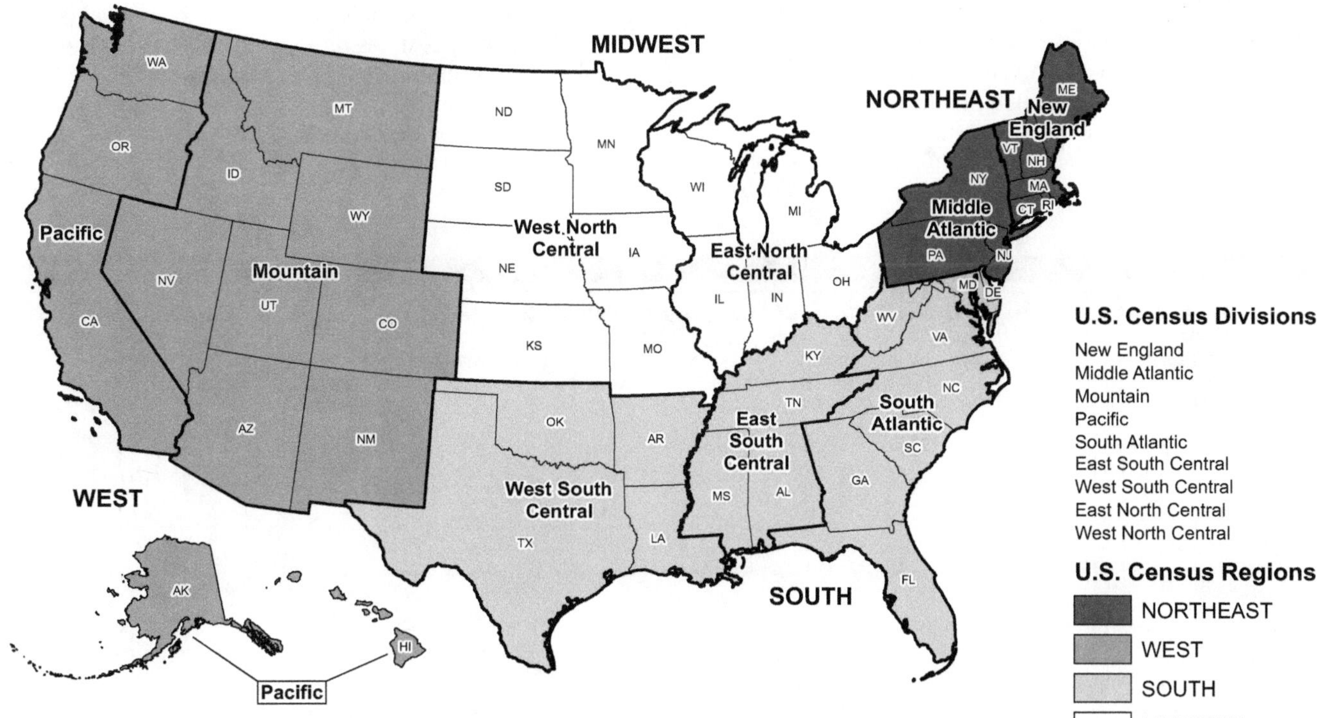

FIGURE 3.16
U.S. States, Census Divisions, and Census Regions
Source: United States Bureau of the Census

suggests that the negative (left) tail of the distribution extends a bit more from the mean than the positive (right) tail of the distribution. This relationship might suggest that larger population states (e.g., California, Illinois, New Jersey, and New York) are growing slower than many small population states in terms of their percent growth. The only states with large populations showing significant growth rates were Texas and Florida. This is a different trend than we've seen in previous editions of the textbook, and perhaps introduces a whole series of hypotheses we might consider for this changing demographic.

The kurtosis values are also generally limited in interpretive value. However, the large positive state-level kurtosis values for both 2010 and 2020 seem to indicate somewhat leptokurtic (peaked) distributions. If you constructed histograms of these data sets, you would probably see the many states with rather low populations creating a peak in the histogram, with a few large population states on the positive tail. The result would represent a rather "peaked" distribution. The census division and census region kurtosis values do not seem to reveal any geographic insights.

COVID-19 Conversation: Descriptive Statistics and Graphic During the COVID-19 Pandemic

As health experts and politicians all over the world explored strategies for dealing with COVID-19, a debate brewed as to the most effective way to protect the public. Rising cases, hospitalizations, and deaths were alarming and required swift decision making to mitigate deaths. Was COVID-19 an equal opportunity virus, or were some portions of the population more at risk than others? In Chapter 1, we began to ask questions regarding the impact of comorbidities, such as obesity, on the severity of COVID-19, and also the regional impact of deaths in the United States. Another question raised was whether children and young adults had similar outcomes as the older population. Understanding who was most vulnerable could hopefully provide insight into effective policies by state and local governments.

States began collecting and presenting data through dashboards. Many of these dashboards were run by state-level GIS specialists, allowing geographers to play an important role in fighting the pandemic. The State of Maryland was an early adopter of GIS and dashboard technology to provide the public with critical information regarding the pandemic. Unfortunately, raw data was not available to the general public, but summaries like those in Table 3.12 show the age breakdown of COVID-19 deaths in Maryland. Using the class intervals and frequencies, we can calculate the weighted age of COVID-19 deaths for the state. Perhaps you can explore the dashboards created by your own state to determine the average age of deaths from COVID-19.

According to the US Census Bureau, the average age in Maryland is 38.7. Our calculated weighted mean of 73 years for COVID-19 deaths is much higher. One thing to note is that the results of our weighted mean might be slightly lower than the actual number, as our last class midpoint is 84.5. Certainly, there were people in their nineties who died of COVID-19, but we simply do not have that class breakdown reported by the State of Maryland. Nonetheless, it is evident that the elderly population appears more vulnerable to COVID-19 deaths than the younger population. Public health officials can use this information to discuss policies regarding testing, vaccination, and lockdowns for businesses, schools, and independent and assisted living communities.

We previously discussed the coefficient of variation's usefulness in assessing variability in proportion to the average. In the case of COVID-19, states with a large population will obviously have a greater number and average number of deaths than states with a smaller population. Figure 3.17 shows the coefficient of variation of deaths by month in 2020. For each state, we calculated the number of deaths for each month, and then computed the average and standard deviation. This

TABLE 3.12

Worktable for Calculating Weighted Mean of Maryland COVID-19 Deaths by Age (as of January 2022)

Class interval j	Class midpoint X_j	Class frequency F_j	X_jF_j
0–9	4.5	4	18.0
10–19	14.5	12	174.0
20–29	24.4	62	1,512.8
30–39	34.5	180	6,210.0
40–49	44.5	465	20,692.5
50–59	55.5	1,185	65,767.5
60–69	64.5	2,205	142,222.5
70–79	74.5	3,108	231,546.0
80+	84.5	5,135	433,907.5
TOTAL		12,356	902,050.8

$$\bar{X}_w = \frac{\Sigma X_j F_j}{n} = \frac{902{,}050.8}{12{,}356} = 73.00$$

map is telling us how variable the deaths were in proportion to the average number of deaths within each state. That is, states with higher coefficients of variation had some months that had very high number of deaths, and some months that were lower in proportion to their average. States with lower coefficients of variation were more consistent in the number of deaths throughout the year.

One can see that New York is in the highest category for coefficient of variation, due most likely to being hit so hard in the early days of the pandemic, and eventually leveling off to a more consistent rate. In fact, exploring the data further, we see that 75% of the deaths in New York occurred in April and May. Curiously, we also see some states like Wyoming, Maine, Montana, and the Dakotas with relatively high coefficient of variation rates. Perhaps this is due to a lower populated state having a spike in COVID-19 deaths during a few months in 2020. When exploring this map, it can help us dig deeper into not only the causes, but potentially the impact of COVID-19 on the state. For example, we can further explore those states with high coefficients of variation to determine which months account for the large variation and determine if hospitalizations were ever at critical levels, or if particular events or seasonality may have had an impact on the spikes occurring in those months. In the case of Wyoming, there were no COVID-19 deaths in March of 2020 as the pandemic had not yet reached its borders, while over 80% of the deaths occurred in November and December, illustrating the movement of the pandemic over time.

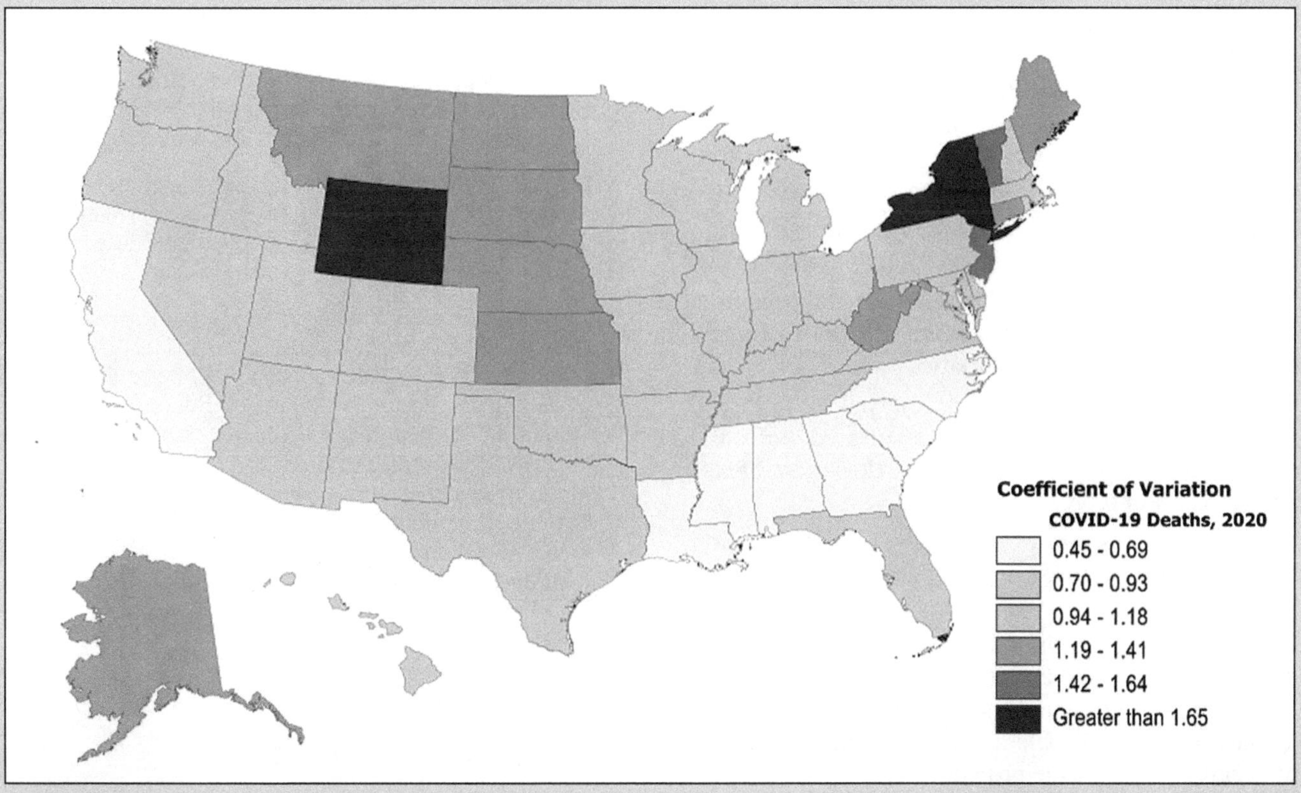

FIGURE 3.17
Coefficient of variation of Covid deaths by month in 2020.

Key Terms

absolute and relative measures, *47, 49*
average deviation (mean deviation), *45*
bimodal, multimodal, unimodal, *43*
boxplot (upper and lower whisker), *44*
class midpoint, *42*
coefficient of variation (coefficient of variability), *49*
histogram (frequency polygon), *40*
individual point distribution (individual value plot), *44*
interquartile range, *44*
Jenks natural breaks method of classification, *47*
Kurtosis (leptokurtic, mesokurtic, platykurtic), *52*
least squares property of the mean, *45*
measures of central tendency:
  mode, *40*
  median, *41*
  mean (arithmetic mean), *41*
modal class and crude mode, *40*
Modifiable Areal Unit Problem (MAUP):
  grouping or zoning problem, *54*
  scale or level of aggregation problem, *53*
ogive, *41*
outliers, *43*

quantiles, *44*
skewness (negative and positive skew), *52*
standard deviation, *45*
standard deviation breaks method of classification, *47*
symmetric, *43*
variance, *45*
weighted mean, *42*

Major Goals and Objectives

If you have mastered the material in this chapter, you should now be able to:

1. Define the basic descriptive measures of central tendency (mode, median, and mean), explain the advantages and disadvantages of each, and select the most appropriate measure given a particular frequency distribution or particular geographic situation.
2. Define the basic descriptive measures of dispersion (range, quantiles, mean deviation, standard deviation, variance) and explain the characteristics of each.
3. Recognize whether a geographic problem or situation requires weighted or unweighted descriptive statistics.
4. Understand the concept of relative variability and its value in comparing different spatial patterns directly by using the coefficient of variation.
5. Understand the characteristics, as well as both the advantages and disadvantages associated with the classification methods of standard deviation breaks and Jenks natural breaks.
6. Define the measures of shape or relative position (skewness and kurtosis) and recognize their potential value in descriptive analysis in geography.
7. Recognize the potential effects of locational data on descriptive statistics. Possible influences include: (a) impact of external boundary delineation, (b) the Modifiable Areal Unit Problem (MAUP), and (c) the spatial aggregation or scale problem.

References and Additional Reading

Bailey, T. C. and A. C. Gatrell. *Interactive Spatial Data Analysis.* London: Longman, 1995.

Lembo, A., Lew, M., Laba, M., and Baveyye, P. "Use of Spatial SQL to Assess the Practical Significance of the Modifiable Areal Unit Problem," *Computer and Geosciences*, Elsevier Press, Vol. 2, No. 2 (2006): 270–274.

Openshaw, S. *The Modifiable Areal Unit Problem.* Concepts and Techniques in Modern Geography (CATMOG), No. 38. Norwich: Geo Books, 1984.

von Hippel, P. T. "Mean, Median, and Skew: Correcting a Textbook Rule," *Journal of Statistics Education,* Vol. 13, No. 2 (2005).

Wheeler, D. J. "Problems with Skewness and Kurtosis, Part Two," *Quality Digest Daily,* Manuscript 231, 2011.

Looking at other introductory textbooks in quantitative geography and spatial analysis can be a valuable experience. Other texts may be simpler, more difficult, or roughly comparable, but whatever their level of difficulty, they will provide different insights. Expect some "symbol shock," as each statistics text will likely use somewhat different statistical notation. A few introductory textbooks are listed here.

Burt, J. E., G. M. Barber, and D. L. Rigby. *Elementary Statistics for Geographers.* 3rd ed. New York: Guilford Press, 2009.

Clark, W. A. V. and P. L. Hosking. *Statistical Methods for Geographers.* New York: John Wiley and Sons, 1986.

Earickson, R. and J. Harlin. *Geographic Measurement and Quantitative Analysis.* New York: Macmillan, 1994.

Ebdon, D. *Statistics in Geography: A Practical Approach.* 2nd ed. Blackwell Publishing, 1991.

Griffith, D. A. and C. G. Amrhein. *Statistical Analysis for Geographers.* Englewood Cliffs, NJ: Prentice-Hall, 1991.

Rogerson, P. A. *Statistical Methods for Geography: A Student's Guide.* 3rd ed. Thousand Oaks, CA: Sage Publications, Inc., 2010.

Taylor, P. J. *Quantitative Methods in Geography: An Introduction to Spatial Analysis.* Prospect Heights, IL: Waveland Press, 1983.

In addition to textbooks on quantitative methods written primarily for geographers, dozens of introductory statistics textbooks are currently in print. A few popular introductory statistics texts are listed here.

Bluman, A. *Elementary Statistics: A Step by Step Approach.* 10th ed. New York: McGraw-Hill, 2018.

De Veaux, R. D., P. F. Velleman, and D. E. Bock. *Stats: Data and Models.* 5th ed. Boston, MA: Pearson Education, Inc., 2020

Freedman, D., R. Pisani, and R. Purves. *Statistics.* 4th ed. New York: W. W. Norton and Co., 2007.

Kirk, R. E. *Statistics: An Introduction.* 5th ed. Belmont, CA: Thomson Learning, 2008.

Peck, R., C. Olsen, and T. Short. *Introduction to Statistics and Data Analysis.* 6th ed. Boston, MA: Cengage, 2020.

Urdan, T. C. *Statistics in Plain English.* 4th ed. New York, NY: Routledge Academic, 2016.

4

Descriptive Spatial Statistics

In the preceding chapter, a variety of basic descriptive statistics were examined, including the mean, standard deviation, and coefficient of variation. To summarize point patterns, a set of descriptive spatial statistics has been developed that are areal or locational equivalents to these nonspatial measures (Table 4.1). Since geographers are particularly concerned with the analysis of locational data, these descriptive spatial statistics, appropriately referred to as **geostatistics**, are often applied to summarize point patterns (and line patterns) and to describe the degree of spatial variability of some phenomena. Geostatistics can also provide a useful summary of an areal pattern on a choropleth map, if each area on the map can be operationally represented by a point.

Spatial measures of central tendency like central feature, mean center, median center (Euclidean median), and Linear Directional Mean (LDM) are examined in Section 4.1. Each of these measures has characteristic properties and a set of practical geographic applications. Section 4.1 also contains some discussion about the concept of "distance." The most important absolute measure of spatial dispersion is standard distance, the spatial equivalent to standard deviation. Standard distance is discussed in Section 4.2, in addition to the concept of relative dispersion and the Standard Deviational Ellipse (SDE).

Only the most fundamental measures of geographic distributions are discussed in this chapter. We have deliberately kept things simple and straightforward at this

TABLE 4.1

Selected Descriptive Statistics: Traditional (Nonspatial) and Spatial*

	Measures of center and central tendency	Measures of absolute dispersion	Measures of relative dispersion
Traditional (nonspatial) measures	mode median mean	range interquartile range mean (average dispersion) standard deviation variance	coefficient of variation skewness kurtosis
Spatial measures			
Point	central feature mean center median center (Euclidean median)	standard distance standard deviational ellipse (SDE)	relative distance
Line	linear direction mean (LDM)		

* Contents of table limited to only those descriptive statistics discussed in chapters 3 and 4.

stage, with the focus exclusively on elementary descriptive spatial statistics. However, a variety of inferential spatial statistics have become increasingly important with the explosive development of geographic information sciences (GIS) in recent years. These issues will be discussed later, especially in Chapters 13 through 15.

It is important to recognize that even simple descriptive measures of a geographic distribution are usually calculated with a GIS software package. Spreadsheets such as Excel and statistical software packages such as *Minitab*, *SPSS*, and *SAS* can also be used for calculation, but the procedures may be more cumbersome. As geographers, we now have more options and computational flexibility than ever before.

4.1 Spatial Measures of Central Tendency

Consider the scatterplot of seven points shown in Figure 4.1. These points might represent any spatial distribution of interest to geographers—the only stipulation is that the phenomenon can be displayed graphically as a set of points in a two-dimensional Cartesian coordinate system. The directional orientation of the coordinate axes and location of the origin are often both arbitrary.

Central Feature

The **central feature** is the feature (point or area) whose total distance to all other features in the study area is the shortest. If we are evaluating a point pattern, the **central point** is the point with the lowest total distance to all other points. Similarly, if evaluating an area pattern, the **central area** is the area whose total distance to all other areas is the shortest. In either case, this is one simple way of defining the most accessible feature. Only the central point calculation procedure is shown here.

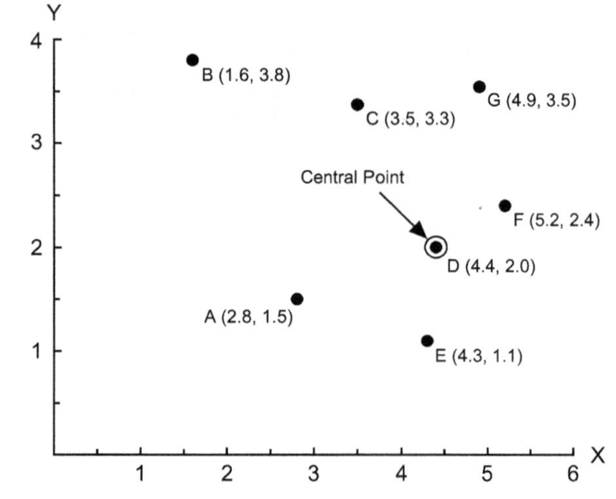

FIGURE 4.1
Graph of Locational Coordinates and Central Point

The worktable for calculating the central point is summarized in Table 4.2. Notice that the definition of "distance" in this example is **Euclidean** (or **straight-line**) **distance**. The Euclidean distance (d_{ij}) separating points i and j is defined by the Pythagorean theorem as:

$$d_{ij} = \sqrt{(X_i - X_j)^2 + (Y_i - Y_j)^2} \qquad \textbf{(4.1)}$$

To refresh your memory, examine Figure 4.2, which illustrates the Pythagorean theorem that states: "In a right-angled triangle, the square of the hypotenuse is equal to the sum of the squares of the other two sides." If the hypotenuse represents the Euclidean distance separating two points (for example, points "A" and "B" in Figure 4.1), the distance between those two points is easily calculated as part of the calculation for the central point (Table 4.2).

The matrix in Table 4.2 shows Euclidean distances for all pairs of points and we can see that point "D" is the one whose total distance from all other points is the lowest with 9.97. By definition, this makes point "D" the central point.

Mean Center

The mean was discussed as an important measure of central tendency in the previous chapter. If this concept of central tendency is now extended to the same set of seven points as shown in Figure 4.1, the **mean center** (or average location), can be determined. Given a coordinate system with the X and Y coordinates of each point deter-

TABLE 4.2

Worktable for Calculating Central Point

Distance matrix between points in figure 4.1

	A	B	C	D	E	F	G	Total Distance
A	0	2.59	1.93	1.68	1.55	2.56	2.90	13.21
B	2.59	0	1.96	3.33	3.82	3.86	3.31	18.87
C	1.93	1.96	0	1.58	2.34	1.92	1.41	11.14
D	1.68	3.33	1.58	0	0.91	0.89	1.58	9.97
E	1.55	3.82	2.34	0.91	0	1.58	2.47	12.67
F	2.56	3.86	1.92	0.89	1.58	0	1.14	11.95
G	2.90	3.31	1.41	1.58	2.47	1.14	0	12.81

Point "D" is the lowest total distance from all other points: 9.97

Euclidean distance from point A to point B:

$$d_{AB} = \sqrt{(X_A - X_B)^2 + (Y_A - Y_B)^2}$$

$$= \sqrt{(2.8 - 1.6)^2 + (1.5 - 3.8)^2}$$

$$= 2.59$$

Locational coordinates: A (2.8, 1.5), B (1.6, 3.8)

mined, the mean center is simply calculated by separately averaging those X and Y values, as follows:

$$\bar{X}_c = \frac{\sum X_i}{n} \quad \text{and} \quad \bar{Y}_c = \frac{\sum Y_i}{n} \qquad \textbf{(4.2)}$$

where: $\bar{X}_c$ = mean center of X
$\bar{Y}_c$ = mean center of Y
X_i = X coordinate of point i
Y_i = Y coordinate of point i
n = number of points in the distribution

The mean center coordinates ($\bar{X}_c = 3.81$ and $\bar{Y}_c = 2.51$) are calculated in Table 4.3 and the mean center location is shown in Figure 4.3.

Just as the nonspatial mean is strongly affected by an outlier or small number of extreme values, the mean center is influenced in a similar way. Suppose, for example, that one additional point with coordinates (15, 13) is included in the previous example (Figure 4.4). The mean center location shifts dramatically from (3.81, 2.51) to (5.21, 3.82), the latter being a coordinate position having larger X and Y coordinates than any of the other seven points. Thus, while the mean center represents an average location, it may not represent a "typical" or "central" location.

The mean center may be considered the **center of gravity** of a point pattern or spatial distribution. Perhaps the most widely known application of the mean center is the decennial calculation of the **geographic "center of population"** by the US Bureau of the Census. This is the point where a rigid map of the country would balance if equal weights (each representing the location of one person) were situated on it. Over the last two centuries, the westward movement of the US population has continued without significant interruption, as reflected in the concomitant westward shift of the center of population (Figure 4.5).

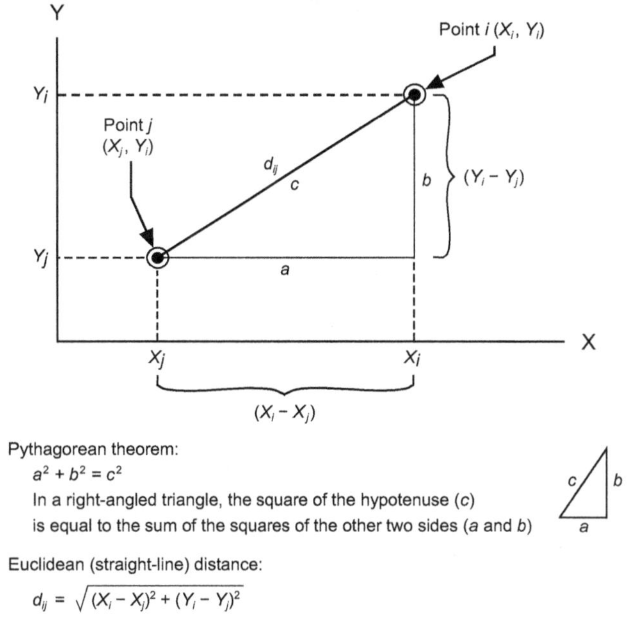

FIGURE 4.2
Calculation of Euclidean Distance (d_{ij}) from Point i to Point j

In the previous seven-point examples, each point is given an equal weight in the central point and mean center calculations—that is, each point is equally important statistically. In many geographic applications, however, points of a spatial distribution could be assigned differential weights. These weights are analogous to frequencies in the calculation of grouped statistics, such as the weighted mean. The points might represent cities and the frequencies the number of people, or the points could be retail store locations and the frequencies could be

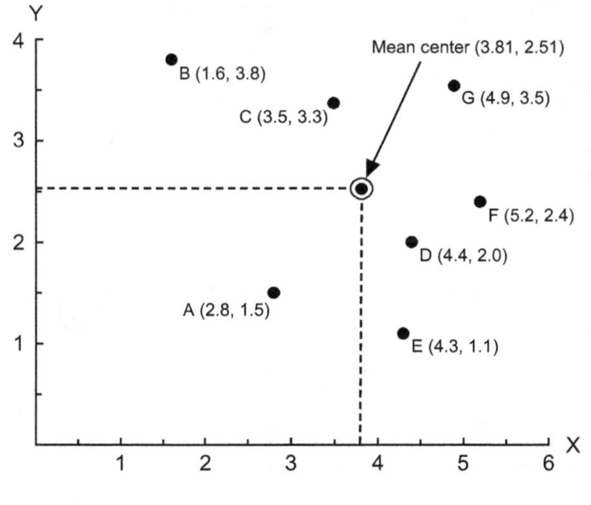

FIGURE 4.3
Graph of Mean Center

TABLE 4.3

Worktable for Calculating Mean Center

	Locational coordinates *	
Point	X_i	Y_i
A	2.8	1.5
B	1.6	3.8
C	3.5	3.3
D	4.4	2.0
E	4.3	1.1
F	5.2	2.4
G	4.9	3.5
$n = 7$	$\sum X_i = 26.7$	$\sum Y_i = 17.6$

$$\bar{X}_c = \frac{\sum X_i}{n} = \frac{26.7}{7} = 3.81 \qquad \bar{Y}_c = \frac{\sum Y_i}{n} = \frac{17.6}{7} = 2.51$$

Mean center coordinates: (3.81, 2.51) *

* See figure 4.3 for graph of locational coordinates and mean center.

volume of sales per store. The **weighted mean center** is defined as follows:

$$\bar{X}_{wc} = \frac{\Sigma f_i \bar{X}_i}{\Sigma f_i} \quad \text{and} \quad \bar{Y}_{wc} = \frac{\Sigma f_i \bar{Y}_i}{\Sigma f_i} \qquad \textbf{(4.3)}$$

where: $\bar{X}_{wc}$ = weighted mean center of X
$\bar{Y}_{wc}$ = weighted mean center of Y
f_i = frequency (weight) of point i

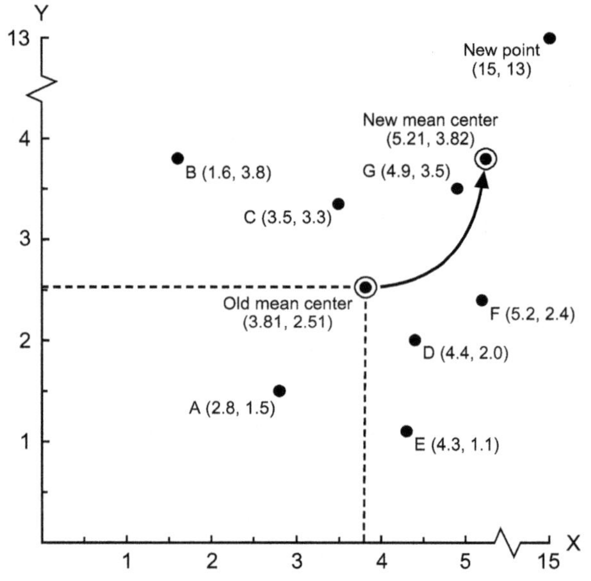

FIGURE 4.4
How an Outlier Might Affect Mean Center Location

Each point in Figure 4.3 is now assigned a weight (Figure 4.6) and the locational coordinates of the weighted mean center are calculated (Table 4.4). These coordinates are somewhat different from the coordinates of the comparable unweighted mean center. The weighted mean center is heavily affected by the relatively large frequency (20) associated with point "B". Gravitation of the center toward a point with an unusually heavy weight will occur even if that point is located peripherally within the spatial distribution.

The mean and mean center both have an important characteristic of locational significance. Recall from the

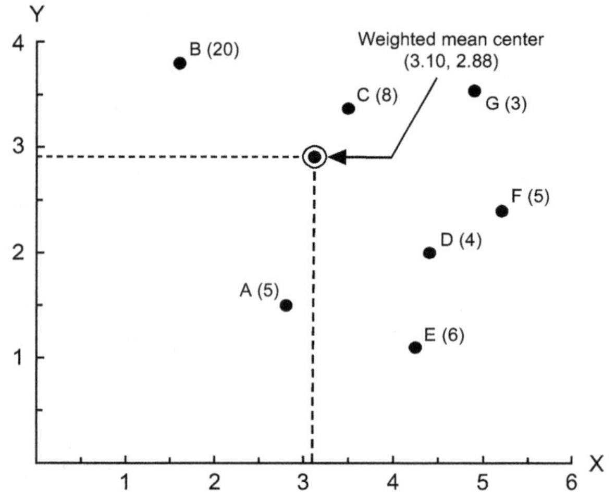

FIGURE 4.6
Graph of Point Locations, Weights (in Parentheses) and Weighted Mean Center

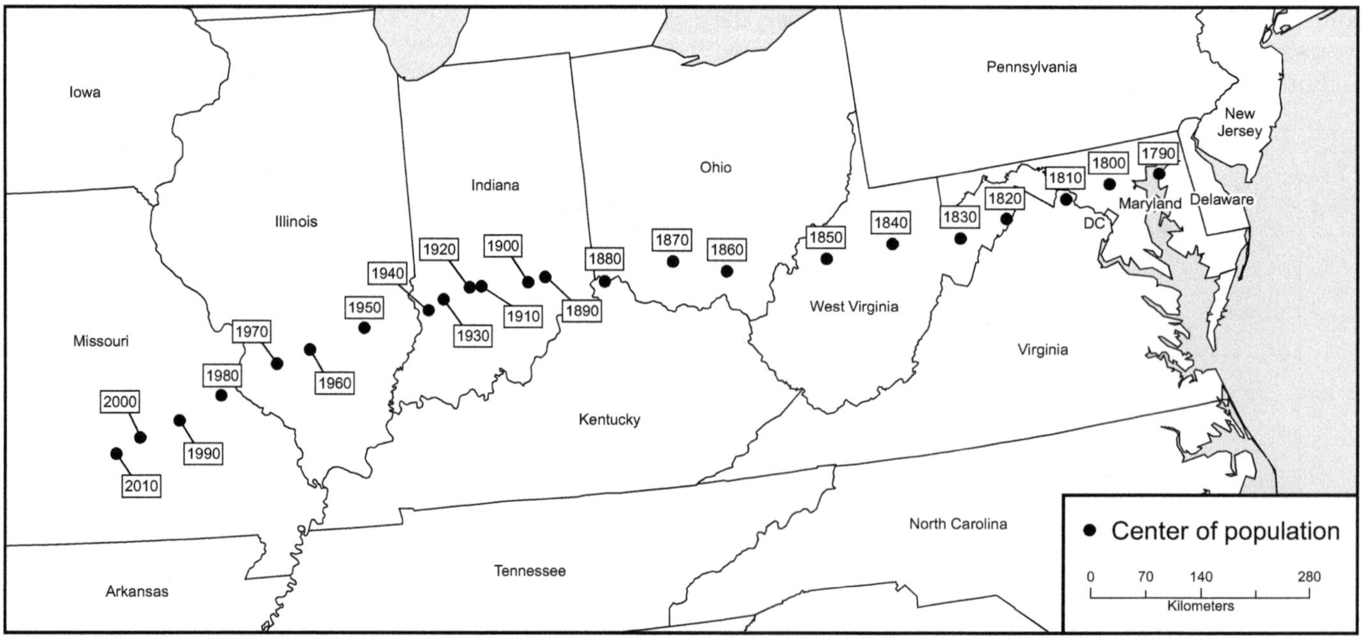

FIGURE 4.5
Geographic Center of U.S. Population, 1790 to 2010
Source: United States Bureau of the Census, 2010

TABLE 4.4

Worktable for Calculating Weighted Mean Center

Point	Locational coordinates * X_i	Y_i	Weight f_i	Weighted coordinates $f_i X_i$	$f_i Y_i$
A	2.8	1.5	5	14.0	7.5
B	1.6	3.8	20	32.0	76.0
C	3.5	3.3	8	28.0	26.4
D	4.4	2.0	4	17.6	8.0
E	4.3	1.1	6	25.8	6.6
F	5.2	2.4	5	26.0	12.0
G	4.9	3.5	3	14.7	10.5

$n = 7$ $\Sigma f_i = 51$ $\Sigma f_i X_i = 158.1$ $\Sigma f_i Y_i = 147.0$

$$\bar{X}_{wc} = \frac{\Sigma f_i X_i}{\Sigma f_i} = \frac{158.1}{51} = 3.10$$

$$\bar{Y}_{wc} = \frac{\Sigma f_i Y_i}{\Sigma f_i} = \frac{147.0}{51} = 2.88$$

Weighted mean center coordinates: (3.10, 2.88) *

* See figure 4.6 for graph of point locations, frequencies, and weighted mean center.

discussion on mean deviation (Section 3.2) that the sum of squared deviations of all observations about a mean is zero. In addition, the sum of squared deviations of all observations about a mean is the minimum sum possible. That is, the sum of squared deviations about a mean is less than the sum of squared deviations about any other number:

$$\bar{X}: \min \Sigma (X_i - \bar{X})^2 \quad \textbf{(4.4)}$$

This important attribute is called the least squares property of the mean.

The mean center is spatially analogous to the mean and has the same least squares property as the mean. That is:

$$(\bar{X}_c, \bar{Y}_c): \min \Sigma (X_i - \bar{X}_c)^2 + (Y_i - \bar{Y}_c)^2 \quad \textbf{(4.5)}$$

In a Cartesian coordinate system based on location, deviations such as $(X_i - \bar{X}_c)$ and $(Y_i - \bar{Y}_c)$ are, in fact, *distances* between points and can be expressed by the Euclidean distance we discussed earlier in the chapter:

$$d_i = \sqrt{(X_i - \bar{X}_c)^2 + (Y_i - \bar{Y}_c)^2} \quad \textbf{(4.6)}$$

Thus, the mean center is the location that *minimizes the sum of squared distances* to all points. This characteristic makes the mean center an appropriate center of gravity for a two-dimensional point pattern, just as the mean is the center of gravity along a one-dimensional number line.

Median Center (Euclidean Median)

For many geographic applications, another measure of "center" is more useful than the mean center. Often, it is more practical to determine the central location that minimizes the sum of *unsquared*, rather than squared, distances. This location, which minimizes the sum of Euclidean distances from all other points in a spatial distribution to that central location, is called the **median center** (X_e, Y_e) or **Euclidean median**. Mathematically, this location minimizes the sum:

$$(X_e, Y_e): \min \Sigma \sqrt{(X_i - X_e)^2 + (Y_i - Y_e)^2} \quad \textbf{(4.7)}$$

Unfortunately, determining coordinates of the Euclidean median is complex methodologically. Computer-based iterative algorithms (step-by-step procedures) must be used to reach a solution. These algorithms evaluate a sequence of possible coordinates and gradually converge on a suitable (often optimal) location for the median center.

A weighted median center is a logical extension of the simple (unweighted) median center, and the same types of algorithmic procedures locate the weighted Euclidean median. The coordinates of the weighted Euclidean median (X_{we}, Y_{we}) will minimize the expression:

$$(X_{we}, Y_{we}): \min \Sigma f_i \sqrt{(X_i - X_{we})^2 + (Y_i - Y_{we})^2} \quad \textbf{(4.8)}$$

The weights or frequencies may represent population, sales volume, or any other feature appropriate to the spatial problem.

The location of the weighted Euclidean median is important to geographers for several practical reasons. For example, a classical problem in economic geography is the so-called Weber problem, which seeks to determine the "best" location for an industry. The optimal location minimizes the total cost of transporting the raw material to the factory and the finished product to the market. The weighted Euclidean median is the location that minimizes these transportation costs.

Perhaps the most extensively developed general application for the median center in geography is public and private facility location. Often an important goal in facility location is minimizing the average distance traveled per person to reach a designated or assigned facility. This efficiency-based objective is equivalent to minimizing the aggregate or total distance people must travel to use the service systemwide. The Euclidean median achieves this goal.

Consider, for example, the problem of locating the site for an urban fire station based on a predicted pattern of fires for the region. Using the past or present pattern of fires as a reasonable estimate of future fires, the optimal central location for the station could be defined as the site that minimizes the total (and hence, average) distance traveled by the fire equipment to reach fires. That location is determined by the Euclidean median.

In another application, suppose location analysts for an exclusive women's apparel chain wish to select an accessible site for a new store. Further, suppose that market analysis indicates that the demographic group most likely to shop in the store is women aged 45–65 who are members of households with incomes greater than $120,000. From the compilation of census tract information in the designated trade area, each tract could be weighted by the number of women having these age and income characteristics. The weighted median center will designate the location that minimizes the total (and average) distance traveled by these women to reach the potential store site.

Extending this procedure to the simultaneous location of multiple facilities within a spatial pattern of demand is known as the "location-allocation" problem or the "multiple facility location" problem. Suppose, for example, city health care planners wish to locate a set of neighborhood medical centers to provide selected types of remedial health care. Not only must a set of medical centers be located, but the potential clientele must be allocated to an appropriate facility, creating "catchment districts" or zones for each center. Problems such as these can be extremely complex and challenging, with both theoretical issues and practical applications receiving considerable attention from geographers.

In recent years, facility location strategies and modeling techniques have become much more sophisticated. The application of Geographic Information Sciences (GIS) to locational decision-making has become routine and expected, and is the procedural method of choice in market analysis and the optimal siting of businesses. Several references at the end of this chapter are worth examining for further information on this subject.

Up to this point, all of the descriptive measures of center have used Euclidean (straight-line) distance. This has been the only definition of "distance." In some geographic situations, however, this may not be the most appropriate way to proceed. For example, in an urban geography study of commercial activity locations (e.g., grocery stores, drug stores, etc.), none of the measures of center considered so far may be very accurate. When travel is confined to a grid or rectilinear street pattern, for example, a straight-line or Euclidean measure of distance underrepresents the actual travel distances and could result in a misplaced center or inaccurately provide a lower measure of dispersion for the facilities being evaluated.

Fortunately, other measures of distance are available. Recall that the formula measuring the Euclidean distance (d_{ij}) of point i (X_i, Y_i) from point j (X_j, Y_j) is:

$$d_{ij} = \sqrt{\left(X_i - X_j\right)^2 + \left(Y_i - Y_j\right)^2}$$

If this Euclidean distance metric is now "generalized" to allow non-Euclidean distance measurement, the result is a **general distance metric**:

$$d_{ij} = \left[\left(X_i - X_j\right)^k + \left(Y_i - Y_j\right)^k\right]^{1/k} \quad \textbf{(4.9)}$$

When the general distance metric (k) equals 2, the formula is conventional Euclidean distance. When k equals 1, however, we are measuring distances in which movement is restricted to a rectangular or grid system. The term **Manhattan distance** describes the restrictive movement typical of travel in the New York City borough of Manhattan. Measuring the distance between points i and j in Manhattan space, where k equals one, gives:

$$d_{ij} = \left|X_i - X_j\right| + \left|Y_i - Y_j\right| \quad \textbf{(4.10)}$$

If you want to travel by car between Columbus, Ohio and Chicago, Illinois, neither Euclidean distance nor Manhattan distance will be very accurate or realistic (Figure 4.7). No major highway or highways allow anything close to straight-line travel between these two cities, so Euclidean distance is a rather poor measure of the effort required (travel time, comfort level while traveling, etc.) to get from the origin to the destination. Following a Manhattan distance route such as the one shown in Figure 4.7 is clearly impractical and probably not possible. However, one of many possible **network distance** routes (traveling along interstate highways through Indianapolis, Indiana) seems reasonable from both travel time and travel convenience perspectives. For many spatial interaction models in geography, some type of network distance measure may be the best alternative for minimizing interaction effort or cost.

Linear Directional Mean

The **linear directional mean** (or simply **directional mean**) identifies the typical or general (mean) direction for a set of lines. That is, the overall trend of a set of lines is determined by calculating the mean angle of the lines. This is easily illustrated by showing the linear directional mean (LDM) as a graphic summary of a set of line data (Figure 4.8).

In some geographic applications you might be interested only in the orientation of the lines. For example, a seismologist working with urban planners in the greater San Francisco area will be concerned only with the mean orientation of the San Andreas and other fault lines. Clearly, the concept of direction does not apply to a fault line. (Although opposing tectonic plates along a transverse fault are certainly moving in different directions!) You could calculate the directional mean of many transportation systems, such as a trail system in a national forest, or a road or railroad system in a region. Since movement can occur in both directions, our only interest would be the average orientation of the system.

In many other geographic applications the direction of each line is of importance and is factored into the measurement. For example, researchers might find it useful to determine the mean direction of such natural phenomena as tornado or hurricane tracks. Environmental planners might be interested in the seasonal migration paths of various deer, elk, or the mean migratory path of Canadian Geese in the Atlantic flyway.

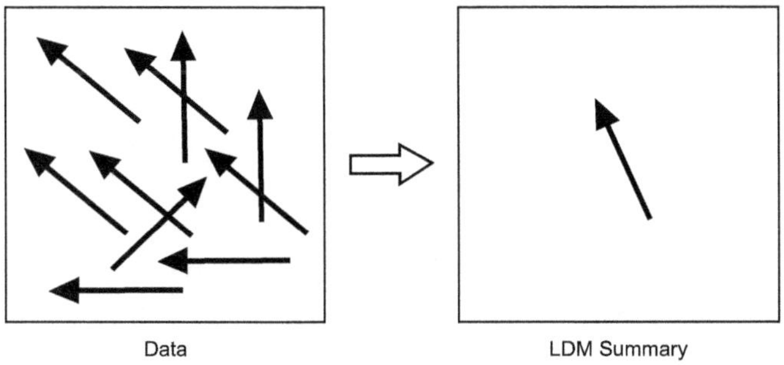

FIGURE 4.8
Linear Directional Mean (LDM) Summarized from Data

Many geographic phenomena change over time, and some applications could focus on temporal changes (possible cycles of change) in mean direction. Does the directional mean of winter storms crossing a particular location (a city in the Midwest, for example) change from decade to decade in any cyclical fashion? Have the seasonal migration paths of elk in the Greater Yellowstone Ecosystem changed due to encroaching development? Also, climatologists have long been interested in looking for repeating cycles that might be associated with some natural phenomenon.

The formula for the linear directional mean (LDM) is:

$$LDM = \arctan \frac{\sum_{i=1}^{n} \sin \theta_i}{\sum_{i=1}^{n} \cos \theta_i} \quad \textbf{(4.11)}$$

where θ_i represents the direction of a set of linear features from a single origin.

When using this application, GIS software is essentially calculating the angle of the resultant vector, the angle that results when each line in the analysis is

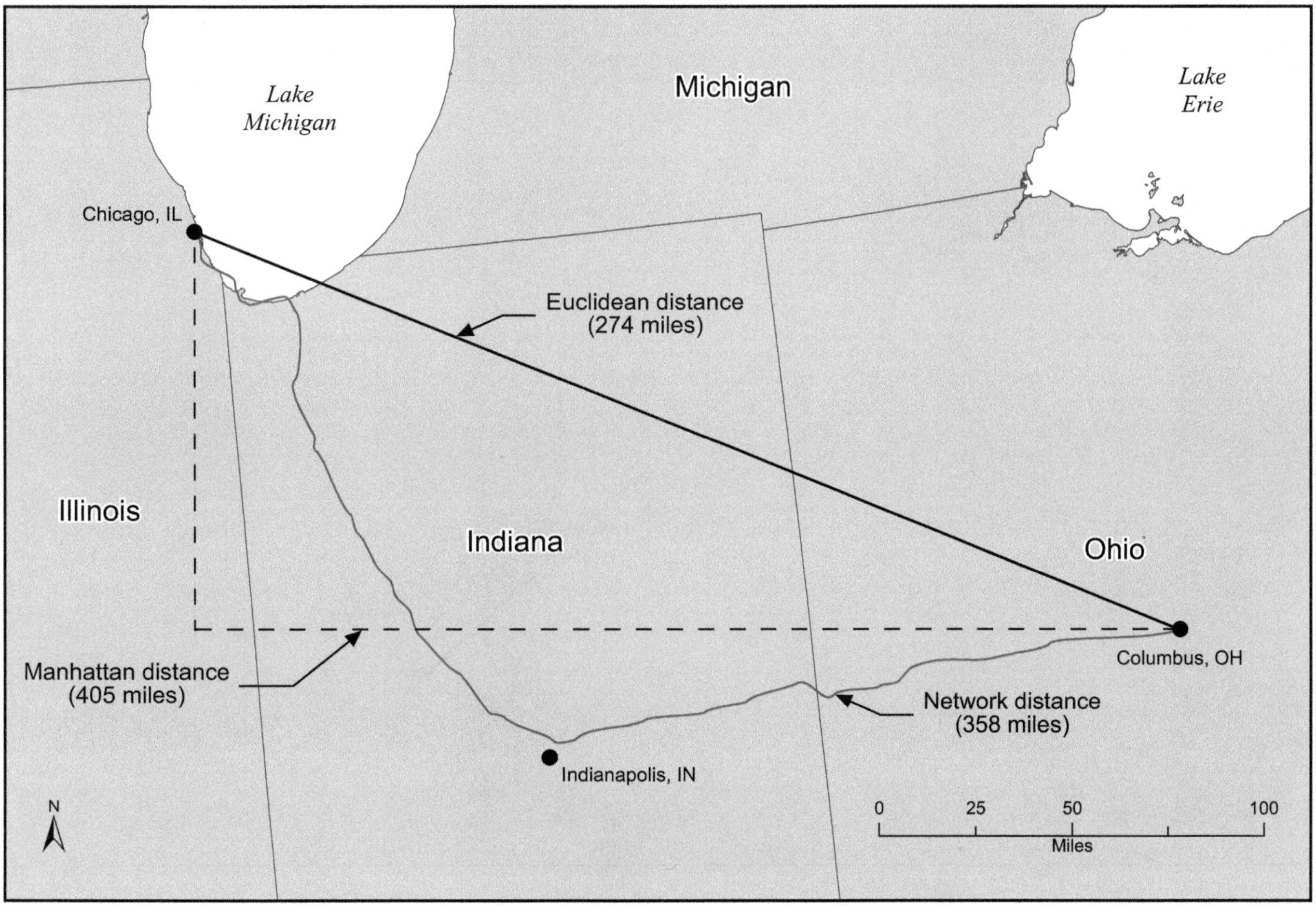

FIGURE 4.7
Measuring "Distance" Between Columbus, Ohio and Chicago, Illinois

treated as a unit vector (length of one unit). The manual calculation of the LDM is nicely illustrated by geographers Jay Lee and David Wong in *Statistical Analysis with ArcView GIS*.

The island territory of Puerto Rico, like many Caribbean countries, is vulnerable to hurricanes. Typically, tropical depressions and storms form over Western Africa and are carried westward by the trade winds to become hurricanes in the Caribbean region. The linear directional mean of all hurricanes and tropical storm tracks that have crossed Puerto Rico from 1852 to 2007 is shown in Figure 4.9. An interesting research hypothesis could be framed around the question of whether the LDM angle changes in any statistically significant way from one time period to another. Since the data are available from the National Climatic Data Center, many interesting research questions could be posed and analyses conducted to pursue this area of inquiry.

4.2 Spatial Measures of Dispersion

Standard Distance

Just as the mean center serves as a locational analogue to the mean, **standard distance** is the spatial equivalent of standard deviation (Table 4.1). Standard distance measures the amount of **absolute dispersion** in a point pattern. After the locational coordinates of the mean center have been determined, the standard distance statistic incorporates the straight-line or Euclidean distance of each point from the mean center. In its most basic form, standard distance (S_D) is written as follows:

$$S_D = \sqrt{\frac{\sum (X_i - \bar{X}_c)^2 + \sum (Y_i - \bar{Y}_c)^2}{n}} \tag{4.12}$$

If Equation 4.12 is modified algebraically, the number of required computations can be reduced considerably.

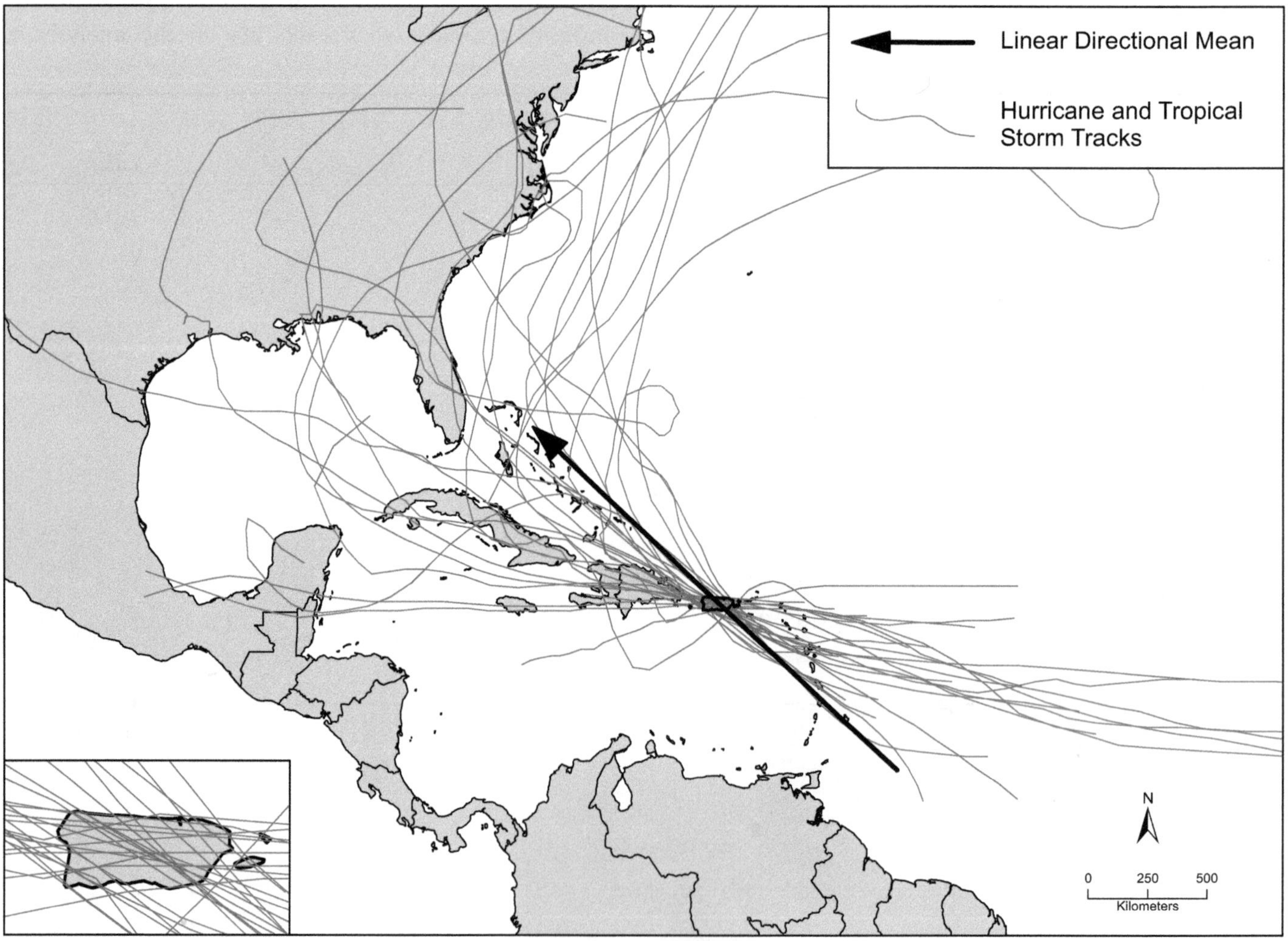

FIGURE 4.9
Linear Directional Mean: Hurricane and Tropical Storm Tracks Crossing Puerto Rico
Source: ESRI Resources Help, Linear Directional Mean

[Note: in practical situations, it is most likely that all computations will be done using a GIS package.]

$$S_D = \sqrt{\left(\frac{\sum X_i^2}{n} - \bar{X}_c^{\,2}\right) + \left(\frac{\sum Y_i^2}{n} - \bar{Y}_c^{\,2}\right)} \quad \textbf{(4.13)}$$

Using the same point pattern as in the earlier examples (Figure 4.1), standard distance is now calculated (Table 4.5) and shown as the radius of a circle whose center is the mean center (Figure 4.10).

Like standard deviation, standard distance is strongly influenced by extreme or peripheral locations. Because distances about the mean center are squared, outliers or atypical points have a dominating impact on the magnitude of the standard distance.

Weighted standard distance is appropriate for those geographic applications requiring a weighted mean center. The definitional formula for weighted standard distance (S_{WD}) is:

$$S_{WD} = \sqrt{\frac{\sum f_i (X_i - \bar{X}_{wc})^2 + \sum f_i (Y_i - \bar{Y}_{wc})^2}{\sum f_i}} \quad \textbf{(4.14)}$$

This may be rewritten in simpler form for computation as:

$$S_{WD} = \sqrt{\left(\frac{\sum f_i (X_i^2)}{\sum f_i} - \bar{X}_{wc}^{\,2}\right) + \left(\frac{\sum f_i (Y_i^2)}{\sum f_i} - \bar{Y}_{wc}^{\,2}\right)} \quad \textbf{(4.15)}$$

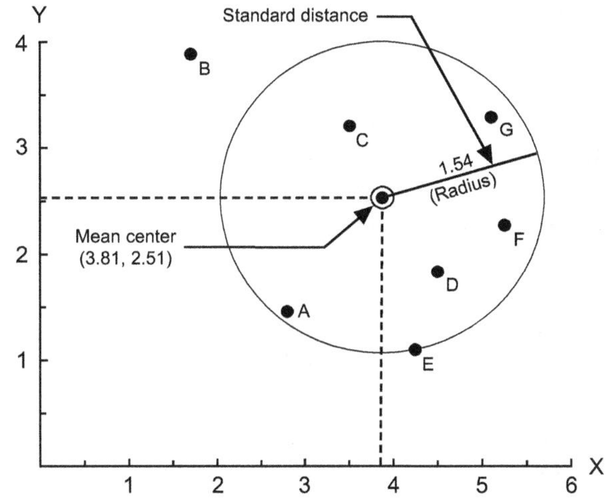

FIGURE 4.10
Graph of Point Locations, Mean Center, and Standard Distance

A weighted standard distance can be computed using the same point pattern as before (Table 4.6). A moderate disparity exists between the relative magnitudes of the unweighted and weighted standard distances (1.54 vs. 1.70). This difference can be primarily explained by point B. Because this point is distant from the mean center and exerts a proportionally greater influence on the standard distance measure with its larger weight, point B causes the weighted standard distance to be larger than the unweighted standard distance.

TABLE 4.5

Worktable for Calculating Standard Distance

Point	Locational coordinates * X_i	Y_i	X_i^2	Y_i^2
A	2.8	1.5	7.84	2.25
B	1.6	3.8	2.56	14.44
C	3.5	3.3	12.25	10.89
D	4.4	2.0	19.36	4.00
E	4.3	1.1	18.49	1.21
F	5.2	2.4	27.04	5.76
G	4.9	3.5	24.01	12.25

$\bar{X}_c = 3.81 \quad \bar{Y}_c = 2.51 \quad \bar{X}_c^{\,2} = 14.52 \quad \bar{Y}_c^{\,2} = 6.30$

$n = 7 \quad \sum X_i^2 = 111.50 \quad \sum Y_i^2 = 50.80$

$$S_D = \sqrt{\left(\frac{\sum X_i^2}{n} - \bar{X}_c^{\,2}\right) + \left(\frac{\sum Y_i^2}{n} - \bar{Y}_c^{\,2}\right)}$$

$$= \sqrt{\left(\frac{111.50}{7} - 14.52\right) + \left(\frac{50.80}{7} - 6.30\right)}$$

$$= 1.54$$

* See figure 4.10 for graph of locational coordinates, mean center, and standard distance.

TABLE 4.6

Worktable for Calculating Weighted Standard Distance

Point	f_i	X_i	X_i^2	$f_i(X_i)^2$	Y_i	Y_i^2	$f_i(Y_i)^2$
A	5	2.8	7.84	39.20	1.5	2.25	11.25
B	20	1.6	2.56	51.20	3.8	14.44	288.80
C	8	3.5	12.25	98.00	3.3	10.89	87.12
D	4	4.4	19.36	77.44	2.0	4.00	16.00
E	6	4.3	18.49	110.94	1.1	1.21	7.26
F	5	5.2	27.04	135.20	2.4	5.76	28.80
G	3	4.9	24.01	72.03	3.5	12.25	36.75

From earlier calculation of weighted mean center:

$\bar{X}_{wc} = 3.10 \quad \bar{Y}_{wc} = 2.88 \quad \bar{X}_{wc}^{\,2} = 9.61 \quad \bar{Y}_{wc}^{\,2} = 8.29$

$\sum f_i = 51 \quad \sum f_i(X_i)^2 = 584.01 \quad \sum f_i(Y_i)^2 = 475.98$

$$S_{WD} = \sqrt{\left(\frac{\sum f_i(X_i)^2}{\sum f_i} - \bar{X}_{WC}^{\,2}\right) + \left(\frac{\sum f_i(Y_i)^2}{\sum f_i} - \bar{Y}_{WC}^{\,2}\right)}$$

$$= \sqrt{\left(\frac{584.01}{51} - 9.61\right) + \left(\frac{475.98}{51} - 8.29\right)}$$

$$= 1.70$$

Standard Deviational Ellipse

While standard distance is a useful measure of dispersion in a point pattern, it does not take the orientation of the pattern into account. Points may be dispersed differently along the X or Y axis, potentially revealing an elliptical trend in the point locations. To capture this trend, the standard distance is calculated separately for the X and Y coordinates using the following formulas:

$$SD_x = \sqrt{\frac{\Sigma(X_i - \bar{X}_c)^2}{n}} \text{ and } SD_y = \sqrt{\frac{\Sigma(Y_i - \bar{Y}_c)^2}{n}} \quad \textbf{(4.16)}$$

SD_x represents the average distance points vary from the mean center on the X axis and SD_y represents the average distance points vary from the mean center on the Y axis. Construction of the standard deviational ellipse begins with creating a cross of dispersion (Figure 4.11). To calculate the cross of dispersion, a line of length SD_x is extended in both directions from the mean center parallel to the X-axis. Similarly, a line of length SD_y is extended in both directions from the mean center parallel to the Y-axis. An ellipse is constructed which encompasses this cross of dispersion. A trigonometric function is used to calculate the angle of rotation for the ellipse. The ellipse is rotated about the mean center so that the distances between the points and both arms of the cross of dispersion are minimized.

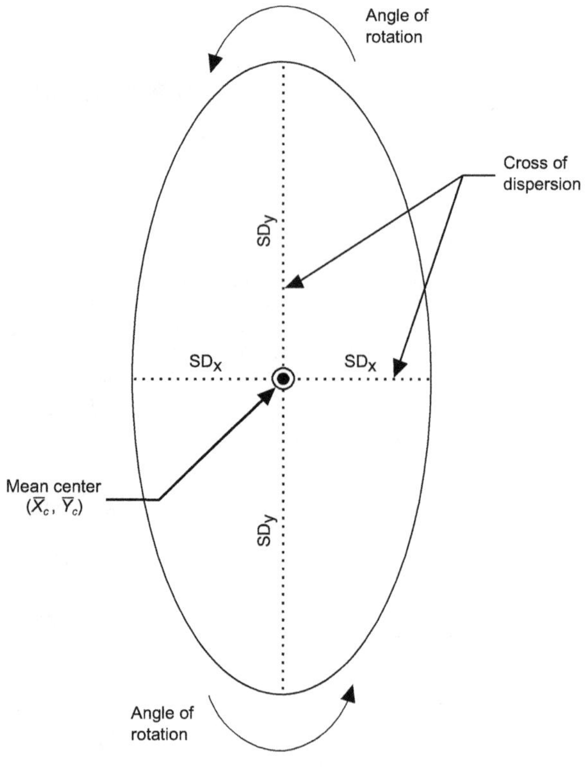

FIGURE 4.11
Standard Deviational Ellipse

TABLE 4.7

Summary Statistics for Standard Deviational Ellipse: Anti-Shipping Activity off the East African Coast

Year	# of incidents	$\bar{X}_c$	$\bar{Y}_c$	SD_x (miles)	SD_y (miles)	Rotation
2009	114	52.53	−1.00	379.7	604.7	45.4
2010	230	52.06	−1.58	415.9	956.3	56.3

Source: National Geospatial-Intelligence Agency (NGA)

The standard deviational ellipse methodology is applied to point patterns representing anti-shipping activities off the east coast of Africa in 2009 and 2010 (Figure 4.12). Anti-shipping activities are hostile actions against ships or mariners and are sometimes referred to as piracy. Motivated by political and economic unrest in the region, the number of incidents of piracy has increased in recent years. The locations of these hostile acts are recorded by the National Geospatial Intelligence Agency. There were 114 recorded incidents in 2009 and 230 incidents in 2010. Given the large number of incidents, it would be difficult even for the trained geographer to discern meaningful patterns from such a large number of points.

The standard deviational ellipse is calculated separately for the 2009 and 2010 incidents. This allows us to explore changes in the magnitude of dispersion and orientation of the pattern from one year to the next. The values summarizing the standard deviational ellipses are shown in Table 4.7 and the ellipses themselves are shown in Figure 4.12. The mean centers and X-axes standard distances for the two distributions are very similar. However, the standard distances for the Y-axes differ by over 350 miles, indicating a more dispersed pattern for the 2010 incidents. One possible explanation for this shift is that the pirates have developed the capability to operate at greater distances from the east coast of Africa. The orientation of the ellipses for both years is remarkably similar. The elongated ellipses suggest a corridor of piracy extending from the coast of Tanzania into the Indian Ocean. In fact, the standard deviational ellipses follow major shipping lanes which extend from the large port of Dar es Salaam, Tanzania to the Middle East and Asia.

Although the standard deviational ellipse is a useful method for identifying the magnitude of dispersion and orientation of a point pattern, two procedural issues must be considered. First is the delineation of the study area boundary. In many geographic problems, a political or administrative boundary defines the study area. If this boundary is itself elongated it may impose an artificial spatial trend on the data. You must consider carefully whether the pattern of the standard deviational ellipse is the result of some underlying spatial process or an artifact of the study area boundary. The second issue is spatial outliers. Both the mean center and standard distance will be impacted by points which are located far from the center

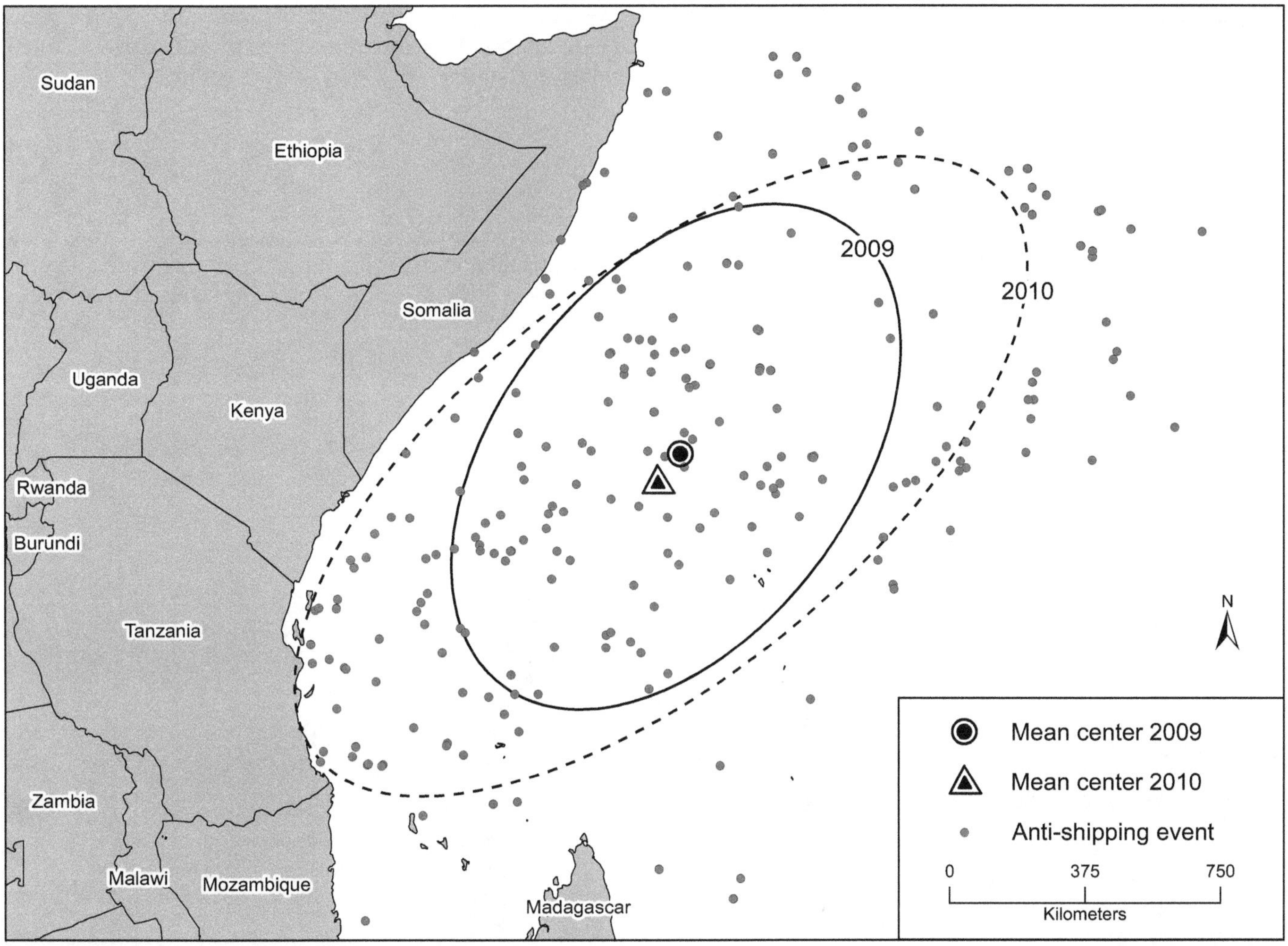

FIGURE 4.12
Standard Deviational Ellipse: Anti-shipping Activity off the East African Coast, 2009 and 2010
Source: National Geospatial-Intelligence Agency (NGA)

of the distribution. The presence of spatial outliers can produce artificially elongated standard deviational ellipses.

The Relative Dispersion Problem

The coefficient of variation (standard deviation divided by the mean) is the nonspatial measure of **relative dispersion** (Table 4.1). Unfortunately, a perfect spatial analogue to the coefficient of variation does not exist for measuring relative dispersion. Although it seems logical to divide the standard distance by the mean center to produce a relative dispersion index, this procedure will not provide meaningful results.

The problem of measuring relative dispersion can easily be demonstrated. Consider a situation where the spatial statistics for the same point pattern are calculated twice using different positions for the coordinate system (Figure 4.13). In Case 1, the X coordinates of each point are three units lower and the Y coordinates one unit lower than in Case 2. Notice, however, that the shift in the coordinate system does not affect standard distance. As a measure of absolute dispersion, standard distance remains unchanged. Conversely, the coordinates of the mean center will change whenever the coordinate system is shifted. Because coordinate system location affects the mean center but not standard distance, a relative dispersion metric based on the ratio of these measures will be meaningless.

Despite these difficulties, some logical estimate of relative dispersion is necessary for spatial measurement. Consider the three point patterns in regions A, B, and C (Figure 4.14, Case 1). The distribution of points in each region has the same amount of absolute dispersion and the same standard distance. However, in small region A, the points have a high degree of relative dispersion, whereas they have a low relative dispersion in region C because the region is larger. The point patterns in regions D and E (Figure 4.14, Case 2) appear to have the same amount of relative dispersion. However, the point pattern in region D has a larger standard distance (absolute dispersion) than region E because of its larger size.

To derive a descriptive measure of relative spatial dispersion, the standard distance of a point pattern must be divided by some measure of regional magnitude. This divisor cannot be the mean center. One possible divisor is the radius (r_A) of a circle with the same area as the region being analyzed. A measure of relative dispersion, called **relative distance**, (R_D), can now be defined:

$$R_D = \frac{S_D}{r_A} \tag{4.17}$$

This relative distance measure allows direct comparison of the dispersion of different point patterns from different areas, even if the areas are of varying sizes.

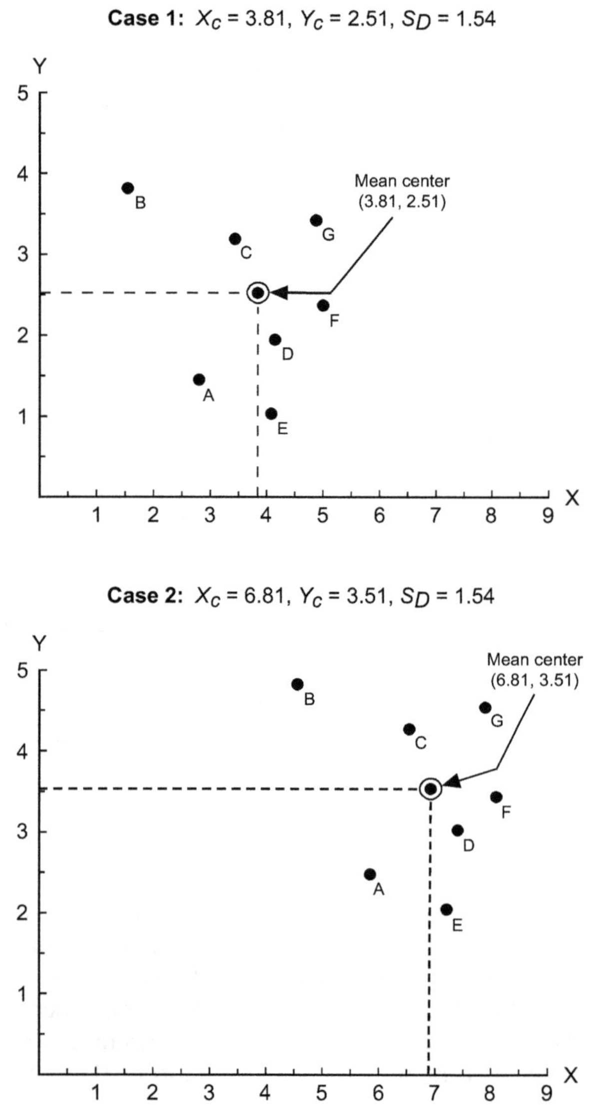

FIGURE 4.13
Arbitrary Placement of Coordinate Axes and Resultant Descriptive Spatial Statistics

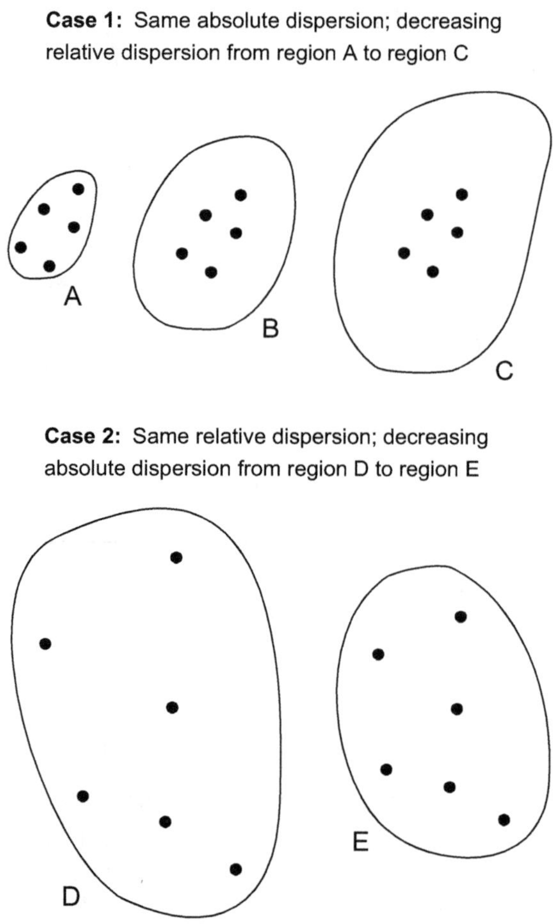

FIGURE 4.14
Comparisons of Absolute and Relative Point Pattern Dispersion

COVID-19 Conversation: Descriptive Spatial Statistics During the COVID-19 Pandemic

In Chapter 3 we looked at the spatial distribution for the coefficient of variation and hypothesized that some areas of the country may have been impacted by a geographic wave of COVID-19 spread over time and space. To further explore this hypothesis, using county-level death figures, we created a map showing locations of the mean center of COVID-19 deaths for each month in 2020. Using the county centroid locations and deaths for each month, we generated the mean center of deaths using the weighted mean center formula illustrated in Table 4.3.

The weighted mean center in the month of February while not shown, was placed in Washington state most likely a result of the nation's first outbreak in a King County, Washington, long term care facility. Figure 4.15 shows the movement of the weighted mean center for subsequent months in 2020. By March, the weighted mean center had moved to eastern Missouri, likely due to the rise in New York City. May saw a further shift eastward, as the rise in cases in New York City still had a large influence on the weighted mean center. There was a short southwest movement from May to June, followed by one of the largest southwest shifts from June to July. The large shift was likely due to declining deaths in the New York metropolitan region along with increasing deaths in the south and west. The southwestern shift continued throughout the summer months as deaths increased in the sunbelt. Further, while waves revisited some areas, New York City never returned to their March highs. The months of October to December showed little movement, as the COVID-19 deaths had more evenly spread throughout the United States.

Although many factors contributed to the ebb and flow of COVID-19 deaths in the United States, a map of the weighted mean center provides insight into the movement of the virus on a month-by-month basis.

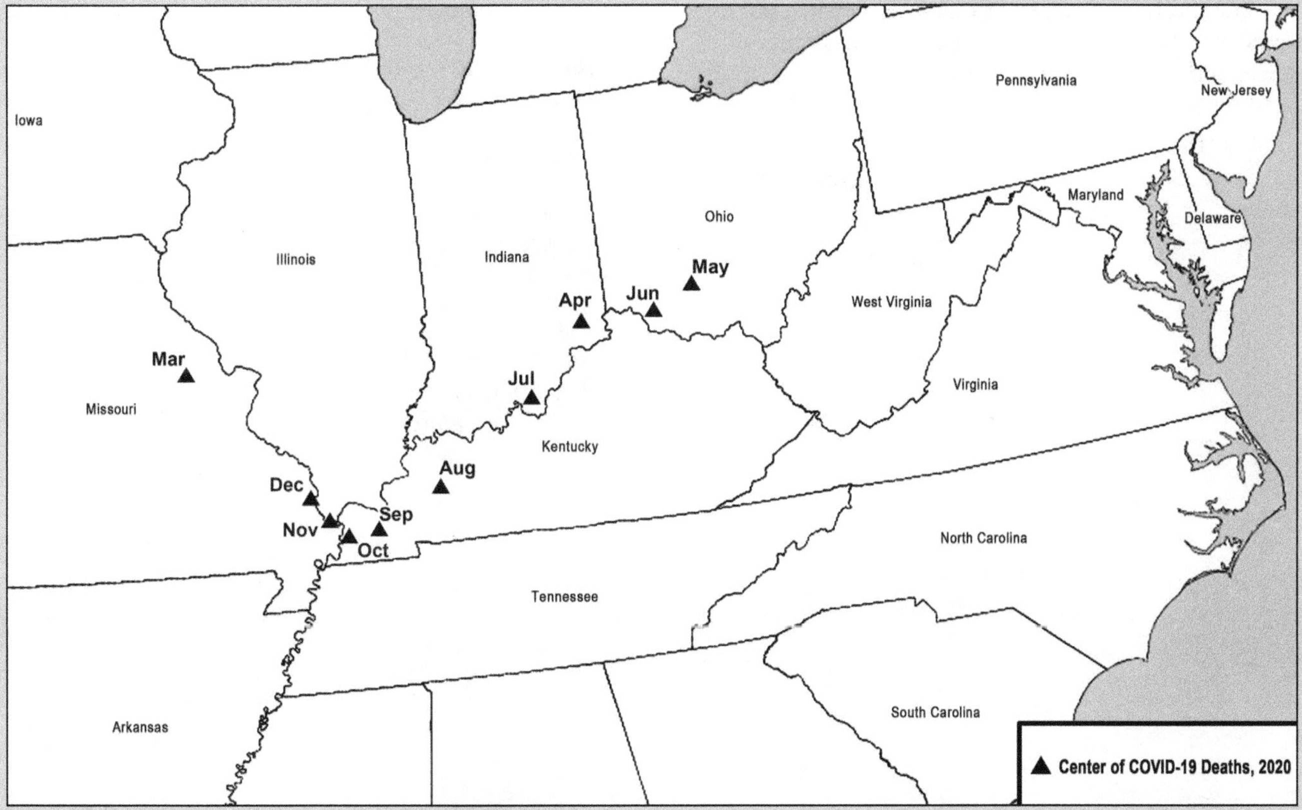

FIGURE 4.15

Key Terms

Major Goals and Objectives

If you have mastered the material in this chapter, you should now be able to:

1. Understand the concept of central tendency as applied in a spatial context and explain the distinctive characteristics of the central point, mean center, and Euclidean median.
2. Define the spatial measures of dispersion (standard distance and relative distance) and recognize their potential applications in geographic problem solving.
3. Recognize when a point pattern in a study region has a large or small absolute dispersion and a large or small relative dispersion.
4. Identify the potential limitations and locational issues associated with the application of descriptive spatial statistics.
5. Understand the concept of "distance metric" and distinguish between Euclidean distance, Manhattan distance, and other distance metrics under various spatial conditions.
6. Explain the goals of the linear directional mean and standard deviational ellipse and suggest practical applications of each.

References and Additional Reading

Abler, R. F., J. S. Adams, and P. R. Gould. *Spatial Organization: The Geographer's View of the World.* Englewood Cliffs, NH: Prentice-Hall, 1971.

Birkin, M. G., G. Clarke, M. Clarke, and A. Wilson. *Intelligent GIS, Location Decisions and Strategies Planning.* New York: John Wiley and Sons, 1996.

Davies, R. and D. Rogers. *Store Location and Store Assessment Research.* New York: John Wiley and Sons, 1984.

Drezner, Z., ed. *Facility Location: A Survey of Applications and Methods.* Heidelberg: Springer Verlag, 1996.

Ebdon, D. *Statistics in Geography.* 2nd ed. Oxford: Basil Blackwell, 1985.

ESRI. "Linear Directional Mean (Spatial Statistics)." Accessed September 29, 2022. https://pro.arcgis.com/en/pro-app/latest/tool-reference/spatial-statistics/linear-directional-mean.htm

Lee, J. and D. W. S. Wong. *Statistical Analysis with ArcView GIS.* New York: Wiley, 2001.

Menecke, E. "Understanding the Role of Geographic Information Technologies in Business: Application and Research Directions." *Journal of Geographic Information and Decision Analysis.* 1 (1997): 44–68.

Mitchell, A. *The ESRI Guide to GIS Analysis. Vol. 2: Spatial Measurements and Statistics.* 2nd ed. Redlands, CA: ESRI Press, 2021.

National Climatic Data Center. Search www.ncdc.noaa.gov for hurricane and tropical storm tracks.

National Geospatial-Intelligence Agency (NGA). Search NGA website for standard deviational ellipse data. www.nga.mil/

Smith, M. J., M. F. Goodchild, and P. A. Longley. *Geospatial Analysis: A Comprehensive Guide to Principles, Techniques and Software Tools.* Leicester: Matador, imprint of Troubador Pub. Ltd., 2009.

Taylor, P. J. *Quantitative Methods in Geography: An Introduction to Spatial Analysis.* Boston: Houghton Mifflin, 1977.

PART 3

The Transition to Inferential Problem Solving

5

Basics of Probability and Discrete Probability Distributions

As geographers, we are vitally interested in exploring and studying spatial patterns found on the earth's physical and cultural landscapes. We develop descriptions and explanations of existing patterns and try to understand the processes that create these distributions. In many cases, we make predictions about future occurrences of geographic patterns and suggest spatial policies. In short, the core of geographic problem solving is the description, explanation, and prediction of geographic patterns and processes.

In earlier chapters, we have focused on ways to classify, describe, and otherwise summarize spatial data. Much of the remainder of the book presents various inferential statistical methods for problem solving in geography. The chapters in this part of the book (Chapters 5–8) deal with probability and sampling and serve as a transition to the inferential statistical techniques that follow.

Chapter 5 provides an introduction to probability and the important role it plays in geographic problem solving. The first section discusses the nature of geographic patterns and the processes that produce them. Included here are geographic examples of deterministic, random, and stochastic processes. This section also introduces basic terms and concepts of probability, including definitions and rules for using probability in geographic analysis.

Sections 5.2 through 5.4 cover certain discrete probability distributions that are important in geographic problem solving. The binomial distribution, discussed in Section 5.2, concerns probability of events having only two possible outcomes and is particularly useful for studying the likelihood of multiple events or trials. Section 5.3 introduces the geometric distribution, which is applied under similar circumstances as the binomial, but is concerned specifically with the timing of the next event. The Poisson distribution, presented in Section 5.4, is used to analyze patterns that are random over time or space. Examples of both spatial and temporal pattern analysis with Poisson are shown in this section. The normal distribution, which is of central importance to all researchers, is the primary topic in the next chapter.

5.1 Basic Probability Processes, Terms, and Concepts

Real-world processes that produce physical or cultural patterns on the landscape are often complex and are not usually totally identifiable. Two general categories are used to describe the nature of these processes-deterministic and probabilistic. **Deterministic processes** create patterns with total certainty, since the outcome can be exactly specified with 100% likelihood. Because of uncertainty and complexity in human behavior and decision-making, virtually no cultural processes are completely deterministic, but some physical processes established through scientific principles fall into this category. For example, the length of time insolation (solar radiation) strikes a point on the earth's surface is determined by both latitude and day of the year. Given these two components, the exact length of daylight and darkness can be determined at any location.

The second category, **probabilistic processes**, concerns all situations that cannot be determined with complete certainty. Given the complex character of most geographic situations, virtually all location-based problems fall into this category. Therefore, while the number of hours of sunlight at a particular location and time is deterministic, the amount of solar energy actually reaching the ground is probabilistic because the amount of cloud cover and particulate matter that absorbs and reflects incoming solar energy as it passes through the atmosphere is changing all the time.

Probabilistic processes can be subdivided into two useful categories. With **random processes**, the same probability is assigned to all outcomes because each outcome has an equal chance of occurring. Typical examples of random processes include drawing a card from a deck, rolling a die, or tossing a coin. Such situations are considered examples of maximum uncertainty. In a geographic context, an ornithologist might develop a GIS model to predict breeding bird occupancy within a wildlife refuge by assigning all nesting locations an equal random probability for occupancy. With **stochastic processes**, the likelihood or probability of any particular outcome can be specified and not all outcomes are equal. For example, an ornithologist may assign a greater probability that a bird will select a nesting location based on how far it is from the edge of the refuge or whether the location is shielded by a dense tree canopy.

One way to illustrate the nature of stochastic processes is to look at the Los Angeles Department of Water and Power's (LADWP) forecasting of serviceability in a water distribution network after an earthquake. The probability of a pipe breaking entails numerous complex factors such as the severity of the earthquake (as measured by peak ground velocity), the pipe material and age, and the characteristics of the soil through which the pipe runs. The combination of these factors determines the likelihood that an individual pipe may break, allowing LADWP to estimate the number and location of broken pipes throughout the system. Because the dynamics of infrastructure performance is complex, particularly when predicting future events, identifying an individual pipe's integrity after an earthquake cannot be specified with certainty. Under these conditions, the agency can only specify the likelihood or probability of each pipe's integrity in the system.

When dealing with this level of uncertainty, a stochastic model is often run numerous times, with each trial yielding a slightly different result. An overall estimate of system serviceability is then made based on all the independent model runs. This approach is known as **Monte Carlo simulation** and is frequently used when there is significant uncertainty in the model parameters.

Some spatial patterns are almost totally unpredictable. For example, the precise location of a tornado touchdown within a region is partly the result of random meteorological processes. Sometimes the size of the study area affects the degree of randomness. The number of tornadoes next year within Kansas can be estimated as a stochastic probability; however, if the focus is narrowed to a small area within Kansas, the exact location of a particular tornado touchdown is random or very close to random.

The study of probability focuses on the occurrence of an **event**, which can usually result in one of several possible **outcomes**. Once all possible outcomes have been considered for an event, probability represents the likelihood of a given result or the chance that an outcome actually takes place. Suppose a simple experiment is conducted by rolling a single standard die having six faces—1, 2, 3, 4, 5, and 6. What is the likelihood of rolling a 6 with a single toss of the die? In this experiment, the event is the roll of the die, and the possible outcomes are the six sides of the die. In effect, rolling the die is taking a sample observation from the infinitely large population of all rolls of that die. The die could be rolled 20 times to collect a sample size of 20. If the die is not "loaded," each outcome or face of the die has an equal chance of occurring and the likelihood of rolling a 6 is one out of six or .167. Of course, the same probability also exists that a 1, 2, 3, 4, or 5 will be thrown. Collecting sample data of outcomes from such experiments (when the sample is only a portion of the population) and then calculating the probabilities of different outcomes is the basis for statistical inference.

Probability can be thought of as **relative frequency**—the ratio between the absolute frequency for a particular outcome and the frequency of all outcomes:

$$P(A) = F(A) \,/\, F(E) \qquad \textbf{(5.1)}$$

where $P(A)$ = probability of outcome A occurring
$F(A)$ = absolute frequency of outcome A
$F(E)$ = absolute frequency of all outcomes for event E

Probabilities can also be interpreted as percentages when the denominator of Equation 5.1 is converted to 100. In the example of the die roll, the probability of a 6 (.167) can also be described as an outcome that has a 16.7% likelihood of occurring.

Some of the basic terms and concepts of probability can be illustrated with a simple geographic example. Suppose a day is classified as wet (W) if measurable precipitation (defined as at least .01 of an inch) falls during a 24-hour period, while the day is termed dry (D) if measurable precipitation does not occur. This is a valid classification, with clearly defined operational definitions of wet and dry. The two categories are **mutually exclusive**, since it is not possible to assign a particular day to more

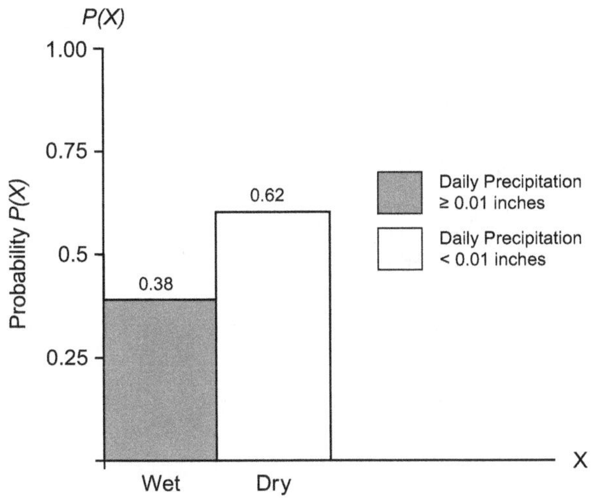

FIGURE 5.1
Relative Frequency of Wet and Dry Days from a 100-Day Period

than one category because the categories of wet and dry do not overlap. By keeping a record of wet and dry days over a 100-day period, absolute frequencies of precipitation can be determined and relative probabilities calculated from the data (Figure 5.1). In this example, 62 days are categorized as dry and 38 as wet. The probability of a wet day occurring $P(W)$ is

$$P(W) = \frac{\text{number of wet days}}{\text{total days}} = \frac{38}{100} = .38 \qquad \textbf{(5.2)}$$

Similarly, the probability of a dry day occurring $P(D)$ is

$$P(D) = \frac{\text{number of dry days}}{\text{total days}} = \frac{62}{100} = .62 \qquad \textbf{(5.3)}$$

Thus, a 38% chance exists that a day will have measurable precipitation and a 62% chance that a day will be dry.

Several rules of probability guide the decision maker. The maximum probability for any outcome is 1.0, indicating total certainty or perfect likelihood of a particular occurrence. The lowest probability for any outcome is 0.0, suggesting no chance of occurrence. Most outcomes have probabilities falling between these maximum and minimum values. Thus, for any outcome (A),

$$0.0 \leq P(A) \leq 1.0 \qquad \textbf{(5.4)}$$

Another straightforward definition concerns the **complement** of an outcome. If (A) contains all those events in which a particular outcome occurs, then the complement of (A), (written as $\bar{A}$), includes all those events where that particular outcome does not occur. The sum of the probabilities of an outcome and its complement is 1.0:

$$P(A) + P(\bar{A}) = 1.0 \qquad \textbf{(5.5)}$$

In the precipitation example, the probability of a wet day is .38, and the probability of a dry day is .62. Since these are complementary outcomes and they are mutually exclusive, their probabilities must total 1.0:

$$P(\overline{W}) = P(D) \qquad \textbf{(5.6)}$$

$$\text{so} \quad P(W) + P(\overline{W}) = P(W) + P(D) = 1.0 \qquad \textbf{(5.7)}$$

Let's return to the example of the die for a moment. With a single toss of the die, it is known that the probability of rolling a 6 is 1/6 or .167. What is the probability of tossing the die twice in succession and getting a 6 both times? These two events, the first roll of the die and the second roll of the die, are **statistically independent** events. Statistical independence exists when the probability of one event occurring is not influenced or affected by whether another event has occurred. The chance of rolling a 6 on the second roll is not at all influenced by what happened on the first roll. Many of the inferential statistics discussed later in the book are based on this property of statistical independence. This means that many of the sampling methods developed in the following chapters must be designed to incorporate this assumption of independence.

If events are independent, the simple *multiplication rule* of probability is applied. For independent events A and B, this rule states

$$P(A \text{ and } B) = P(A) \cdot P(B) \qquad \textbf{(5.8)}$$

Therefore, if event A is the first roll of the die and event B is the second roll of the die,

$$P(A \text{ and } B) = P(A) \cdot P(B) = 1/6 \cdot 1/6 = 1/36 \qquad \textbf{(5.9)}$$

That is,

$$P(A \text{ and } B) = P(A) \cdot P(B) = .167 \cdot .167 = .02778 \qquad \textbf{(5.10)}$$

There is one chance in 36 (somewhat less than a 3% chance) of tossing a die twice in succession and getting a 6 with both tosses.

The multiplication rule of probability can be extended using similar logic. For example, if a die is rolled three times, the likelihood of rolling a 6 on the first roll (event A), another 6 on the second roll (event B), and yet another 6 on the third roll (event C) is

$$\begin{aligned} P(A \text{ and } B \text{ and } C) &= P(A) \cdot P(B) \cdot P(C) \\ &= 1/6 \cdot 1/6 \cdot 1/6 = 1/216 \end{aligned} \qquad \textbf{(5.11)}$$

That is,

$$P(A \text{ and } B \text{ and } C) = P(A) \bullet P(B) \bullet P(C) = .167 \bullet .167 \bullet .167 = .00463 \quad \textbf{(5.12)}$$

The chance of tossing three 6s in succession on a die is very low (less than one-half of one percent.)

Let's consider how these basic probability rules can be applied to a set of US immigration data for fiscal year 2019 (Table 5.1). These data show the number of persons obtaining legal permanent residence by country of birth and state destination. Only the five largest countries of birth and five largest state destinations are shown in the table, but they account for a significant amount of the total immigration. Of all persons obtaining legal permanent residence in the United States in 2019, 32.0% were born in one of these five countries and 57.3% located in one of these five states.

Some categories in the table are mutually exclusive. For example, some immigrants initially settled in California while others first settled in New Jersey. These are mutually exclusive (nonoverlapping) categories because no single person can initially live in more than one state. Similarly, no single person can have more than one country of birth. Within certain portions of the table, however, the categories are not mutually exclusive. For example, a single immigrant could certainly be from Mexico *and* settle in Florida and thus be included in both of these (nonmutually exclusive) categories. In fact, 4,001 such individuals exist.

Suppose your task is to calculate the probability of selecting an immigrant from China *or* India at random from a list of all 1,031,765 persons obtaining legal permanent residence. The concept of randomness requires that each individual on this very lengthy list has an equal chance of being selected, and the selection is not biased in any way. Since these occurrences do not overlap, the *addition rule for mutually exclusive events* can be applied. In general form, this rule is as follows:

TABLE 5.1

Persons Obtaining Legal Permanent Residence Status by Country of Birth and U.S. State Destination, Fiscal Year 2019

Country of birth	State destination					TOTAL (all states)
	CA	NY	FL	TX	NJ	
Mexico	50645	2437	3113	32811	2437	139,120
China	18680	18859	1620	3280	2253	70,863
India	15099	5116	2019	5777	8123	69,162
Philippines	24082	2361	2320	2525	2321	58,173
Dominican Republic	172	26249	3900	275	8444	53,870
TOTAL (all countries)	208446	147999	107276	87750	56920	1,042,625

Note: Five largest countries of birth listed in order by row. Five largest state destinations listed in order by column. Grand total obtaining legal permanent status: 1,042,625.

Source: U.S. Department of Homeland Security, Yearbook of Immigration Statistics, 2019

$$P(A \text{ or } B) = P(A) + P(B) \quad \textbf{(5.13)}$$

where $P(A \text{ or } B)$ = probability of event A or event B
$P(A)$ = probability of event A
$P(B)$ = probability of event B

If event A is the selection of a person from China and event B is the selection of a person from India, then the probability of selecting an immigrant from one or the other of these countries is:

if $P(A) = 60{,}029 \;/\; 1{,}031{,}765 = .0580$ **(5.14)**

and $P(B) = 69{,}162 \;/\; 1{,}031{,}765 = .0670$ **(5.15)**

then $P(A \text{ or } B) = .0580 + .0670 = .1250$ **(5.16)**

Using similar logic for more than two events, the addition rule for mutually exclusive events can be generalized. For example, with three mutually exclusive events—A, B, and C—the addition rule becomes

$$P(A \text{ or } B \text{ or } C) = P(A) + P(B) + P(C) \quad \textbf{(5.17)}$$

If we identify event A as the selection of an immigrant from China, event B as the selection of an immigrant from India, and event C as the selection of an immigrant from the Philippines, then the probability of selecting a person from any one of these three countries is their sum, according to the addition rule:

if $P(A) = 60{,}029 \;/\; 1{,}031{,}765 = .0581$ **(5.18)**

and $P(B) = 69{,}162 \;/\; 1{,}031{,}765 = .0670$ **(5.19)**

and $P(C) = 43{,}478 \;/\; 1{,}031{,}765 = .0421$ **(5.20)**

then $P(A \text{ or } B \text{ or } C) = P(A) + P(B) + P(C)$
$= .0581 + .0670 + .0421 = .1672$ **(5.21)**

When randomly selecting a single individual from all persons obtaining legal permanent residence status during fiscal 2019, the probability of choosing a person from China, India, or the Philippines is .1672 (16.72%).

Suppose your task is to determine the probability of selecting a person from Mexico *or* an immigrant settling in California, *or both*. These are certainly not mutually exclusive events, as many Mexicans obtained legal permanent residence in California in 2019. In fact, according to Table 5.1, that number is known to be 50,645. For this situation, an addition rule for nonmutually exclusive events must be applied:

$$P(A \text{ or } B) = P(A) + P(B) - P(A \text{ and } B) \quad \textbf{(5.22)}$$

If event A is a person from Mexico, and event B is a person settling in California, then the following is known:

if $P(A) = 139{,}120 / 1{,}031{,}765 = .1348$ **(5.23)**

and $P(B) = 208{,}446 / 1{,}031{,}765 = .2020$ **(5.24)**

and $P(A \text{ and } B) = 50{,}645 / 1{,}031{,}765 = .049$ **(5.25)**

then $P(A \text{ or } B) = P(A) + P(B) - P(A \text{ and } B)$
$= .1348 + .2020 - .049 = .2877$ **(5.26)**

Note that the 50,645 Mexicans settling in California have been counted twice; those immigrants are included in the numerator of both $P(A)$ and $P(B)$. Therefore, this outcome $P(A \text{ and } B)$ needs to be subtracted to avoid double counting or duplication. Thus, if a single person is randomly selected from a list of all 1,031,765 persons obtaining legal permanent residence in the United States in fiscal year 2019, there is a .2859 probability (28.59% chance) of selecting someone from Mexico, or someone settling in California, or both.

Suppose you already know that a particular individual has settled in New York (event B), and the probability that this individual is from El Salvador is to be calculated. In this case, determining the *conditional probability* $P(A \mid B)$ is the appropriate strategy:

$$P(A \mid B) = P(A \text{ and } B) / P(B)$$

where $P(A \text{ and } B) = 26{,}249 / 1{,}031{,}765 = .0254$

and $P(B) = 147{,}999 / 1{,}031{,}765 = .1434$

then $P(A \mid B) = .0252 / .1434 = .1757$

Many other rules of probability exist for more complex applications. However, because this discussion of probability is only introductory, you should examine the sources listed at the end of the chapter for more thorough treatments of the subject.

The probability of outcomes in certain problems follows consistent or typical patterns. Such patterns, called **probability distributions**, relate closely to frequency distributions, such as the histogram and ogive diagrams shown earlier. In a frequency distribution, the number of occurrences appears on the vertical axis, and in a probability distribution, the probability of occurrence is displayed on the vertical axis. In both cases, the horizontal axis shows the actual outcomes, occurrences, or values of the variable being studied.

Recall that variables are termed discrete or continuous depending on whether the values occur as distinct whole numbers (discrete) or decimal values (continuous). Probability distributions for discrete outcomes are termed **discrete probability distributions**, whereas those for outcomes that can occur at an infinite number of points are termed **continuous probability distributions**. In this chapter, discussion is focused on three widely used discrete distributions—binomial, geometric, and Poisson.

5.2 The Binomial Distribution

The **binomial** is a discrete probability distribution associated with events having only two possible outcomes. Binary outcomes are often described in a zero-one, yes-no, or success-failure format and many geographic situations fit a binary framework. Consider these examples: a location either has measurable precipitation over a 24-hour period or it does not, a person is either a male or a female, a river is either above or below flood stage, a respondent to an opinion poll either favors or opposes an issue. In each of these situations, only one of two outcomes is possible, assuming that an undecided or uncertain result is not possible. The general shape of the binomial distribution is shown in Figure 5.2.

The binomial distribution is especially useful in examining probabilities from multiple events or trials, each of which has only two possible outcomes. Geographers are often interested in evaluating sequences of these events or trials-phenomena that either occur or don't occur at some location through time. In many of these situations, the probability of a single event is determined quite easily, perhaps using historical data. For example, the probability of a river in Bangladesh reaching flood stage during a given year may be .40. Thus, on average, flooding occurs four years out of ten. With this information, the binomial distribution may be used to determine the probability of multiple flood events occurring over other time periods (perhaps 15 years out of the next 30). The assumption is that a river reaching flood stage during a given year is an event independent of whether the river reached flood stage in other years.

When using binomial probabilities, the focus is on one of the two possible outcomes, termed the *given* out-

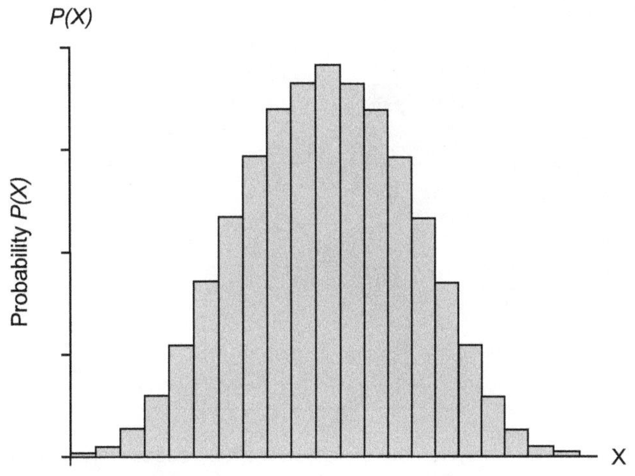

FIGURE 5.2
The Binomial Distribution

come (X). The binomial distribution is shown mathematically as follows:

$$P(X) = \frac{n!\, p^X q^{n-X}}{X!\,(n-X)!} \quad \textbf{(5.27)}$$

where n = number of events or trials

p = probability of the given outcome in a single trial

q = $\bar{p} = 1 - p$ or the probability of the *other* outcome in a single trial

X = number of times the given outcome occurs within the n trials

$n!$ = n factorial:
if $n > 0$, $n! = [n(n-1)(n-2)\ldots(2)(1)]$
if $n = 0$, $n! = 1$
for example, $5! = (5)(4)(3)(2)(1) = 120$

To summarize, the binomial distribution is applicable for those geographic problems in which the following conditions are met:

- The objective is to determine the probability of multiple (n) independent events or trials, and the number of trials is fixed.
- Each event or trial has two possible outcomes, one termed the given outcome (X), with associated probability p, and a complementary or *other* outcome having probability $q = \bar{p} = 1 - p$.
- The probabilities p and q must remain stable or consistent over the duration of the study period or over successive trials through time—that is, the probability of a given outcome is the same for each trial.

The best way to illustrate the characteristics and practical uses of the binomial probability is with a geographic example. Suppose a potential vegetable grower is seeking a location to start a business. One of the key variables in site selection is the probability of adequate precipitation at that site, thereby reducing the necessity of expensive irrigation. Suppose the potential grower has learned from other growers in the immediate region that at least three inches of precipitation are needed during the growing season to avoid irrigation. The grower estimates that irrigation can be afforded only one year in five to make a sufficient profit. Precipitation data collected for a potential site show that in 21 of the last 25 years rainfall exceeded three inches during the growing season. The historical record therefore suggests that the probability of a given year requiring supplemental irrigation is 4/25 or 0.16. What is the probability that the grower can meet the requirement of having to irrigate only one year in five or not having to irrigate at all?

Over the five-year period the farmer faces six possible outcomes, from none of the five years requiring irrigation up to all five years requiring irrigation. However, only two of the six outcomes (no years and one year)

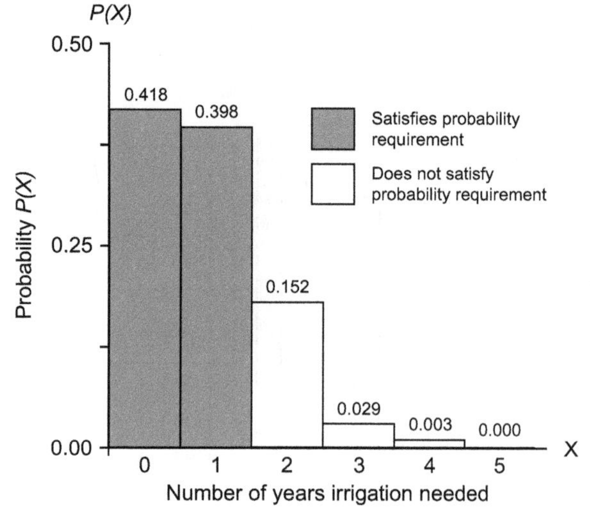

FIGURE 5.3
Probability of Vegetable Owner Needing Irrigation for 0 to 5 Years

TABLE 5.2

Critical Values and Binomial Probabilities of Suitable Outcomes for Vegetable Owner

$$P(X) = \frac{n!\, p^x q^{n-x}}{X!\,(n-X)!}$$

Critical Values:

$X = 0, 1, 2, 3, 4, 5$
$n = 5$ years
$p = 4/25 = .16$
$q = (1 - p) = .84$
$n! = (5)(4)(3)(2)(1) = 120$

When $X = 0$: $P(0) = \frac{5!(.16)^0(.84)^5}{0!(5)!} = \frac{120(1)(.418)}{1(120)} = .418$

When $X = 1$: $P(1) = \frac{5!(.16)^1(.84)^4}{1!(4)!} = \frac{120(.16)(.498)}{1(24)} = .398$

When $X = 2$: $P(2) = \frac{5!(.16)^2(.84)^3}{2!(3)!} = \frac{120(.026)(.593)}{2(6)} = .152$

When $X = 3$: $P(3) = \frac{5!(.16)^3(.84)^2}{3!(2)!} = \frac{120(.004)(.706)}{6(2)} = .029$

When $X = 4$: $P(4) = \frac{5!(.16)^4(.84)^1}{4!(1)!} = \frac{120(.001)(.84)}{24(1)} = .003$

When $X = 5$: $P(5) = \frac{5!(.16)^5(.84)^0}{5!(0)!} = \frac{120(.000)(1)}{120(1)} = .000$

TABLE 5.3

Binomial Probabilities of Vegetable Owner Needing Irrigation Over a Five-Year Period

Number of years out of five	Binomial probability	Type of Outcome
0	.418	Suitable
1	.398	Suitable
2	.152	Unsuitable
3	.029	Unsuitable
4	.003	Unsuitable
5	.000	Unsuitable
TOTAL	1.000	

Total probability of suitable outcomes = .816
Total probability of unsuitable outcomes = .184

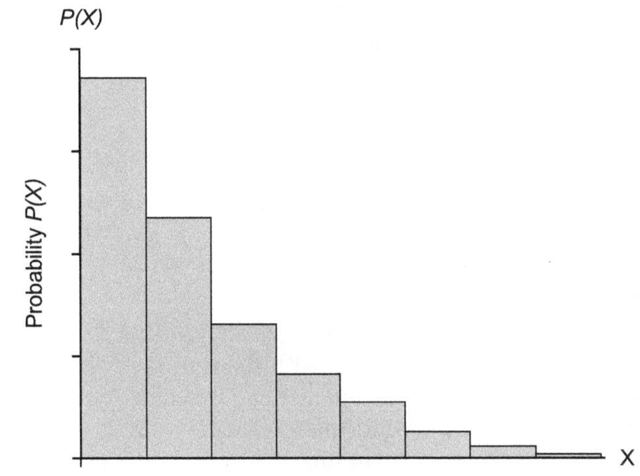

FIGURE 5.4
The Geometric Distribution

would result in a profitable situation for the grower; each of the other four outcomes would be too costly. Thus, the probability that the grower will meet the profit requirement is found by summing the probabilities for the two suitable outcomes. The critical values and calculations of the binomial probabilities for all possible outcomes are shown in Table 5.2. The binomial probabilities for the problem are also listed in Table 5.3 (with suitable or unsuitable outcome) and plotted in Figure 5.3.

The vegetable grower would have a .418 probability of needing no irrigation and a .398 chance of needing one year of irrigation during the next five years. By adding the binomial probabilities for no years and one year of irrigation (.418 + .398) we conclude that the grower has an 81.6% chance of meeting the profitability requirement at this potential site during the five-year interval. This level of risk may or may not be acceptable to the grower.

5.3 The Geometric Distribution

The geometric distribution is another discrete probability distribution. While the binomial distribution determines the probability associated with a number of given outcomes or events ("successes"), the narrower focus of the geometric distribution is on the exact number of consecutive trials necessary to observe the given outcome or event for the first time. For example, a geographer might be interested in the timing of the next category three hurricane at a particular location. As another example, you might want to estimate the probabilities associated with the timing of the next major earthquake (say, above 7.0 on the Richter scale) along the San Andreas Fault in the San Francisco area. Questions like these could be asked: "What is the probability that the next big quake in this area will occur after 2022?" or "I've just signed a yearly contract for an apartment located very close to the San Andreas Fault and want to know the likelihood of the next major earthquake occurring next year." The general form of a geometric distribution is shown in Figure 5.4.

In conclusion, the geometric distribution applies to geographic problems in which the following conditions are met:

- The objective is to determine the number of consecutive independent trials needed to observe a given event (success) for the first time. Alternatively, the objective is to determine the number of failures that will occur before the first success.
- As with the binomial distribution, each event or trial has two possible outcomes (success/failure), each with a stable probability of occurrence.
- The geometric probability is "memoryless." That is, if a success has not yet occurred in a sequence of trials, the number of past trials in the sequence does not change the probability distribution of the number of trials that remain before a success occurs. For example, if an unbiased coin has been tossed ten consecutive times and all tosses have been heads, the probability of tails on the next toss remains unchanged at .50. Quite simply, the past doesn't influence the future (a characteristic that many gamblers fail to grasp!).

The usefulness of the geometric distribution is best shown with an actual geographic application. The Susquehanna River Basin drains 27,510 square miles of New York, Pennsylvania, and Maryland—where about 4 million people live (Figure 5.5). The Basin has a history of flooding. In fact, the main stem of the Susquehanna has experienced 15 major floods since record-keeping began around 1800, with the latest incident occurring quite recently (Table 5.4). This averages out to a major flood every 14 years.

Some past flooding has been catastrophic. The flooding from tropical storm Agnes in 1972 caused record-breaking damage in the Susquehanna River

Basin. Seventy-two people died and damage topped $2.8 billion—about $14.3 billion in today's dollars. It was the nation's most costly disaster until Hurricane Katrina struck New Orleans and the Gulf Coast in 2004. Future flooding of the Susquehanna has the potential to be even more devastating than past events. Of the 1,400 communities in the Basin, an estimated 1,160 still have residents who live in flood-prone areas. Obviously, flood prediction, management, and protection are of utmost concern for these people.

The geometric distribution can be applied to determine probabilities associated with the timing of the next major flood along the Susquehanna. It seems logical to want flood probability information for each single year in the future and also a cumulative probability estimate (cumulative distribution function—CDF) showing when the next major flood will most likely occur.

According to the historical record, the probability a major flood will occur in the first year of analysis is $p = .07$. It is assumed that the timing of each major flood is an independent event, not influenced by the timing of previous floods—certainly a reasonable assumption. [You might ask if this is really a correct assumption: if climate change is significant, the frequency of flood events could be changing!] If events are truly independent, the probability of a flood not occurring in the first year is $(1 - p) = .93$.

TABLE 5.4

Major Floods in the Susquehanna River Basin Since 1800

1810		1964	
1865		1972	Tropical Storm Agnes
1889		1975	Tropical Storm Eloise
1894		1996	Basinwide Flash Flood
1935		2004	Tropical Storm Ivan
1936	St. Patrick's Day	2006	
1946		2011	Tropical Storm Lee
1955	Hurricanes Connie & Diane		

Source: Susquehanna Flood Forecast and Warning System

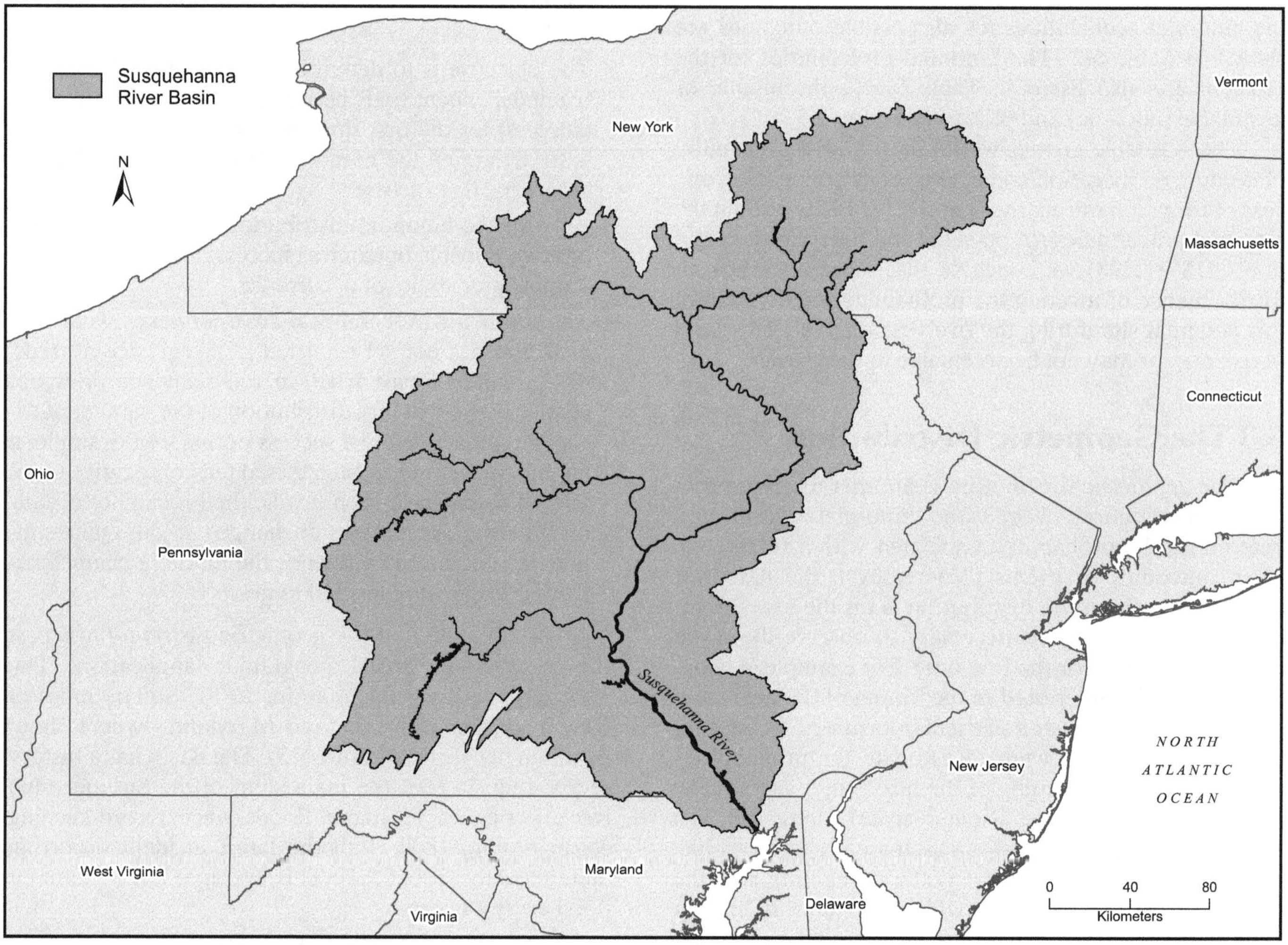

FIGURE 5.5
The Susquehanna River Basin

The probability of a flood occurring in the second year (if none occurred in the first year) is $(1 - p)\ (p) = (.93)\ (.07) = .0651$. The probability of a flood occurring in the third year (if no flood occurred in years one or two) is $(1 - p)\quad (1 - p)\ (p) = (1 - p)^2\ (p) = (.93)\ (.93)\ (.07) = .06$. These expressions are applications of the multiplication rule of probability for independent events (see Equation 5.8). Continuing this logic, the probability of a flood occurring in the year k (if no flood occurred in any prior year—1, 2, 3 . . . $k - 1$) is $(1 - p)^{k-1}\ (p) = (.93)^{k-1}\ (.07)$. This last expression corresponds to $(k - 1)$ consecutive years of no flood, followed by a flood in year k.

The application of the geometric probability distribution to predict the likelihood of possible flooding in the Susquehanna River Basin over the next 10 years can be summarized both graphically and in table form. The regular decline in flood probability is shown directly in Figure 5.6 and in the middle column of Table 5.5. Year 1 has the greatest likelihood of the next major flood occurrence (.07), Year 2 the second largest probability of the next flood, and so on. The rightmost column of Table 5.5 lists the **cumulative distribution function (CDF)** probabilities. If past flooding frequencies continue into the future, then the next major flood has more than a 50% chance of occurring (52.96%, to be precise) sometime in the next five years. For residents on a Susquehanna River Basin floodplain, this must be viewed as an alarmingly short time period.

The average amount of time until the next major flood is $(1\ /\ p)$ years: that is $(1\ /\ .07) = 14.28$. However,

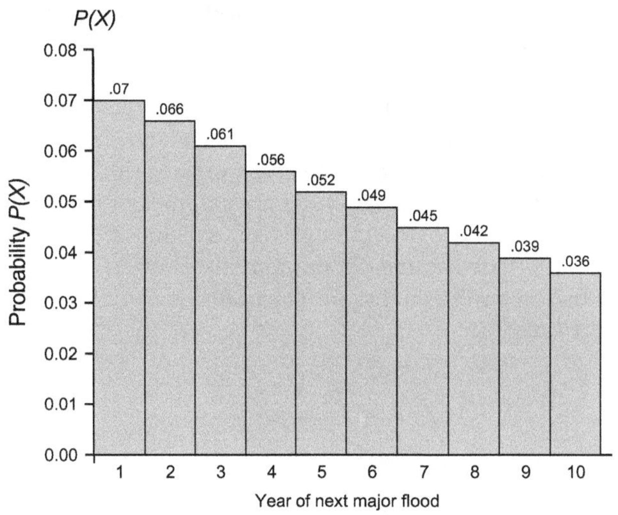

FIGURE 5.6
Geometric Distribution: Timing of the Next Major Flood on the Susquehanna River

over the last couple of centuries the sequencing of floods has been highly variable. No major floods occurred over a 55-year period from 1810 to 1865, and Basin residents were given another lengthy break from flooding between 1894 and 1935. Of course, major floods can also strike in rapid succession. Serious flooding events devastated the region in consecutive years (1935 and 1936) and over the last decade or so there have been three notable incidents (2004, 2006, and 2011). This irregular sequencing of floods is reflected by the rather large standard deviation of 13.77 years. How should this standard deviation value be interpreted? Recall that standard deviation measures the "typical" or "standard" difference of an observation from the mean. Since the mean of this geometric distribution is 14.28 years and the standard deviation is 13.77 years, considerable variability can be expected in the amount of time that will elapse from one flood incident to the next. This high level of uncertainty and unpredictability must be very frustrating to emergency preparedness personnel and local government officials that are constantly fighting to get funding for flood protection. In times of economic difficulties, it is much easier to argue for budgetary needs that are more predictable, such as police and fire protection or education.

TABLE 5.5

Geometric Distribution: Probabilities Associated with the Next Major Flood in the Susquehanna River Basin

Number of trials (years) $p\ (X = k)$	Probability the next major flood will occur this year (probability mass function) $(1 - p)^{k-1}\ p$	Cumulative probability that next major flood will occur this year (cumulative distribution function) $1 - (1 - p)^k$
1	$(.93)^0\ .07 = .07$	$1 - (.93)^1 = .07$
2	$(.93)^1\ .07 = .066$	$1 - (.93)^2 = .135$
3	$(.93)^2\ .07 = .061$	$1 - (.93)^3 = .196$
4	$(.93)^3\ .07 = .056$	$1 - (.93)^4 = .252$
5	$(.93)^4\ .07 = .052$	$1 - (.93)^5 = .304$
6	$(.93)^5\ .07 = .049$	$1 - (.93)^6 = .353$
7	$(.93)^6\ .07 = .045$	$1 - (.93)^7 = .398$
8	$(.93)^7\ .07 = .042$	$1 - (.93)^8 = .44$
9	$(.93)^8\ .07 = .039$	$1 - (.93)^9 = .48$
10	$(.93)^9\ .07 = .036$	$1 - (.93)^{10} = .516$

Mean = $1\ /\ p = 1\ /\ .07 = 14.28$

Variance = $(1 - p)\ /\ p^2 = (1 - .07)\ /\ .07^2 = 189.80$

Standard Deviation = $((1 - p)\ /\ p^2)^{0.5} = ((1 - .07)\ /\ .07^2)^{0.5} = 13.77$

5.4 The Poisson Distribution

Some probability problems in geography involve the study of events that occur repeatedly and randomly over either time or space. For example, the placement of calls to an emergency response dispatcher might be considered random over a short period of time. At certain spatial levels, multiple occurrences of weather-related phenomena (e.g., thunderstorms, tornadoes, and hurricanes) may occur with little spatial predictability. Geographers also study various cultural entities whose patterns may be the result of random processes. In instances where events occur repeatedly and at random (i.e., independent of past or future events), the **Poisson probability distribution** can be used to analyze how frequently an outcome occurs during a certain time period or across a particular area. Other geographic applications of Poisson involve the analysis of existing frequency count data to determine if a random distribution exists.

Poisson probabilities are calculated using this general formula:

$$P(X) = \frac{e^{-\lambda}(\lambda^X)}{X!} = \frac{\lambda^X}{e^{\lambda}(X!)} \quad \textbf{(5.28)}$$

where X = frequency of occurrence
λ = mean frequency
e = base of the natural logarithm (approximately 2.71828)
$X!$ = X factorial

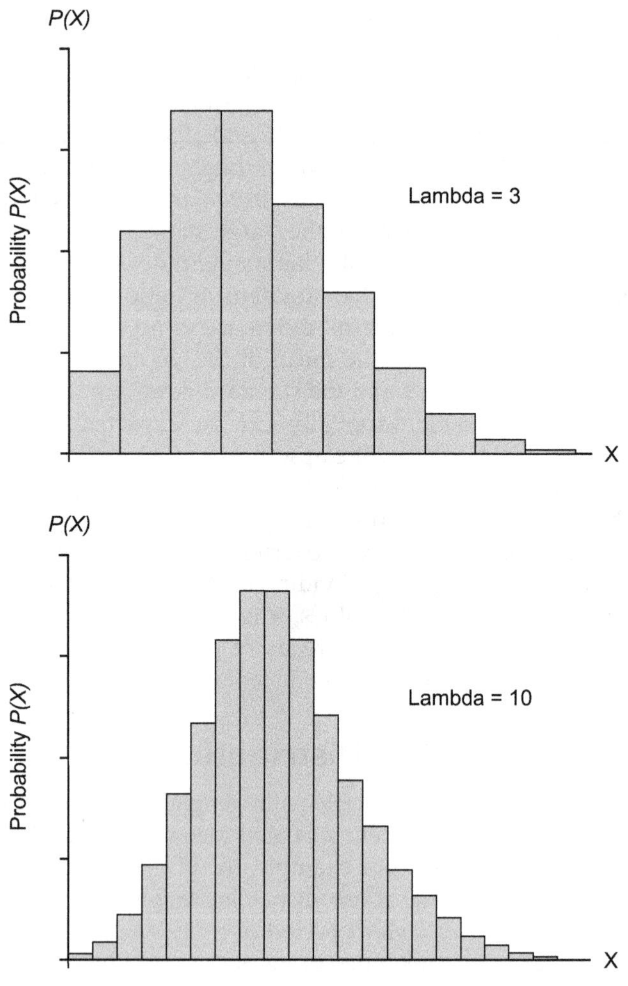

FIGURE 5.7
The Poisson Distribution

The general form of the Poisson distribution is a function of lambda, the mean frequency. Figure 5.7 shows the shape of the distribution with mean frequencies of 3 and 10.

In summary, the Poisson distribution is appropriate for geographic problems that meet the following criteria:

- The objective is to determine how frequently an outcome occurs where events occur repeatedly and at random over time or across space. That is, data are counts of events. In other instances, the objective is to analyze a set of existing frequency count data to determine if it is randomly distributed through time or to analyze a point pattern to determine if it is randomly distributed across space.
- The time period being analyzed is divided into discrete units (e.g., years) or the study area being analyzed is divided into discrete areal subdivisions (e.g., quadrats).
- The frequency distribution of the occurrence of multiple events by discrete time unit or the frequency distribution of a point pattern by quadrat is estimated under the assumption that a random process is operating. Alternatively, an existing frequency distribution is analyzed to determine if it has been generated by a process operating randomly through time or across space.

As with the binomial and geometric distributions, the most effective way to demonstrate the usefulness of the Poisson distribution is by looking at practical geographic applications.

Analyzing Geographic Data through Time

This first application involves snowfall prediction over time—specifically, predicting the average seasonal frequency of separate and unique "crippling heavy snowfalls" (which we define as a snowfall event of at least 10 inches) in the very snowy city of Flagstaff, Arizona.

The city of Flagstaff is located in north central Arizona. At 6,900 feet, surrounded by ponderosa pine forest and south of mountains that reach well over 12,000 feet, Flagstaff enjoys a four-season climate. However, at this elevation, Flagstaff receives a lot of snow, with a seasonal average of 97 inches.

Both positive and negative outcomes become associated with very heavy snowfall incidents. In addition, important budget and management policy issues must be made before the winter season even starts. Flagstaff is a winter sports area and substantial economic activity depends on adequate snowfall. A successful season is determined by the quality and quantity of snowfall, the timing of seasonal precipitation, its amount, and the temperature. Conversely, very heavy snowfalls can be economically crippling in a number of ways. Extremely high seasonal totals are possible, as in winter of 1973, when nearly 200 inches fell. Local public officials must consider the likelihood of crippling snow events that have the potential to paralyze transportation systems and jeopardize safe mobility.

Obviously, planners need to budget for major snow events in advance. For example, how much annual money should be allocated to road salt purchase, or how many snow removal vehicles should be at the ready (in inventory). Should planners stay with the current number of snow-removal vehicles (which can handle the 5-inch snowstorm adequately), or should two additional vehicles be purchased, which public service officials estimate would be sufficient to handle most 10-inch events? Local planners routinely face such snow-related decisions that require them to balance cost with level of preparedness.

If we examine the snowfall record over a recent 30-year period (1971–2000), Flagstaff experienced 54 separate crippling heavy snowfall events of at least 10 inches—an average of 1.8 events per winter. With this mean frequency value of $\lambda = 1.8$, we can calculate all Poisson probabilities.

TABLE 5.6

Poisson Distribution: Expected Number of Days per Winter with Heavy Snowfall (Greater than 10 Inches), Flagstaff, Arizona

$$P(X) = \frac{e^{-\lambda}(\lambda^x)}{X!} = \frac{\lambda^x}{e^{\lambda}(X!)}$$

Critical Values:

X = frequency of occurrence = 0, 1, 2, 3, . . . , 6
λ = mean frequency per winter = 1.8 (historical record shows 54 separate heavy snowfall events over 30 years)
e = 2.718 (base of the natural logarithm)
$e^{\lambda} = 2.718^{1.8} = .16533$

When $P(0) = \frac{(2.718^{-1.8})(1.8^0)}{0!} = \frac{(.16533)(1)}{(1)} = .16533$

When $P(1) = \frac{(2.718^{-1.8})(1.8^1)}{1!} = \frac{(.16533)(1.8)}{1} = .29759$

When $P(2) = \frac{(2.718^{-1.8})(1.8^2)}{2!} = \frac{(.16533)(3.24)}{2} = .26883$

When $P(3) = \frac{(2.718^{-1.8})(1.8^3)}{3!} = \frac{(.16533)(5.832)}{6} = .16070$

. . .

When $P(6) = \frac{(2.718^{-1.8})(1.8^6)}{6!} = \frac{(.16533)(34.012)}{720} = .00781$

Since both e and λ are constants, the expression $e^{-\lambda}$ in Equation 5.28 is also a fixed value ($e^{-\lambda} = 2.72^{-1.8} = .16533$). Using this constant $e^{-\lambda}$ value, Poisson probabilities are calculated for a series of heavy snowfall frequencies ($X = 0, 1, 2, 3, 4$) and presented in Table 5.6.

If heavy snowfalls occur randomly with a mean of 1.8 per winter, the likelihood of no such snowfall during a given winter is .16533. Extending this logic to a period of 100 winters, about 16 or 17 of those winters should be free of very heavy snowstorms. The most likely occurrence is one very heavy snowfall per winter, which should occur at a probability of .29759, or almost 30 winters out of 100 (Figure 5.8). Amazingly, there is more than a 7% chance (probability of .072315) that a future winter in Flagstaff will have 4 different crippling heavy snow events in which over 10 inches of snow will fall!

Analyzing Geographic Data over Space

The United States Department of Agriculture's Forest Service maintains the Fire Statistics System (FIRESTAT), which contains over 500,000 wildfire records collected by the six federal land management agencies that occurred between 1910 and 2021. Users can query, research, and download this data for statistical and other geographic analyses.

In 2012, a fire in Gila National Forest, New Mexico, became the largest wildfire in the state's history, burning over 200,000 acres. Suppose, for example, foresters in New Mexico are concerned about the frequency of very large wildfires (greater than 20,000 acres burned). Because of the uncertain nature of human influences, land use patterns, and weather events, the number of large wildfires varies greatly from year to year. Due to the financial pressures associated with fighting very large fires, foresters may want to know the probability of expe-

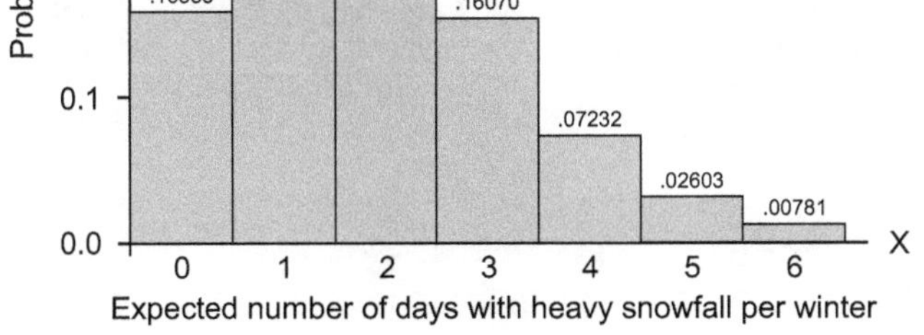

FIGURE 5.8
Poisson Distribution: Expected Number of Days per Winter with Heavy Snowfall (Greater than 10 Inches), Flagstaff, Arizona

riencing more than 3 fires exceeding 20,000 acres in a given year. The current occurrence of wildfires is independent of past or future occurrences and can be considered random. Therefore, Poisson is the valid probability model for describing wildfire frequency across the State of New Mexico.

During the 30 years between 1980 and 2010, 19 years were completely free of very large wildfires, while 9 wildfires occurred in 1 of the 30 years. Since a total of 32 large wildfires occurred during the period, the mean number of large wildfires per year was 1.06.

The wildfire probability can be calculated for any frequency of occurrence. For example, since the data cover 30 years, and 19 of these years did not have a very large wildfire, the resultant probability of no large wildfires occurring in a year is 19/30 or 0.63. Thus, given this set of data, New Mexico has a 63% chance of avoiding a very large wildfire (greater than 20,000 acres burned) in any given year.

If the process producing the pattern is truly random, the probabilities of occurrence will follow the Poisson distribution. The mean frequency is the average number of large wildfires per time period or geographic area. Knowing only this mean value is sufficient to allow all Poisson probabilities to be calculated:

$$P(X) = \frac{e^{-\lambda}(\lambda^X)}{X!} = \frac{\lambda^X}{e^{\lambda}(X!)} \qquad \textbf{(5.28)}$$

where X = frequency of occurrence
λ = mean frequency
e = base of the natural logarithm (approximately 2.71828)
$X!$ = X factorial

Like the binomial distribution, Poisson requires discrete or integer outcomes (X) to represent the number of frequency of occurrences. Unlike the binomial, however, the Poisson distribution is not binary. In the wildfire example, discrete outcomes represent the different number (frequency counts) of wildfires occurring annually. The mean outcome or average number of wildfires per year (λ) is 1.06. Since both e and λ are constants, the expression e^{λ} in Equation 5.28 is also a fixed value ($e^{\lambda} = 2.72^{1.06} = 2.88$). The exponential function on a calculator or computer is needed for this calculation. Using this constant e^{λ} value, Poisson probabilities are calculated for a series of wildfire frequencies ($X = 0, 1, 2, 3, \ldots, 9$) and presented in Table 5.7.

If large wildfires occur randomly with a mean of 1.06 per year, the likelihood of no large wildfire over 20,000 acres occurring in a given year is 0.35. In other words, extending this logic to a period of 100 years, about 35 of those years should be free of very large wildfires. The most likely occurrence is one very large wildfire per year, which should occur at a probability of .37 or about 37 years out of 100 (Figure 5.9). A geographer can then compare the expected probabilities with the observed probabilities in Table 5.7 to determine whether the occurrence of large wildfires is a series of random events.

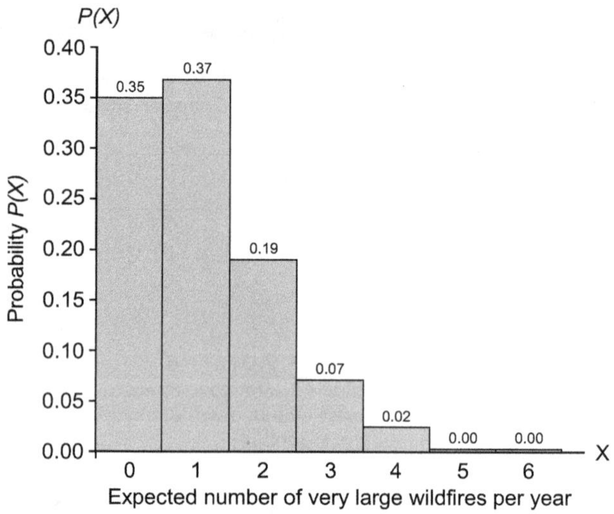

FIGURE 5.9
Poisson (Expected) Probabilities for Number of Very Large Wildfires in New Mexico per Year

TABLE 5.7

Observed and Expected (Poisson) Frequencies for Number of Very Large Wildfires in New Mexico per Year

Number of very large wildfires per year	Observed frequency (years)	Total wildfires	Observed probability of occurrence	Poisson probabilities	Expected frequencies (years)
0	19	0	0.63	0.35	11.09
1	4	4	0.13	0.37	11.75
2	2	4	0.07	0.19	6.23
3	2	6	0.07	0.07	2.20
4	1	4	0.03	0.02	0.58
5	1	5	0.03	0.00	0.12
9	1	9	0.03	0.00	0.00
TOTAL	30	32	1.00 (100%)	1.00 (100%)	31.97

Source: U.S. Geological Survey, Federal Wildland Fire Occurrence Website

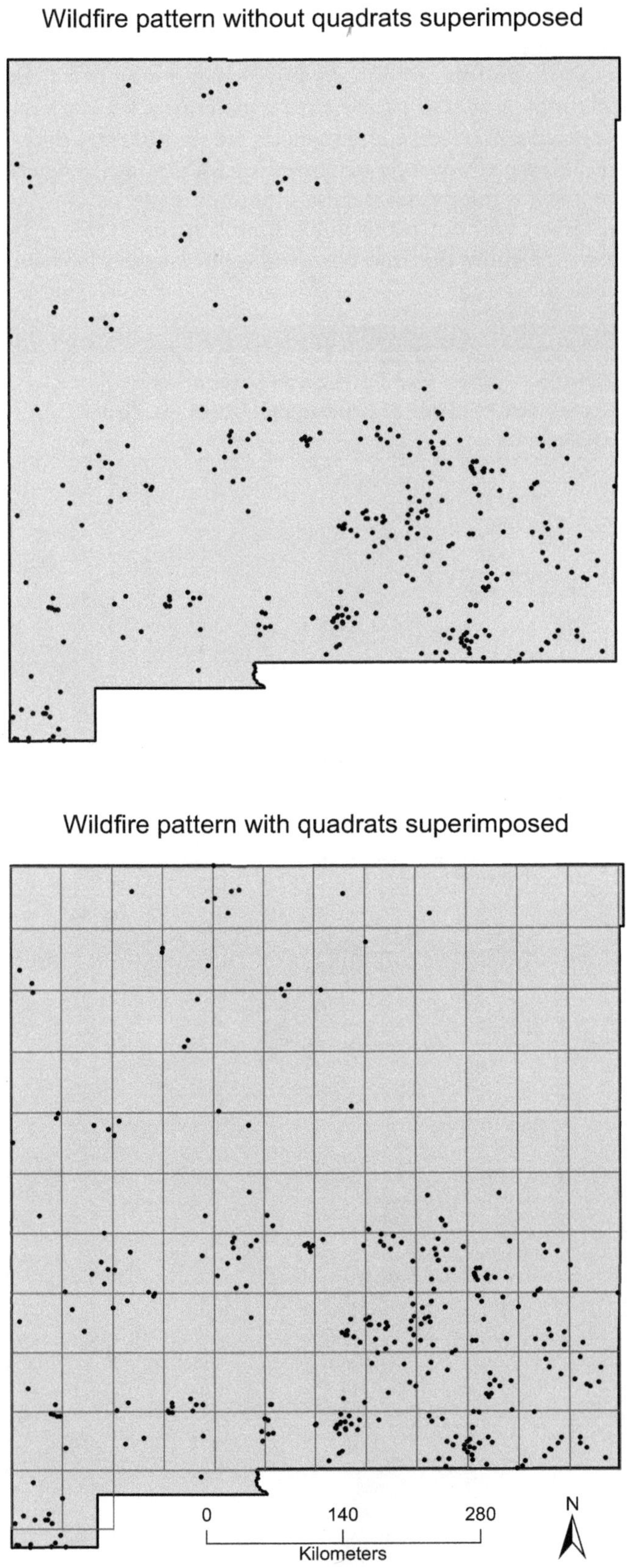

FIGURE 5.10
Spatial Pattern of Very Large Wildfires (over 20,000 Acres) in New Mexico, 1980 – 2009
Source: U.S. Geological Survey, Federal Wildland Fire Occurrence Website

In a spatial application of Poisson, the geographic region under study is divided into spatial areas, usually a series of regular-sized square cells known as *quadrats.* The number of occurrences of an item or phenomenon being studied is recorded for each of the quadrats covering the study area. Given the mean number of occurrences per quadrat or cell across the region, the Poisson distribution shows the probability of a quadrat containing a certain frequency of occurrences.

We can now ask the question: is this set of very large wildfires across New Mexico randomly distributed? To calculate the expected Poisson pattern of large wildfire occurrences that would be found under an assumption of randomness, a set of quadrats is placed over the study area (Figure 5.10).

Because the southern boundary of New Mexico is somewhat irregular, some of the quadrats near the border would lie outside the study area. Therefore, the quadrats were placed in a manner to maximize the number of quadrats that would fit in the state and also cover the greatest amount of points. Of the 326 wildfires, only 10 are not included in a quadrat.

Which cells should be included in the analysis? A procedural method must be selected that is as objective, consistent, and unbiased as possible. One logical way to proceed would be to include only those cells with more than half of their area in New Mexico. If Poisson probabilities for the wildfire pattern are determined in this fashion, 122 quadrats (containing 316 of the 326 large wildfire locations on the map) are analyzed, representing almost 97% of the large wildfires that occurred in New Mexico over the 30-year period.

The frequency of wildfires is determined for each cell by counting the number of points inside each quadrat. This is easily accomplished using the overlay operations in a GIS. The observed frequencies for wildfire occurrence by cell range from a low of 0 to a high of 19 (Table 5.8). In those cells experiencing a very large wildfire, the most frequent occurrence or mode is one point per quadrat, found in 23 cells. The average cell frequency (λ)—total number of wildfires (316) divided by the number of cells (122)—is 316/122 or 2.59 (Table 5.9). Thus, for this set of quadrats, an average cell contains 2.59 points (wildfires).

The probability of very large wildfire occurrence under an assumption of randomness is determined from the Poisson equation (Equation 5.28) using the mean cell frequency. The calculations for zero, one, and two wildfires per cell are shown in Table 5.8, and Figure 5.11 shows the Poisson probabilities graphically. To compute the expected cell frequencies, each Poisson probability is multiplied by the total number of cells. For example, since the Poisson probability of a cell having 4 wildfires is .14, 14% of the 122 cells or 17 cells should have 4 wildfires (Table 5.9). The largest frequency expected for a random pattern of wildfires is 2 points per cell, which should occur in 30.69 cells of the study area. This maximum expected value is consistent with the mean cell frequency of 2.59 points per quadrat.

The observed frequencies do not appear to match the calculated Poisson frequencies. Major discrepancies occur between the observations. Notably, under a random distribution we would expect 9 cells to have zero fires, and 30 cells to have two fires. These expected frequencies are very different from the 50 observed frequencies with zero fires and the 14 observed frequencies with two fires. Inferential statistics can be used to determine whether the expected and observed frequencies are significantly different. This procedure is discussed in Section 13.1, where the focus is inferential statistics for point patterns.

Even without using inferential statistics, Table 5.10 seems to show that the observed and expected frequen-

TABLE 5.8

Observed Frequency of Very Large Wildfires per Cell in New Mexico

Number of very large wildfires per cell	Observed frequencies	Total points	Observed probabilities of wildfires per cell
0	50	0	0.410
1	23	23	0.189
2	14	28	0.115
3	7	21	0.057
4	7	28	0.057
5	2	10	0.016
6	3	18	0.025
7	2	14	0.016
8	2	16	0.016
9	3	27	0.025
10	1	10	0.008
11	1	11	0.008
13	2	26	0.016
14	1	14	0.008
16	2	32	0.016
19	2	38	0.016
TOTAL	122	316	1.00 (100%)

TABLE 5.9

Critical Values and Poisson Probabilities of Expected Wildfire Frequency per Cell for Three Outcomes

$$P(X) = \frac{\lambda^X}{e^{\lambda}(X!)}$$

Critical Values:

X = 0, 1, 2, 3, 4, . . ., 17, 18

N = 122 cells

f = total frequency = 316 wildfires

λ = mean frequency per cell = f/N = 316/122 = 2.59

e = exponential (approximately 2.72)

e^{λ} = $2.72^{2.59}$ = 13.35

$$P(0) = \frac{2.59^0}{13.35(0!)} = \frac{1}{13.35} = .07$$

$$P(1) = \frac{2.59^1}{13.35(1!)} = \frac{2.59}{13.35} = .19$$

$$P(2) = \frac{2.59^2}{13.35(2!)} = \frac{6.7081}{26.7} = .25$$

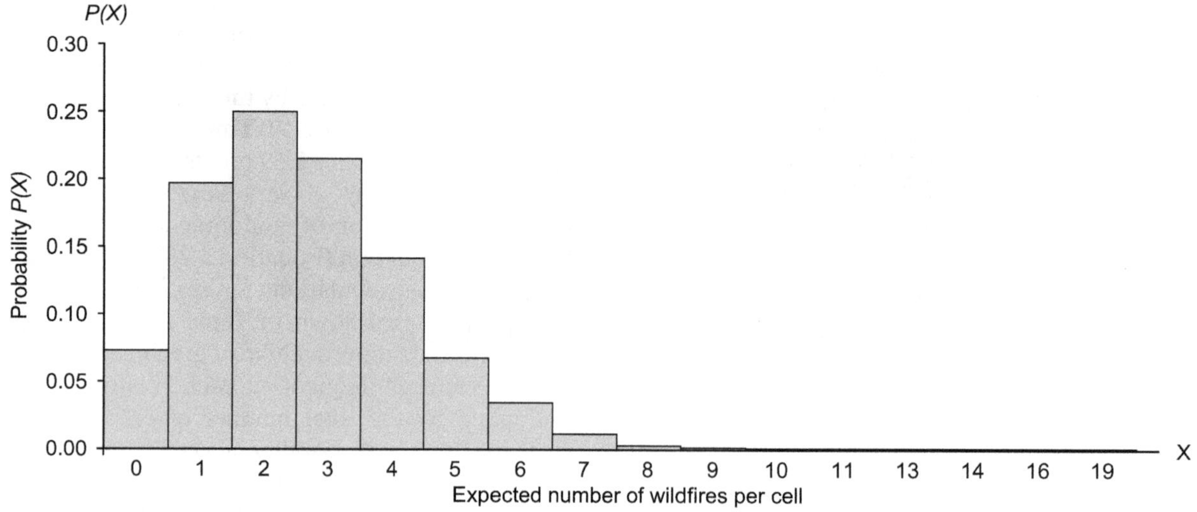

FIGURE 5.11
Poisson (Expected) Probabilities for Number of Very Large Wildfires per Cell

TABLE 5.10

Observed and Expected Frequencies (Poisson) for Wildfires per Cell

Number of very large wildfires per cell	Observed frequencies	Expected frequencies of wildfires per cell	Expected frequencies
0	50	0.08	9.15
1	23	0.19	23.70
2	14	0.25	30.70
3	7	0.22	26.50
4	7	0.14	17.16
5	2	0.07	8.89
6	3	0.03	3.84
7	2	0.01	1.42
8	2	0.00	0.46
9	3	0.00	0.13
10	1	0.00	0.03
11	1	0.00	0.01
13	2	0.00	0.00
14	1	0.00	0.00
16	2	0.00	0.00
19	2	0.00	0.00
TOTAL	122	0.99	121.99

cies are quite different. Therefore, the distribution of large wildfires throughout New Mexico is not random. This finding makes intuitive sense, since some areas of the state are devoid of forests and would have no chance of having a forest fire. Also, different management practices might prohibit human interaction with some forests, while encouraging the use of other forests for activities such as camping. If human activity contributes to wildfire frequency, then excluding people from certain areas will have an influence on the locations of large wildfires.

The geographer must make several methodological decisions when using quadrats to examine spatial point patterns. In addition to deciding how to handle quadrats partially outside the study area, researchers must consider the important issue of quadrat size. How would the Poisson probabilities have differed if more quadrats of smaller size had been placed over the pattern of wildfires? How does a decision to include quadrats only partially inside the study area affect results?

Key Terms

binomial distribution, *83*
complement, *81*
cumulative distribution function, *87*
deterministic processes, *79*
discrete and continuous probability distributions, *83*
event and outcome, *80*
factorial, *84, 88, 90*
geometric distribution, *85*
Monte Carlo simulation, *80*
mutually exclusive, *80*
Poisson distribution, *88*
probabilistic processes, *80*
randomness and random processes, *80*
relative frequency, *80*
rules of probability, *81*
statistical independence, *81*
stochastic processes, *80*

Major Goals and Objectives

If you have mastered the material in this chapter, you should now be able to:

1. Understand the general nature of deterministic and probabilistic processes in geography.
2. Recognize the distinction between stochastic and random probabilistic processes.
3. Explain the concept of randomness and recognize the characteristics of a random experiment.
4. Explain the probability concept of relative frequency and recognize applications in geography.
5. Understand the basic rules of probability.
6. Understand the characteristics of the binomial, geometric, and Poisson probability distributions and identify potential applications of each in geography.

References and Additional Reading

Bulmer, M. G. *Principles of Statistics.* Edinburgh: Oliver and Boyd, 1967.
Evans, M. J., and J. S. Rosenthal. *Probability and Statistics: The Science of Uncertainty.* 2nd ed. New York: W. H. Freeman and Co., 2009.

Hereford, R. *Climate Variation at Flagstaff, Arizona—1950–2007.* US Geological Survey Open-File Report 2007-1410, 17 pp., 2007.

National Climatic Data Center. *United States Climate Normals, 1971–2000.* Contains Flagstaff, AZ annual snowfall data.

Rogerson, P. A. *Statistical Methods for Geography: A Student's Guide.* 5th ed. Thousand Oaks, CA: Sage Publications, Inc., 2020.

Susquehanna Flood Forecast and Warning System. www.susquehannafloodforecasting.org. Search "flood-history" for list of past major floods.

Unwin, D. *Introductory Spatial Analysis.* London: Methuen, 1981.

US Department of Homeland Security. *Yearbook of Immigration Statistics: 2010.* Washington, DC, US Department of Homeland Security, Office of Immigration Statistics, 2011. Data source for immigration data in Table 5.1.

US Forest Service. Fire Statistics System (FIRESTAT). https://data.fs.usda.gov/geodata/edw/datasets.php?xmlKeyword=National%20USFS%20Fire%20Occurrence

6

Continuous Probability Distributions and Probability Mapping

6.1 The Normal Distribution

The most generally applied probability distribution for geographical problems is the **normal distribution**. When a set of geographical data is normally distributed, many useful conclusions can be drawn, and various properties of the data can be assumed. The normal distribution provides the basis for sampling theory and statistical inference, both of which are discussed in later chapters. The discussion in this section shows how probability statements are made from normally distributed data sets.

Although a normal distribution is fully described with a rather complex mathematical formula, it can be generally understood by a simple graph showing the frequency of occurrence on the vertical scale for the range of values displayed on the horizontal axis (Figure 6.1). Since the values on the horizontal axis are not restricted to integers, the normal curve is an example of a continuous distribution. The most striking feature of the normal curve is its symmetry; the lower (left-hand) and upper (right-hand) ends of the frequency distribution are balanced. This symmetric pattern of values in a normal distribution means that no skewness exists in the data. The central value of the data represents the peak or most frequently occurring value. In a normally distributed set of data, this position corresponds to all three measures of central tendency—the mean, median, and mode.

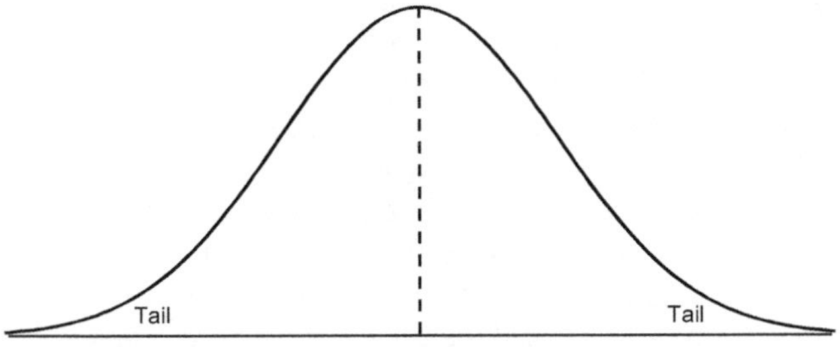

FIGURE 6.1
The Normal Distribution

A normal curve is sometimes referred to as "bell-shaped" in appearance. This characteristic shape represents a frequency distribution with an intermediate amount of kurtosis. The tails of a normal distribution are the portions of the curve with the lowest frequencies located furthest from the center. Note that the frequency of values in a set of normally distributed data declines in a gradual manner in both directions away from the mean.

Because of its particular shape and mathematical definition, the normal distribution is very useful in making probability statements about actual outcomes in many different situations. For example, if the amount of precipitation a location receives in a year is normally distributed over a series of years, the probability of this site receiving a given amount of precipitation can be calculated. The normal distribution provides the theoretical basis for sampling and statistical inference—a primary focus of the rest of the book.

The way in which areas are distributed under the normal curve provides the basis for making probability estimates. The total area under the normal curve represents 100% of the outcomes, therefore we can determine the percentage of values within any portion of the curve along the horizontal axis. For example, due to the symmetry of the normal distribution, 50% of the values must lie under the curve and to the right of the central or mean value. Since the normal curve is also a probability distribution, a value taken from a normal distribution has a .50 probability of falling above the mean.

Given the symmetric form of the normal curve, it is clear that 50% of all values are greater than the mean. This methodology is used to determine percentages for other intervals under the normal curve. Using integral calculus, statisticians can calculate areas under the normal curve. A simple alternative is to use a table of normal values, showing the proportion of total area under any part of the normal curve. This table is derived mathematically from a theoretical normal distribution. In virtually all statistical software packages, the area under any portion of the normal curve is calculated automatically, given the relevant parameters.

To use the table of normal values, data must be standardized. On a standardized scale, each observation is assigned a **standard score** (also called a **normal deviate**), which indicates how many standard deviations separate a particular value from the mean of the distribution. Standard scores can be either positive or negative. For units of data greater than the mean, the corresponding standard scores are positive, and for values less than the mean, standard scores are negative. A standard score of plus 1 represents a value that is one standard deviation *above* the mean, whereas a score of minus 1 is one standard deviation *below* the mean. The mean corresponds to a standard score or normal deviate of 0. The larger the standard score (in either the positive or negative direction), the farther the value lies above or below the mean.

For any standard score, the table of normal values provides a probability that can be interpreted in two ways. First, the standard score gives the probability of a value falling between the mean and that standard score location for a set of normally distributed data. Second, multiplying the probability by 100, it shows the percentage of all values in the normal distribution that lie between the mean and the standard score. Thus, the table of normal values provides information to determine the area under the normal curve for any interval.

Consider these examples using the table of normal values (Appendix, Table A). The probability value associated with a standard score of 1.0 is .3413 (or 34.13%) (Figure 6.2). In a normally distributed set of data, approximately 34% of the values lie between the mean and one standard deviation above the mean. Since the normal curve is symmetric, 34.13% of the values also lie between the mean and one standard deviation *below* the mean. Therefore, by combining these two areas, approximately 68% of all values in a normal distribution lie within one standard deviation on either side of the mean.

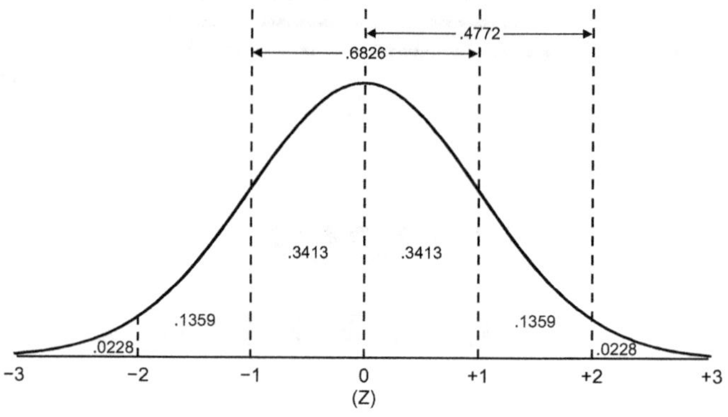

FIGURE 6.2
Selected Areas of the Normal Curve

Similarly, the probability of a value lying between the mean and a standard score of 2 is .4772. Thus, almost 48% of values in a normally distributed set of data lie between the mean and two standard deviations above the mean, and somewhat more than 95% of the values are within two standard deviations on either side of the mean. The remaining 5% of the values are in the two tails of the distribution, where the standard score is either greater than plus 2.0 or less than minus 2.0.

Although the table of normal values provides probabilistic information on a standardized scale, data can be measured on various scales. To apply the normal probabilities to specific sets of data, values must first be converted from their original units of measurement (X) to a standardized (Z) scale without units of measurement. In this process, data values are represented by their relative position in comparison to the mean:

$$Z_i = \frac{X_i - \overline{X}}{s} \tag{6.1}$$

where Z_i = Z-score or standard score for the ith value
X_i = observation i
$\overline{X}$ = mean of the data
s = standard deviation of the data

The numerator of Equation 6.1 shows the deviation of observation i from the mean of the data set. This deviation is then divided by the standard deviation of the distribution. The resulting Z-value (or standard score) can be interpreted as the number of standard deviations above or below the mean an observation is located. If the value under consideration is greater than the mean, the deviation is positive and Z will be greater than zero, while if the observation is less than the mean, the deviation and resulting Z-score will be negative. If X equals the mean, there is no deviation from that value and the Z-score will be zero.

Any set of data can be converted from its original units of measurement into the corresponding set of standardized values. However, to estimate probabilities using the table of normal values, the original data must be normally distributed. If the data is not normal, it is invalid to use the table of normal values to estimate probabilities. Techniques for testing whether a sample has been drawn from a normally distributed population are presented in Section 12.1.

The 40-year data set of annual precipitation for Washington, DC, presented in earlier chapters is used again to make probability estimates. Probability questions can be stated in a couple of different ways. One general type of question we could ask is: "Given a particular data value, what is the corresponding probability?" A second general type of probability question we could ask is: "Given a particular probability value, what is the corresponding data value?" We'll now take a look at both of these approaches.

For the first approach, what is the probability of annual precipitation in Washington, DC, exceeding 48 inches? These 40 years of data are normally distributed, with a mean of 39.95 inches and a standard deviation of 7.5 inches. We can estimate probabilities associated with a data value using a simple three-step procedure.

Step 1: Calculate the standard score:

The standard score corresponding to 48 inches is calculated, where $\overline{X} = 39.95$ and $s = 7.5$

$$Z_i = \frac{X_i - \overline{X}}{s} = \frac{48.0 - 39.95}{7.5} = \frac{8.05}{7.5} = 1.07 \qquad \textbf{(6.2)}$$

Thus, a precipitation level of 48 inches is 1.07 standard deviations *above* the mean precipitation of 39.95 inches (Figure 6.3).

Step 2: Determine the probability from the normal table:

Using the table of normal values, the *Z*-score of 1.07 corresponds to a probability level of .3577. Thus, almost 36% of the values under the normal curve lie between the mean and 1.07 standard deviations. In other words, in about 36 years out of 100, precipitation in Washington, DC, should fall between 39.95 and 48 inches.

Step 3: Evaluate the probability value:

Although the table of normal values always determines probabilities for areas under the curve in relation to the mean, the actual probability being sought may represent a different part of the curve. For this problem, the answer lies in the shaded portion of the curve above (to the right of) the *Z*-score for 1.07. Because the proportion of the total area under the normal curve above the mean is .5000, the correct answer is found by subtracting the probability in Step 2 from .5000 (.5000 – .3577 = .1423). Therefore, in 14 years out of 100, annual precipitation in Washington, DC, should exceed 48 inches.

For the second approach, what amount of annual precipitation in Washington, DC, is likely to be exceeded with a probability of .90 (that is, 9 years out of 10)? To answer this question, the three-step methodology is altered.

Step 1: Determine the probability from the normal table:

As shown on Figure 6.4, precipitation in Washington, DC, will be exceeded 90% of the time at the position indicated by the shading. The total shaded area represents .90 of the total area under the curve, while the portion to the left of the mean is .40. Therefore, we must determine the *Z*-score corresponding to a probability of .40.

Step 2: Calculate the standard score:

According to the table of normal values, a probability of .40 lies between a *Z*-score of 1.28 (when $Z = 1.28$, the probability is .3997) and a *Z*-score of 1.29 (when $Z = 1.29$, the probability is .4015), but is considerably closer to $Z = 1.28$. Since the location is below (or less than) the mean, the value is *minus* 1.28. Thus, the precipitation that occurs at least 90% of the time (9 years out of 10) is 1.28 standard deviations below the mean.

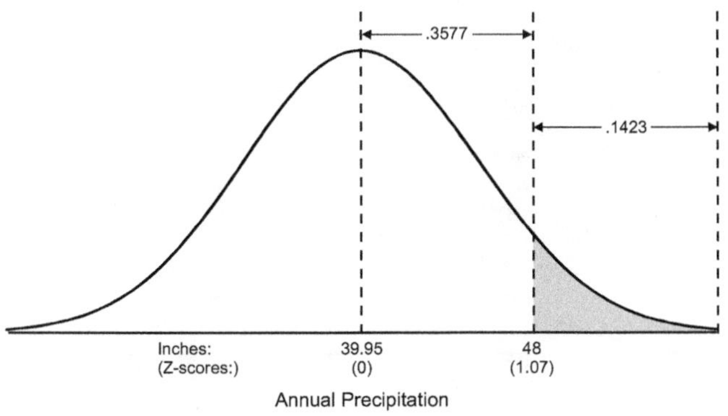

FIGURE 6.3
Probability of Annual Precipitation in Washington, DC, Exceeding 48 Inches

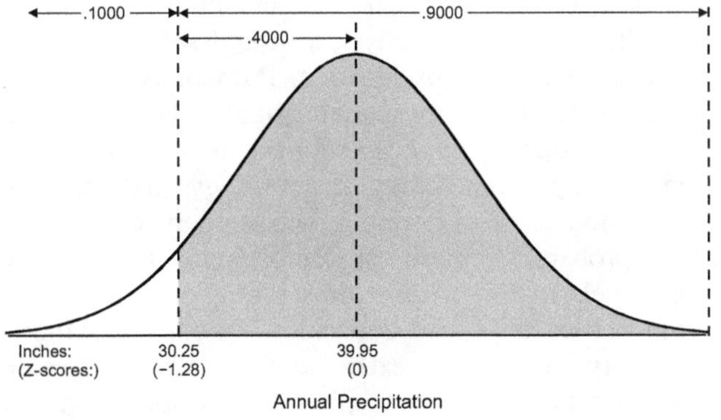

FIGURE 6.4
Amount of Precipitation Exceeding Nine Years Out of Ten in Washington, DC

Step 3: Calculate the Precipitation Value (X_i):

Using Equation 6.1, the precipitation value corresponding to a Z-score of minus 1.28 is determined:

$$-1.28 = \frac{X_i - 39.95}{7.50} \quad \textbf{(6.3)}$$

and

$$X_i = -1.28(7.50) + 39.95 = 30.35 \text{ inches} \quad \textbf{(6.4)}$$

Therefore, 9 years out of 10, precipitation in Washington, DC, will likely exceed 30.35 inches, and 1 year out of 10, precipitation should be less than 30.35 inches.

In summary, the normal curve is a very useful probability distribution in geography. For normally distributed data, the proportions of total area under the normal curve are fixed, allowing a variety of probability statements to be made. The normal curve serves as the basis for sample estimation and inferential statistics, discussed in upcoming chapters. You might want to determine the likelihood of a future heavy snowfall or drought using precipitation or snowfall data from your hometown, similar to what we have done for Washington, DC.

6.2 Probability Mapping

The previous section discussed how to make probabilistic statements about an event at a single location. We can extend this technique to calculate probability estimates for multiple locations distributed across a region. If probability data are plotted on a map, the resulting information represents a **probability map** or "probability surface," showing spatial variation in the variable under consideration.

One very important practical constraint must be carefully considered when constructing probability maps—the variable you are mapping must be continuously distributed across space. Since virtually no human phenomena are continuously distributed, (e.g., population, number of murders) probability map applications are limited to natural or physical variables where a value can be measured at any location. Probability maps can be constructed for many spatial variables. The technique seems most directly applicable for the analysis of natural phenomena in climatology, meteorology, and environmental studies. For example, geographers could construct probability maps of atmospheric particulates, ozone levels, major winter storms, timing of killing frosts, or the pH levels of acid deposition. In some instances, probability mapping is extended to selected topics in human geography, such as disease, unemployment, or poverty. However, extreme care must be taken because virtually all variables in human geography are not distributed continuously over space.

Suppose annual precipitation data are collected for a set of cities in the United States, and the mean and standard deviation of values at each location are calculated. Assuming all data sets are normal, the level of precipitation exceeded 9 years out of 10 could be determined for each city using the technique discussed in the previous section. A probability map is produced by assigning each probability value to its map location and by connecting equal probability values with isolines. Just as contours show elevation patterns in an area, the probability surface would show the spatial pattern of precipitation probability. Many GIS software products have the capability to automatically generate isolines from a point pattern. However, selection of the actual probabilities is the job of an educated geographer.

You might be asking: what additional information is provided by a map of precipitation probability that is not provided by a simple annual precipitation map? Probability maps consider *both* the central tendency and variability of the data *at each location*. In fact, the key advantage of probability maps over maps of central tendency (e.g., a map of average annual precipitation) is their consideration of the variability of values at each location.

The following example illustrates the importance of variability in the construction of probability maps. Consider two cities, A and B. They have equal average annual precipitation (50 inches), but very different levels of variability around those averages (Figure 6.5). Note the contrast between city A and city B with regard to the minimum annual precipitation expected 9 years out of 10. City A has little precipitation variability from year to

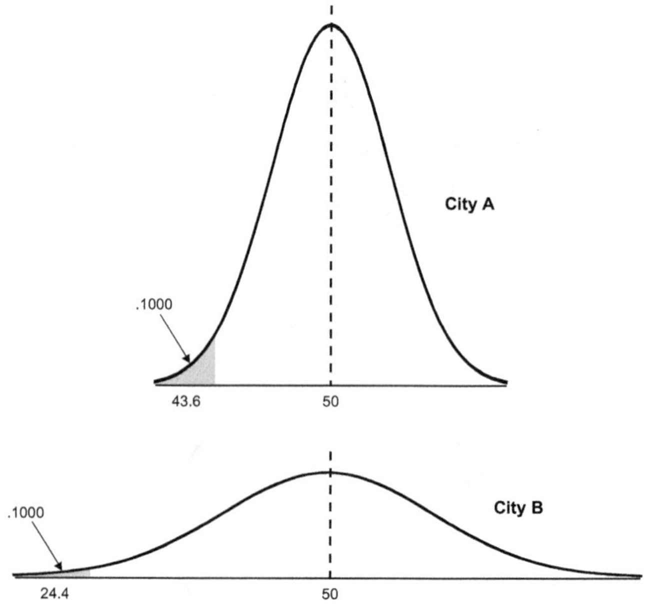

FIGURE 6.5
Comparison of Expected Precipitation in Two Cities with the Same Average Annual Precipitation, but Very Different Precipitation Variability

year and the minimum expected precipitation is only slightly lower than the mean (43.6 vs. 50). By contrast, city B has great variability in precipitation over time, making the minimum expected precipitation considerably lower than the mean (24.4 vs. 50). When mapping the minimum annual precipitation expected 9 years out of 10 for many different cities, the resulting probability map might have a spatial pattern that differs greatly from that of average annual precipitation. Such maps may be more useful in identifying areas that are at greater risk for repeated flooding or drought. Finally, note that a conventional map of average annual precipitation is comparable to a probability map showing the amount of precipitation expected 5 years out of 10 or 50% of the time.

Example: Annual Heating Degree Day Pattern across the Conterminous United States

A *heating degree day* (HDD) is a surrogate measure for the amount of heating energy needed to produce "comfortable" indoor warmth for a home or business in cold or cooler climates. It is a widely used energy management index for measuring yearly heating energy demand. Generally, the heating requirements for a building at a specific location are directly proportional to the number of HDDs at that location. A cooling degree day (CDD) is the opposite of a heating degree day; the CDD measurement reflects the amount of air-conditioning energy needed to cool a structure during the warmer summer months.

There are many ways to calculate heating degree days—the more detailed the temperature data, the more accurate the HDD calculation. One widely used approximation method simply compares the average daily temperature at a site (which is the mean of the high and low temperatures for that particular day) with a base temperature. The average daily temperature is subtracted from a base temperature of 65 degrees Fahrenheit. If temperature is measured in degrees Celsius, a base temperature of 18.3 is used. If the HDD statistic is negative, no heating degree days are recorded, and the assumption is that heating will not be needed on that day. If the difference is positive, the HDD value represents the number of HDDs needed that day. These daily values totaled over a year define the annual number of heating degree days at that location.

We have collected 40-year records of annual heating degree day values from over 100 weather stations across the conterminous United States, excluding Alaska and Hawaii. These are all level-one weather stations, generally located at large city airports. Our goal was to have a spatial pattern of weather stations that is fairly evenly distributed and not clustered in just one part of the country.

Using the means and standard deviations calculated from the 40-year HDD data, we calculated the number of HDDs expected to be exceeded 1 year out of 10 using the standard score formula (Equation 6.1). The need for this much energy for heating is fairly rare, as a winter this cold is expected to occur only once per decade. The calculation for Lexington, Kentucky, indicates that a heating degree day level of 5,168.6 will be exceeded 10% of the time (Figure 6.6).

Following this procedure, values for all cities are calculated and placed on a probability map of the United States, with isolines (much like contours) drawn connecting locations having the same estimated number of HDDs 1 year out of 10. A similar process could be followed for any other probability level. For example, a probability map could show how many HDDs occur 1 year out of 20 (.05 probability) or 3 years out of 10 (.30 probability).

The probability map of HDDs (Figure 6.7) roughly reflects the spatial pattern of solar energy or heat received across the United States. Since the map portrays a situation expected to occur 1 year out of 10, it provides a reasonable spatial estimate of the amount of energy needed for that once-in-a-decade cold winter.

The general east-west trend of isolines emphasizes the importance of latitude on the distribution of heat. This influence is demonstrated by the regular south-to-north increase in heating degree days and winter heat need (e.g., from Louisiana to Minnesota).

Where isolines dip southward, weather station(s) at a higher elevation probably influenced the pattern, since locations with higher elevations generally have cooler temperatures all year

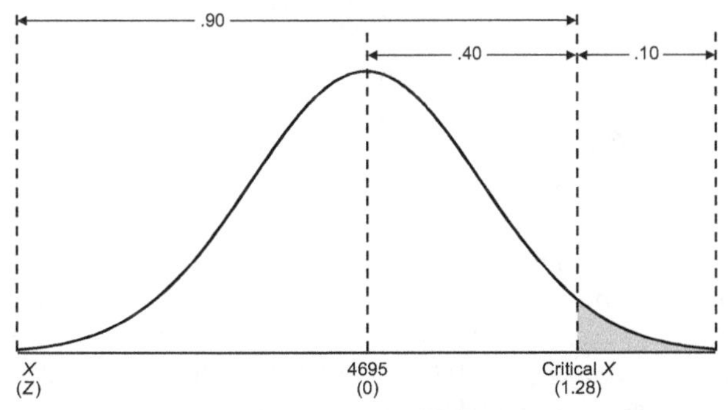

Step 1: Calculate $\bar{X}$ and s: $\bar{X} = 4695$, $s = 370$

Step 2: Use Z table to determine standard score when $p = .10$ Z will be 1.28

Step 3: Calculate the critical heating day value (X)

$$Z = \frac{X - \bar{X}}{s} \text{ then } X = \bar{X} + (Z)(s)$$

$$= 4695 + (1.28)(370)$$

$$= 5168.6 \text{ (Critical } X\text{)}$$

Lexington, KY, will exceed 5168.6 heating degree days one year out of ten.

FIGURE 6.6
Calculation of Heating Degree Days Exceeded One Year Out of Ten for Lexington, KY

(continued)

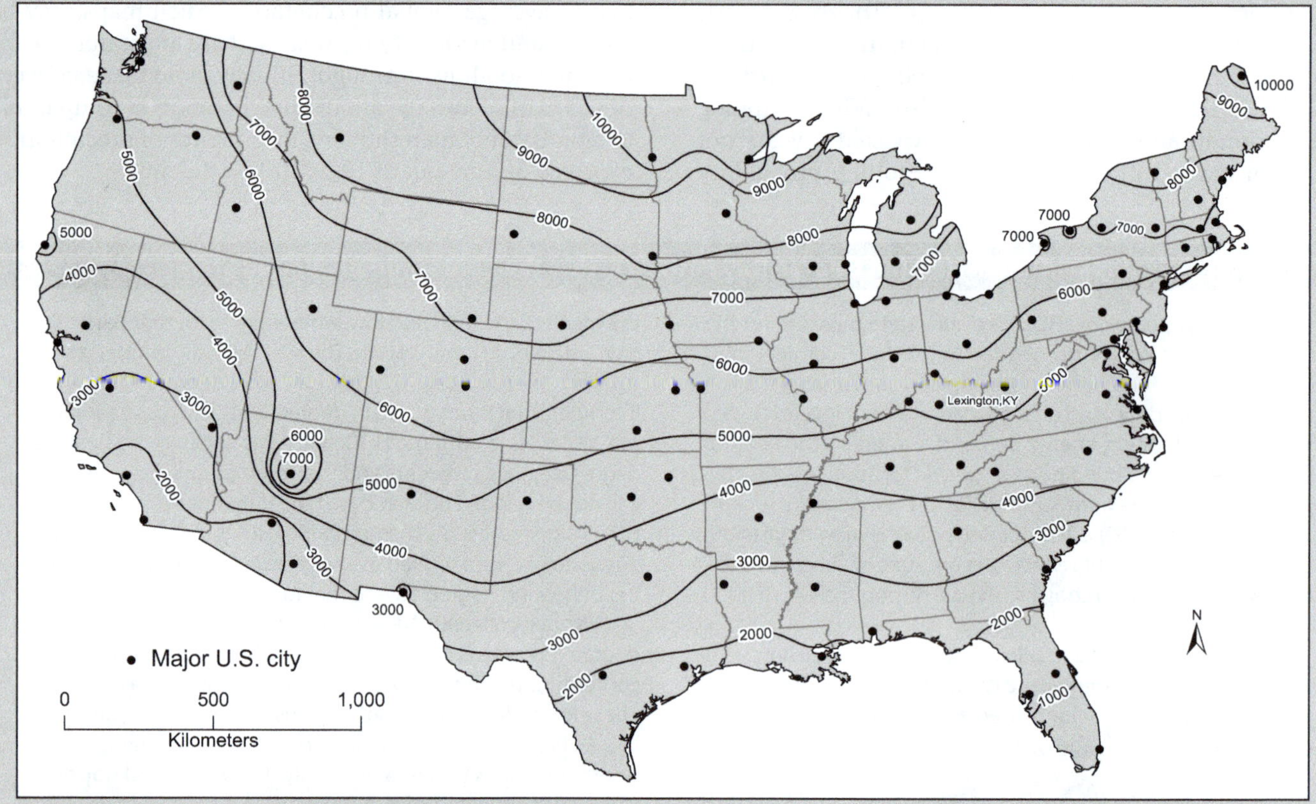

FIGURE 6.7
Probability Map: Number of Heating Degree Days Expected One Year Out of Ten
Source: National Climatic Data Center (NCDC)

long, but especially in the winter months. For instance, in Figure 6.7 you can see a slight southward dip over the Appalachian Mountains of West Virginia and western Virginia.

Perhaps the dominant feature on the probability map is the higher HDD values found in much of the western third of the country. Compared to corresponding latitudes in the eastern two-thirds, HDD values throughout the Rocky Mountain region are somewhat higher. Again this is mostly due to the higher elevations in that area. You might also note the moderating influence of the Pacific Ocean along the California, Oregon, and Washington coastline, keeping the number of HDDs needed during winter cold spells lower than locations at comparable latitudes several hundred miles inland.

The heating degree day pattern becomes complicated in places with abrupt elevation changes. A steep contour gradient is especially notable in places like Arizona where the significant decline in elevation from Flagstaff to Phoenix results in a complex regional isoline pattern. Flagstaff has an elevation of nearly 7,000 feet and requires lots of winter heating energy. The 1-in-10-year annual heating degree day value of Flagstaff is 7948. Phoenix has a lower elevation of about 1,100 feet, and the city's location in the Sonoran Desert results in a limited need for artificial heating, even in the winter months (the annual HDD value exceeded 1 year in 10 is only 1623.

Interpretation of the HDD pattern nationwide illustrates a common problem associated with isoline mapping. The isoline pattern is influenced by the density and placement of control points from which the isolines are drawn. If data from additional weather stations were used, the isolines that make up the probability surface could be located more precisely.

We need to caution you about several problems associated with the HDD statistic. For a variety of reasons, the results are sometimes less accurate than many people believe, even though the underlying theory behind these measures is solid. For example, using a one-size-fits-all base temperature may be inaccurate because different structures heat to different temperatures and average heat gain will vary significantly from building to building, even with the same energy input. Also, most structures are heated to ideal comfort levels on an irregular and intermittent basis. Clearly, home heating and business heating needs fluctuate in different complicated ways (business hours versus overnight, weekdays versus weekends, etc.). Yet another problem has to do with properly monitoring internal temperatures that often vary considerably at different internal locations within a building. However, it is also true that a well-designed and carefully monitored HDD data set can be right on target and allow the implementation of an excellent energy conservation program.

Example: Last Spring Frost Pattern in the Southeast United States

Recall that one of the examples of geographic problem solving we looked at in Chapter 1 dealt with the timing of the average last spring frost in the southeastern United States (see Figure 1.6). We now return to this data set in the context of drawing an isochronal probability map showing the likelihood of a late spring frost occurring at any particular location for any particular date that we want to select.

Using the means and standard deviations calculated from the date of last spring frost data, the probability of the last spring frost occurring on or after April 1 is calculated for each of the 76 weather stations. The calculation for Waycross, Georgia (Figure 6.8) shows there is a .1660 probability (that is, a 16.6% chance) that Waycross will have a late spring frost on or after April 1.

A probability map is usually effective in showing a general regional pattern, but is often less likely to be effective at portraying changes in probability over shorter distances or at the micro-level. We have already seen with the HDD application that the precision of probability map isoline placement is directly affected by the density of the point pattern from which the isolines are interpolated.

A similar problem occurs with the last spring frost probability map. The spatial pattern of average last spring frost dates across the southeast is complex (Figure 6.9). In places like western North Carolina and northern Georgia where the Piedmont plateau transitions into the Blue Ridge Mountains, average last spring frost dates vary considerably over relatively short distances, partly due to substantial elevation changes and associated local climatological factors. Frost prevalence is clearly affected by local topography. Cool air settles at the bottom of slopes because it is heavier than warm air. This allows frost pockets to form in valleys where the cool air is trapped. In general, however, higher elevations have colder temperatures. Combine these influences and you are likely to see frost damage at the bottom of slopes and on the higher-altitude hilltops, while the sloped hillsides are frost-free.

Many local conditions affect the timing and severity of frost. These include: (1) the presence of certain weather conditions (clear skies and calm winds) that sometimes allows development of an inversion layer (colder air near the ground with warmer air above this trapped cold layer); (2) compass orientation of the site (aspect); (3) the degree of inclination (slope); and (4) barriers to cool air drainage, such as narrow valleys and enclosed basins.

In addition, any frost climatology data from official weather stations are based on air temperature measurements taken about five feet above ground level, from weather station shelters or platforms. It is not unusual to discover that ground surface temperatures are 4–6°F lower than shelter readings.

In summary, the timing of the last spring frost at any particular site is the result of both general factors like those hypothesized in Chapter 1 (latitude, elevation, and distance from the coast) and local climatological conditions such as those just discussed. The cumulative result of all these regional and localized factors is a probability map with isochrones changing direction frequently over fairly short distances. The spatial complexity of the last spring frost pattern will be discussed further when we construct regression models to predict the date for each of the weather stations.

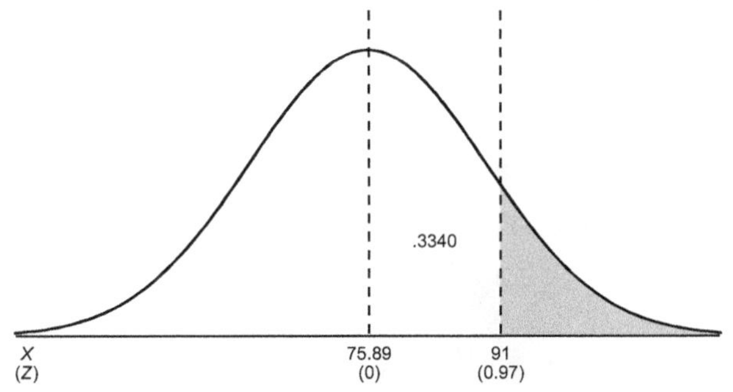

Step 1: Calculate $\bar{X}$ and s: $\bar{X} = 75.89$, $s = 15.63$ } Julian Date

March 17 is Julian day 76

March 31 is Julian day 90

April 1 is Julian day 91

Step 2: Calculate the Z score when $X = 91$ (April 1)

$$Z = \frac{91 - 75.89}{15.63} = 0.97$$

Step 3: Determine the area under the normal curve associated with $Z = 0.97$. This area is .3340. Therefore the associated probability is .5000 − .3340 = .1660. There is a 16.6% chance of a last spring frost on or after April 1 in Waycross, GA.

FIGURE 6.8
Calculation of Probability that Last Spring Frost Occurs on or After April 1 in Waycross, GA

(continued)

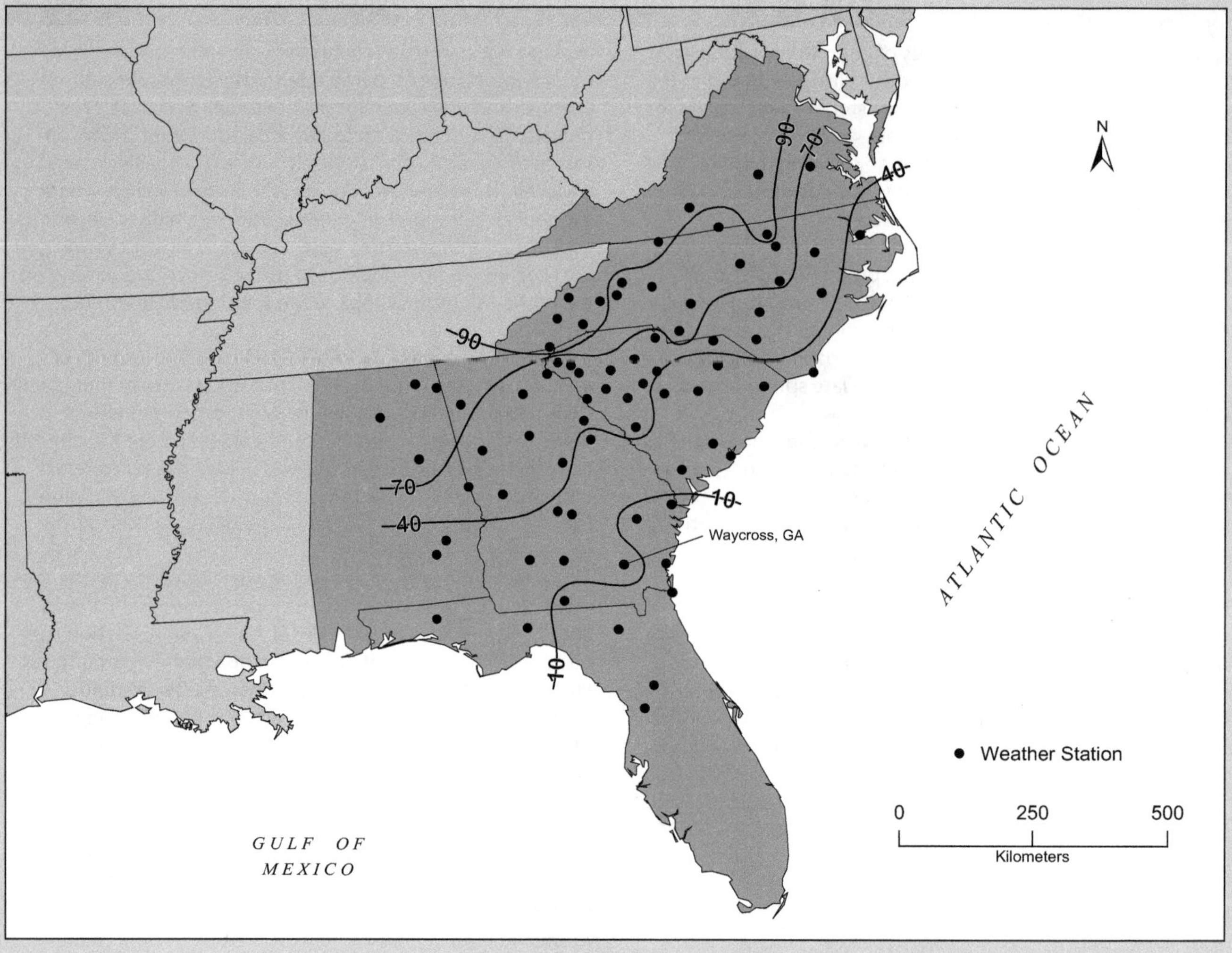

FIGURE 6.9
Probability Map: Likelihood of Last Spring Frost Occurring on or After April 1 in Portion of Southeast United States
Source: Parnell, 2013

Key Terms

normal distribution, *95*
probability map, *98*
standard score (normal deviate), *96*

Major Goals and Objectives

If you have mastered the material in this chapter, you should now be able to:

1. Explain the nature of the normal distribution, know how to use the table of normal values, and identify potential applications in geography.
2. Understand the calculation and interpretation of standard (*Z*) scores.
3. Describe the technique of probability mapping and recognize the types of spatial variables for which probability maps can be constructed.

References and Additional Reading

Energy Lens, Energy Management Software. *Degree Days—Handle with Care!* www.energylens.com/articles/degree-days.

National Climatic Data Center (NCDC). *Climatography of the United States,* No. 20, 1971–2000. Source for heating degree data.

Parnell, D. B. *A Climatology of Frost Extremes Across the Southeast United States, 1950–2009.* Unpublished manuscript, Salisbury, MD: Salisbury University, 2013.

Ristinen, R. A., J. J. Kraushaar, and J. T. Brack. *Energy and the Environment.* 4th ed. Hoboken, NJ: John Wiley and Sons, 2022.

7

Basic Elements of Sampling

Sampling was discussed briefly in Chapter 1 as an essential component of the scientific research process. In many aspects of geographic research and problem solving, statistical techniques are incorporated into that research procedure. In fact, a basic knowledge of sampling procedures and methodology is valuable, whatever the area of geographic study.

How does sampling fit into the overall scheme of statistical analysis and scientific research? Earlier chapters include some brief comments about the nature of these relationships. The ultimate aim of inferential statistics is to generalize about certain characteristics of a large group, based on information obtained from a portion (often a very small portion) of that large group. That is, characteristics regarding a statistical **population** (the universe of all individuals) are *inferred* from information obtained from a **sample** (a subset or portion of individuals selected from the population for detailed analysis). These concepts are further clarified in the chapter on probability, where the importance of randomness and random processes is discussed in the context of making generalized probability statements about a large population (the conceptual notion of an infinite number of die rolls) using only sample observations (such as a few rolls of a die).

This chapter introduces some basic sampling concepts and discusses the important roles of sampling in geographic problem solving and research. Without a properly drawn sample, valid generalizations or statistical inferences about the population may not be possible. Section 7.1 presents an overview of the advantages of sampling, the sources of sampling error, and the steps involved in a well-designed sampling procedure. The various types of probability sampling are reviewed in Section 7.2. All probability samples contain an element of randomization, but simple random sampling is often not the best choice of sample design. We will examine some useful alternatives to simple random sampling, including systematic, stratified, cluster, and hybrid sample designs. Section 7.3 discusses the circumstances under which spatial sampling is necessary and reviews different types of spatial sampling. Special attention is given to point sampling designs.

7.1 Sampling Concepts

Sampling is an essential skill in virtually all areas of geographic study. A biogeographer interested in the spatial pattern of environmental change associated with high-intensity recreation use in a national park cannot examine conditions everywhere in the park. A representative sample of study sites needs to be selected for detailed analysis of these human-environment relations. An urban geographer examining the variation of housing quality in a metropolitan area selects a sample of homes for detailed study. A behavioral geographer conducting research on natural hazards distributes questionnaires to a representative sample to learn public attitudes toward alternative flood-management policies along the flood plain of a river. A medical geographer concerned with neighborhood variations in the use of hospital emergency rooms as primary care centers uses sampling to select the neighborhoods and the hospitals to include in the study.

These examples illustrate the geographer's use of sampling in both spatial and nonspatial contexts. Spatial sampling takes place when the biogeographer selects *locations* to examine environmental change in the national park. Likewise, the neighborhoods where the medical geographer chooses to examine patterns of hospital use also constitute a spatial sample. On the other hand, the geographer conducting the study on attitudes toward natural hazards along the river flood plain may have selected individuals from a nonspatial list of households in the area. The urban geographer conducting the study of housing quality could have taken a sample of homes from either a nonspatial list (e.g., tax rolls) or a spatial source (e.g., a map).

Advantages of Sampling

A variety of both practical and theoretical reasons make sampling preferable to complete enumeration or census of an entire population. These are some of the advantages of sampling:

- *Sampling is a necessity in many geographic research problems.* If the population being studied is extremely large (or even theoretically infinite), completing a total enumeration is not possible. If you are studying the reasons why families in the United States change residence, sampling is the only alternative—all those who move cannot possibly be contacted and surveyed individually. A geographer analyzing global changes in the nature and spatial extent of tropical rain forests will find it impossible to have total spatial coverage, since the number of locations in a rain forest is infinite.
- *Sampling is an efficient and cost-effective method of collecting information.* An appropriate amount of data concerning the population can be obtained and analyzed quickly with a sample. Not only is sample information collected in less time, but sampling also keeps expenditures lower and logistical problems to a minimum. The overall scale of effort (time, cost, personnel, logistics, etc.) becomes realistic with sampling as opposed to an examination of the entire population.
- *Sampling can provide highly detailed information.* In geographic problems where in-depth analysis is necessary, only a small number of individuals or locations can be included in the study. These few elements in the sample could then be closely scrutinized with the collection of a comprehensive set of information. A study of shopping behavior patterns, for example, might require numerous questions about the number of shopping trips, locations visited while shopping, attitudes about alternative stores, as well as demographic or socioeconomic information concerning household members. Such detailed analysis is obtained only through sampling.
- *Sampling allows repeated collection of information quickly and inexpensively.* Many geographic research problems require detailed information collected over a specific period of time or focus on spatial changes that occur rather quickly. For example, if you are studying the attitudes of citizens living in a barrier island community toward alternative coastal zone management strategies, you may want to follow a sample of residents as pertinent legislation moves through the political process. Their attitudes are likely to change over time, especially if a hurricane hits the island during the study period! An urban geographer analyzing the spatial pattern of growth may wish to focus on the views of residents in a neighborhood before, during, and after the construction of a nearby shopping mall. With dynamic situations such as these, sampling is required; information from all individuals in the population cannot possibly be collected without unreasonable effort and cost.
- *Sampling can provide a high degree of accuracy.* With sampling, an acceptable level of quality control is assured. Complete and accurate questionnaire returns are more likely to be obtained if a small number of well-trained personnel conduct all of the interviews. This procedure may provide more accurate results than a complete census requiring a larger number of personnel, some of whom may not be fully trained. The 2010 U.S. Census of Population once again illustrated the difficulties of acquiring accurate information from everyone. Considerable controversy developed because many people living in America were literally not counted.

Despite the many clear and obvious advantages of statistical sampling, the decision to use sampling (and how to use sampling) in a specific practical situation is often controversial with multifaceted implications. Real-world difficulties and political policy issues can arise concerning the implementation of sampling. In addition, many citizens are not aware of the advantages of sampling over total enumeration and will not be strong advocates of sampling.

Sources of Sampling Error

A central goal of sampling is to derive a truly *representative* set of values from a population. A representative sample accurately reflects the actual characteristics of the population without bias. To ensure an unbiased representative sample, an element of **randomness** is always incorporated into the sample design procedure. However, just having randomness built into a sampling plan does not guarantee an unbiased sample, for many other **sources of sampling error** are possible.

Unfortunately, it is usually not possible to know with absolute certainty whether a sample is totally representative. The very nature of sampling often means that everything cannot be known about the population containing the sample. Because only sample data are available, some uncertainty is associated with sample estimates, and some sampling error occurs. If you are taking a sample, you must try to minimize sampling error given various practical constraints such as cost or time.

The measurement concepts of precision and accuracy help categorize the many sources of sampling error. The results from a small sample may not be very exact or precise. Increasing the sample size permits a more exact estimate (Figure 7.1, line A) and can reduce imprecision. Larger samples, however, are invariably more costly and time-consuming to obtain (Figure 7.1, line B). Satisfactory resolution of this difficult trade-off between sample precision and the effort required to sample is important in most real-world problems involving sampling.

Sampling inaccuracy is a more complex issue than sampling precision. Systematic bias can enter sampling in many different ways. Some inaccuracies are the result of problems with the sampling procedure itself. For example, the elements or individuals in a population could be "mismatched" with the set of elements or individuals from which a sample is taken. An urban geographer studying new home construction patterns might use a list of building permits as a source of data. If this list contains home renovations as well as new home construction, some of the renovations might be erroneously included in the new home construction sample.

Selecting an improper sampling design is another potential source of inaccuracy. The complexity of a problem might call for a more sophisticated approach, rather than a simple random design. In other sampling problems, the method of data collection may be inappropriate. For example, a geographer studying population mobility might decide to use a mail questionnaire to keep costs down when telephone interviews would provide more accurate results.

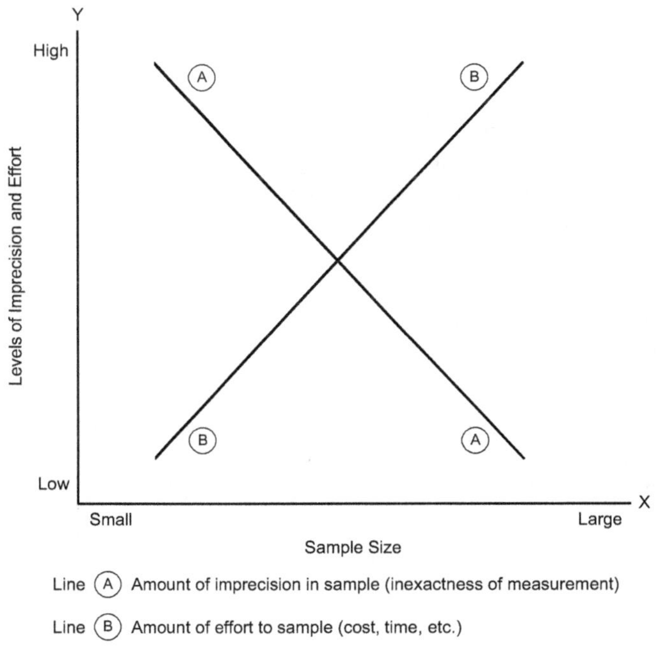

FIGURE 7.1
How Change in Sample Size Affects Level of Imprecision and Effort to Sample

Other types of inaccuracy may result from some operational, logistic, or personnel problem not directly connected to the actual sampling procedure. Inconsistencies in collecting field data or interviewing could adversely affect the sample results. Errors could be made in the editing, recording, or tabulating of information. Even forces of nature beyond the control of the researcher could bias results. For example, if you are examining land use patterns along the flood plain of a large river you could find the entire project in jeopardy if severe flooding occurs during the study period.

Simply stated, the quality of statistical problem solving in geography depends heavily on samples that are properly designed and collected. It is important to develop procedures carefully to reduce sampling error to a tolerable or acceptable level. Most sources of error can be reduced substantially (and perhaps even avoided or eliminated) if all steps of the sampling procedure are carefully planned and evaluated *before* the full set of sample data is collected and analyzed.

The American Community Survey

The government of the United States has a long history of information gathering about Americans. Starting in 1790, as required by the U.S. Constitution and with funds authorized by congress, the Census Bureau has conducted a national census every 10 years. One original mandated purpose was to determine the population size of each state, thereby allocating the proper number of congressional seats in the House of Representatives according to population size.

Over time, the amount of census information collected has increased dramatically, reflecting the growing need for data concerning the demographic and socioeconomic characteristics of our population. As we move into the 21st century, the "long form" of the census (given to a sample of housing units) has been replaced by the American Community Survey (ACS). The ACS has been described as the most substantial change in the decennial census in more than 60 years. The ACS is a nationwide monthly survey designed to provide communities with precise and accurate demographic, social, economic, and housing data annually. With full implementation of the ACS sampling process, we can now obtain a continual stream of updated information about localities, congressional districts, and states. It is hoped that this will revolutionize the way in which all levels of government plan and evaluate their programs. In addition, small businesses, large corporations, and individuals can utilize ACS data to estimate the sales potential of services and products and develop strategies for starting or expanding a business. For example, a restaurant could use demographic and economic data from the ACS to determine the best location and available workforce to start a new franchise. In addition to these benefits, the Federal government uses the survey to allocate more than $400 billion in annual funding to states and localities for education, health care, roads and numerous other programs.

Steps in the Sampling Procedure

If faced with a geographic research problem in which the collection of sample information is necessary, you need to follow a number of steps (Table 7.1 and Figure7.2). Collecting sample data is *not* the first action taken. You must first anticipate and resolve various problems that might cause sampling error or other difficulties. Numerous safeguards or checks should be incorporated into the sampling procedure whenever necessary. The characteristics and relevant issues at each step of a well-designed sampling procedure are now summarized.

Step 1: As descriptive statements and inferential hypotheses are formulated in scientific research, variables in the problem must be defined both conceptually and operationally. In sampling, the population and area "targeted" for study must be defined. The **target population** is the complete set of individuals from which information is collected, whereas the **target area** is the entire region or set of locations from which information is gathered. Precise delineation of the target population and target area is not always a simple task. Suppose, for example, you are planning to study the spatial variation of public attitudes regarding several proposed revitalization projects in the central business district (CBD) of a large metropolitan area. What set of individuals should make up the target population? Where should you place the target area boundaries? Arguments could be made in support of any of the following alternatives:

- all city residents active in local civic groups
- all city landowners
- all city residents
- all city and suburban residents
- all area residents plus nonresident visitors

Step 2: After conceptually defining the target population and target area, an operational sampling frame(s) is created. A **sampling frame** is defined as the practical or operational structure that contains the entire set of elements from which the sample is drawn. This structure is sometimes a comprehensive list of individuals (nonspatial) and sometimes the boundary line delimiting the extent of the study area (spatial). Sometimes both spatial and nonspatial sampling frames are components of the same problem. A **sampled population** is the set of all individuals contained in the sampling frame from which the sample is actually drawn. A **sampled area** is the set of all locations within the study area boundary line that delimits the spatial sampling frame from which the sample is actually drawn.

TABLE 7.1

Steps in the Sampling Procedure

Step 1: Conceptually Define Target Population and Target Area

- **Target population:** the complete set of individuals from which information is to be collected.
- **Target area:** the entire region or set of locations from which information is to be collected.

Step 2: Designate Sampled Population and Sampled Area from Sampling Frame

- **Sampling frame:** the practical or operational structure that contains the entire set of elements from which the sample will actually be drawn.
- **Sampled population:** the set of individuals contained in the sampling frame, from which the sample is actually drawn.
- **Sampled area:** the set of all locations within the study area boundary line that delimits the spatial sampling frame, from which the sample is actually drawn.

Step 3: Select Sampling Design

- Probability sampling preferred over nonprobability sampling.
- Types of probability sampling include random, systematic, stratified, cluster, and hybrid designs.
- Spatial and nonspatial variations in sampling design exist.

Step 4: Design Research Instrument and Operational Plan

- Methods of data collection include direct observation, field measurement, mail questionnaire, personal interview, telephone interview.
- Establish protocols for handling all problems or situations that can be anticipated in the sampling procedure.
- Complete miscellaneous logistic and procedural tasks in the preparation of sample taking.

Step 5: Conduct Pretest

- Complete trial run or pilot survey of sample data collection method.
- Correct all discovered problems that could lead to sampling error.
- Pretest results may be used to determine sample size.

Step 6: Collect Sample Data

- Consistency in collection methods and procedures is essential.
- Ensure overall high level of quality control.

Why is it important that you distinguish between target population and sampled population in nonspatial sampling? The CBD revitalization example illustrates this distinction. Suppose the population you want to target is *all metropolitan area residents*, both in the city and in adjacent suburbs. Operationally, however, this target population is not listed in any single comprehensive source. We must construct a sampling frame that lists the intended target population as completely as possible.

Many possible sources of data could provide an operational definition for this sampling frame, yet each presents certain challenges. Customer lists from electric utilities would not include all households or all residents at those households listed. Some people in apartments or other group accommodations would be omitted from the sampling frame. Telephone listings will also not provide a complete enumeration of the target population. Lower-income residents without phones would be underrepresented in the sample, as would students in dormitories, residents in group quarters, and those with unlisted numbers. Landline phones and cell phones may not be available together on any list. The result is that the sampled population often cannot duplicate the target population because the sampling frame used to delineate the sampled population is not complete.

Suppose you want to update past studies concerning land use and land cover change in the "Greater Yellowstone Ecosystem" (GYE) (Figure 7.3) and would like to collect sample data from various locations within the area. The GYE contains Yellowstone and Grand Teton National Parks, 7 national forests, over 20 other federal and state jurisdiction areas (such as Wind River Indian Reservation), and surrounding private lands. The area supports threat-

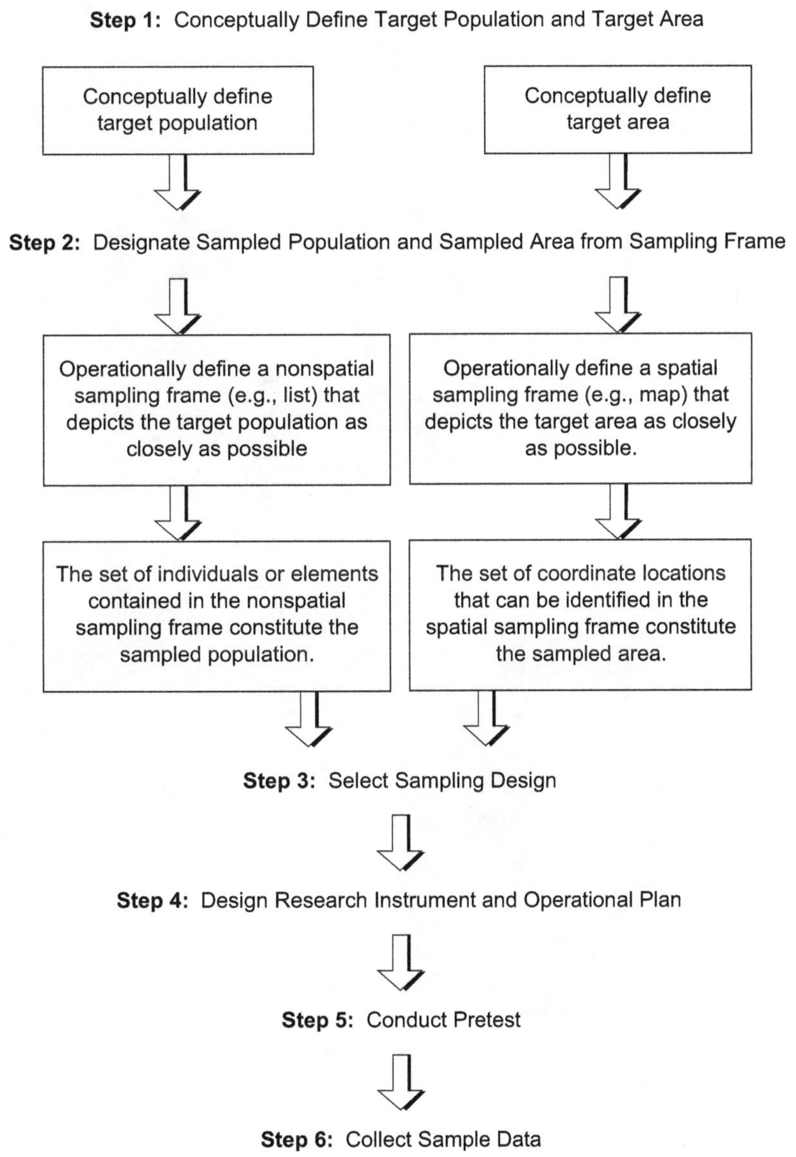

FIGURE 7.2
Steps in the Sampling Procedure

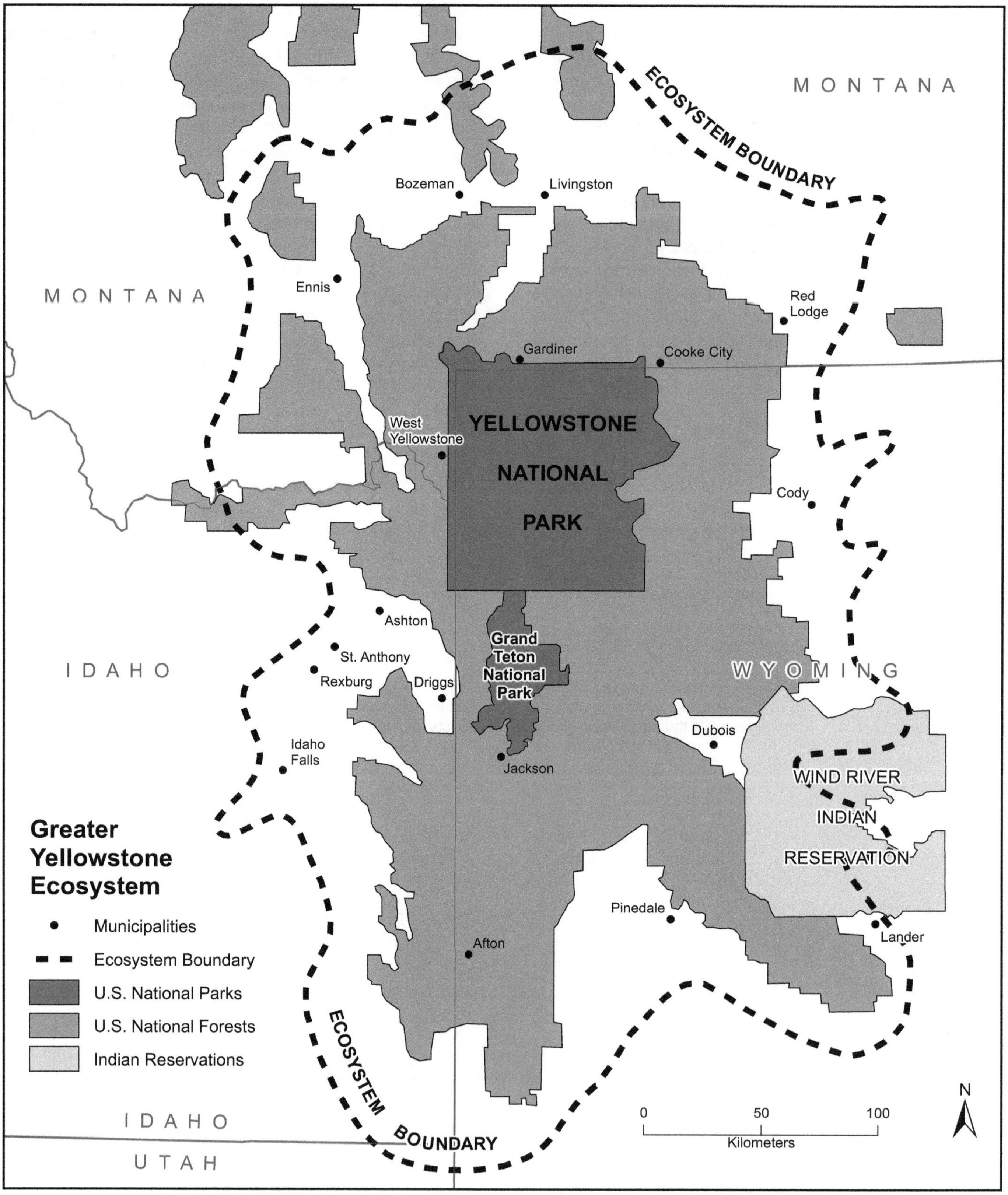

FIGURE 7.3

The Greater Yellowstone Ecosystem

Source: Greater Yellowstone Coalition (Adapted from Greater Yellowstone Ecosystem Map) ened species such as grizzly bear and free-roaming populations of elk and bison. The surrounding private lands include a number of towns whose residents traditionally relied on natural resource industries such as farming, ranching, logging, and mining. In recent years, however, the local economies of communities in the GYE have shifted from primary sector activities to an economy based on land development, real estate, and recreational businesses.

With such a diverse and complex study area, a spatial sampling procedure is going to be difficult. Suppose your conceptual target area is the ecosystem boundary shown in Figure 7.3. You would like the actual, operationally-defined sampled area to correspond as closely as possible to this target area. However, there are likely to be almost insurmountable problems with gaining access to all selected sample locations within the GYE. Entry onto some ranches, resorts, and other types of private property will surely be prohibited, and access may also be denied onto some public lands leased by lumber and energy companies. To what extent will the inability to sample from these portions of the target area bias the results and weaken the accuracy of the study? These questions must be carefully considered before continuing further.

An effective sampling frame is characterized by a target population that matches the sampled population and a target area that matches the sampled area as closely as possible. In the problem concerning CBD revitalization, what is the nature and probable extent of sampling error that will occur if a telephone survey (sampled population) is used to elicit opinions of all metropolitan area residents (target population)? In the Yellowstone example, how much bias is injected into the sampling procedure if a considerable portion of private land in the region is excluded from analysis? Whenever facing mismatches like these, only qualified or conditional statements are possible, and you must clearly explain the sources of systematic bias and estimate their magnitude.

Step 3: Selecting an appropriate sample design method is crucial to the success of the entire sampling procedure. Sampling design is the way in which individuals or locations are selected from the sampling frame.

A fundamental distinction can be made between **nonprobability** and **probability sampling designs**. Nonprobability sampling is subjective because the selection of sampled individuals or locations is based on personal judgment. These samples are therefore sometimes called "judgmental" or "purposive samples." Criteria for sample selection might include:

- personal experience and background knowledge;
- convenience or better access to a nonrandom selection of individuals or locations;
- use of only those who volunteer or respond to the survey instrument, such as a questionnaire in a newspaper; and
- selection of a single nonrandom study area or sample region (case study) in which to conduct research.

Excellent descriptive results are sometimes derived from a nonprobability sampling procedure. For example, a case study approach may provide considerable insight on variable interaction within the study area. However, the underlying problem with all nonprobability approaches to sample selection is that no valid inferences can be made from the sampled elements to any wider group in the population or to any other areas. In other words, nonprobability sampling limits the generalizations that can be inferred from the research findings.

By contrast, the nature of probability sampling is more objective and is closely associated with scientific research. In a probability sample, each individual or location that could be selected from the sampling frame has a known chance (or probability) of being included in the actual sample. The advantage of probability sampling is that we can estimate the amount of sampling error. Because of the importance of probability sampling in geographic research, Section 7.2 is devoted entirely to this topic.

Step 4: Many procedural matters must be resolved when sampling in real-world geographic situations. A key task is the selection and design of an appropriate measurement instrument. Possible formats for data collection include direct observation, field measurement (especially in areas of physical geography), mail questionnaire, personal interview, and telephone interview. To select the most appropriate format, the nature of the research problem must be evaluated carefully. Even after the method of data collection has been selected, problems in instrument design may remain. In survey design, each question must be carefully worded, all possible responses to a question must be considered in advance, and the questions must be sequenced properly. Questionnaire design is complicated because many different considerations are involved. For those interested in further details regarding questionnaire design, see the references at the end of the chapter.

Other procedural decisions and logistic arrangements may be necessary, depending on the nature of the sampling problem. Preliminary site reconnaissance may be necessary, and interviewers and other personnel may need to be hired and trained. If not anticipated, any number of difficulties can plague the researcher and even create bias in the sample, undermining its inferential power. Discussing the wide variety of circumstances that might arise is not possible, as every geographic problem is likely to have its own peculiarities. However, a list of some common problems includes:

- low response to a survey or questionnaire;
- unexpected response resulting from ambiguity or misunderstanding;
- environmental or political change in the sampled area; and
- unqualified or incompletely-trained interviewers.

Step 5: Before extensive time and money are expended on the main data collection effort, a small-scale pretest should be conducted to assess both the strengths and deficiencies in the research instrument design and operational plan. In a field study, the pretest should reveal problems with instrument calibration or malfunction, site access problems, and other logistical difficulties. Problems in a survey or questionnaire format may include non-response, improperly-worded questions, and poorly-sequenced questions. A quality pretest can identify and correct difficulties in research design and data collection.

A good pretest or pilot survey can also help determine the ultimate sample size that needs to be taken. Too small a sample will not yield meaningful results, whereas an unnecessarily large sample will waste time, money, and effort. Pretest information can help you estimate the sample size needed to achieve a certain level of precision in the main sample, and this topic is discussed in Section 8.3.

Step 6: The collection of sample data is a major task, where high levels of quality control must be maintained and well-considered data management procedures carefully followed. Consistency in all aspects of data collection and processing is absolutely essential. In this step, the benefits of the careful design of the research instrument and planning (step 4) become evident, and the improvements made in pretesting (step 5) will expedite the major data collection effort significantly.

7.2 Types of Probability Sampling

In probability sampling, each individual or item that could be selected from the sampling frame has a known chance (or probability) of being included in the sample. This important advantage occurs because a randomization component is incorporated in some known way into the sample design. If sample data are used in an inferential manner (either for estimation of population parameters or for hypothesis testing), then an element of randomness must be built into the sample design procedure. All types of probability sampling contain this characteristic of randomness and avoid the subjectivity of nonprobability sampling.

Simple Random Sampling

Randomization is fundamental to all probability sampling, and a **simple random sample** is the most basic way to generate an unbiased, representative cross-section of a population. In a simple random sample, every individual in the sampling frame has an equal chance of being included. All types of probability sampling contain this characteristic of randomness and avoid the subjectivity of nonprobability sampling. In addition, the selection of elements should be independent of one another; meaning, the selection of one element does not affect the selection of another element.

Traditionally, geographers and statisticians used a random number table to obtain random numbers for a sample. The process of using a random number table is tedious, and generally unnecessary, as random numbers are easily obtained using computer technology. As an example, desktop applications like Excel have a random function, = *rand()*, built in, while computer languages like Visual Basic and JavaScript, use *rnd()* and *random()*, respectively. These functions will return a random value between 0 and 1. Another option is to use web pages such as http://www.random.org/integers/ to generate random integers. Therefore, in this edition of our textbook we have skipped the discussion of the random number table and recommend more modern approaches.

As an example, suppose university personnel want to determine the detailed opinions of students regarding which activities to provide in a planned new student center. Because time is limited, only a small sample of students is surveyed. From a student body of 8,500, a simple random sample of 25 students is selected. Each student must have an equal chance of being chosen (that probability is 25/8,500 = .0029, or about 3 per 1,000). Since the target population is 8,500, each member of that population can be assigned a number between 1 and 8,500. Whether using a spreadsheet or a computer program, the random number generated between 0 and 1 can be multiplied by the target population (in this case 8,500). The formula in Excel would look like:

= rand()*8500

This formula would be entered 25 times in the spreadsheet to obtain 25 randomly generated numbers. Of course the number calculated would be a floating point value (with decimal precision), so you can easily round the number off to obtain an integer value.

Systematic Sampling

Systematic sampling is a widely-used design that often simplifies the selection process. A **systematic sample** makes use of a regular sampling interval (k) between individuals selected for inclusion in the sample. A "1-in-k" systematic sample is generated by randomly choosing a starting point from among the first k individuals in the sampling frame, then selecting every kth individual from that starting point. For example, if a sample of 25 ($n = 25$) is taken from a population of 500 ($N = 500$), the sampling interval (k) would be N/n ($k = 500/25 = 20$; a "1-in-20" sample). Care must be taken to ensure that no nonrandom pattern or sequencing is present in the list from which every kth individual is selected. Otherwise, bias may be introduced into the sample process.

A systematic method of sampling is generally less cumbersome to apply than simple random sampling. Systematic sampling provides a relatively quick way to derive a large size sample and obtain more information at a reasonable cost. As a result, it is used in many practical contexts. Government agencies (such as the United States

Census Bureau and Statistics Canada) and political polling firms (such as Gallup or Harris) routinely apply systematic samples when detailed analyses or follow-up studies are needed. To estimate the quality of a product coming off an assembly line, factory management could have a more detailed inspection of every 100th item. Every 20th visitor to a park could be asked to complete a survey regarding park services or every 50th taxpayer in a city could be asked about alternative funding or planning projects. In market analysis, the editors of a magazine could send a detailed survey or questionnaire concerning existing and proposed features to every 10th subscriber.

Stratified Sampling

In many geographic sampling problems, the target population or target area is separated into different identifiable subgroups or subareas, called "strata." If sample units in the different strata are expected to provide different results, such "target subdivision" is logical. With **stratified sample** design, the effect of certain possible influences can be controlled. Taking a simple random sample from each class or stratum makes the fullest possible use of available information and increases the precision of sample estimates.

Case 1:
Stratify by place of residence

Campus dormitory	Off-campus housing
1. ____	1. ____
2. ____	2. ____
3. ____	3. ____
etc.	etc.

Case 2:
Stratify by year in school

Freshman	Sophomore	Junior	Senior
1. ____	1. ____	1. ____	1. ____
2. ____	2. ____	2. ____	2. ____
3. ____	3. ____	3. ____	3. ____
etc.	etc.	etc.	etc.

Case 3:
Stratify by place of residence and year in school

	Freshman	Sophomore	Junior	Senior
Campus dormitory	1. ____ 2. ____ 3. ____ etc.	1. ____ 2. ____ 3. ____ etc.	1. ____ 2. ____ 3. ____ etc.	1. ____ 2. ____ 3. ____ etc.
Off-campus housing	1. ____ 2. ____ 3. ____ etc.	1. ____ 2. ____ 3. ____ etc.	1. ____ 2. ____ 3. ____ etc.	1. ____ 2. ____ 3. ____ etc.

Note: Random samples of varying sizes taken from each stratum

FIGURE 7.4
Alternative Stratified Sample Designs: Student Opinion Survey

Stratified sample designs may be either **proportional (constant-rate)** or **disproportional (variable-rate)**. In a proportional stratified sample, the percentage of the total population in each stratum matches the proportion of individuals actually sampled in that stratum as closely as possible. For example, suppose 20% of all residents in a city are apartment dwellers. Proportional representation of the apartment stratum in a stratified sample design requires that 20% of the sampled individuals be apartment dwellers. Suppose further that this housing study specifically involves a resident survey on rent control. Because apartment dwellers would be particularly affected by the proposed legislation and their opinions important to decision makers, city council members might want their views to be represented more heavily. If such a disproportional (or variable-rate) stratified sampling design is appropriate, the apartment resident stratum would be **oversampled** (with a larger than proportional sample size), and residents not living in apartments would be **undersampled** (with a smaller than proportional sample size). An analogous practical sampling strategy is to maintain proportional representation of both the apartment and non-apartment strata, but weight the apartment dweller responses more heavily when calculating sample statistics.

For the opinion survey on the new student center, a simple random sample would probably yield less precise results than would a well-designed stratified sample. A university has a diverse student population, and different student subgroups would likely have varying opinions on activities for a student center. Several stratified sample designs are possible (Figure 7.4). For example, those living in campus dormitories may have significantly different priorities than do commuters living off-campus (Figure 7.4, Case 1). Also, the views of freshmen may differ from those of sophomores, juniors, and seniors (Figure 7.4, Case 2). Stratifying *both* by place of residence *and* year in school might prove most useful, resulting in a composite sample design structure with eight strata (Figure 7.4, Case 3).

If the views of each student are considered equally important, a proportional stratification is applied. However, if the views of dormitory residents are considered more important, that group could be oversampled in a disproportional stratified sampling scheme. In both instances, if results are expected to vary by stratum, a stratified sample is preferable to a simple random sample of the same size. That is, stratification should be applied if background knowledge or logic suggests that it would be beneficial and practical.

Cluster Sampling

For some geographic problems, **cluster sampling** is most appropriate and may be more efficient or cost-effective than random, systematic, or stratified sampling. A cluster sample is derived by first subdividing the target population or target area into mutually exclusive and exhaustive categories (Figure 7.5). An appropriate number of categories (clusters) are selected for detailed analysis through random sampling. Two alternatives exist:

- all individuals within each cluster are included in the sample, making a total enumeration or census within that cluster;
- a random sample of individuals is taken from each cluster.

The latter option is sometimes called a "two-stage" cluster sample because a random process is used twice—once to select clusters and then again to select sampled individuals within each cluster. The actual approach used depends on the circumstances and practical sampling conditions. The complete cluster sample is the composite of these selected clusters of observations.

Cluster sampling is generally an effective design option when practical or logistic problems make other choices more expensive, difficult, or time-consuming. In some situations, a complete sampling frame is not available, but partial information can be obtained. This might occur in an urban geography study where total enumerations are available for many city blocks, but no complete list of all city residents exists. A cluster could be defined as all (or many) homes on a city block. In other geographic problems, the population may be widely dispersed, resulting in high travel costs and major logistical problems. In these circumstances, non-cluster design options (e.g., simple random sample) would not be practical. The choice of a spatially-contiguous or concentrated cluster keeps the costs of obtaining the necessary sample to a minimum and permits an adequate sample size to be generated with reasonable effort.

A total enumeration cluster approach makes it easier to obtain large sample sizes (option 1 in Figure 7.5). In the student opinion survey, interviewing all students in a randomly-selected number of dormitories or sampling all students in a number of general education classes would result in an appropriately large sample size.

For cluster sampling to be most effective, however, the individuals within each cluster should be as different or heterogeneous as possible. This will make the cluster sample observations representative of the entire sampled population and avoid systematic bias. If the clusters are internally similar or homogeneous in nature, stratified sampling may be a better alternative. Therefore, the appropriateness of clusters in a geographic research problem must be evaluated carefully. In this context, the total enumeration cluster sample of a dormitory or general education class may not be appropriate because these clusters could be too homogeneous (i.e., a dorm may only house freshmen).

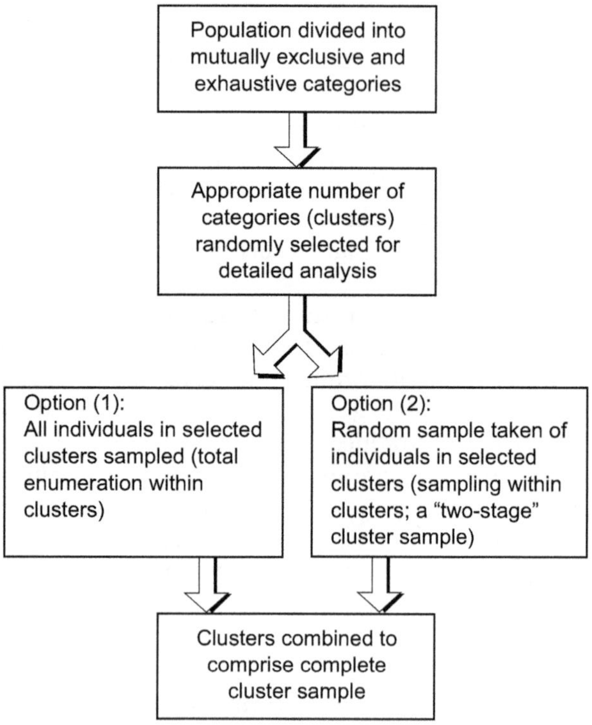

FIGURE 7.5
Steps in Cluster Sampling

Combination or Hybrid Sampling

Choosing a sampling design is seldom a simple, straightforward matter. Decision makers should consider cost, time, and convenience, as well as various practical problems unique to the specific situation. Common sense and experience are also important in selecting a sample design.

In many cases, the simplicity of a simple random approach offers an important advantage. However, practical circumstances may make the use of *any* simple type of sampling (random, systematic, stratified, or cluster) difficult or unwieldy. When practical conditions dictate, a **combination sample or hybrid sample** may be most appropriate. The following experience illustrates how practical realities influence the selection of sample design.

Some time ago, members of a geography department were asked to conduct a survey of air passengers enplaning at the nearby regional public airport. The survey had three general objectives: (1) establish the airport's service area; (2) determine selected passenger characteristics; and (3) provide various recommendations on how to improve or expand airport services. Since a complete census of all passengers was impractical, a survey sampling procedure was devised. Selection of a proper sample design was critical to ensure that the passengers being surveyed accurately represented the entire population of passengers enplaned at the airport. A number of statistical requirements and practical realities limited the sampling design choice:

- ensuring that each passenger was selected at random with known probability of inclusion;
- guaranteeing representative coverage over the different seasons of the year and different days of the week;
- working within economic constraints, to keep both the researchers' transportation costs and the number of hours required by airport personnel to administer the survey at a reasonable level; and
- selecting a convenient sample design that would produce appropriate accuracy, allow airport personnel to closely supervise the procedure, and result in minimal administrative errors.

As things developed, it turned out that a combination or hybrid sample design was best able to meet both the statistical requirements and practical constraints. A year-long survey period was chosen, and an initial survey date (cluster) was selected at random. All flights on this calendar day were sampled (total enumeration within clusters) rather than distributing questionnaires to individual passengers on scattered individual flights (sampling within clusters). Succeeding survey dates were systematically spaced every 13th day after the initial date. This sequencing produced a total of 28 survey days or clusters, seven in each season of the year, with each day of the week represented once each quarter. Thus, all seasons were proportionally represented, as were all weekdays and weekend days in each season. By systematically spacing the sample clusters every 13th day, we assumed that no bias was introduced for either the season of the year or day of the week. If a holiday happened (randomly) to be selected as a sample day, it was included in the study. In fact, to exclude a holiday purposely would have introduced an unwanted systematic bias into the analysis.

The overall result was a type of *random systematic cluster* sample. Randomness existed because the initial survey date was randomly chosen, surveying passengers every 13th day added a systematic component, and giving all passengers questionnaires on the selected survey dates created a cluster effect.

7.3 Spatial Sampling

The types of sample design we have considered up to this point have not been explicitly spatial. Under certain circumstances, however, spatial sampling is necessary. Spatial sampling is applied when using a map of a continuously-distributed variable (such as vegetative cover, soil type, or pH of surface water) and a sample of locations is being selected from this map. If you are conducting fieldwork and need to select sample site locations within a defined target area, then you are going to need to apply a spatial sampling procedure. The use of GIS technology makes it relatively easy to select objects randomly. For example, if you need to randomly select objects that are already mapped, you can apply the same technique used to develop random numbers. In this case, you could select features in a GIS dataset by randomly selecting the appropriate feature ID numbers (assuming the feature ID is labeled 1 to n). If on the other hand you need to identify site locations that are not currently mapped, you can randomly generate X, Y coordinates that correspond to the bounding box of your study area. The randomly generated X, Y coordinates can be placed in a list, imported into the GIS, and mapped. These locations can then be exported to a GPS device to allow navigation to the appropriate sites.

Spatial sampling from maps or other spatial sampling frames may involve point samples, line samples (traverses), or area samples (quadrats) (Figure 7.6). Of these three types of spatial sampling, geographers use point sampling most frequently, and the various concepts involving spatial sampling procedures are illustrated most easily through point sampling designs. Therefore, this section focuses on point sampling, and those wanting details on quadrat and traverse sampling are referred to more specialized texts and references.

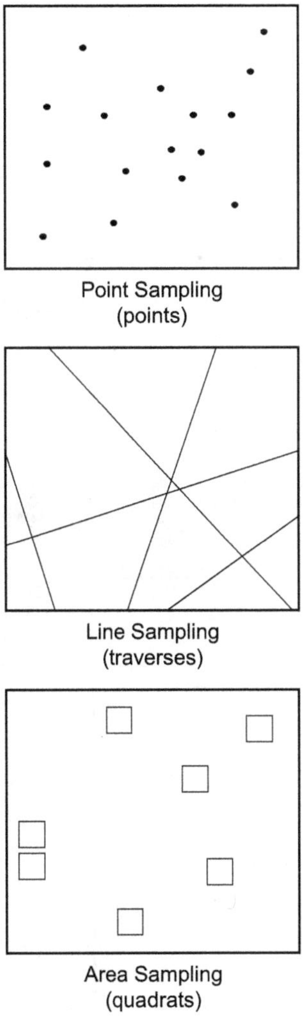

FIGURE 7.6
Spatial Sampling: Points, Lines, and Areas

For all types of spatial sampling, a spatial sampling frame (such as a map) must be constructed. This frame must include a coordinate system that allows clear identification of locations within the sampled area. For point sampling from a map, designation of the (X, Y) Cartesian coordinate pair will identify a unique location for each point.

In a **simple random point sample** (Figure 7.7, Case 1), two numbers (an X and Y coordinate pair) are selected to designate each point. To locate n sample points in a sampled area for a study, $2(n)$ random numbers must be drawn from a random numbers table or other source. The resulting pattern of points may be uneven, with some portions of the sampled area seemingly underrepresented, and others appearing overrepresented.

As an example, consider three different random distributions of 24 points on an agricultural field (Figure 7.8). Each sample distribution is clearly different, with some apparent clustering of points on parts of the field while other parts of the study area have few or no sample points. This spatial unevenness is to be expected, of course, because these are all random samples. In theory, if we took an infinite number of 24-point samples, and plotted them all on the same map, we would have points rather evenly distributed over the entire field.

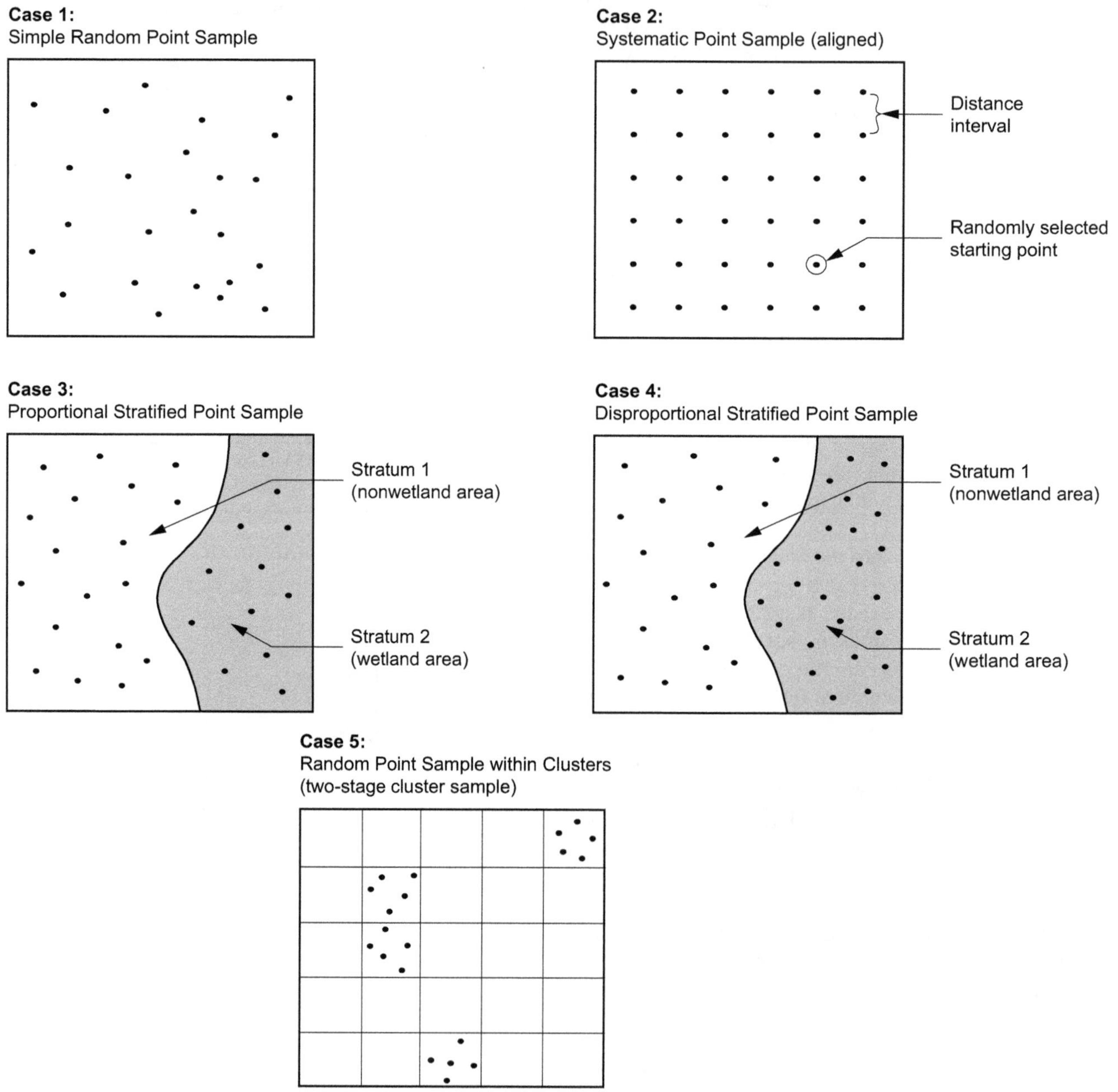

FIGURE 7.7
Types of Spatial Point Sampling

Systematic point sampling is a convenient way to avoid the problems possible with an uneven distribution of points across the study area. Systematic point sampling uses a "distance interval, similar to a regular sampling interval (*k*) from a list of individuals in a nonspatial context (Figure 7.7, Case 2). A "starting point" is randomly selected, then all other points are located using the distance interval to space them evenly across the entire sampled area. Only one point must be randomly located; the subsequent placement of all other points in the systematic pattern is determined automatically by the distance interval. This approach has the advantage of offering representative, proportional coverage of the entire sampled area. Bias will enter the sample design when a spatial regularity or periodicity exists in the distribution of some phenomenon that happens to match the distance interval. Systematic point sampling is often used in geographic research, particularly when dealing with environmental and resource problems where data are continuously distributed across an area.

Just as in nonspatial cases, **stratified point sampling** in spatial sample design may be either proportional (constant-rate) or disproportional (variable-rate). The approach selected depends on the circumstances of the problem. Suppose you are investigating environmental degradation in a region that includes particularly vulnerable tidal and nontidal wetlands. In Figure 7.7 (Cases 3 and 4), stratum 1 shows the nonwetland portion of the spatial sampling frame (comprising 60% of the total sampled area), while stratum 2 identifies wetlands (the other 40%). Suppose proportional representation of both strata is desired. This would dictate that 60% of the sample points be placed in non-wetland locations and 40% on wetland sites. If a total of 30 sample points is sufficient, 18 points should be in non-wetland locations, and 12 points in wetlands (Figure 7.7, Case 3). Suppose instead that particular attention needs to be focused on possible environmental problems in the wetlands area. In this situation, the wetlands stratum needs more detailed monitoring or "oversampling," while maintaining adequate coverage of the non-wetland portion of the sampled area. The disproportional stratified design (Figure 7.7, Case 4) shows twice the intensity of sample points (24 rather than 12) in the wetland stratum.

In spatial point sampling, a **cluster design** has the important advantage of keeping travel costs and other logistical problems to a minimum because the points of a cluster are located together (Figure 7.7, Case 5). This feature makes point sampling within clusters particularly attractive in geographic projects having an extensive study area. In many field studies, a cluster approach may be the only practical alternative.

Although two methods of detailed analysis within clusters are available in nonspatial cluster sampling (Figure 7.5), only the "two-stage" cluster approach is appropriate in spatial cluster sampling. In the spatial sampling of a variable continuously distributed across an area, an infinite number of points could be selected in each subarea. As a result, total enumeration is virtually impossible in a spatial context.

However, a cluster approach to point sampling is not always advisable. If the phenomenon being studied varies from one section of the sampled area to another, it may be

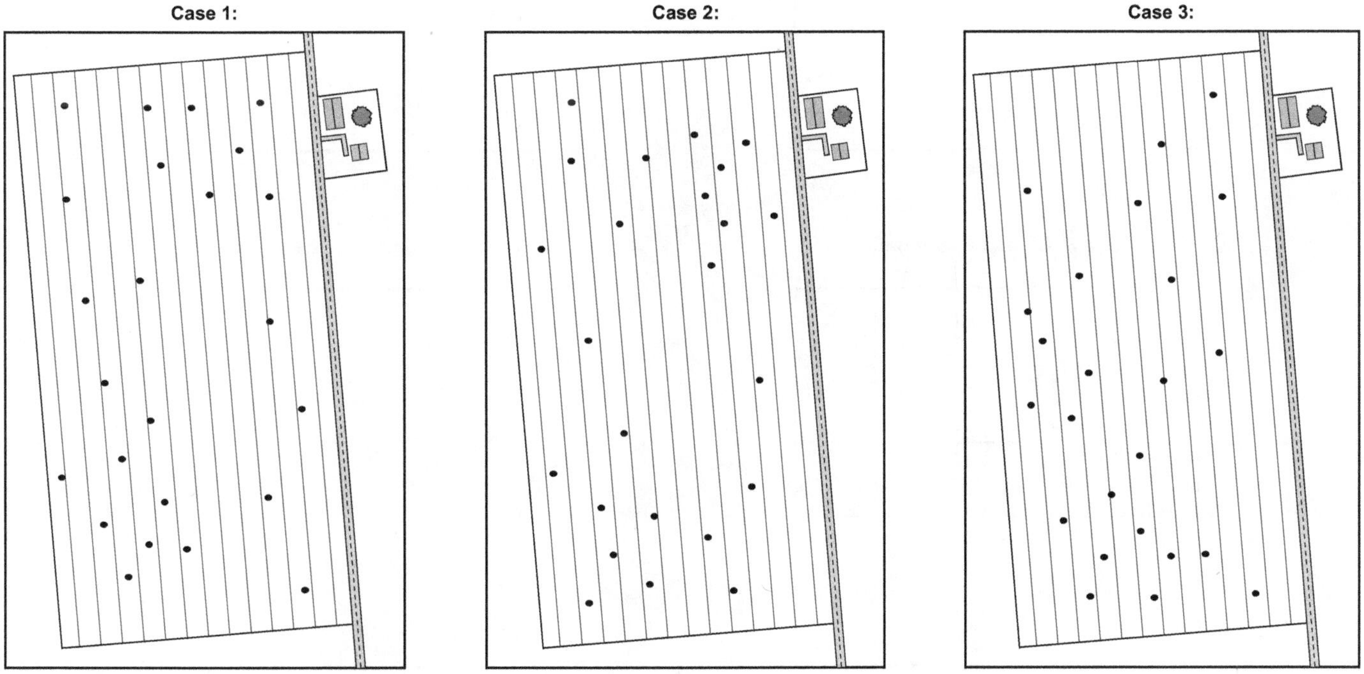

FIGURE 7.8
Three Different Random Samples of 24 Points on the Same Agricultural Field

necessary to ensure that some point locations are selected from all sections or subareas. Cluster sampling excludes substantial parts of the study area, and such uneven areal representation can be a disadvantage in many geographic problems. As a result, the practical advantages of convenience and reduction in travel cost often have to be balanced against unrepresentative spatial bias.

In addition to simple point sampling, a number of combination or hybrid point sampling designs are also possible (Figure 7.9). A composite sample design with its complex procedures is only worthwhile if it is likely to improve the accuracy of the sample in estimating the population.

Perhaps the most widely used hybrid sampling model is the **stratified systematic unaligned point sample**. First suggested by Berry and Baker (1968), this approach includes the following steps (Figure 7.9, Case 1):

1. Place a regular grid system over the entire study area, taking care to ensure that grid size is at an appropriate level of spatial resolution for the problem.
2. Select a random number for each row and column in the gridded area, where the row number specifies the horizontal position of points in that row and the column number specifies the vertical position of points in that column.
3. Locate a single point in each grid using the appropriate row and column positions.

This procedure provides proportional representation of all segments of the sampled area, yet avoids possible problems with regularities or periodicities in the spatial pattern that can be encountered with an aligned systematic point sample (Figure 7.7, Case 2). The needed sample points can also be generated fairly easily. As a result of these advantages, a stratified systematic unaligned point sample is a good choice in many realistic geographic situations.

Other combination or hybrid sampling models can be designed, but should be used only if they are likely to improve the accuracy or precision of the sampling

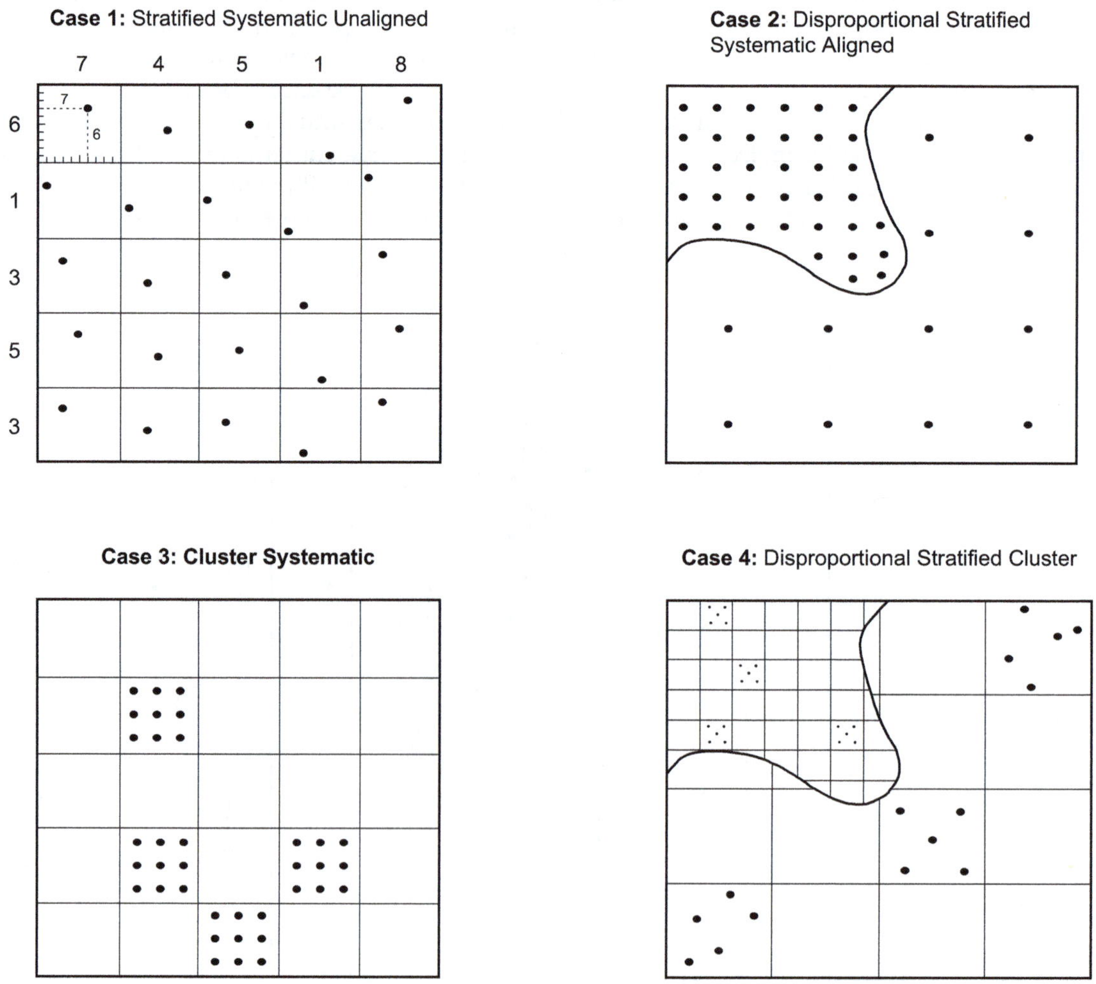

FIGURE 7.9
Selected Types of Composite or Hybrid Point Sampling

problem. Geographers try to create models in such areas as climatology, resource management, and migration to replicate or duplicate reality as closely as possible. Similarly, in geographic sampling problems, the sample design must reflect the character of the situation. In this way, precise and accurate estimates can be obtained with reasonable effort. Other hybrid point sample design options include: "disproportional stratified systematic aligned" (Figure 7.9, Case 2); "cluster systematic" (Case 3); and "disproportional stratified cluster" (Case 4).

Are any of these hybrid designs really feasible? Clearly the answer is yes. Suppose you are concerned with monitoring changes in both the spatial distribution and degree of intensity of nitrogen and phosphorus levels in a bay receiving agricultural runoff. In the narrow estuaries of the bay, runoff problems are likely to be more severe and will need careful monitoring. On the other hand, no portion of the bay should be left totally unmonitored. What type of spatial point sampling procedure should be used when locating monitor buoys in the bay? A disproportional stratified systematic aligned design might be the most practical alternative (Figure 7.10). Disproportional stratification places an appropriate density of nutrient monitoring stations in the necessary locations. Systematic placement allows efficient collection of water samples, since the buoys are aligned regularly for pick-up by boat. This sample design is "tailor-made" to fit the practical circumstances of this geographic problem.

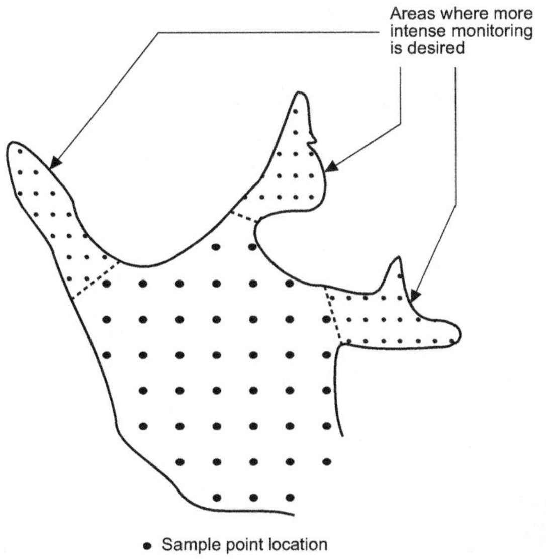

FIGURE 7.10
Location of Nutrient Monitoring Stations on a Bay: Disproportional Stratified Systematic Aligned Sample

Key Terms

basic types of sampling:
- simple random, *110*
- systematic, *110*
- stratified, *111*
- cluster, *111*

combination or hybrid sample designs, *112*
oversampling and undersampling, *115*
population, *103*
probability and nonprobability sampling designs, *109*
proportional and disproportional stratified sampling, *111*
randomness, *104*
sample, *103*
sampled population and area, *106*
sampling frame, *106*
sources of sampling error, *104*
stratified systematic unaligned point sample, *116*
target population and area, *106*
total enumeration within clusters and two-stage cluster sampling, *112*
types of spatial sampling:
- simple random point sample, *114*
- systematic point sample, *115*
- stratified point sample, *115*
- cluster point sample, *115*

Major Goals and Objectives

If you have mastered the material in this chapter, you should now be able to:

1. Understand the various advantages of sampling in contrast to a complete enumeration or census of an entire population.
2. Identify possible sources of sampling error in a geographic problem or situation.
3. Explain the basic objectives and practical uses of the American Community Survey.
4. Understand the steps in the general sampling procedure, from the conceptual definition of target population and target area to the collection of sample data.
5. Define and distinguish between these sampling terms: target population and target area, sampled population and sampled area, and sampling frame.
6. Explain the characteristics of the important types of probability sampling—including simple random, systematic, stratified, cluster, and combination or hybrid designs.
7. Select an appropriate sampling design (one able to meet both the statistical requirements and practical circumstances) when presented with a "geographic situation" or practical problem that requires sampling.
8. Explain the characteristics (including both advantages and disadvantages) of the various types of simple and composite point sampling.

References and Additional Reading

American Community Survey (ACS). United States Census Bureau. www.census.gov/programs-surveys/acs/

American Statistical Association, Survey Research Methods Section. *What is a Survey? How to Plan a Survey, How to Collect Survey Data*. Accessed January 9, 2014. http://www.amstat.org/sections/srms/

Berry, B. J. L. and A. M. Baker. "Geographic Sampling," in *Spatial Analysis: A Reader in Statistical Geography,* edited by B. J. L. Berry and D. F. Marble. Englewood Cliffs, NJ: Prentice-Hall, 1968.

Cochran, W. G. *Sampling Techniques.* 3rd ed. New York: John Wiley and Sons, 1977.

Dixon, C. J. and B. Leach. *Sampling Methods for Geographical Research*. Concepts and Techniques in Modern Geography, CATMOG 17. Norwich, England: Geo Abstracts, 1978.

Greater Yellowstone Coalition. *The Greater Yellowstone Ecosystem*. www.greateryellowstone.org

Henry, G. *Practical Sampling*. Newbury Park, NJ: Sage, 1990.

Lohr, S. L. *Sampling: Design and Analysis*. Pacific Grove, CA: Brooks/Cole, 1999.

McGrew, J. C. and R. A. Rosing. *Salisbury-Wicomico County Regional Airport Passenger Survey.* Salisbury: Delmarva Advisory Council, 1979.

Parmenter, A. W., et. al. "Land Use and Land Cover Change in the Greater Yellowstone Ecosystem: 1975–1995," *Ecological Applications.* Vol. 13, No. 3 (Jun. 2003): 687–703.

Salant, P. and D. A. Dillman. *Practical Sampling*. Newbury Park, NJ: Sage, 1990.

Scheaffer, R. L., W. Mendenhall, R. L. Ott and K. G. Gerow. *Elementary Survey Sampling*. 7th ed. Boston: Brooks/Cole, 2012.

Thompson, S. K. *Sampling*. 3rd ed. New York: John Wiley and Sons, 2012.

Williams, B. *A Sampler on Sampling*. New York: John Wiley and Sons, 1978.

8

Estimation in Sampling

The primary objective of sampling is to make inferences about the population from which a sample is taken. More specifically, sample statistics are used to estimate population parameters such as the mean, total, and proportion. When sample statistics represent a larger population accurately, they are considered unbiased estimators.

This chapter considers both the theory and practice of estimating from samples. The basic terminology and the theoretical concepts that underlie sample estimation are discussed in Section 8.1. Distinction is made between point estimation and interval estimation, and the nature of the sampling distribution of a statistic is explained. Also discussed is the importance of the central limit theorem when inferring sample results to a population. Confidence intervals that indicate the level of precision for a population estimate can be determined for any desired sample statistic. Sections 8.2 and 8.3 include a variety of confidence interval equations. Material is organized by type of sample (random, systematic, and stratified) and by parameter (mean, total, and proportion). A series of examples illustrates the procedure for constructing confidence intervals around each parameter using the different sampling methods.

To save time and effort, geographers often want to know—*before* taking a full sample—how large a sample is needed for a particular research problem. Issues and methods that determine minimum necessary sample size before taking the full and complete sample are discussed in Section 8.4. Examples illustrate how sample size is established for problems using the mean, total, and proportion.

8.1 Basic Concepts in Estimation

Point Estimation and Interval Estimation

In statistics, a basic distinction is made between **point estimation** and **interval estimation**. The concept of point estimation is relatively straightforward—a statistic is calculated from a sample and then used to estimate the corresponding population parameter. With probability sampling, the "best" (unbiased) point estimate for a population parameter is the corresponding sample statistic. Therefore, the best point estimate for the population mean (μ) is the sample mean ($\bar{X}$); the best point estimate for the population standard deviation (σ) is the sample standard deviation (s), and so on (Table 8.1).

Note that the denominator for calculating sample standard deviation is ($n - 1$) rather than (n). This slight adjustment makes the sample standard deviation a less-biased estimator of the population standard deviation, particularly for smaller samples. With larger sample sizes (larger than 30 or so), the difference between dividing by (n) or dividing by ($n - 1$) is insignificant. However, using (n) with a smaller sample would result in an estimate that under-represents the magnitude of the true standard deviation in the population.

How precise are sample point estimators? How close (or distant) from the true population parameter is the calculated sample statistic? Because probability sampling involves some uncertainty, it is unlikely that a sample statistic will exactly equal the true population parameter. What can be determined, however, is the likelihood that

TABLE 8.1

Sample Statistics as Best Point Estimators of Population Parameters

Descriptive statistic	Population parameter	Sample statistic *	Calculating formula
Mean	μ	$\bar{X}$	$\frac{\sum_{i=1}^{n} X_i}{n}$
Standard deviation	σ	s	$\sqrt{\frac{\sum_{i=1}^{n}(X_i - \bar{X})^2}{n-1}}$
Total	τ	T	$N(\bar{X}) = N\frac{\sum X_i}{n}$
Proportion **	ρ	p	$\frac{x}{n}$

* best point estimator
** x = number of units sampled having a particular characteristic
n = total number of units sampled

a sample statistic is within a certain range or interval of the population parameter. The determination of this range is the basis for interval estimation. A **confidence interval** or bound represents the level of precision associated with the population estimate. Its width is determined by: (1) the sample size; (2) the amount of variability in the population; and (3) the probability level or level of confidence selected for the problem. We will now explore these ideas in some detail.

The Sampling Distribution of a Statistic

Suppose a random sample of size n is drawn from a population, and the mean of that sample ($\bar{X}$) is calculated. Now suppose a second sample of size n (totally independent of the first sample) is drawn and its mean calculated. If this process is repeated for many similar-sized independent samples in a population, the frequency distribution of this set of sample means can be graphed (Figure 8.1). This curve is referred to as the **sampling distribution of sample means**.

A sampling distribution can be developed for any statistic, not just the mean. After many independent samples of size (n) are drawn from a population, the statistic of interest (i.e., mean, total, and proportion) is calculated for each sample, and the distribution of the sample statistics graphed. The resulting frequency distribution has a shape, a mean, and a certain amount of variability (as reflected by the standard deviation and variance). These general characteristics of sampling distributions are important, but for now the focus is on particular characteristics of sample means.

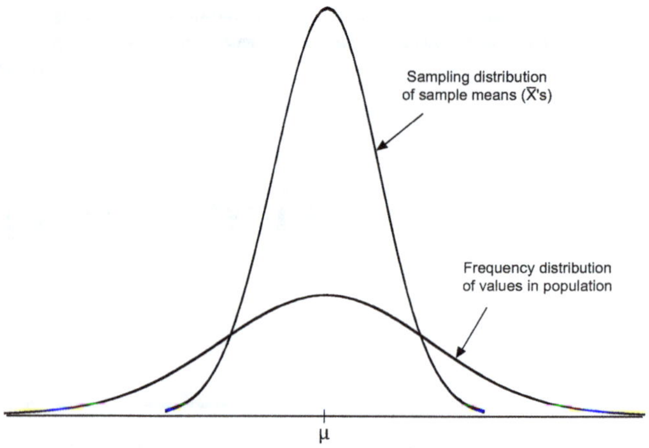

FIGURE 8.1
Sampling Distribution of Sample Means and Frequency Distribution of Population Values

The Central Limit Theorem

When the sample statistic is the mean, its frequency distribution has a particular set of features that are of vital theoretical importance in sampling and general statistical inference. These features are summarized in the **central limit theorem (CLT)**:

Understanding how the central limit theorem works is essential for understanding why we can make inferences about a population using only a sample. Let's take a look at each of the three primary ideas in the CLT theoretically and then consider a simplified example.

The first idea states that the frequency distribution of sample means will have a mean of μ, the mean of the total population. Suppose we have a population of 1000 students and decide to sample five students to estimate the amount of the average student loan. One of our samples might include these student identification numbers (27, 252, 516, 717, and 927), allowing us to compute a sample mean, $\bar{x}_1$. Suppose we decide to take another

Central Limit Theorem

Suppose all possible random samples of size (n) are drawn from an infinitely large, normally-distributed population having a mean = μ and a standard deviation = σ. The frequency distribution of these sample means ($\bar{X}s$) will have:

1. a mean of μ (the population mean);
2. a normal distribution around this population mean; and
3. a standard deviation of $\sigma / \sqrt{n}$.

This particular set of characteristics applies only to the distribution of sample means and not to the distribution of other sample statistics.

sample of five students; this would yield yet another set of students for which we can calculate a sample mean, $\bar{x}_2$. We now have two sample means, and can use that to generate a mean of both samples, $(\bar{x}_1 + \bar{x}_2)/2$. Theoretically, we can continue to take an infinite number of random samples of five students, and the average of these samples would be equal to the population average. In an extreme example, it is possible that one of our samples selects five students with very high student loans. Therefore, $\bar{x}_1$ would be much higher than the population average for student loans. However, it is equally possible that one of our samples selects five students with very low student loans, and thus $\bar{x}_2$ would be much lower than the population average for student loans. If we averaged these two extreme samples, the result would be a number closer to the population mean.

The second idea states that the frequency distribution of sample means will be normally distributed around the true population mean. In reality, each sample of five students will likely have some students whose loan amount is larger than the population average, and some students whose loan amount is smaller than the population average. Therefore, the average of each sample of 5 students will tend to be closer to the mean. As we accumulate these five-student samples, we will have many more sample means that are closer to the population mean than farther away. If this frequency distribution is plotted, it will look increasingly normal in shape as the number of samples increases.

The third idea provides insight into the variability of the sample means. It states that the standard deviation of the sampling distribution of means is equal to the population standard deviation divided by the square root of the sample size, $\sigma/\sqrt{n}$. This measure of standard deviation is called the **standard error of the mean** ($\sigma_{\bar{X}}$):

$$\sigma_{\bar{X}} = \frac{\sigma}{\sqrt{n}} \quad \textbf{(8.1)}$$

The standard error of the mean indicates how much a typical sample mean is likely to differ from the true population mean. Quite simply, standard error is a basic measure of the amount of **sampling error** in a problem.

Recall that the standard deviation is a descriptive statistic that measures the average amount that a single observation differs from the mean. In our example, we are measuring how much an individual sample of 5 students differs from the true population mean. This difference is dependent upon the amount of variability in the population and the size of the sample. As an illustration, suppose the student loans of all students range anywhere from $12,000 to $14,000. That is a small range of only $2,000. Any sample we take will tend to be close to the mean. However, if the range of student loans is between $0 and $215,000, it is very likely that a sample will be farther from the mean because we have a range of loan values of $215,000. Therefore, samples taken from populations with a small standard deviation have a lower standard error than samples taken from populations with a higher standard deviation.

Similarly, we know intuitively that larger samples tend to provide better estimates of the population than smaller samples. Thus, increasing the sample size will improve the estimate of the population parameter. Therefore, mathematically, an increase in n in the denominator will cause the standard error (sampling error) to be reduced.

Example: Sampling from an Experimental Agricultural Plot

While the discussion above makes intuitive sense from a mathematical standpoint, we are interested in determining if the concept actually works in a practical sense. To test the practicality of the central limit theorem, we present an illustration using some real world data.

The experimental agricultural plot shown in Figure 8.2 contains 7240 yield values. The average yield was 123.1 bu/ac with a standard deviation of 38.5. Furthermore, the histogram portion of Figure 8.2 shows that the data had a moderate skew of 1.63.

Suppose we didn't actually know the average and standard deviation for the population, but were only able to obtain samples. Therefore, to *theoretically* test the claims of the central limit theorem, we can generate 2000 random samples with each of these different sizes (5 yield locations, 10 yield locations, and 40 yield locations), and calculate the average, standard deviation, and skew. For the remainder of this discussion, keep in mind that we did not include an infinite number of samples, but rather capped our number of samples at 2000.

Table 8.2 shows the descriptive statistics for the original 7240 yield locations as well as the computed average and standard deviation of the respective sampling distributions using sizes of 5, 10, and 40 samples. You can see that as the sample size increased from 5 sample values (average yield of 122.5) to 40 sample values (average yield of 122.2), the average yield value for the sampling distribution was closer to the actual value of 122.1. Thus, even with skewed data, the theoretical sampling distribution has a mean of μ (in reality, we are slightly off because we only included 2000 samples).

Figure 8.3 shows a series of sampling distributions for the three tests. While the original data of 7240 points had a skew of 1.63, the sampling distribution of 2000 samples using 5 yield values reduced the skew to 1.09, and thus caused the sampling distribution to appear more normally distributed. We are looking at these changes in skew because a reduced skew (as related to larger sample size) is an indicator of a distribution closer to normal. Using a larger sample size of 10 yield values further reduced the skew to 0.61, and the sample size of 40 reduced the skew even more to 0.36. This series of figures

(continued)

TABLE 8.2

Descriptive Statistics for Population (Size 7,240) and Samples of Size 5, 10, and 40

Samples	Mean	Standard Deviation	Theoretical standard error of the mean $(\sigma / \sqrt{n})$
Original 7,240 values	122.1	38.5	
5 sample values	122.5	17.9	17.21
10 sample values	122.3	13	12.174
40 sample values	122.2	6.8	6.08

illustrate that as a larger sample size is used to estimate the average yield, the sampling distribution appears more normal—even if the original data was skewed to begin with!

Revisiting Table 8.2 also shows the average yield, the actual computed standard deviation of the sampling distributions, and the theoretical standard error of the mean using the formula $\sigma/\sqrt{n}$. In our example, *s* represents the population standard deviation of 38.5, and *n* represent the sample sizes we investigated (5, 10, and 40 respectively).

For the 2000 samples of 5 yield values, the theoretical standard error of the mean is computed as $17.2 = 38.5/\sqrt{5}$, which is very close to the actual value of 17.9. Similarly, the theoretical standard error computed with 40 yield values is $6.08 = 38.5/\sqrt{40}$, which is also similar to the theoretical value of 6.8. Once again, even with skewed data, the theoretical standard error of the mean is very similar to the actual standard deviation of the sampling distribution, especially when the sample size increases.

In reality, we will never take an infinite number of samples, nor will we even have the luxury of taking 2000 samples of 40 data points. Often the best we can do is to obtain a single sample of a certain number of data points. However, as our practical example shows, all samples will theoretically (and practically) fall within the normal sampling distribution. Therefore, if we only take a single sample, it is safe to assume that the sample will fall somewhere within the normal distribution. Further, given the size and standard deviation of the sample, we can reasonably determine the confidence interval of our sample (the probability that the true population falls somewhere within a specified range).

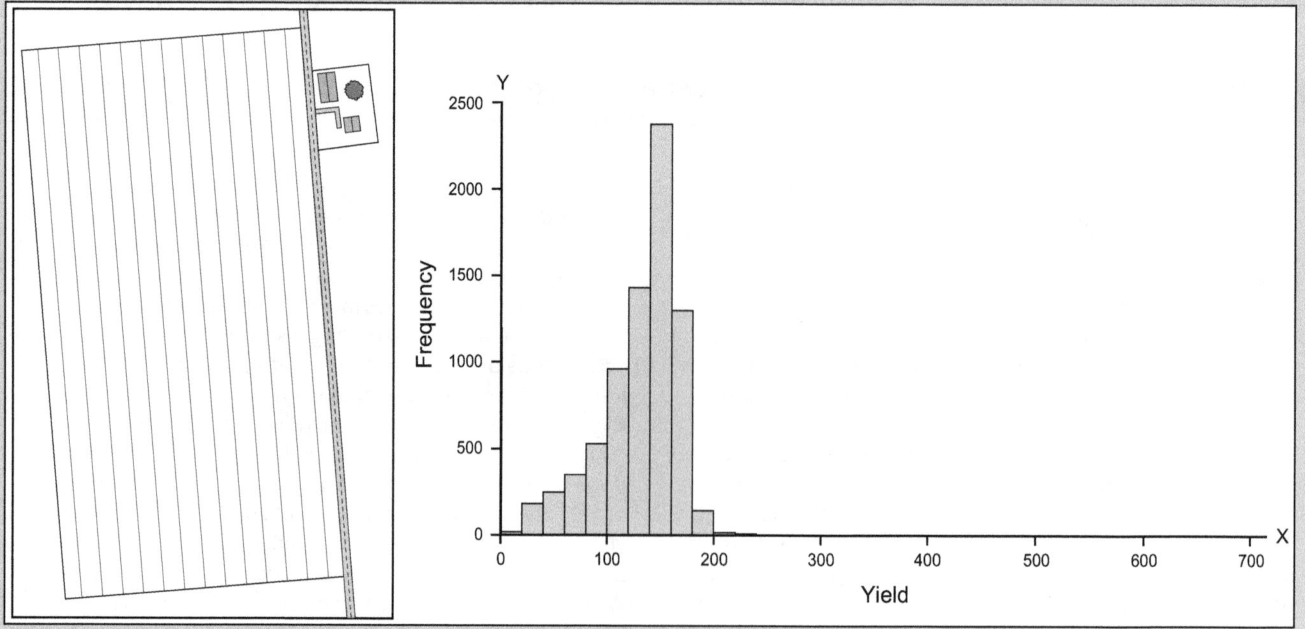

FIGURE 8.2

An Experimental Agricultural Plot and Histogram Showing Distribution of 7,240 Yield Values

Source: Lembo, A., Lew, M., Laba, M., Baveyye, P. 2006

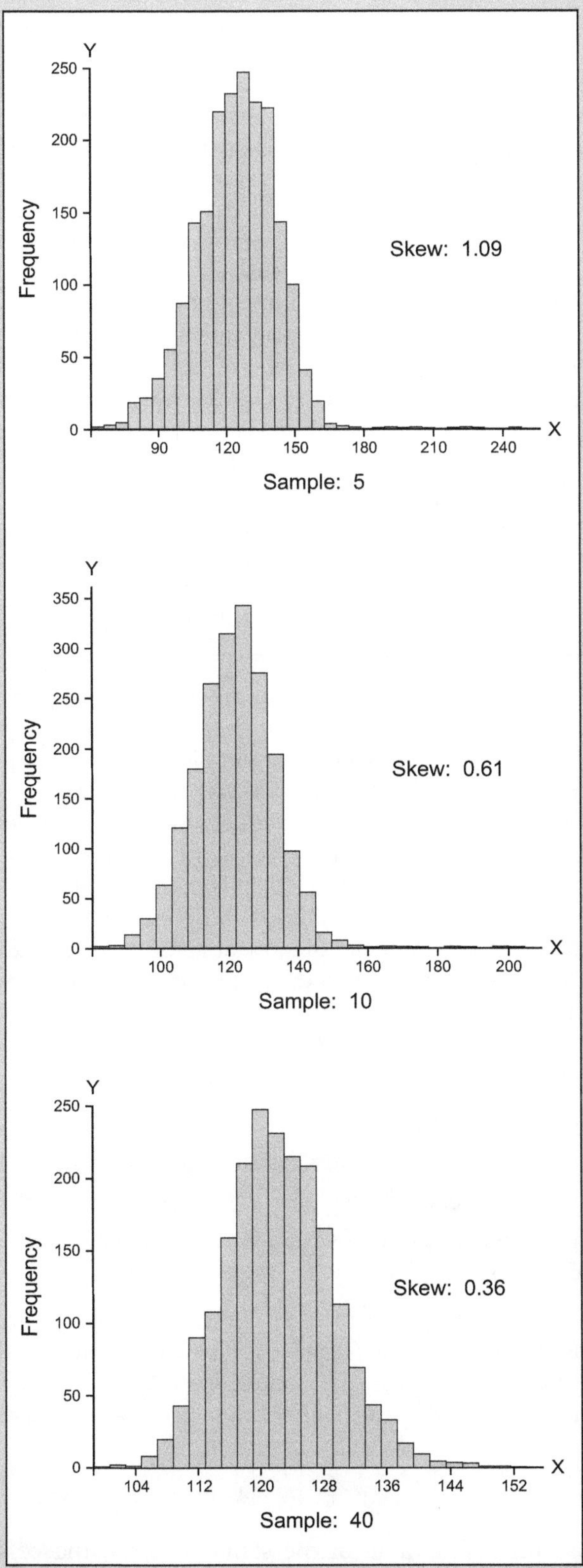

FIGURE 8.3
Histograms Showing Sampling Distributions of Size 5, 10, and 40. (While the distribution appears left-skewed, the positive skewness measure arises due to a few high outliers that are difficult to see.)

8.2 Confidence Intervals and Estimation

Suppose you want to place a confidence interval about a sample mean with 90% certainty that the interval range contains the actual population mean. The general formula for a confidence interval is:

$$\bar{X} \pm Z\sigma_{\bar{X}} \quad \textbf{(8.2)}$$

where $\bar{X}$ = sample mean
Z = Z-value from the normal table
$\sigma_{\bar{X}}$ = standard error of the mean

Because the central limit theorem establishes the normality of the distribution of sample means, Z-values from the normal table (Appendix, Table A) provide the proper probabilities for defining the confidence interval. Given a desired certainty of 90%, what is the corresponding Z-value or area under the normal curve? That is, what Z-value from the normal table is associated with the confidence interval having a 90% likelihood of containing the true population mean? The desired Z-value represents the situation where 90% of the total area under the curve is encompassed by the confidence interval and 45% of this total area is on either side of the mean (Figure 8.4).

If you look at the table of normal values in Appendix A, you can see that when the area "A" equals 4495, the Z-value is 1.64 and when the area "A" equals 4505, the Z-value is 1.65. If we follow the convention of rounding up if midway between two numbers, a Z-value of 1.65 is used for this confidence interval estimate. Therefore, we can be 90% confident that the true population mean lies within the confidence interval defined by:

$$\bar{X} \pm 1.65\sigma_{\bar{X}}$$

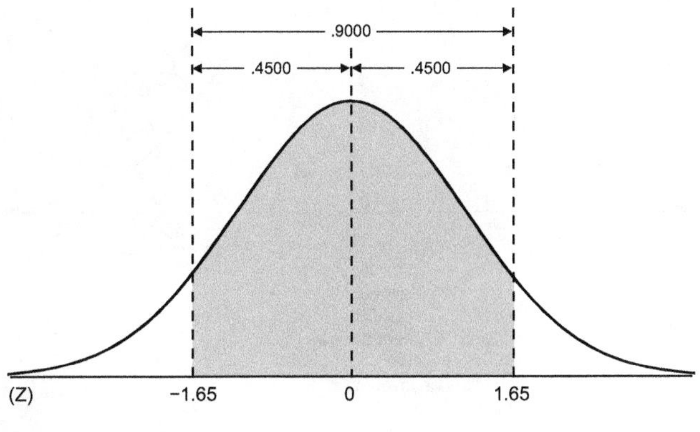

FIGURE 8.4
Standard Scores (Z-Values) Associated with a 90 Percent Confidence Interval

The confidence interval contains expressions that are added to and subtracted from the mean to define the **upper and lower bounds** of the interval:

$$\text{Upper bound} = \bar{X} + 1.65\sigma_{\bar{X}}$$

$$\text{Lower bound} = \bar{X} - 1.65\sigma_{\bar{X}}$$

Recall that the entire area under the sampling distribution curve of Figure 8.1 represents the set of *all* possible sample means that could be drawn from the original population. The shaded area in the center of the sampling distribution (Figure 8.4) shows the location of 90% of all sample means that could be drawn from the population. Suppose 10 actual samples are taken from this population and the location of each sample mean is plotted (Figure 8.5). Of the 10 sample means shown $(\bar{X}_1 \text{ to } \bar{X}_{10})$, 9 of them (.90 probability) are within the confidence interval. This result is expected because the interval was constructed with a 90% chance of containing μ. Of course, when taking a single sample, the location of μ is not known. However, a .90 probability exists that the interval or bound placed around that single sample mean *does* contain the true population mean. Conversely, a .10 probability exists that the confidence interval placed around the single sample mean *does not* include μ. The probabilities that an unusually large sample mean (well above μ) or an unusually small sample mean (well below μ) could be drawn are represented graphically by the two unshaded tails of the sampling distribution. Note that by chance, the fifth sample mean $(\bar{X}_5)$ falls below the lower bound of the confidence interval around μ.

Several terms are used when making interval estimates in sampling. The **confidence level** refers to the probability that the interval surrounding a sample mean *encompasses* the true population mean. This confidence level probability is defined as 1 – α. The **significance level** refers to the probability that the interval surrounding a sample mean *fails to encompass* the true population mean. The significance level is denoted by α and equals the total sampling error. Because error is equally likely in either direction from μ, the probability of the sample mean falling into either tail of the distribution is α/2.

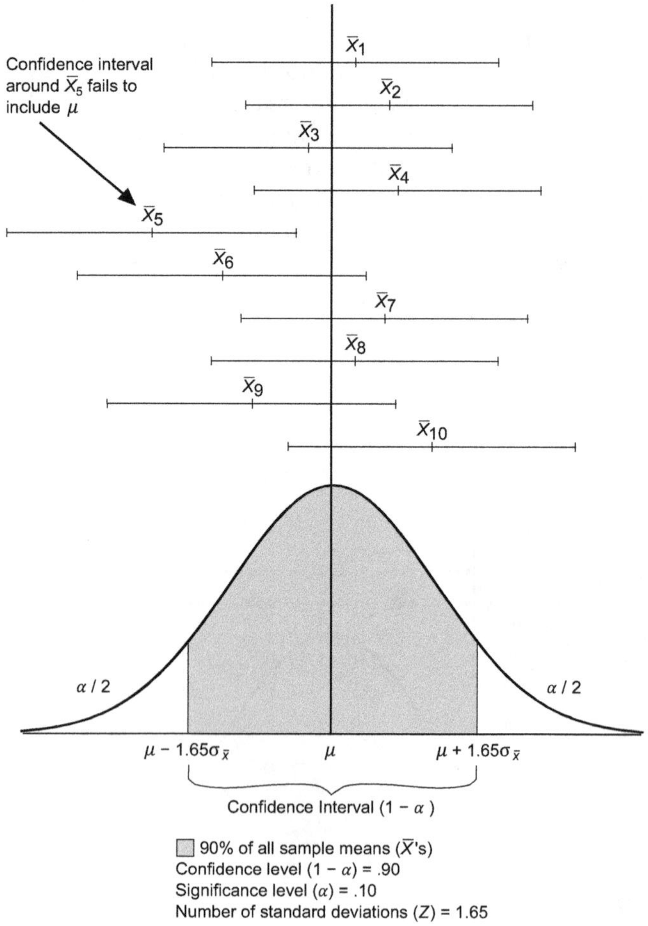

FIGURE 8.5
Distribution of Sample Means and the Confidence Interval Concept

General Procedure for Constructing a Confidence Interval

The best way to learn about the construction of confidence intervals is to present a simple example in some detail. Suppose that a random sample of 50 commuters in a metropolitan area revealed that their average journey-to-work distance was 9.6 miles. Moreover, a recent study has determined that the standard deviation of journey-to-work travel distances for this metropolitan area is approximately 3 miles. What is the confidence interval around this sample mean of 9.6 that guarantees with 90% certainty that the true population mean is enclosed within that interval?

The confidence interval for μ is calculated from the following values:

- the sample mean ($\bar{X} = 9.6$)
- the population standard deviation ($\sigma = 3$)
- the sample size ($n = 50$)
- the Z-value associated with a 90% confidence level ($Z = 1.65$)

The single best estimate of the true population mean (μ) is the sample mean ($\bar{X} = 9.6$). This sample mean is the statistic around which the confidence interval is placed. From Equation 8.1, the confidence interval is defined as

$$\bar{X} \pm Z\sigma_{\bar{X}}$$

Substituting the value of the standard error, the confidence interval equation becomes

$$\bar{X} \pm Z\frac{\sigma}{\sqrt{n}} \qquad \textbf{(8.3)}$$

Inserting all values for the journey-to-work problem:

$$\overline{X} \pm Z\frac{\sigma}{\sqrt{n}} = 9.6 \pm 1.65\frac{3}{\sqrt{50}} = 9.6 \pm 0.70$$

This interval ranges from a lower bound of 8.90 (9.6 – .70) to an upper bound of 10.30 (9.6 +.70). It can now be concluded with 90% certainty that the mean journey-to-work distance for all commuters is between 8.90 and 10.30 miles.

When dealing with confidence intervals, various situations arise. Sometimes sample sizes are small and often population standard deviations are unknown. Proper decisions must be made about how to proceed under these circumstances.

What level of confidence should be used? A confidence interval can be created for any desired level of confidence ($1 - \alpha$). The most commonly accepted and widely used confidence levels are .99, .95, and .90. The corresponding likelihoods of making a sampling error (also called the significance level, α) when using these confidence levels are .01, .05, and .10. These "conventional" levels of confidence are generally considered rigorous enough to guarantee acceptably low probabilities of sampling error.

The impact of various confidence levels on significance levels, number of standard errors, and confidence interval characteristics are now summarized for the journey-to-work problem (Table 8.3). Notice what happens when the confidence level increases from .80 to .99. First, the significance level decreases from .20 to .01, reducing the sampling error for the problem. Also, the number of standard errors (Z) increases from 1.28 to 2.58, thereby widening the confidence interval range and width. Thus, higher levels of confidence result in wider confidence intervals and less precise estimates, but lower sampling error. The investigator has the responsibility for deciding how to trade off confidence level against level of precision.

What if the population standard deviation (σ) and variance (σ^2) are unknown? In the previous example, it was assumed that the population standard deviation and variance were known, while the population mean (μ) was unknown. This assumption is unrealistic and impractical: in most situations, if the population mean is unknown, the standard deviation and variance of the population are also unknown. With an unknown population variance, the sample variance (s^2) provides the best estimator and is the statistic inserted into the standard error formula.

Recall the standard error formula (Equation 8.1):

$$\sigma_{\overline{X}} = \frac{\sigma}{\sqrt{n}}$$

Squaring both sides to convert standard deviations into variances:

$$\sigma_{\overline{X}}^2 = \frac{\sigma^2}{n} \quad \textbf{(8.4)}$$

If the population variance (σ^2) is unknown, the sample variance (s^2) is substituted:

$$\sigma_{\overline{X}}^2 = \frac{\sigma^2}{n} = \frac{s^2}{n} \quad \textbf{(8.5)}$$

To get this expression into an effective form to place confidence intervals (bounds) around point estimates, one final step is needed—take the square root of both sides.

$$\sigma_{\overline{X}} = \sqrt{\frac{s^2}{n}} \quad \textbf{(8.6)}$$

This standard error formula is the expression used for the many confidence interval equations that follow.

What if the sample size is small? The journey-to-work example incorporated the standard normal deviate (Z) into the confidence interval calculation. However, Z is valid only if the sample size is greater than 30. With smaller samples, the confidence interval equation must be altered, and the standard *Z-value* must be replaced by the corresponding value from the student's *t* distribution. This modification is necessary because the sample standard deviation (s) is not always accurate when the sample size

TABLE 8.3

The Relationship Between Confidence Level and Confidence Interval for the Journey-to-Work Example

Confidence Level ($1 - \alpha$)	Significance Level (α)	Number of standard errors (Z)	Number of standard errors ($\overline{X} \pm Z\sigma_{\overline{X}}$)	Confidence Interval	
				Range	Width
.80	.20	1.28	9.6 ± .54	9.06 to 10.14	1.08
.90*	.10	1.65	9.6 ± .70	8.90 to 10.30	1.40
.95	.05	1.96**	9.6 ± .83	8.77 to 10.43	1.63
.98	.02	2.33	9.6 ± 1.00	8.60 to 10.60	2.00
.99	.01	2.58	9.6 ± 1.09	8.51 to 10.69	2.18

* Calculations for ($1 - \alpha$) = .90 are detailed in situation 1.
** A Z-value of 2.00 is often used for 1.96 for convenience.

is smaller than 30. In some situations, it may be impractical or impossible to have a sample size as large as 30. If the value of s appears to be inaccurate because of the small sample size, logic demands that a wider interval be presented as the confidence interval estimate for μ to avoid introducing bias. This wider interval allows an equivalent amount of confidence in the result that the true population mean lies within the interval.

Whenever the sample size is less than 30, the confidence interval formula is

$$\bar{X} \pm t\sigma_{\bar{X}} \quad \textbf{(8.7)}$$

Like the normal distribution, the t distribution is symmetric and bell-shaped. The exact shape of the t distribution depends on the sample size: as n approaches 30, the value of t approaches the standard normal (Z) value. Two pieces of information are needed to use the t table (Appendix, Tables C and D):

TABLE 8.4

Confidence Interval Equations for Random and Systematic Samples*

Part 1: Population parameter — mean (μ)	
Best point estimate:	$\hat{\mu} = \bar{X} = \frac{\sum X_i}{n}$
Standard error of the point estimate (sampling error):	$\hat{\sigma}_{\bar{X}} = \sqrt{\frac{s^2}{n}\left(\frac{N-n}{n}\right)}$ where $s^2 = \frac{\sum(X_i - \bar{X})^2}{n-1}$
Confidence interval (bound) around the point estimate:	$\bar{X} \pm Z\hat{\sigma}_{\bar{X}} = \bar{X} \pm Z\sqrt{\frac{s^2}{n}\left(\frac{N-n}{N}\right)}$
Part 2: Population parameter — total (τ)	
Best point estimate:	$\hat{\tau} = T = N\bar{X} = \frac{N\sum X_i}{n}$
Standard error of the point estimate (sampling error):	$\hat{\sigma}_{\tau} = \sqrt{N^2\left(\frac{s^2}{n}\right)\left(\frac{N-n}{n}\right)}$
Confidence interval (bound) around the point estimate:	$T \pm Z\hat{\sigma}_{\tau} = T \pm Z\sqrt{N^2\left(\frac{s^2}{n}\right)\left(\frac{N-n}{N}\right)}$
Part 3: Population parameter — proportion (ρ)	
Best point estimate:	$\hat{\rho} = p = \frac{x}{n}$
Standard error of the point estimate (sampling error):	$\hat{\sigma}_p = \sqrt{\left(\frac{p(1-p)}{n-1}\right)\left(\frac{N-n}{N}\right)}$
Confidence interval (bound) around the point estimate:	$p \pm Z\hat{\sigma}_p = p \pm Z\sqrt{\left(\frac{p(1-p)}{n-1}\right)\left(\frac{N-n}{N}\right)}$

* Exclude the finite population correction, $\left(\frac{N-n}{N}\right)$, from the confidence interval equations if $n/N < .05$. Replace Z with the corresponding t if $n < 30$.
N = population size; n = sample size, $\bar{X}$ = sample mean, s^2 = sample variance. [In part 3 only: p = sample proportion].

- the desired significance level (α) – common levels are $\alpha = .10, .05$, and $.01$.
- the number of degrees of freedom (df), where df is defined as one less than the sample size (df = $n - 1$). An explanation of degrees of freedom is included in Chapter 9 in the context of inferential statistics.

For smaller sample sizes (30 or less), the t value will always be slightly larger than the Z-value at the corresponding level of significance, resulting in a slightly larger confidence interval, or bound, on the error of estimation. For example, when placing a 95% confidence interval ($\alpha = .05$) around a sample statistic when $n = 30$,

TABLE 8.5

Confidence Interval Equations for Stratified Samples*

Part 1: Population parameter — mean (μ)	
Best point estimate:	$\hat{\mu} = \bar{X} = \frac{1}{N}\sum_{i=1}^{m} N_i\bar{X}_i = \frac{1}{N}[N_1\bar{X}_1 + \ldots + N_m\bar{X}_m]$
Standard error of the point estimate (sampling error):	$\hat{\sigma}_{\bar{X}} = \sqrt{\frac{1}{N^2}\sum_{i=1}^{m} N_i^2\left(\frac{s_1^2}{n_1}\right)\left(\frac{N_i - n_i}{N_i}\right)}$
Confidence interval (bound) around the point estimate:	$\bar{X} \pm Z\hat{\sigma}_{\bar{X}} = \bar{X} \pm Z\sqrt{\frac{1}{N^2}\sum_{i=1}^{m} N_i^2\left(\frac{s_1^2}{n_1}\right)\left(\frac{N_i - n_i}{N_i}\right)}$
Part 2: Population parameter — total (τ)	
Best point estimate:	$\hat{\tau} = T = \sum_{i=1}^{m} N_i\bar{X}_i = [N_1\bar{X}_1 + \ldots + N_m\bar{X}_m]$
Standard error of the point estimate (sampling error):	$\hat{\sigma}_{\tau} = \sqrt{\sum_{i=1}^{m} N_i^2\left(\frac{s_i^2}{n_i}\right)\left(\frac{N_i - n_i}{N_i}\right)}$
Confidence interval (bound) around the point estimate:	$T \pm Z\sigma_{\tau} = T \pm Z\sqrt{\sum_{i=1}^{m} N_i^2\left(\frac{s_i^2}{n_i}\right)\left(\frac{N_i - n_i}{N_i}\right)}$
Part 3: Population parameter — proportion (ρ)	
Best point estimate:	$\hat{\rho} = p = \frac{1}{N}\sum_{i=1}^{m} N_i p_i = \frac{1}{N}\sum_{i=1}^{m} N_i\frac{x_i}{n_i} = \frac{1}{N}\left[N_1\frac{x_1}{n_1} + \ldots + N_m\frac{x_m}{n_m}\right]$
Standard error of the point estimate (sampling error):	$\hat{\sigma}_p = \sqrt{\frac{1}{N^2}\sum_{i=1}^{m} N_i^2\left(\frac{p_i(1 - p_i)}{n_i - 1}\right)\left(\frac{N_i - n_i}{N_i}\right)}$
Confidence interval (bound) around the point estimate:	$p \pm Z\hat{\sigma}_p = p \pm Z\sqrt{\frac{1}{N^2}\sum_{i=1}^{m} N_i^2\left(\frac{p_i(1 - p_i)}{n_i - 1}\right)\left(\frac{N_i - n_i}{N_i}\right)}$

* Exclude the finite population correction $\left(\frac{N-n}{N}\right)$ and all $\left(\frac{N_i - n_i}{N_i}\right)$ from the confidence interval equations if $n/N < .05$. Replace Z with the corresponding t if $n < 30$.
m = number of strata; subscript i refers to stratum i; N_i = size of population stratum i; n_i = size of sample from stratum i; s_i^2 = variance of sample i; N = total population of all strata ($N = N_1 + N_2 + \ldots + N_m$). [In part 3 only: p_i = sample proportion with characteristic of stratum i].

$Z = 1.96$, while $t = 2.04$. This larger confidence interval for t is to be expected because the smaller the sample, the larger the sampling error and the greater the uncertainty that the sample precisely represents the population from which it is drawn.

What if the sample size is large relative to the population size? You may sometimes encounter the fortunate situation where your sample size is relatively large compared to the population size. This is a fortunate situation because having a relatively large sample drawn from a relatively small population allows you to add what is called the **finite population correction (fpc)** to the confidence interval formula. Including the fpc when calculating the confidence interval allows you to narrow the confidence interval width and derive a more precise sample. Since narrowing the confidence interval width is always a good thing, the fpc should be included in the confidence interval formula whenever possible.

All of the confidence interval formulas are shown in Tables 8.4 and 8.5, and the last portion of each confidence interval formula is the fpc component:

$$\text{fpc} = \left(\frac{N-n}{N-1}\right) \quad \textbf{(8.8)}$$

where: fpc = finite population correction
N = population size
n = sample size

The **sampling fraction** of a problem is defined as the ratio of sample size to population size: (n/N). Whenever this sampling fraction is large, the finite population correction should be included in the confidence interval equation. In general, the fpc value is included in population estimation equations when the ratio of sample size to population size exceeds 5% ($n / N > .05$). In the next section of this chapter (Section 8.3) the only situation that includes the fpc is situation 6.

Whenever the sampling fraction is less than 5%, the fpc portion of the confidence interval formula should not be included, as it will have only a negligible factor. In the next section, the fpc is not included in the confidence interval equations for situation 1–5.

All of the components are now in place to calculate various confidence intervals. However, deciding which one to select depends on two factors:

- Different *sample types* require different confidence interval equations. The following discussion of confidence interval calculation will be limited to three sample types—random, systematic, and stratified.
- Different *population parameters* require different confidence interval equations. The following discussion will be restricted to three parameters—mean, total, and proportion.

These different sample types and population parameters are combined to create the six basic situations shown in Tables 8.4 and 8.5 with regard to confidence intervals. Each of these basic situations is illustrated with a geographic example. Other situations focusing on additional sampling designs (such as cluster, composite, or hybrid samples) are discussed in advanced texts that cover sampling more extensively. If you want to find out more about sample estimation, refer to the readings at the end of the chapter.

8.3 Geographic Examples of Confidence Interval Estimation

The following examples are all placed in a very practical geographic context. A realistic set of scenarios has been created utilizing representative values from the 2010 Census of Population and other reliable sources. Each of the following six "situations" is designed to be a practical problem or issue that could occur in any community. With each situation, the emphasis is on the geographic characteristics of the problem and related policy considerations. Following one of the main themes of the book, our hope is to demonstrate the strong linkages between informed geographic decision-making and the use of statistics.

According to the 2010 U.S. Census of Population, the fictitious city of Middletown (Figure 8.6) has a population of 20,340 and is the county seat of Clinton County (104,500 total population). For various administrative and economic reasons, Middletown has collected official statistics for the last few decades from each of their four districts: Northside (pop. 2,444), Central (pop. 9,498), Easton (pop. 4,910) and Southside (pop. 3,528). Just outside the Middletown city limits is the planned community of Parkwood Estates (pop. 438), whose residents will be deciding in the next election whether to approve or reject annexation into Middletown (Figure 8.7).

Situation 1: Random or Systematic Sample—Estimate of Population Mean

Like many other communities, a critical concern of many Middletown residents is the rapidly rising cost of household energy expenditures. According to a recent Residential Energy Consumption Survey (RECS) conducted by the U.S. Energy Information Administration, the average annual energy expenditure per household nationwide is nearly $2,000. Local officials suspect energy costs in Middletown may be even higher than the national average. Community officials and a number of citizen organizations have been supporting various federal and state voluntary initiatives to promote energy conservation and reduce energy consumption. These programs include improved (more energy-conserving) building codes, improved appliance standards and incentives, expanded weatherization assistance, and retrofit incentives (with

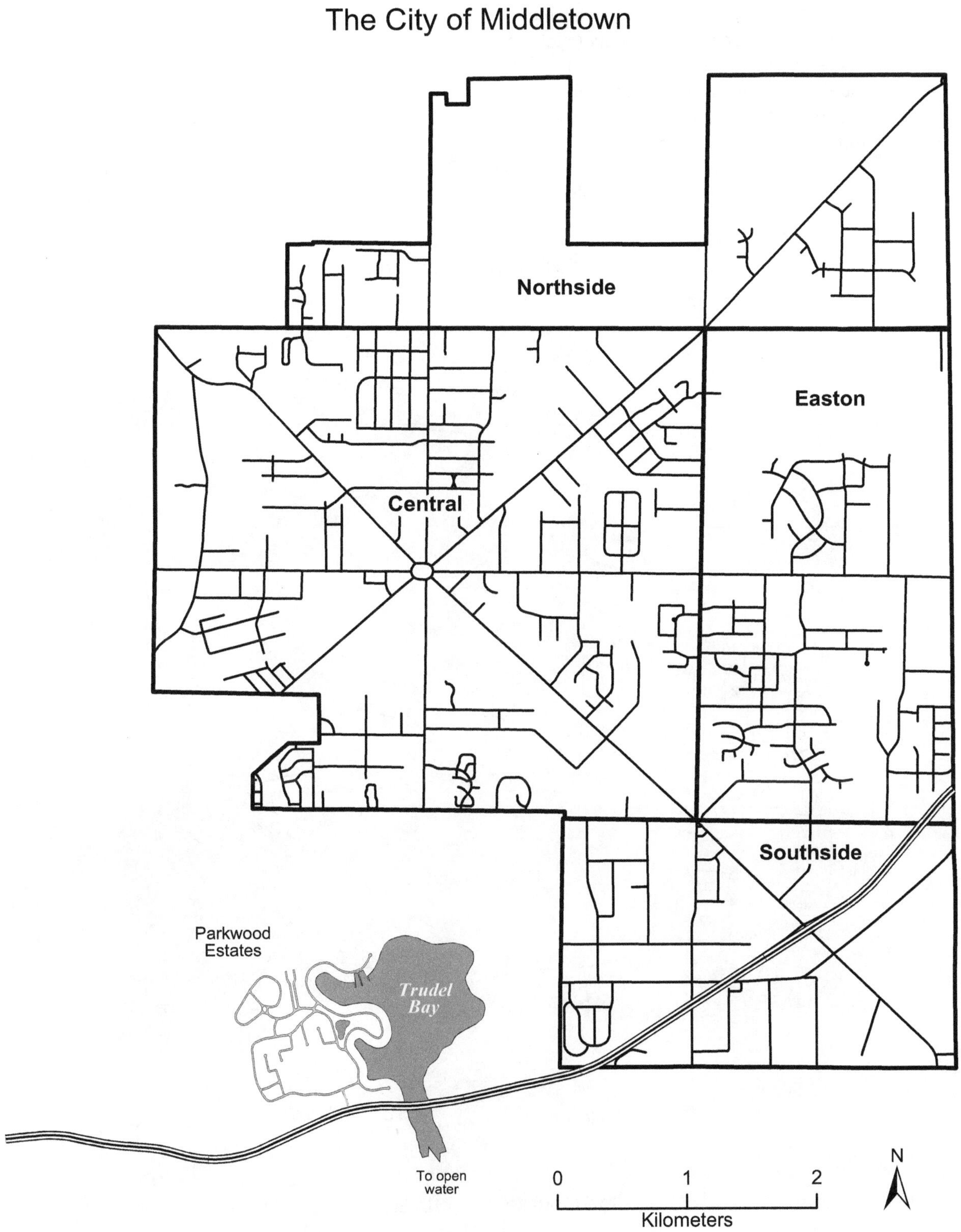

FIGURE 8.6
The City of Middletown

increased equipment standards). To date, however, these community efforts have had negligible results.

As part of a newly-created local residential energy conservation program, Middletown officials want to estimate the average annual energy expenditures per household. The hope is that this information, along with more detailed monitoring of energy cost components (such as energy expense by specific appliance) will allow them to recommend specific energy reduction strategies for local households that will best reduce their overall energy costs.

To estimate and place bounds on the average annual energy expenditures per household, Middletown planners take a simple random or systematic sample of 25 households. From the sample data, the sample mean and variance are $\bar{X}$ = \$1,810 and s^2 = 38,416. They wish to determine the confidence interval that contains the true average annual energy expenses per household for Middletown, with a 90% certainty ($1 - \alpha = .90$). The relevant values for this problem are:

$$\bar{X} = 1{,}810 \quad s^2 = 38{,}416 \quad N = 8{,}306 \quad n = 25 \quad \alpha = .10$$

Since $n < 30$, t rather than Z should be used in the confidence interval equation (the top portion of Table 8.4). Since $\alpha = .10$ and degrees of freedom $= n - 1 = 24$, the value from the t table is 1.71. Because the sampling

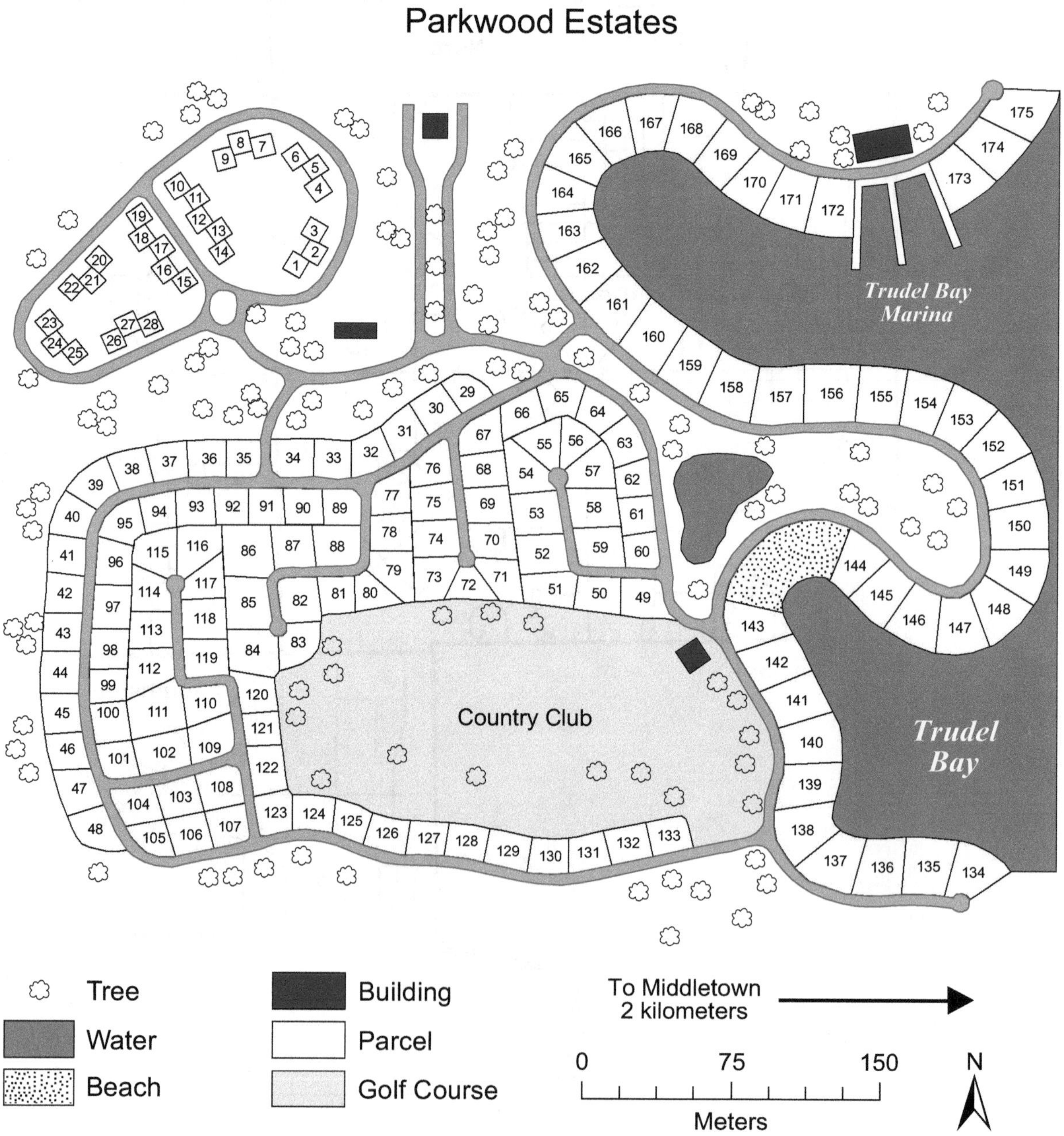

FIGURE 8.7
The Planned Community of Parkwood Estates

fraction (n / N) is less than .05, the finite population correction can be ignored. The confidence interval in this problem is calculated as

$$\bar{X} \pm t\sqrt{\frac{s^2}{n}} \qquad \textbf{(8.9)}$$

$$\bar{X} \pm t\sqrt{\frac{s^2}{n}} = 1,810 \pm 1.71\sqrt{(38,416/25)} = 1,810 \pm 67.03$$

Middletown planners are 90% certain that the true mean energy expenditures per household falls within the interval from 1,742.97 to 1,877.03.

This is a fairly narrow confidence interval, but it can be made even more precise if decision-makers think the additional precision will allow them to make even better energy policy recommendations to Middletown households. Two strategies are available to narrow the interval: (1) lower the confidence level from .90 and/or (2) increase the sample size above 25. The latter strategy requires more effort, but the larger sample size will permit the confidence interval to be narrowed.

Suppose the decision is made to increase the number of households surveyed from 25 to 250. If the level of confidence is kept at .90, the resulting interval should be considerably more precise. Suppose the mean and variance of this larger survey of households are calculated from the sample data as $\bar{X}$ = 1,821 and s^2 = 36,864. Now, with $n > 30$, Z rather than t is used in the confidence interval formula. Also, with α = .10, the value from the normal table becomes Z = 1.65. Once again, the finite population correction (fpc) is not useful because the sampling fraction n / N is still less than .05. The revised confidence interval is calculated as

$$\bar{X} \pm Z\hat{\sigma}_{\bar{X}} = \bar{X} \pm Z\sqrt{\frac{s^2}{n}} \qquad \textbf{(8.10)}$$

$$\bar{X} \pm Z\sqrt{\frac{s^2}{n}} = 1,821 \pm 1.65\sqrt{\frac{36,864}{250}} = 1,821 \pm 20.04$$

With this larger sample, planners are now 90% certain that the true mean household energy costs in Middletown (μ) falls within the very narrow interval from \$1,800.96 to \$1,841.04.

Situation 2: Random or Systematic Sample—Estimate of Population Total

Middletown planners can now use the best point estimate of average annual energy expenditures per household to determine the total amount of energy costs for all residential households in their community. The best estimate of the population total $(\hat{\tau})$ is the sample total (T), which is the sample mean $(\bar{X})$ multiplied by the population size (N):

$$\hat{\tau} = T = N\bar{X} \qquad \textbf{(8.11)}$$

For the Middletown problem, if the smaller sample size of (n = 25) is being used, the best estimate of the population total is

$$\hat{\tau} = T = 8,306(1,810) = 15,033,860$$

Based on this best point estimate, Middletown households spend slightly more than \$15 million annually to meet their collective residential energy needs. Using the equations in the middle portion of Table 8.4, the confidence interval around the sample total can be determined by:

$$T \pm Z\sqrt{N^2\left(\frac{s^2}{n}\right)} \qquad \textbf{(8.12)}$$

As in situation 1, the sampling fraction is less than .05 and the fpc can be ignored. Also, remember that when the sample size is less than 30, t rather than Z should be used. Therefore, when α = .10 and t = 1.71, the confidence interval for the total household energy use in Middletown is calculated as

$$T \pm t\sqrt{N^2\left(\frac{s^2}{n}\right)} = 15,033,860 \pm 1.71\sqrt{(8306)^2\left(\frac{38416}{25}\right)}$$
$$= 15,033,860 \pm 556,767.79$$

Planners are 90% certain that the true total amount of energy costs in all community residential households (τ) is found within the interval from \$14,477,092 to \$15,590,627. Even though the sample size was only 25 households, the interval appears narrow enough to allow Middletown planners to assess the total energy costs for the community.

Situation 3: Random or Systematic Sample—Estimate of Population Proportion

The recent nationwide meltdown in the real estate market has placed extraordinary pressures on communities to meet their financial obligations. As homeowners default on mortgages, the number of properties on the tax roll diminishes, along with the revenues generated by those taxes. Not only is housing foreclosure a problem, but many residential properties with a mortgage are now finding themselves "underwater" as the value of their homes has declined. This means many homeowners are now experiencing negative equity, as the amount owed on their mortgage is greater than the current value of their home.

The spatial pattern of foreclosures in the United States is very highly concentrated. According to RealtyTrac, more than half of the nation's foreclosures in 2008 took place in only 35 counties. This dramatic clustering is a sign that the financial crisis that devastated the US economy probably began with collapsing home loans in only a few scattered places. A few of the 35 "crisis counties" were in the already depressed areas in and around Cleveland and Detroit. But nearly all of the most severely stressed places were concentrated in Southern California, Las Vegas, South Florida, Phoenix, and Washington. These are locations where home values impressively increased in the years 2000–2007, but then plummeted dramatically later in the decade.

The nationwide pattern of negative equity and "underwater" households (those who owe more in a mortgage than the value of their house) is much more widely dispersed. The most recent data available as this is written (the fourth quarter of 2011) estimates that over 11 million residential properties were in a condition of negative equity. That is nearly 23% of all residential properties carrying a mortgage. The states with the highest negative equity percentages are Nevada (where 61% of all mortgaged properties are underwater), Arizona (48%), Florida (44%), Michigan (35%), and Georgia (33%).

Though not in one of these "problem states," officials from fictitious Middletown are concerned that their percentage of households with mortgages currently "underwater" is greater than the national average. From the tax records and other sources, it is known that there are 8,306 occupied housing units in town. If Middletown follows the national average, slightly over 68% of all occupied housing units are owner-occupied and nearly 32% are renter-occupied. Middletown officials also know that about 70% of all owner-occupied housing units nationwide are carrying a mortgage.

Armed with this general background information, planners estimate that nearly 4,000 Middletown owner-occupied housing units have a mortgage. The goal is to take a random or systematic sample of these housing units to estimate what proportion are currently "underwater." Eighty-five owner-occupied housing units having a mortgage are sampled, and 25 of these are found to have negative equity.

The best point estimate of the population proportion $(\hat{\rho})$ is the sample proportion (p), which is the number in the sample having the specified characteristic (x) divided by the total sample size (n).

$$\hat{\rho} = p = \frac{x}{n} \tag{8.13}$$

The best estimate of the proportion of Middletown owner-occupied households that are underwater is

$$\hat{\rho} = p = \frac{x}{n} = \frac{25}{85} = .294$$

Using the equations in the bottom portion of Table 8.4, the confidence interval around this population estimate of the proportion is calculated as

$$p \pm Z\sqrt{\left(\frac{p(1-p)}{n-1}\right)} \tag{8.14}$$

As in situations 1 and 2, the sampling fraction is less than .05 and the fpc can be ignored.

To enclose the true population proportion with 95% confidence ($1 - \alpha = .95$ and $Z = 1.96$), the confidence interval is:

$$.294 \pm 1.96\sqrt{\left(\frac{.294(.706)}{85-1}\right)} = .294 \pm .096$$

Middletown planners can conclude with 95% certainty that ρ, the true proportion of owner-occupied housing units with a mortgage that currently has negative equity, is in the interval from .198 to .390. This is a very wide confidence interval, and would undoubtedly be a disappointing result for Middletown officials. Their choices for narrowing the confidence interval width are to increase the sample size, reduce the confidence level (perhaps to 80%), or both.

Situation 4: Stratified Sample—Estimate of Population Mean

Members of the Board of Education in fictitious Clinton County wish to evaluate the "geographic competencies" of all high school juniors in the County school district. One group of students has completed an introductory geography course taught in several of the County high schools, whereas another group of students has not. School officials suspect that the geographic competencies of these two groups of juniors are quite different. If a simple random sample is taken of all juniors, either of the groups could be over- or under-represented in the sample. Therefore, a stratified sample design with proportional representation of each group of students is appropriate.

To evaluate students' geographic skills, school officials decide to use the Secondary-Level Geography Test designed by the National Council for Geographic Education (NCGE). This test evaluates a student's basic geographic knowledge in three areas: geographic skills, physical geography, and human geography. Of 2,160 total juniors in the school district, 480 have completed the geography course and 1,680 have not. School officials have only 90 test booklets from NCGE and must restrict their sample to 90 students. The general procedure for estimating the mean from a stratified sample is shown in the top portion of Table 8.5 and the worktable showing the best estimate and confidence interval for the test scores of all Clinton County students is shown in Table 8.6.

Notice that the overall best estimate of the true population mean is the sample mean of 54.11. This 54.11 value is actually the weighted mean of the two sample means (64.36 for the 480 students who have taken the geography course and 51.18 for the 1,680 students who have not taken the course). Because proportional representation of these two student groups is necessary to have an unbiased sample, it is not surprising that the much larger group of students who did not complete the geography course would influence the overall result much more (with a heavier weighting) than the group of students completing the course. However, we can be assured that this stratified sample result is more likely to be closer to the true population mean (which is unknown) than would be a simple random sample of the same size.

Situation 5: Stratified Sample—Estimate of Population Total

The Middletown City Council has just held a public meeting presenting the results of the energy expenditures sample survey. Recall from situations 1 and 2 that best estimates and confidence intervals were placed around the average annual energy costs per household and the total annual energy costs for all residential units in Middletown. The vast majority of citizens attending the meeting expressed concern about the limited usefulness of the sample results.

A number of criticisms were aired, but several complaints seemed to predominate: (1) the sample of households was too small; (2) the best point estimates seemed

TABLE 8.6

Confidence Interval Estimate of Stratified Sample Mean: Secondary-Level Geography Test Results for Clinton County Students

Task: Estimate the mean test score of all high school juniors on the Secondary-Level Geography Test, based on a stratified sample, and place a 95% ($\alpha = 0.5$) confidence interval (bounds) around the estimate.

Stratum 1 Students who have taken introductory geography courses		**Stratum 2** Students who have not taken introductory geography courses	
$N_1 = 480$	$n_1 = 20$	$N_2 = 1680$	$n_2 = 70$
$\bar{X}_1 = 64.36$	$s_1^2 = 65.61$	$\bar{X}_2 = 51.18$	$s_2^2 = 90.25$

The best estimate of the true population mean (μ) is the stratified sample mean ($\bar{X}$):

$$\hat{\mu} = \bar{X} = \frac{1}{N}\sum_{i=1}^{m} N_i \bar{X}_i$$

$$= \frac{1}{2160}[480(64.34) + 1680(51.18)] = 54.11$$

The confidence interval around $\bar{X}$ is:

$$\bar{X} \pm Z\hat{\sigma}_{\bar{X}} = \bar{X} \pm Z\sqrt{\frac{1}{N^2}\sum_{i=1}^{m} N_i^2\left(\frac{s_i^2}{n_i}\right)\left(\frac{N_i - n_i}{N_i}\right)}$$

Exclude the finite population correction $\left(\frac{N_i - n_i}{N_i}\right)$ as $(n/N) < .05$ for each stratum:

$$\bar{X} \pm Z\sqrt{\frac{1}{N^2}\sum_{i=1}^{m} N_i^2\left(\frac{s_i^2}{n_i}\right)}$$

$$= 54.11 \pm 1.96\sqrt{\frac{1}{2160^2}\left[(480)^2\left(\frac{65.61}{20}\right) + (1680)^2\left(\frac{90.25}{70}\right)\right]}$$

$$= 54.11 \pm 1.96$$

With 95% certainty, the mean test score of all high school juniors, μ, *is* within the interval from 52.21 to 56.01.

to be reasonable, but the confidence intervals around those estimates were so wide that it seemed to make the predictions basically useless (this criticism was particularly aimed at the total expenditures of all households in town on energy needs); (3) some people were concerned that the sample very likely did not reflect the complexity of the types of housing units in Middletown. Citizens know that the city has a wide variety of housing types which probably have very different energy needs (single-family detached, single-family attached, large and small apartments, and even some mobile homes) and the simple random sample procedure used in the energy expenditures survey didn't account for this at all; and (4) other people were upset that the sample might not have adequately represented their neighborhood (recall that Middletown has four districts—Northside, Central, Easton, and Southside) and the different demographic profiles that exist in each district which would certainly be reflected in various housing characteristics (year of construction, total square feet of floor space, owner-occupied versus renter-occupied, and so on). For example, one person living in the Central district pointed out that her neighborhood was mostly rental units, with lots of apartments and relatively few single-family homes of any kind. A resident in Easton mentioned that his district had the city's only large mobile home park.

TABLE 8.7

Confidence Interval Estimate of Stratified Sample Total: Energy Expenditures for Middletown Households

Stratum 1 Northside	Stratum 2 Central	Stratum 3 Easton	Stratum 4 Southside
$N_1 = 873$	$N_2 = 4{,}112$	$N_3 = 1{,}964$	$N_4 = 1{,}357$
$n_1 = 21$	$n_2 = 99$	$n_3 = 47$	$n_4 = 33$
$\bar{X}_1 = 1{,}910$	$\bar{X}_2 = 1{,}680$	$\bar{X}_3 = 1{,}840$	$\bar{X}_4 = 1{,}885$
$s_1 = 194$	$s_2 = 173$	$s_3 = 192$	$s_4 = 181$
$s_1^2 = 37{,}636$	$s_2^2 = 29{,}929$	$s_3^2 = 36{,}864$	$s_4^2 = 32{,}761$

$N = 8{,}306 \qquad n = 200 \qquad \bar{X} = 1{,}816 \qquad m = 4$

$$\text{BPE} = T = \sum_{i=1}^{m} N_i\bar{X}_i = 873(1{,}910) + 4{,}112(1{,}680) + 1{,}964(1{,}840) + 1{,}357(1{,}885) = 14{,}747{,}295$$

If 90% confidence level, then $Z = 1.65$ (no finite population correction)

$$T \pm Z\hat{\sigma}_T \sqrt{\sum_{i=1}^{m} n_i^2\left(\frac{s_i^2}{n_i}\right)}$$

$$14{,}747{,}295 \pm 1.65\sqrt{(873)^2\left(\frac{37{,}636}{21}\right) + (4{,}112)^2\left(\frac{29{,}929}{99}\right) + (1{,}964)^2\left(\frac{36{,}864}{47}\right) + (1{,}357)^2\left(\frac{32{,}761}{33}\right)}$$

$$14{,}747{,}295 \pm 1.65\sqrt{136{,}588{,}750{,}000 + 511{,}145{,}280{,}000 + 302{,}527{,}720{,}000 + 182{,}819{,}050{,}000}$$

$$14{,}747{,}295 \pm 1.65\sqrt{1{,}133{,}080{,}800{,}00}$$

$$14{,}747{,}295 \pm 1.65\,(1{,}064{,}463)$$

$14{,}747{,}295 \pm 1{,}756{,}364$ 90% confidence from 12,990,931 to 16,503,659

With 90% certainty, total energy expenditures for all Middletown households is within the interval from $12,990,931 to $16,503,659.

After considerable discussion, it was decided that if any practical policies were ever going to be implemented, another energy expenditures survey was needed. All steps in the sampling procedure were reexamined and a new sampling protocol was designed. A district-level proportional (constant-rate) stratified sample procedure was selected, with the number of households in each stratum (district) to be proportionally represented. It was also decided to increase the total sample size to 200 households. The general procedure for estimating the total energy costs from a stratified sample is shown in the middle portion of Table 8.5 and the associated worktable calculations are summarized in Table 8.7.

The best point estimate of total annual energy costs for all residential units in Middletown is $14,747,295. We can be 90% certain that the true amount of energy

TABLE 8.8

Confidence Interval Estimate of Stratified Sample Proportion: Parkwood Residents Opinion Regarding Annexation into Middletown

	Stratum 1 Apartments	Stratum 2 Homes not on water	Stratum 3 Homes on water
	$N_1 = 28$	$N_2 = 105$	$N_3 = 42$
	$n_1 = 8$	$n_2 = 30$	$n_3 = 12$
Opinion regarding annexation	(X) Yes = 5	(X) Yes = 22	(X) Yes = 4
	No = 3	No = 8	No = 8

$N = 175$ $n = 50$ (X) Yes = 31 No = 19 $m = 3$

$$P_1 = \frac{X_1}{n_1} = \frac{5}{8} = .625 \qquad P_2 = \frac{X_2}{n_2} = \frac{22}{30} = .733 \qquad P_3 = \frac{X_3}{n_3} = \frac{4}{12} = .333$$

$$\text{BPE} = \frac{1}{N}\sum_{i=1}^{m} N_i p_i = \frac{1}{175}[28(.625) + 105(.733) + 42(.333)] = .620$$

If 90% confidence level, then Z = 1.65 (include fpc, as $\left(\frac{n}{N}\right) > .05$)

$$p \pm Z\hat{\sigma}_p = p \pm Z\sqrt{\frac{1}{N^2}\sum_{i=1}^{m} N_i^2\left(\frac{p_i(1-p_i)}{n_i-1}\right)\left(\frac{N_i-n_i}{N_i}\right)}$$

$$.620 \pm 1.65\sqrt{\frac{1}{175^2}\left[(28)^2\left(\frac{(.625)(.375)}{7}\right)(.714) + (105)^2\left(\frac{(.733)(.267)}{29}\right)(.714) + (42)^2\left(\frac{(.333)(.667)}{11}\right)(.714)\right]}$$

$$.620 \pm 1.65\sqrt{\frac{1}{30{,}675}[(784)(.03348)(.714) + (11{,}025)(.00675)(.714) + (1{,}764)(.0202)(.714)]}$$

$$.620 \pm 1.65\sqrt{\frac{1}{30{,}675}[18.7413 + 53.1350 + 25.4418]}$$

$$.620 \pm 1.65\sqrt{.003173}$$

$$.620 \pm 1.65\,(.0563293)$$

$.620 \pm .0929$ confidence interval from .5271 to .7129

consumed by all residential units in the community is within the confidence interval from $12,990,931 to $16,503,659. This is certainly a much more precise estimate than was obtained in situation 2, which was limited by the small and simple random sample.

Situation 6: Stratified Sample—Estimate of Population Proportion

Yet another issue is facing some of the citizens of Clinton County. The planned community of Parkwood Estates is located just outside the city limits of Middletown (Figures 8.6 and 8.7). Residents of Parkwood Estates will have the opportunity in the next election to decide whether they want to be annexed into the City of Middletown. Although both positive and negative attitudes have been expressed about this decision, no specific information has been collected from Parkwood residents regarding their opinions about annexation.

It is recommended that a survey of Parkwood residents be taken to ascertain their views about this question. Specifically, planners would like to know what proportion of Parkwood residents are in favor of annexation. The entire community contains 175 residential units—too many to interview with the limited amount of time, money, and personnel available. Based on knowledge of the area, it is known that the community has 28 apartment units (Karla Court), 105 homes on lots without direct water access, and 42 lots with direct water access. The water-accessible properties are located in a separate part of the development along the shoreline of Trudel Bay. It is felt that a proportional stratified sample is appropriate because it is likely that residents of these three portions of Parkwood Estates might have different opinions regarding annexation. It is common knowledge that the three groups have different demographic and economic profiles (income, age, square footage of property, etc.). Clearly it is important to ensure that none of the three groups is over- or under-represented in the sample.

There is enough time and personnel to collect opinions about annexation from a stratified sample of 50 households. The general procedure for estimating the proportion from a stratified sample is shown in the bottom portion of Table 8.5 and the calculations are summarized in Table 8.8.

Because our sample (50 households) is greater than 5% of the population (175 households), it is advantageous to include the finite population correction (fpc) in the confidence interval estimation. Remember, including the fpc when the sampling fraction is large will improve the precision of our sample estimates—a desirable outcome.

The results indicate that 62.0% of the Parkwood Estates households sampled are in favor of annexation into Middletown. However, the confidence interval around this best estimate is rather wide, ranging from 52.71–71.29%. The somewhat imprecise confidence interval is due to the divergent opinions regarding annexation between the three portions of the development. Unlike the other two strata, homeowners on the water (lakefront) seem to be opposed to annexation. Notice that only four of twelve households sampled are in favor of annexation.

Despite the differences of opinion within Parkwood Estates, we can be very confident that annexation into Middletown will be approved when put to the vote. The entire range of the 90% confidence interval is above 50% approval of annexation, so it is quite unlikely that annexation will be defeated. In fact, the lower end of the confidence interval is well above 50% (at 52.7%), and only 5% of the range of possible values is below the estimated confidence interval, so we can be well over 95% confident that annexation will be approved when voted upon by Parkwood residents.

8.4 Sample Size Selection

In problems using sampling, geographers often want to determine the minimum sample size needed to make sufficiently precise estimates *before* the complete sample is actually taken. Taking a sample larger than necessary wastes both time and effort. Some of the major factors to consider in selecting sample size are:

- the type of sample (random, stratified, etc.);
- the population parameter being estimated (mean, total, proportion);
- the degree of precision (width of confidence interval that can be tolerated);
- the level of confidence to be obtained for the estimate.

As sample size increases, key trade-offs occur. At a particular confidence level (.95, for example), increasing the sample size provides greater precision and narrows the confidence interval width around the population estimate. Similarly, at a particular level of precision (e.g., estimating the population proportion within .03 of its true value), increasing the sample size will raise the level of confidence that the estimate is within the selected interval.

Unfortunately, a larger sample generally requires more time and effort. In many practical sampling problems, you are likely to have multiple conflicting objectives. You must take a sample large enough to achieve the desired precision level and confidence interval width, but *simultaneously* avoid taking too large a sample, which wastes time and effort and provides estimates more precise than necessary. When samples get much larger than needed, the extra effort yields smaller and smaller incremental improvements in precision. In other words, the extra effort is more costly and yields less "bang for the buck."

This section illustrates how an appropriate sample size is determined from random sampling for the three basic population parameters—mean (μ), total (τ), and proportion (ρ).

Sample Size Selection—Mean

Suppose your task is to determine the minimum sample size needed to place the population *mean* estimate within a desired confidence interval around the true population mean. After choosing a confidence interval having an acceptable range, the issue becomes finding the appropriate sample size needed to place the sample statistic within that range. Simply stated, how large a sample is needed to get within the stated range of the actual population mean?

Recall that the confidence interval for μ is $\bar{X} \pm Z\hat{\sigma}_{\bar{X}}$, where $\hat{\sigma}_{\bar{X}}$ is the standard error of the mean (sampling error). Suppose E (for **Error**) represents the amount of sampling error we are willing to tolerate. That is, for any particular problem, a maximum acceptable difference separates the sample mean statistic from its population mean:

$$E = Z\hat{\sigma}_{\bar{X}} = Z\sqrt{\frac{\sigma^2}{n}} \tag{8.15}$$

The magnitude of tolerable sampling error depends on the circumstances of the problem. In some applications, a considerable amount of error can be tolerated, making E a large number. In other instances, only a limited amount of error can be accepted, so E must be a smaller number. For example, a political party taking a survey many months before an election only wants a "rough estimate" of a candidate's popularity and is willing to accept an E value of large magnitude. Just a few days before the election, however, the party wants a more precise estimate of the expected election outcome, so a small-sized E value is desired.

The tolerable sampling error (E) is equivalent to one-half the width of the confidence interval (Figure 8.8).

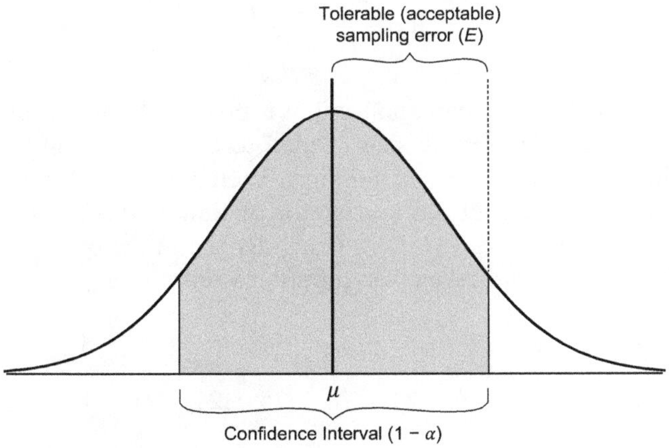

FIGURE 8.8
Relationship Between Confidence Interval and Tolerable (Acceptable) Sampling Error (E)

After some algebraic manipulation, the sample size (n) needed to estimate μ with a certain level of precision or tolerable error (E), at a chosen level of confidence (Z), can be expressed as

$$E = Z\sqrt{\frac{\sigma^2}{n}}$$

$$E^2 = Z^2\left(\frac{\sigma^2}{n}\right) \tag{8.16}$$

$$nE^2 = Z^2\sigma^2$$

$$n = \frac{Z^2\sigma^2}{E^2} = \left(\frac{Z\sigma}{E}\right)^2$$

If the population standard deviation (σ) is unknown, the sample standard deviation (s) is substituted:

$$n = \left(\frac{Zs}{E}\right)^2 \tag{8.17}$$

The required minimum sample size is directly related to the desired level of confidence (Z) and variability in the sample (s), but inversely related to the degree of error (E) you are willing to tolerate. In the political example, this relationship suggests that sample size will need to be increased in the second polling taken just a few days before the election to ensure that sampling error (confidence interval range) is reduced to an acceptable level.

In most real-world examples, the population standard deviation is not known. The sample standard deviation value is usually substituted to determine an appropriate sample size. A value for s is best derived from a pretest or preliminary sample. However, s could also be obtained from some previous study, or, if no better alternative exists, s could be an "educated guess" based on past experience.

Since a population parameter is being estimated, at least 30 observations should be included in a pretest or preliminary sample. You should return to the general population to draw additional sample units if needed. These new units can then be combined with the original sample units to create a single larger sample that meets the size requirement. If this "return" procedure is used, it is considered a **two-stage sampling design**.

Example of Sample Size Selection—Mean

Middletown planners wish to estimate the *mean* annual energy expenditures per household and be 90% confident that their estimate will be within $50 of the true population mean (μ). What is the minimum number of households that must be surveyed to ensure this degree of precision at this selected confidence level?

(continued)

Suppose a preliminary sample of 30 households is drawn, and *s* is calculated as $210. Since the population standard deviation is unknown, Equation 8.18 is used:

$$n = \left(\frac{Zs}{E}\right)^2 = \left(\frac{(1.65)(210)}{50}\right)^2 = 48.02$$

Therefore, a random sample of at least 49 households should be taken to ensure the desired degree of precision at the 90% confidence level. It is always better to include an extra observation or two above the absolute minimum necessary sample size, so rounding up to 49 households is appropriate. Since this result was obtained from a preliminary sample of 30 households in a two-stage sampling design, only 19 additional households need be contacted to complete the study satisfactorily.

Sample Size Selection—Total

The minimum sample size needed to make an interval estimate of a population *total* within a certain tolerable error level (E) can also be determined. The confidence interval for τ is $T \pm Z\hat{\sigma}_T$, where $\hat{\sigma}_T$ is the sampling error or standard error of the total:

$$E = Z\hat{\sigma}_T = Z\sqrt{N^2\left(\frac{\sigma^2}{n}\right)} \quad \textbf{(8.18)}$$

Algebraic manipulation isolates (n), the minimum sample size needed to estimate τ with a selected level of tolerable error (E) at a chosen confidence level (Z):

$$n = \left(\frac{NZ\sigma}{E}\right)^2 \quad \textbf{(8.19)}$$

If the population standard deviation (σ) is unknown, the sample standard deviation (s) is substituted:

$$n = \left(\frac{NZs}{E}\right)^2 \quad \textbf{(8.20)}$$

Once again, the population standard deviation is seldom known and must be estimated with the sample standard deviation, which generally requires a pretest or preliminary sample.

Example of Sample Size Selection—Total

Middletown planners also want to estimate the *total* amount of energy expenditures for all residential households in the community. Suppose they wish to estimate total energy costs within $500,000 and be 90% confident with that level of precision and sampling error (*E*). What is the minimum number of households that must be surveyed?

From a pretest or preliminary survey of 30 households, the standard deviation of the annual energy expenditures per household is calculated as $215. Because the population standard deviation is not known, the sample size needed to meet the requirements of the problem is calculated from Equation 8.20:

$$n = \left(\frac{NZs}{E}\right)^2 = \left(\frac{(8{,}306)(1.65)(215)}{500{,}000}\right)^2 = 34.71$$

A random sample of at least 35 households should be taken to estimate the total energy cost within $500,000 and be 90% confident in that level of precision. Since 30 households were already surveyed in the preliminary survey, only another 5 households need to be contacted in the second stage of the survey design.

Sample Size Selection—Proportion

To estimate a population *proportion* within a certain allowable level of error (E), the minimum sample size can also be calculated in advance of full sampling. Again, a pretest or preliminary survey can be used to estimate the population proportion (ρ) from the sample proportion (p). The confidence interval for ρ is $p \pm Z\hat{\sigma}_p$, where $\hat{\sigma}_p$ is the sampling error or standard error of the proportion:

$$E = Z\hat{\sigma}_\rho = Z\sqrt{\frac{\rho(1-\rho)}{n-1}} \quad \textbf{(8.21)}$$

You can isolate the minimum sample size (n) algebraically as:

$$n = \frac{Z^2\rho(1-\rho)}{E^2} \quad \textbf{(8.22)}$$

If the population proportion (ρ) is unknown, the sample proportion (p) is substituted:

$$n = \frac{Z^2 p(1-p)}{E^2} \quad \textbf{(8.23)}$$

Unlike the mean and total, it is possible to determine the minimum sample size needed to estimate a population proportion *without* taking a pretest or preliminary sample. Note that the numerator of Equation 8.23 contains the product $p\,(1 - p)$. Consider the range of values this product can take given different values of p:

p	.1	.2	.3	.4	.5	.6	.7	.8	.9
$p(1-p)$	.09	.16	.21	.24	.25	.24	.21	.16	.09

The maximum product of .25 occurs when $p = .5$ and $(1 - p) = .5$. Inserting .25 into Equation 8.24, the result is the largest minimum sample size needed under conditions of maximum uncertainty and represents a "worst case scenario." This value, .25, will *always* provide a large enough minimum sample size, thereby making this a popular sampling strategy:

$$\begin{aligned} n &= \frac{Z^2 p(1-p)}{E^2} \\ &= \frac{Z^2(.5)(.5)}{E^2} \\ &= \frac{Z^2(.25)}{E^2} \\ &= \left(\frac{1}{4}\right)\frac{Z^2}{E^2} \end{aligned} \tag{8.24}$$

Example of Sample Size Selection—Proportion

Middletown planners wish to estimate the *proportion* of owner-occupied housing units that are carrying a mortgage and have negative equity. They want to be 90% certain their sample statistic is within .10 (10%) of the true population proportion. To determine in advance the minimum sample size needed when estimating a population proportion (p), two options are possible:

(1) *Preliminary survey taken.* Suppose a preliminary survey of 30 housing units reveals that 30% of Middletown owner-occupied households with a mortgage are currently "underwater". Since the population proportion (p) is unknown, $p = .30$ is used with Equation 8.24 to determine the necessary sample size:

$$n = \frac{Z^2 p(1-p)}{E^2} = \frac{(1.65)^2(.30)(.70)}{(.10)^2} = 57.2$$

Thus, a random sample of at least 58 households should be taken to estimate the proportion of Middletown households that are in a condition of negative equity within .10 (10%) and be 90% confident in a result that precise. Since 30 households have already been surveyed, another 28 observations must be taken in the second stage of the survey design.

(2) *Preliminary survey not taken.* If no preliminary survey is taken, the maximum product $p(1-p) = .5(.5) = .25$ may be used in a "worst case" situation (Equation 8.24):

$$n = \frac{Z^2 p(1-p)}{E^2} = \frac{(1.65)^2(.25)(.25)}{(.10)^2} = 68.1$$

In this case, a random sample of at least 69 households should be taken to achieve the desired level of precision. A larger required minimum sample size should always be expected with this option.

Key Terms

bounds of a confidence interval (upper and lower), *124*
central limit theorem, *120*
confidence interval, *120*
confidence level, *124*
E (tolerable or acceptable sampling error), *137*
finite population correction (fpc) and fpc rule, *128*
interval estimation, *119*
point estimation, *119*
sampling distribution of sample means, *120*
sampling fraction, *128*
significance level, *124*
standard error of the mean (sampling error), *121*
two-stage sampling design, *137*

Major Goals and Objectives

If you have mastered the material in this chapter, you should now be able to:

1. Understand the basic concepts in point estimation and interval estimation, including the sampling distribution of a statistic and the central limit theorem.
2. Recognize the conditions when it is appropriate to apply the finite population correction.
3. Explain the terms "confidence interval," "significance level," and "confidence level" and the relationship between them.
4. Understand the procedure for constructing a confidence interval (bounds around the point estimate) when appropriate sample statistics are provided.
5. Understand clearly the factors involved in selecting the exact equation used to place a confidence interval around a population estimate, including type of sample taken and population parameter being estimated by the sample statistic.
6. Know how to apply the appropriate equations to calculate confidence intervals around population estimates for various geographic problems.
7. Identify the factors that need consideration when selecting sample size for a geographic problem and know how to determine the appropriate minimum sample size when presented with a geographic situation.

References and Additional Reading

Many of the references from Chapter 7 are also relevant here.

If you are interested in exploring further the geographic examples mentioned in this chapter, the following are good places to start.

- To examine the farm data presented in both Chapter 3 and this chapter, see the Lembo, A., et.al. reference in Chapter 3.

- For more information about the Residential Energy Consumption Survey (RECS), a comprehensive nationwide survey on household energy use, see the U.S. Energy Information Administration website: www.eia.gov.
- RealtyTrac is a real estate information company and an online marketplace for foreclosed and defaulted properties in the U.S. www.realtytrac.com. They constantly update the spatial patterns of foreclosures and negative equity properties.
- Additional information regarding geography competency tests is available at the National Council for Geographic Education (NCGE) website. In particular, it is worthwhile examining the Geography-for-Life material: https://ncge.org/teacher-resources/national-geography-standards/
- Considerable data is available regarding housing vacancies and home ownership rates from the U.S. Census Bureau at: www.census.gov/housing/hvs

PART 4

Inferential Problem Solving in Geography

9

Elements of Inferential Statistics

In the last chapter we introduced various concepts concerning estimation in sampling. We learned that a primary objective of sampling is to infer some characteristic of the population based on statistics derived from a sample of that population. Sample statistics are used to make point estimates of population parameters such as the mean, total, and proportion. In addition, to determine the level of precision of these point estimates, a confidence interval, or bound, is placed around the sample statistic, making it possible to state the likelihood that a sample statistic is within a certain range or interval of the population parameter. In this chapter, these ideas are extended to a form of statistical inference known as **hypothesis testing**. Using these inferential procedures, we are able to reach statistical conclusions for a wide variety of problems.

The practical application and value of inferential statistics are best understood in the context of the scientific research process. As outlined and discussed in Chapter 1 (Figure 1.1), the formulation of hypotheses and their testing through inferential statistics play a central role in the development of the science of geography. For example, hypothesis evaluation may lead to the refinement of spatial models and the development of laws and theories. In addition, conclusions from inferential testing often contribute toward the advancement of scientific research.

Methods of hypothesis testing are introduced in this chapter with examples of one-sample difference tests. A properly created sample is essential in the successful application of inferential statistics. We sometimes need to verify whether a particular sample is truly typical or representative of the population from which it is drawn. If a sample is used in further analysis or research and *does not* represent the original population accurately, then future results obtained using this sample may be flawed. For example, suppose community planners want to sample the local population to determine the level of support for constructing a new vocational-technical school. Suppose the views of upper- and lower-income residents differ significantly on this issue. If the sample happens to represent upper-income residents disproportionately, survey results are likely to be flawed because the views of lower-income residents are under-represented. Therefore, a one-sample difference test needs to be used early in the planning process to verify that the sample opinions are truly representative, thereby avoiding subsequent problems.

In other cases, we might be interested in comparing known statistics from a local sample to population parameters that are available from some external source. The issue here is to show that the sample taken from the local population differs significantly (or does not differ significantly) from some given external population. For example, suppose you want to know whether pollution levels from a local sample of sites differ significantly from national or state EPA standards (i.e., the external population) for those pollutants. If the sample of local pollution levels meets the standards, then federal or state penalties are not imposed on the community.

If the population parameter and the sample statistic are not significantly different, you can be confident that the sample is truly representative of the target population. In such cases, the chosen sample is adequate for further analysis. On the other hand, if the population parameter and sample statistic are significantly different, the sample may be inaccurate or otherwise deficient. Excessive sampling error may produce an unrepresentative sample. Human error may occur in one of the steps taken to produce the sample.

Alternatively, a particular sample may represent the situation described in the last chapter (Figure 8.5), where the mean of sample five $(\bar{X}_5)$ falls into a tail of the sampling distribution of means. That illustration demonstrates how a sample statistic calculated from a properly drawn random sample has a 10% chance of being quite different from the actual population parameter value.

An example helps clarify the need for testing the difference between a sample and population mean. Suppose local officials want to determine the attitudes of community residents toward constructing a new public swimming pool. The advisory committee wants to select a 5% sample of families at random for a telephone survey. However, because families with children are more likely to want this facility, the committee needs to make sure that their sample accurately represents the family structure in the community.

One way to measure the success of the committee's objective is to determine the number of children in each family selected in the sample. Using these data, the mean family size for the sample could be computed. This sample mean could then be compared to the average family size for the target population found in either recent census data for the community or perhaps in information from a school census. If the mean family size for the sample is not significantly different than the mean for the target population, then the committee can have confidence that their sample is truly representative of family structure in the area and can confidently base other analysis off of this sample. The critical results of the survey can then be analyzed with the knowledge that the sample is not biased.

In Section 9.1, two complementary methods of hypothesis testing are presented. The classical-traditional approach provides a solid, logical foundation for understanding hypothesis testing. The *p*-value (or *prob*-value) method of hypothesis testing builds on this solid foundation yet provides additional useful information concerning the research problem. An example problem from the previous chapter is used to illustrate these methods. The one-sample difference of means *t* test is applied as the hypothesis testing terminology and procedure is introduced.

Section 9.2 introduces the one-sample difference of proportions test. Then in Section 9.3 the discussion focuses on the circumstances or geographic problems for which inferential testing is appropriate. In addition, some issues that influence the selection of the proper inferential statistical test are briefly discussed.

9.1 Terms and Concepts in Hypothesis Testing: One-Sample Difference of Means Test

Classical or **traditional hypothesis testing** involves a formal multi-step procedure that leads from the statement of hypothesis to a conclusive statement (decision) regarding the hypothesis (Table 9.1). Hypothesis testing allows us to make inferences about the magnitude of one or more population parameters, based on sample statistics estimating those parameters. Hypotheses regarding a population parameter are evaluated using sample information, and a conclusion is reached (at some preselected significance level) about the hypotheses. Because of the nature of sampling, a measurable probability can always be assigned to the conclusions reached through statistical hypothesis testing. This logic should seem very familiar, as it is based directly on the sampling estimation and confidence interval themes we discussed in Chapter 8.

Recall in the first scenario from the previous chapter (random or systematic sample—estimate of population mean) that planners in the city of Middletown estimated the average annual energy expenditures per household. More specifically, they estimated the mean energy costs per household from two random samples—a smaller sample having only 25 households and a larger sample of 250 households. Both samples were taken from a population of 8,306 Middletown households, and a confidence interval was placed around both sample estimates. The smaller 25-household sample used the *t* table to estimate the width of the confidence interval. This procedure resulted in a 90% confidence interval of \$67.03 placed around a mean household expenditure of \$1,810. In the larger 250-household sample the mean household energy expenditure was \$1,821 and the associated *Z*-value for 90% confidence produced a considerably more precise interval of \$20.04:

$$\text{smaller sample } (n=25): \quad \bar{X} \pm t\hat{\sigma}_{\bar{X}} = 1810 \pm 67.03$$

$$\text{larger sample } (n=250): \quad \bar{X} \pm Z\hat{\sigma}_{\bar{X}} = 1821 \pm 20.04$$

From the 2009 Residential Energy Consumption Survey conducted by the Energy Information Administration, Middletown officials have learned that the aver-

TABLE 9.1

Steps in Classical/Traditional Hypothesis Testing

Step 1:	State null and alternate hypothesis
Step 2:	Select appropriate statistical test
Step 3:	Select level of significance
Step 4:	Delineate regions of rejection and nonrejection of null hypothesis
Step 5:	Calculate test statistic
Step 6:	Make decision regarding null and alternate hypothesis

age household in the United States annually spends nearly \$2,000 ($\mu = 2000$). They want to determine if average household energy costs in their community are similar to (or different from) this national figure. To answer this question, a formal hypothesis testing procedure is established and appropriate terminology and concepts regarding statistical testing are introduced.

Steps in the Classical or Traditional Hypothesis Testing Procedure

Step 1: State the Null and Alternate Hypotheses: Two complementary hypotheses of interest are the **null hypothesis** (denoted H_0) and the **alternate** or **alternative hypothesis** (denoted H_A). Consider the formulation of hypotheses concerning the mean of a population (μ). The typical claim is that μ is equal to some value, μ_H (for **hypothesized mean**). This claim of equality is called the null hypothesis, and takes the general form:

$$H_0: \mu = \mu_H$$

The null hypothesis can also be stated:

$$H_0: \mu - \mu_H = 0$$

In the latter form, attention is focused on H_0 as a statement of "no difference" between μ and μ_H, which is the case if ($\mu - \mu_H$) equals 0 (or null). The null hypothesis statement always includes the equal sign. We should note that even though the more "interesting" result might be to demonstrate that differences exist, the null hypothesis always focuses on the assumption that no differences exist.

The converse of the null hypothesis is the alternate hypothesis, H_A. The alternate hypothesis expresses the conditions under which H_0 is to be rejected and can be viewed as a positive statement of difference. The two hypotheses are mutually exclusive, for if H_0 is rejected, H_A is accepted. The alternate hypothesis takes one of two forms, depending on how the research problem is structured. In some problems, the form of H_A is **non-directional** (or **two-tailed**), while in others it is **directional** (or **one-tailed**), but H_A always consists of an inequality indicating the conditions under which H_0 is rejected. So, for example, each of the following is a valid form of H_A:

$$H_A: \mu \neq \mu_H \text{ (non-directional)}$$

$$H_A: \mu < \mu_H \text{ (directional)}$$

$$H_A: \mu > \mu_H \text{ (directional)}$$

The selection of a specific form of H_A depends on how the hypothesized difference is stated. The directional and non-directional formats offer two possibilities. In our example, we know that the nationwide average annual energy cost per household is \$2,000. We also know that in the smaller Middletown sample ($n = 25$), the average energy cost per household was estimated to be \$1,810. If our goal is to determine whether the Middletown average is different than the national average, we would state the alternate hypothesis in a non-directional form:

$$H_0: \mu = 2{,}000$$

$$H_A: \mu \neq 2{,}000$$

These statements hypothesize that mean annual energy cost for Middletown households is no different from the national average of 2,000. Obviously, \$1,810 is a different amount of money than \$2,000. However, our consideration in statistical testing is whether the magnitude of the difference can be explained by the inherent natural variability of the data. If the Middletown mean is close to this population mean, the likely conclusion is that the null hypothesis *should not* be rejected. Conversely, if the Middletown mean is not close to the national population mean, the likely conclusion is that H_0 *should* be rejected. By expressing H_A in this form, the direction of difference between household energy costs in Middletown and the United States is not important. That is, the Middletown average annual household energy expenditure could be either greater than *or* less than the national figure. If H_0 is rejected, the *only* conclusion that can be drawn is that the difference in household energy cost is significant; no conclusion on the direction of that difference is possible.

For some situations, the alternate hypothesis can provide more specific information about the hypothesized direction of difference:

$$H_0: \mu = 2{,}000$$

$$H_A: \mu < 2{,}000$$

or

$$H_A: \mu > 2{,}000$$

In addition to determining whether a significant difference exists, the direction of that difference (expressed in H_A) is also specified. If H_A: $\mu < 2{,}000$, then rejection of H_0 indicates that the average annual energy cost of Middletown households is significantly lower than the national average. This would be a reasonable alternate hypothesis to test, for example, if Middletown has a mild year-round climate or an accessible location making energy transportation costs relatively low. If H_A: $\mu > 2{,}000$, then rejection of H_0 indicates that the average Middletown household energy cost is significantly higher than the national average. This might be a logical alternate hypothesis if Middletown has an energy-demanding climate or a relatively inaccessible location resulting in high energy transport costs.

With classical hypothesis testing, Middletown planners must decide which form of H_A to use. The critical issue is whether *a priori* knowledge exists about any direction of difference in energy use between Middletown and households nationwide. If Middletown planners have no preconceived rationale for believing their household energy cost is larger or smaller than the national average, the non-directional format would be appropriate.

In classical hypothesis testing the conclusion is to either reject or not reject the null hypothesis. *Because this decision is based on a single sample,* we can measure the chance or probability of making an incorrect decision or reaching a wrong conclusion. Error comes from two possible sources (Table 9.2):

Type I Error: A decision could be made to reject the null hypothesis as false when it actually is true. For our Middletown example, it could be concluded that the difference between the national average household energy cost and the sample of Middletown households is significant, when actually no difference exists. The likelihood of this sort of error occurring (Type I error) is equivalent to the significance level (α), discussed in Chapter 8.

Type II Error: Conversely, a decision could be made to not reject the null hypothesis when it actually is false. For the Middletown example, it could be concluded that their household energy cost does not differ from the national average, when a significant difference really exists. The likelihood of this error (Type II) occurring is beta (β).

The logic of hypothesis testing operates on much the same principle as judicial decision making in a court of law (Table 9.3). In hypothesis testing, the null hypothesis is presumed correct until rejected or proven otherwise. Similarly, in court, a defendant is presumed innocent until proven guilty. In the judicial system, convicting a person who is truly not guilty is considered a more serious error than freeing a guilty person. Similarly, in hypothesis testing, rejecting H_0 as false when it is actually true (a Type I error) is considered more serious than not rejecting H_0 when it is actually false (a Type II error). In a court of law, when reasonable doubt exists about the guilt of a defendant, the jury reaches a verdict of not guilty. The defendant is *not* declared "innocent," but rather the verdict is "not guilty." Similarly, if reasonable doubt exists about rejecting the null hypothesis, you should not reject it. For this reason, the significance level (α) of most problems is kept relatively low (at a level such as .05 or .01), to minimize the chances of a serious Type I error.

TABLE 9.2

Possible Decisions in Classical/Traditional Hypothesis Testing

Decision from hypothesis testing	Null hypothesis in reality: True	Null hypothesis in reality: False
Reject H_0 as false	Type I error (prob. = α)	Correct decision (prob. = $1-\beta$)
Do not reject H_0	Correct decision (prob. = $1-\alpha$)	Type II error (prob. = β)

TABLE 9.3

Possible Decisions in a Court of Law

Decision from jury	True situation (unknown): Not guilty	True situation (unknown): Guilty
Guilty	Incorrect decision	Correct decision
Not guilty	Correct decision	Incorrect decision

Step 2: Select the Appropriate Statistical Test: The second step in classical hypothesis testing is selecting the appropriate statistical test for your problem. As will be explained near the end of this chapter, a logical and convenient way to categorize the many statistical tests available is by the type of question asked and the assumptions that are met. Because we are comparing our one sample against a population parameter, it is sufficient for now to state that the appropriate statistical test for the Middletown household energy cost problem is a **one-sample difference of means test:**

$$Z \text{ or } t = \frac{\bar{X}-\mu}{\sigma_{\bar{X}}} = \frac{\bar{X}-\mu}{\sigma/\sqrt{n}} \tag{9.1}$$

where Z or t = test statistic
$\bar{X}$ = sample mean
μ = population mean
$\sigma_{\bar{X}}$ = standard error of the mean
σ = population standard deviation
n = sample size

You should select Z only if the sample size is greater than or equal to 30 *and* you know the population standard deviation (a fairly rare occurrence); select t if the sample size is less than 30 *or* if the population standard deviation is unknown (the usual situation), in which case you must substitute the sample standard deviation.

In the chapter introducing continuous probability distributions (Chapter 6), the concept of a Z-score or standard score is presented. The one-sample difference of means test is structurally similar to the Z-score equation (Table 9.4). The standardized Z-score of a value in a set of data may be interpreted as the number of standard deviations a value is above or below the mean. Using similar logic, the Z-model or t-model one-sample difference of

TABLE 9.4

Comparable Logic: *Z*-Score of an Observation and *Z*-Value (or *t*-Value) of a Sample Mean

Z-score of a value (i) in a set of data	*Z*-value or *t*-value of a sample mean in a frequency distribution of sample means
$Z_i = \dfrac{X_i - \bar{X}}{s}$	$Z \text{ or } t = \dfrac{\bar{X} - \mu}{\sigma_{\bar{X}}}$ where $\sigma_{\bar{X}} = \dfrac{s}{\sqrt{n}}$

means test statistic measures the number of standard errors a sample mean lies above or below the hypothesized population mean.

For the Middletown household energy expenditure examples, the population standard deviation (σ) is not known (this is the usual circumstance). Recall from Chapter 8 (Table 8.1) that the sample standard deviation, s, is a proper estimator of σ when σ is unknown, but a slight adjustment is needed. Thus, substituting s for σ in Equation 9.1 gives:

$$Z = \frac{\bar{X} - \mu}{\sigma_{\bar{X}}} = \frac{\bar{X} - \mu}{s/\sqrt{n}} \quad (\text{if } n \geq 30) \qquad \textbf{(9.2)}$$

or

$$t = \frac{\bar{X} - \mu}{\sigma_{\bar{X}}} = \frac{\bar{X} - \mu}{s/\sqrt{n-1}} \quad (\text{if } n < 30) \qquad \textbf{(9.3)}$$

For illustrative purposes, suppose Middletown officials must use the smaller 25-household sample in which the average annual energy expenditure is \$1,810 ($\bar{X}$ = 1,810) because the larger sample of 250 households has not yet been collected. This means that t rather than Z is used.

Step 3: Select the Level of Significance: The next task in classical hypothesis testing is to place a probability on the likelihood of a sampling error. As mentioned earlier, committing a Type I error and rejecting a null hypothesis as false when it is actually true is generally considered serious. Therefore, the usual procedure is to select a fairly low significance level (α) such as .10 or .05. Then, the conclusion is specified in terms of the level of significance of the result. In classical hypothesis testing, a null hypothesis may be rejected at the .05 level, which is the same as saying the statistical test is significant at the .05 level. This would mean that there is a 5% chance that a Type I error has occurred, and it is only 5% likely (1 chance in 20) that the null hypothesis has been improperly rejected because of random sampling error.

For many geographic research problems, an extremely stringent significance level is not necessary. In general, the choice of significance level depends on the nature of the problem and the effects of the decision. With some problems, the consequences of sampling error may be severe, and in these instances, a very low significance level is required. With the Middletown household energy cost, using a commonly accepted, "conventional" significance level such as α = .05 provides sufficient precision.

Step 4: Delineate Regions of Rejection and Non-rejection of Null Hypothesis: Once a significance level has been selected, this value is used to create the regions of **rejection and non-rejection** of the null hypothesis (Figure 9.1). For our example, the total area in which H_0 is rejected, as represented by the significance level, encompasses 5% (α = .05) of the area under the curve. This area of rejection can be distributed in one of two ways. In Case 1, the alternate hypothesis is non-directional, so the shaded rejection area of H_0 is distributed equally between the two tails of the curve. With this **two-tailed** format and α = .05, 2.5% (α / 2 = .025) of the total area is in each of the rejection regions or tails of the distribution. In Case 2, the alternate hypothesis is directional (**one-tailed**), so the shaded rejection region of H_0 is placed entirely on one tail of the distribution. In this diagram, the rejection region happens to be on the right tail, but the placement of the rejection region depends on the form of H_A. In both the two-tailed and one-tailed cases, the unshaded area under the curve delineates test statistic values where H_0 is not rejected. This area of non-rejection is 95% (1 – α = .95) of the total area under the curve. When α = .05 and H_A is non-directional, each of the two tails contains α /2 or .025 of the area under the curve (Figure 9.2). This leaves .475 of the area on *each* side of the distribution (1 – α = .95 in total) in the non-rejection region.

One-Sample Difference of Means *Z* or *t* Test

Primary Objective: Compare a random sample mean to a population mean for difference

Requirements and Assumptions:

1. Random sample
2. Population from which sample is drawn is normally distributed
3. Variable is measured at interval or ratio scale

Hypotheses:

$H_0 : \mu = \mu_H$ (where μ_H is the hypothesized mean)
$H_A : \mu \neq \mu_H$ (two-tailed)
$H_A : \mu > \mu_H$ (one-tailed) or
$H_A : \mu < \mu_H$ (one-tailed)

Test Statistic:

$$Z \text{ or } t = \frac{\bar{X} - \mu}{\sigma_{\bar{X}}}$$

If sample size ≥ 30, use *Z*; if sample size < 30, use *t*

The next task is to determine the critical *t*-table values that delimit the boundaries separating the rejection and non-rejection regions. To do this precisely, we must further explain the concept of **degrees of freedom**, which was briefly introduced in Chapter 8. The degree of freedom indicates the number of values in a calculation that are free to vary. Suppose, for example, we know that the mean of a data set having four numbers is 10 and the numbers are 7, 8, 14, and some unknown number. To determine the mean, we add all the values and divide by the number of observations, resulting in the formula $(7 + 8 + 14 + x) / 4 = 10$, where x is the unknown number. With some algebra we can calculate that $x = 11$. In this simple example, the first three numbers are allowed to vary, but when they (and the mean—the known statistical parameter in this case) are fixed, the value of the fourth number cannot vary and has no freedom to move. This example therefore has 3 degrees of freedom.

Suppose now that we have a sample size of 20. If we know that the sample has a mean of 10, but do not know the values of any of the observations, then there are 19 degrees of freedom. All numbers must add up to a total of 200: (20 × 10). Once we know the magnitude of 19 of the 20 numbers, the magnitude of the 20th number is fixed (not free to vary), so we have 19 degrees of freedom.

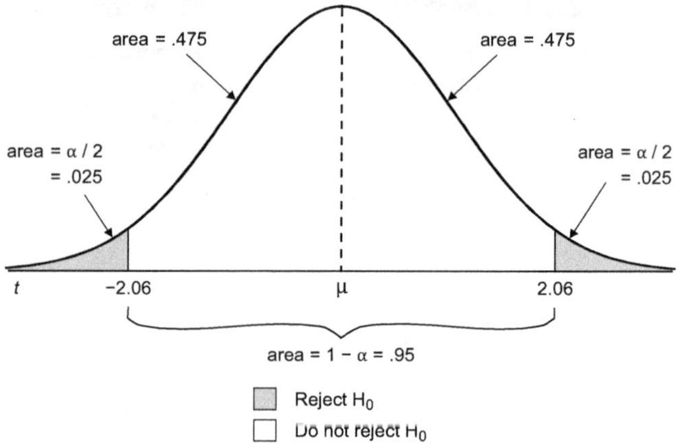

FIGURE 9.2
Normal Distribution Values Associated with a Significance Level (α) = .05: Two-Tailed Case

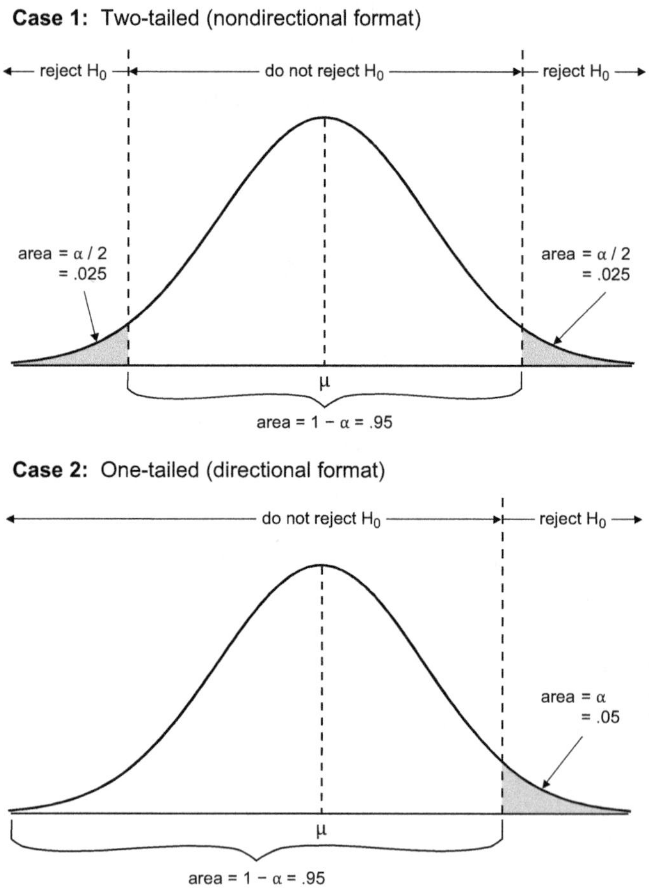

FIGURE 9.1
General Regions of Rejection and Non-rejection of Null Hypothesis: Significance Level (α) = .05

Degree of freedom is a vital component when using the *t*-table to determine the critical test statistic value. There are many *t*-table distributions, and the one selected for a particular problem depends on the sample size. If our sample size is n, then the number of degrees of freedom is $n - 1$. For example, a sample of size 20 requires use of the *t*-table row with 19 degrees of freedom.

We return now to our problem of estimating average annual energy expenditure for Middletown households from a sample of 25. In the *t*-table (Appendix, Table D), the value that corresponds with a 0.95 level of confidence at 24 $(n - 1)$ degrees of freedom is 2.06. Therefore, in this problem, a *t*-value of 2.06 defines the boundary separating the rejection and non-rejection regions, and a calculated *t*-value (test statistic) must have an absolute value of less than 2.06 to keep the null hypothesis from being rejected.

The following decision rule identifies and summarizes the regions of rejection and non-rejection of the null hypothesis for $\alpha = .05$ and H_A non-directional:

Decision rule:

If $t < -2.06$ or if $t > 2.06$, reject H_0
Conversely, if $-2.06 \leq t \leq 2.06$, do not reject H_0

Step 5: Calculate the Test Statistic: At this step of the hypothesis testing procedure, sample data are evaluated using the test statistic. In the Middletown sample, mean household annual energy cost was \$1,810, with a sample standard deviation of \$196. Substituting these sample

TABLE 9.5

Summary of Classical/Traditional Hypothesis Testing: Middletown Household Energy Cost Example

Step 1:	$H_0{:}\mu = 2000$ and $H_A{:}\mu \neq 2000$
Step 2:	One-sample difference of means t test selected as test statistic
Step 3:	$\alpha = .05$
Step 4:	If $t < -2.06$ or if $t > 2.06$, reject H_0 If $-2.06 \leq t \leq 2.06$, do not reject H_0
Step 5:	Calculate t (from random sample) = −5.00
Step 6:	Since $t < -2.06$, reject H_0

statistics into the difference of means t test gives the following:

$$t = \frac{\bar{X} - \mu}{s/\sqrt{n-1}} = \frac{1810 - 2000}{196/\sqrt{24}} = -5.00 \tag{9.4}$$

The Middletown sample mean is 5.00 standard errors *below* the U.S. average. If household energy cost in Middletown had been more than the national norm of 2000, the calculated test statistic would have been positive. This calculated value is now compared with the critical t-table values determined in step 4 to reach a final decision.

Step 6: Make Decision Regarding Null and Alternate Hypotheses: All the information is now available to make a decision regarding rejection or non-rejection of the null hypothesis. The calculated test statistic is $t = -5.00$. The critical t-table values are −2.06 and 2.06. Since the calculated test statistic is not between the critical table values ($-2.06 \leq t \leq 2.06$) the null hypothesis *should be rejected.* The conclusion is that mean annual household energy cost in Middletown differs significantly from the mean annual household energy cost nationally. The steps in the classical hypothesis testing procedure for the Middletown example are summarized in Table 9.5.

P-Value or *Prob*-Value Hypothesis Testing Procedure

The formal multistep procedure of classical hypothesis testing provides a logical basis and excellent theoretical underpinning for all inferential decision making. For example, all of the terminology just presented (null and alternate hypotheses, one-tailed and two-tailed tests, Type I and Type II errors, etc.) is fully valid and effectively carries over to the p-value method of hypothesis testing. However, the usefulness of the results from classical analysis is limited in some important ways. First, a specific significance level must be selected to delineate the regions of rejection and non-rejection of the null hypothesis. This *a priori* selection of α is often arbitrary and may lack a clear theoretical basis. A significance level of .05 or .10 is often chosen because they are the "conventional" probabilities commonly provided in statistical tables. Second, the final decision regarding the null and alternate hypotheses is binary in nature—either H_0 is rejected or not rejected at that arbitrary significance level. This type of conclusion provides only limited information about the calculated test statistic. We rarely use classical hypothesis testing today for statistical problem solving; instead, the p-value approach is commonly employed in scientific research.

The more flexible p-value method of hypothesis testing provides additional valuable information. With this approach, the *exact* significance level associated with the calculated test statistic value is determined. That is, the ***p*-value** is the exact probability of getting a test statistic value of a given magnitude or larger, *if* the null hypothesis of no difference is true. This can generally be interpreted as the probability of making a Type I error. Unnecessary pre-selection of a "standard" significance level (α) is avoided, and decisions to reject or not reject H_0 at that arbitrary level need not be made.

In the Middletown household energy cost example, the decision rule was to reject the null hypothesis if $t < -2.06$ or $t > 2.06$ and to not reject H_0 if $-2.06 \leq t \leq 2.06$. The calculated test statistic value of $t = -5.00$ led to the conclusion that the mean annual energy cost of Middletown households was significantly different than mean national household energy cost, and the null hypothesis was rejected.

This conclusion is of limited use. All that has been decided is that we can be at least 95% certain that the Middletown household annual average energy cost is different from the national average. However, knowing the exact significance level associated with the specific sample mean or calculated test statistic value of −5.00 would be more informative. If the null hypothesis had been rejected on the basis of the 1810 sample mean, what would be the exact significance level and likelihood that a Type I error had been made? The precise probability of making a Type I error cannot be determined with the classical approach.

We can assess the Middletown sample more informatively using the p-value approach. The critical values that separate the regions of rejection and non-rejection of the null hypothesis are now based on the location of the particular sample mean ($\bar{X} = 1810$) relative to the population mean ($\mu = 2000$). The region of non-rejection is centered on $\mu = 2000$, and the absolute difference between $\bar{X}$ and μ, $|\bar{X} - \mu| = 190$, is used to establish the width of the interval on either side of μ. Thus the non-rejection region has an upper bound of 2190 ($\mu + 190$) and a lower bound of 1810 ($\mu - 190$). The rejection regions occupy the extremes of the distribution, outside the upper and lower bounds of the confidence interval.

The proportion of the total area under the normal curve area lying within the rejection region(s) represents the p-value, and most statistical software packages complete these steps:

1. The test statistic for t is calculated.
2. The probability or relative area under each tail of the normal curve is determined for that t-value.
3. The rejection area is determined by subtracting the probability (of step 2) from .5000.
4. This area is doubled if a non-directional (two-tailed) alternate hypothesis is used.

This procedure is completed with the Middletown example of mean annual household energy expenditure. The difference of means t test statistic is calculated:

$$t = \frac{\overline{X} - \mu}{s/\sqrt{n-1}} = \frac{1810 - 2000}{196/\sqrt{24}} = -5.00 \quad \textbf{(9.5)}$$

This statistical test value is then used to determine a probability or area under the t-model curve. When $t = -5.00$ and the degrees of freedom is 24, the Student's t distribution table (Appendix, Table C) does not list a value because it is so far out in probability that it is beyond the table. You may safely assume that the p-value is much less than .005. We should note that computer software will often report the exact p-value of .000, representing a .000% chance of a Type I error. This is often due to rounding or truncating, as p-values cannot actually be zero. Therefore, one can replace $p = .000$ as $p < .001$. We will use the computer convention of .000 throughout the remainder of the book.

How can this p-value be interpreted? If the decision is made to reject H_0, the significance level equals the p-value or <.001, representing a <.001% chance of a Type I error. In this situation, the likelihood of making an error is extremely low, so the clear and logical decision is to reject the null hypothesis. In the Middletown energy cost example, a p-value of .000 can also be interpreted this way: you could take 1000 independent random samples of 25 Middletown households, but not even 1 time out of 1000 would you expect to get a test statistic value having an absolute magnitude of 5.00, *if* there really is no difference between Middletown and national average annual household energy expenditures.

Looked at from another perspective, the p-value is a probabilistic measure of the belief or conviction that the decision not to reject the null hypothesis is correct. For example, a p-value relatively close to 1 indicates a high degree of trust in the validity of the null hypothesis, whereas a p-value very close to 0 suggests that little or no faith should be placed in the null hypothesis. In the Middletown example, the p-value of .000 indicates that the null hypothesis should be rejected, and the Middletown mean annual household energy cost of \$1810 is significantly different from the nationwide average annual household energy cost of \$2000.

Use of the p-value approach, however, does not provide an excuse to interpret results subjectively or avoid making decisions. Just because computers produce exact p-value instantaneously, you are still responsible for interpreting the statistical and geographic meaning of the p-values in a consistent and rational way. P-values report what the sample data reveal about the credibility of a null hypothesis, but still demand the same stringent rules of inference as required in classical hypothesis testing.

Since the p-value method offers many advantages over classical hypothesis testing, *all statistical tests presented in the remainder of this text will use the p-value method.* In the statistical analysis of all the geographic problems that follow, associated p-values will be reported without showing all steps in their manual calculation.

9.2 One-Sample Difference of Proportions Test

Hypotheses can also be formed to study the difference between a sample proportion and a population proportion. The test statistic for a difference of proportions test is sometimes called the Z test for proportions or one-proportion Z test. Like the difference of means Z test, the normal distribution is used.

In one of the examples from the last chapter (random or systematic sample—estimate of proportion), Middletown planners estimated the proportion of all owner-occupied housing units in the community that are "underwater." Remember this term refers to homeowners experiencing negative equity, with the amount still owed on their mortgage being greater than the current value of their home. Eighty-five Middletown owner-occupied housing units with current mortgages were sampled, and 25 of them were "underwater," with negative equity. This is a sample proportion of .294.

In the last chapter, we were interested in placing a confidence interval around this best sample estimate. Now, we want to compare the Middletown proportion of owner-occupied housing units that are underwater with the national proportion. Therefore, a one-sample difference of proportions test is appropriate, since we have one sample proportion (from Middletown) to compare with a population (the nationwide proportion). RealtyTrac reports that, nationwide, 22.8% (.228) of all owner-occupied housing units carrying a mortgage are underwater in the fourth quarter of 2011 ($\rho_H = .228$). Middletown officials are concerned that the proportion of housing units underwater in their community (ρ) *is greater than* the national proportion. Since the direction of difference is stated, this is a one-tailed or directional alternate hypothesis. The corresponding null hypothesis would be that the Middletown proportion of housing units underwater *is not greater than* the nationwide proportion. Quite simply, we are asking if the Middletown sample proportion (.294) is greater than the national proportion (.228).

That is:

$$H_0: \rho = \rho_H$$

$$H_A: \rho > \rho_H \text{ (directional)}$$

The appropriate test statistic is the one sample difference of proportions test:

$$Z = \frac{p - \rho}{\sigma_p} \quad \textbf{(9.6)}$$

where σ_p = standard error of the proportion
The standard error of the proportion (σ_p) is the standard deviation of the sampling distribution of proportions:

$$\sigma_p = \sqrt{\frac{p(1-p)}{n}} = \sqrt{\frac{(.228)(.772)}{85}} = .0455 \quad \textbf{(9.7)}$$

therefore:

$$Z = \frac{p - \rho}{\sigma_p} = \frac{.294\text{-}.228}{.0455} = \frac{.066}{.0455} = 1.45$$

If the null hypothesis is rejected when Z = 1.45, the exact associated significance level and p-value is .0735 (Figure 9.3). This indicates a 7.35% chance that a Type I error has been made if H_0 is rejected. This is a borderline situation. There is a 92.65% chance that Middletown does indeed have a greater proportion of housing units "underwater" than the national proportion, but it is not a totally definitive result. How Middletown officials decide to respond to this statistical outcome, and whether they decide to develop any housing policy or program are questions they have to resolve for themselves.

The various one sample difference tests are related, which we can see in the similar structure of the one sample test statistics (Table 9.6). A general format is common to each of these difference tests. Note that the numerator of each test statistic is the difference between the sample statistic and the hypothesized population parameter value. The denominator is the standard deviation of the sampling distribution, which is also referred to as the standard error of the sample statistic.

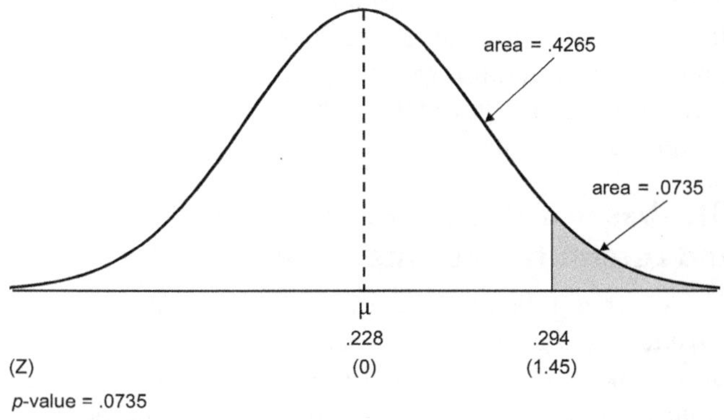

FIGURE 9.3
One-Sample Difference of Proportions Test: Probability that Middletown Proportion of Households "Underwater" is Greater than the National Proportion

TABLE 9.6

The Similar Structure of One-Sample Difference Tests

Difference Test	Test Statistic
Difference of means* large sample (n > 30)	$Z = \frac{\bar{X} - \mu}{\sigma_{\bar{X}}} = \frac{\bar{X} - \mu}{s/\sqrt{n}}$
Difference of means* small sample (n ≤ 30)	$t = \frac{\bar{X} - \mu}{\sigma_{\bar{X}}} = \frac{\bar{X} - \mu}{s/\sqrt{n-1}}$
Difference of proportions	$Z = \frac{\rho - p}{\sigma_p} = \frac{\rho - p}{\sqrt{p(1-p)/n}}$

* Population standard deviation (σ) is unknown.

Best estimate of $\sigma = s = \sqrt{\frac{\Sigma(X_i - \bar{X})^2}{n-1}}$

General Format of One-Sample Difference Test:

$$Z \text{ or } t = \frac{\text{sample statistic} - \text{hypothesized population parameter value}}{\text{standard deviation of sample statistic (standard error)}}$$

One-Sample Difference of Proportions Test

Primary Objective: Compare a random sample proportion to a population proportion for difference

Requirements and Assumptions:

1. Random sample
2. Variable is organized by dichotomous (binary) categories

Hypotheses:

$H_0: \rho = \rho_H$ (where ρ_H is the hypothesized proportion)
$H_A: \rho \neq \rho_H$ (two-tailed)
$H_A: \rho > \rho_H$ (one-tailed) or
$H_A: \rho < \rho_H$ (one-tailed)

Test Statistic:

$$Z \text{ or } t = \frac{p - \rho}{\sigma_p}$$

9.3 Selected Issues in Inferential Testing

So far in this chapter, attention has been focused on a related set of one sample difference tests. However, there are numerous other types of inferential tests that can be applied to help solve a wide variety of geographic problems. Before continuing, however, some basic questions need to be addressed: Under what general circumstances are inferential statistics appropriate? More specifically, can or should an inferential test be used when one has a total population (as in a GIS-based data base) and not a sample from which to infer to a population? How does one determine which of the many available inferential techniques is appropriate and should be selected for a particular geographic problem? That is, what are the dimensions and choices that need to be made, and should the questions and answers be in any sequence? Under what circumstances should a parametric or nonparametric statistical test be chosen and are there situations where it doesn't matter or situations where both might be applied?

All inferential techniques have certain general characteristics and assumptions in common. As stated earlier, the general goal of hypothesis testing is to make inferences about the magnitude of one or more population parameters based on sample estimates of those parameters. When applying any inferential test statistic it is assumed that a *random sample* has been drawn from a population. Often, however, other unbiased types of sampling (such as systematic sampling) are also valid for inferential hypothesis testing. Recall from Chapters 7 and 8 that an element of randomness must be included in the sample procedure, whatever specific type of sample is selected. If multiple samples are required for a particular problem, it is always assumed that *each sample is drawn separately and independently.* Suppose a geographer wishes to compare the average size of pebbles on two beaches. This would be done by taking a random sample of pebbles from one beach and a separate and independent random sample of pebbles from the other beach. (An exception to the assumption of independence occurs if a matched-pair or dependent sample difference test is used, examples of which will be seen in Chapter 10).

Artificial versus Natural Samples and Inferential Testing

Beyond these general characteristics, geographers have actively debated the circumstances for appropriate use of inferential testing in problem solving. One dimension of the discussion centers on the difference between artificial and natural sampling. In an **artificial sample**, the investigator draws an unbiased, representative sample from a statistical population and then is able to infer certain characteristics about the population based on the sample data. No one questions the appropriateness of inferential techniques in geographic problems or situations in which a proper artificial sample has been taken.

Opinion differs on whether inferential statistics should be applied on geographic data sets not obtained from artificial sampling. Geographers often wish to analyze spatial patterns or data sets that comprise complete enumerations or total populations. In these situations, are inferential procedures ever appropriate? Some geographers suggest that inferential statistics may be permitted when using a data set considered a "natural sample."

In **natural sampling**, the "natural" or real-world processes that produce the spatial pattern under analysis contain random components. Suppose a geographer wishes to analyze a data set showing the pattern of all hurricane "landfalls" along a coastline during the last century. The landfall site of a hurricane is partly a function of prevailing global water currents and wind patterns (nonrandom or systematic influences), but is also affected by a complex variety of meteorological processes associated with that particular hurricane (random influences). It is possible to argue then that the observed pattern of hurricane landfall sites is actually a "natural sample" from the population of all possible landfall locations that could have occurred.

The pattern of state-level 2000–2010 population change (shown in Figure 1.7) is another example of a natural sample. This choropleth map pattern is partly the result of such nonrandom or systematic factors as climate and job opportunities related to the location of natural resources and partly the result of numerous individual and family decisions to migrate from one state to another (some of which could be random).

If inferential procedures are applied to natural samples such as these, *the results must be interpreted with extreme caution.* Descriptive summaries of natural samples are certainly appropriate, but you must take care to avoid making improper inferential statements. Issues regarding the application of inferential statistics to natural samples are quite complex and continue to generate considerable discussion and controversy in applied statistics. Those interested in pursuing these arguments further (particularly from the geographer's perspective) are directed to the references at the end of this chapter.

GIS-based Total Enumeration and Inferential Testing

Another related issue that geographers face when considering the use of inferential statistics, especially those making extensive use of GIS technology, is differentiating between a sample and a total enumeration within a data set. You will recall that the purpose of inferential statistics is to make inferences about a population based on a statistical sample. Inferential tests allow us to understand the degree to which a sample reflects the characteristics of the total population, taking sampling error into account.

Oftentimes, especially in GIS-based applications, you might already have what might be considered a total population. For instance, a county-wide GIS may have the assessed value of every property in the county. Similarly, a weather monitoring station will have the rainfall totals of every day of the year, and not just a sample of days. Finally, some data like that illustrated for life expectancy may only include a single data value for each country—in many cases it is unclear if these data represent a total enumeration or a sample.

When a total enumeration of data exists, one can certainly compare the descriptive values. That is, the annual total rainfall in one city may be 1.24" higher than the total rainfall in another city, or the average assessed value of property in one community is $1,267 higher than the assessed value of property in another community. A non-geographic example is the test scores illustrated in Table 8.6. In reality, the high school probably maintains records for every student's exam. Therefore, descriptive statistics regarding the average test scores can be computed using the entire population.

In cases where the total population is known, you might simply decide to evaluate the descriptive statistics and perform no additional tests. The simple reason for this is because the total population provides us with the actual descriptions, whether it be property values, rainfall, or test scores. Also, inferential statistics assume that there is a sample with a corresponding sampling error. Using the total population masked as a sample violates assumptions of the inferential tests.

Nonetheless, simply knowing that one city has 1.24 more inches of rainfall than another city, or that one community's property value is $1,267 more than another community may be unsatisfying for certain policy decisions. In many cases, the geographer is really asking whether the differences between two groups are explained by the natural variation in the data itself. For example, when the average property values are around $500,000 in two communities, a difference of $1,267 may not represent a significant difference. In this case, the differences in the values may simply be explained by the natural variation in the property values throughout both communities.

When faced with the common situation of a GIS dataset having what is considered the total population, you have a few reasonable options:

1. **Ignore the use of inferential statistics and make extensive use of descriptive statistics**. Remember that descriptive statistics can tell us a significant amount about one or more datasets. As previously discussed, the coefficient of variation (CV) provides great power in comparing two different datasets while also considering the variability in the data.
2. **Take a sample of the larger dataset to allow the use of inferential statistics**. Oftentimes, GIS datasets have tens or hundreds of thousands of records as in the case of a countywide parcel map with assessed values. A geographer can sample a smaller subset from that dataset and perform inferential statistics. The advantages of this approach are that you will have a sample, the sampling error will be known theoretically through application of the Central Limit Theorem, and inferential tests for differences can be conducted. While that solves the sampling problem, it raises other concerns regarding the wisdom in throwing out what others would consider perfectly good data. In our own review of this question we found that statisticians were rather evenly split on the merit of this approach.
3. **Treat the population as though it is a sample and perform inferential tests**. One might argue that many of the attributes stored in a GIS have inherent error built in. For example, the amount of rain captured at a monitoring station has some inherent error caused by wind direction or other environmental conditions, while the assessed value of a property may be subject to an assessor's apparent mood that particular day. Similarly, for some data, like life expectancy or obesity discussed in Chapter 1, it is impossible to say that every person who has died in a country had their age recorded, or every person in a state has been evaluated for obesity.

Obviously there are many issues that a geographer using GIS data must consider when attempting to apply statistical inferences. At a minimum, you should feel comfortable exploring descriptive statistics. If you choose to apply an inferential test based on the above scenarios, take great care in considering the applicability of the test for the given data. In most cases, if an inferential test is used, you can evaluate the results to make thoughtful observations but refrain from a strict adherence to a *p-value* to represent statistical significance.

Selecting a Suitable Inferential Test

We need to know more than the general characteristics of a particular problem to select an appropriate inferential test. Because of the idiosyncratic nature of geographic problems, selecting the single best statistical procedure is often difficult. To complicate matters, the same geographic problem can be structured or organized in different ways, or data can be collected and correctly analyzed statistically in various ways. You should consider this discussion as providing general guidelines, not a rigid set of fixed rules.

However, determining the structure of a problem and the organization of data in that problem will almost certainly direct you to a particular technique or set of appropriate techniques. To help you make an appropriate selection, we provide a logical organizational framework in which to place each of the many inferential techniques presented in this text.

The fundamental dimensions of our organizational framework are shown in Table 9.7. The only specific inferential tests covered so far in the text are the one-sample difference tests presented earlier in this chapter, and they are placed in the appropriate cell of the table. At the present time, all of the other cells in this table are left blank. After we have presented a number of other inferential tests in the following chapters, this table is provided again in the final (epilogue) chapter with all inferential tests covered in the text placed in their proper location.

The columns of the table represent the *type of question* being asked or being investigated. At the most basic level, statistical inference is concerned with estimating differences, similarities, or relationships between a sample (or

TABLE 9.7

Organizational Structure for Selection of Appropriate Inferential Test

Level of measurement	Questions about differences: One sample	Dependent sample (matched-pairs)	Two samples	Three or more (k) samples
Nominal scale (categorical variables)				
Ordinal scale				
Interval/ratio scale	One-sample difference of means One-sample difference of proportions test			

	Questions about relationships: Questions about strength and direction of relationship between two variables	Questions about form or nature of relationship between one dependent and one independent variable	Questions about form or nature of relationship between one dependent variable and two or more independent variables
Nominal scale (categorical variables)			
Ordinal scale			
Interval/ratio scale			

	Questions about explicitly spatial relationships
Nominal scale (categorical variables)	
Ordinal scale	
Interval/ratio scale	

set of samples) and a hypothesized value of a population parameter (or set of parameters). However, problems are organized or structured differently according to the type of difference, similarity, or relationship being examined.

The first four columns in the top portion of Table 9.7 refer to various types of questions about differences. For example, the problems considered earlier in this chapter were concerned with the magnitude of difference between a single sample and a hypothesized population parameter. These problems are collectively referred to as one-sample difference tests (first column in top portion of Table 9.7). In certain special cases, the data set under investigation consists of one sample of observations collected for two or more different variables or at two or more different time periods. In these circumstances, a dependent-sample (matched-pairs) difference test is appropriate (second column in top portion of Table 9.7). Other inferential tests consider the magnitude of difference between two or more independently-drawn random samples, and hence these tests are known as two-sample, three-sample, or *k*-sample difference tests (columns 3 and 4 in the top portion of the table).

The columns in the middle portion of Table 9.7 respectively refer to: questions about the strength and direction of relationship between two variables; questions about the form or nature of relationship between one dependent and one independent variable; and questions about the form or nature of relationship between one dependent variable and multiple (two or more) independent variables. Notice that the types of questions asked in the columns of the middle portion of the table are fundamentally different than the columns in the top portion of the table. Rather than asking if a sample is significantly different from a population or if two or more samples are significantly different from one another, the focus in the middle segment of the table is on how similar (strong or weak) the relationships are between two or more variables and a description of the form or nature of those relationships.

The bottom portion of the table considers questions about patterns and relationships that are explicitly spatial. Specially designed tests are available to analyze a point pattern or area pattern to determine if it is random or significantly different from random (more clustered or more dispersed than random). Geographers face a number of particular concerns with the statistical analysis of data that are explicitly spatial, and an entire section of the text (Part Five) deals with inferential spatial statistics.

The rows of Table 9.7 represent different *levels of measurement*. Depending on the nature of the geographic problem under investigation, data may be scaled at the nominal (categorical), ordinal, or interval/ratio level of measurement. Selection of an appropriate inferential test depends in part on this level of measurement.

Statisticians traditionally divide inferential techniques into two categories: parametric and nonparametric. Inferential tests that require knowledge about population parameters and make certain assumptions about the underlying population distribution are termed **parametric tests**. For example, in the one sample difference of means tests (Z or t) and the one sample difference of proportions test (Z), population parameters such as μ, σ, and ρ are included in the test statistic formulas. A commonly-applied assumption is that the population is normally distributed with mean μ and standard deviation σ. Other assumptions that apply for some multi-sample parametric techniques will be discussed as needed.

Another group of statistical tests requires no such knowledge about population parameter values and has fewer restrictive assumptions concerning the nature of the underlying population distribution. This group of tests is termed **nonparametric** or **distribution-free**.

Parametric tests *require* sample data measured on an interval/ratio scale, because their test statistics use parameters such as the mean or standard deviation—descriptive statistics that can only be calculated from interval/ratio data. Recall, a mean or standard deviation should not be calculated from nominal or ordinal data. Nonparametric tests, on the other hand, do not require such "interval/ratio" statistics. Some nonparametric tests are specifically designed to be used with ordinal data, whereas others are designed to be applied effectively to nominal or categorical data (or combinations thereof). For data measured at a nominal or ordinal scale, a nonparametric test can be applied.

Different strategies may be possible, however, with data at the interval/ratio scale:

1. **Run only a parametric test.** This approach is appropriate if there is virtually no doubt that the requirements and assumptions needed to use the test have all been met.
2. **Run only a nonparametric test.** If there is reason to believe that one or more of the parametric test assumptions is moderately or severely violated, then the results from running the parametric test are likely to be invalid or inaccurate. Therefore, the set of interval/ratio data may be converted ("downgraded") to an ordinal or nominal scale, and a corresponding appropriate nonparametric test can be run.
3. **Run both a parametric and nonparametric test**. If there is uncertainty about the degree to which requirements and assumptions are violated or an assumption is violated only slightly (for example, a sample is drawn from an underlying population distribution that is only slightly non-normal), then it may be appropriate to run both a parametric and a comparable nonparametric test.

Additional insights may be gained by running both tests on the same data and comparing the resulting *p*-values. This "pairing" strategy may be particularly useful when conducting an exploratory or investigative geographic analysis. This idea is demonstrated several times in later chapters.

The number of geographic problems that can be examined using inferential statistics is limitless. In the remainder of the text, inferential techniques are applied to a number of real-world geographic situations. Each inferential technique or set of techniques is presented using a common format:

1. Presentation of the rationale, purposes, and objectives of the technique. A set of appropriate geographic problems that could be solved using the technique are listed, and an overview of the key assumptions and required conditions is provided.
2. Presentation of basic formulas and computations associated with the technique.
3. Discussion of the example geographic problem under examination and the application of the inferential technique. This step has been considerably strengthened in this edition of the text, as we constantly strive to include more practical geography in support of the use of statistics.
4. Evaluation of the use of the technique for a particular geographic problem and discussion of both inferential issues and geographic factors that might have affected the test results.

Key Terms

artificial and natural samples, *152*
classical (traditional) hypothesis testing, *144*
degrees of freedom, *148*
hypothesis testing, *143*
hypothesized mean, *145*
null and alternate (alternative) hypotheses, *145*
one-sample difference of means test, *146*
one-sample difference of proportions test, *150*
one-tailed (directional) hypothesis, *145*
parametric and nonparametric (distribution-free) tests, *155*
p-value hypothesis testing, *149*
rejection and non-rejection regions, *147*
two-tailed (non-directional) hypothesis, *145*
Type I and Type II errors, *146*

Major Goals and Objectives

If you have mastered the material in this chapter, you should now be able to:

1. Explain null and alternate hypotheses, directional and nondirectional hypothesis testing formats, Type I and Type II errors, rejection and non-rejection regions, degrees of freedom, significance levels, and test statistics.
2. Recognize when it is appropriate to use a directional or nondirectional alternate hypothesis.
3. Understand and apply the steps necessary to test hypotheses using the classical/traditional approach.
4. Understand and apply the *p*-value approach to hypothesis testing and evaluate its advantages over the classical approach.
5. Know when to apply the difference of means Z or t test or the difference of proportions Z test in comparing a sample statistic to a population parameter.
6. Recognize the difference between artificial and natural sampling approaches.
7. What are the statistical options available when faced with the situation of having a GIS dataset considered to be a total population?
8. Understand the issues involved in selecting the appropriate inferential statistical test. Issues include: type of question investigated, level of measurement used, and whether a parametric and/or nonparametric test is appropriate.

References and Additional Reading

Burt, J. E., G. M. Barber, and D. L. Rigby. *Elementary Statistics for Geographers.* 3rd ed. New York: Guilford Press, 2009.
Court, A. "All Statistical Populations are Estimated from Samples," *Professional Geographer*, Vol. 24 (1972): 160–161.
Gould, P. R. "Is Statistix Inferens the Geographical Name for a Wild Goose?" *Economic Geography* (Supplement), Vol. 46 (1970): 439–450.
Hays, A. "Statistical Tests in the Absence of Samples: A Comment." *Professional Geographer*, Vol. 37 (1985): 334–338.
Marzillier, L. F. *Elementary Statistics.* Dubuque, IA: Wm. C. Brown, 1990.
Maxwell, N. P. "A Coin-Flipping Exercise to Introduce the *P*-value." *Journal of Statistics Education*. Vol. 2, No. 1 (1994).
Meyer, D. R. "Geographical Population Data: Statistical Description, Not Statistical Inference." *Professional Geographer.* Vol. 24 (1972a): 26–28.
Meyer, D. R. "Samples and Populations: Rejoinder to 'All Statistical Populations are Estimated from Samples.'" *Professional Geographer.* Vol. 24 (1972b):161–162.
Siegel, S. and N. J. Castellan. *Nonparametric Statistics for the Behavioral Sciences.* 2nd ed. New York: McGraw-Hill, 1988.
Summerfield, M. "Populations, Samples, and Statistical Inference in Geography." *Professional Geographer.* Vol. 35 (1983): 143–149.

10

Two-Sample and Dependent-Sample (Matched-Pairs) Difference Tests

The previous chapter introduced the process of hypothesis testing using both the classical and *p*-value approaches. The focus in that chapter was on the application of *one-sample* difference tests, where a sample statistic was compared to a population parameter using both means and proportions. Geographers encounter many other situations where the objective is to determine whether a significant difference exists between *two samples.* If the sample statistics are significantly different, you can infer that the samples were drawn from truly different populations. However, if the sample statistics are not significantly different, it can be concluded that they were drawn from a single population.

This chapter is organized by the type of statistic being compared (mean or proportion) and the nature of the relationship between the samples (independent or dependent). Sections 10.1 and 10.2 expand on the methods discussed in the previous chapter for testing for differences between *two independent* samples. Independent samples occur when the items collected in the first sample are unrelated to (independent of) the items collected in the second sample. In Section 10.1, differences between two sample means are tested for statistical significance using both parametric and nonparametric procedures. Section 10.2 considers the two-sample difference of proportions test, which is designed to compare two sample proportions for significant difference.

In other problems, the samples may not be independent. For example, you might measure each subject or location twice—*before* and *after* some type of treatment or event. The before and after measurements constitute a *matched-pair* of observations and are considered *dependent samples.* In other geographic situations, two variables or indicators for the same sample of observations or locations are compared to see if the samples of matched-pairs of observations are significantly different. Asking two types of questions of the same sample of participants could also constitute a matched pair. Parametric and nonparametric methods for comparing differences between dependent samples are examined in Section 10.3.

10.1 Two-Sample Difference of Means Tests

Geographers use two primary methods to compare and test means from two independent samples for significant difference. The basic distinction separating the two methods is the requirements for parametric and nonparametric tests. If the sample data and population distributions meet the requirements of parametric testing, the appropriate independent sample difference of means procedure would be either the *Z* or *t* test. For these methods, the sample data are measured on an interval-ratio scale and the samples are drawn from sufficiently large and normally distributed populations. Conversely, if the sample data are measured at the ordinal scale or if the samples are too small or drawn from populations that are clearly not normal, a nonparametric procedure is required. For two-sample difference of means problems that do not meet parametric requirements, the Wilcoxon Rank Sum *W* test or Mann-Whitney *U* test is used.

Two-Sample Difference of Means *Z* or *t* Test

Suppose you want to compare surface water runoff levels in two different environmental settings: (1) a natural landscape with little or no modification by human activity; and (2) an exurban setting partially modified with the development of several residential subdivisions and their access roads. You might hypothesize that the presence of artificial surfaces like concrete and asphalt affects the volume of surface water runoff, making the amount different in the exurban versus natural settings. To test for significant differences in runoff, you could take spatially random samples measuring surface runoff levels from both settings. The question being asked is whether the mean volume of runoff from the natural locations differs significantly from the mean volume of runoff at the exurban locations. If the two sample means are significantly different, you can logically infer that the samples were taken from two distinct populations.

In another possible application, suppose you want to determine if current home values differ between two adjacent residential subdivisions built at the same time and with similar construction costs, but with very different design principles. One of the subdivisions used cluster zoning of homes on smaller lots, including such amenities as a tennis court, baseball diamond, playing field/village common, and some open space. The other subdivision used a conventional "cookie cutter" approach to design with larger private lots and no community recreational amenities or open space. Twenty years later, if a random sample of homes is taken from each subdivision, you could determine if the properties in the cluster-zoned development appreciated in value at a faster rate than those in the conventional subdivision.

As another example, suppose you want to compare fertility patterns in two different Chinese provinces. The first is mostly an urban coastal province heavily exposed to the one-child policies of the government over previous years, whereas the second is a rural interior province with a large ethnic population that experienced less pressure from central government officials regarding the one-child policy. Suppose independent random samples of women aged 30 to 34 are chosen from each of the two provinces, and the number of births for each woman recorded. The question can then be asked: is there a significant difference in the mean number of children per woman between these two provinces?

The null hypothesis when testing two independent sample means for significant differences has the form H_0: $\mu_1 = \mu_2$, which is equivalent to H_0: $\mu_1 - \mu_2 = 0$. The alternate hypothesis is stated in two ways: H_A: $\mu_1 \neq \mu_2$, when the direction of difference is not hypothesized (two-tailed format) or as H_A: $\mu_1 > \mu_2$ (or H_A: $\mu_1 < \mu_2$), when the direction of difference is hypothesized (one-tailed format).

Which of the two forms should be used in the preceding examples? In the water runoff problem, a one-tailed (directional) format seems appropriate because natural surfaces in an undeveloped landscape logically absorb more water, allowing more moisture to soak directly into the ground, resulting in a *lower* volume of water runoff. Conversely, precipitation falling on artificial surfaces in the exurban setting might not be absorbed as readily, resulting in *higher* volumes of runoff into gutters and drains along residential streets. In the subdivision design problem, a two-tailed (nondirectional) format is suggested if no logical, pre-established basis exists for hypothesizing that either design type would produce a faster home value appreciation rate. In the problem examining patterns of Chinese fertility, a one-tailed format is appropriate because a particular direction of difference makes sense; it is logical to hypothesize that women in the coastal province have *fewer* children on average than women in the more isolated interior province due to the increased one-child policy pressure placed on coastal communities.

Similar to the one-sample difference of means procedure, the parametric two-sample difference of means test can take different forms depending on the size of the samples and nature of the population variances. If the populations are normally distributed with known variances and the sample sizes are large (n_1 and n_2 greater than 30), the sampling distribution for the difference of means follows the normal (Z) distribution, and the test statistic is

$$Z = \frac{\bar{X}_1 - \bar{X}_2}{\sigma_{\bar{X}_1 - \bar{X}_2}} \qquad \textbf{(10.1)}$$

where $\bar{X}_1$ = mean of sample 1

$\bar{X}_2$ = mean of sample 2

$\sigma_{\bar{X}_1 - \bar{X}_2}$ = standard error of the difference of means

and

$$\sigma_{\bar{X}_1 - \bar{X}_2} = \sqrt{\frac{\sigma_1^2}{n_1} + \frac{\sigma_2^2}{n_2}} \qquad \textbf{(10.2)}$$

Following the general format for difference tests, the numerator of the two-sample test statistic in Equation 10.1 shows the actual or observed difference between the two sample means. Recall from Chapter 8 that the sample mean $(\bar{X})$ is the best estimate of the population mean (μ). Similarly, in this situation involving two samples, the difference of sample means $(\bar{X}_1 - \bar{X}_2)$ is the best estimate of the difference of population means ($\mu_1 - \mu_2$).

The denominator of Equation 10.1 is the standard error of the difference of means. This standard error expression is an estimate of how much difference between $\bar{X}_1$ and $\bar{X}_2$ is expected to occur due to sampling and is considered a measure of the expected sampling error. Think of it this way—if two independent random samples are taken from the very same population, the two sample means would likely be somewhat different, just because of the random effects of sampling. Thus, the standard error of the difference of means is an estimate of the expected difference that should occur simply by chance.

What factors influence the size of the test statistic in Equation 10.1? The magnitude of the *actual or observed* difference in sample means (the numerator) is divided by the magnitude of the *expected* difference in sample means (the denominator). If the actual difference in sample means is considerably greater than the expected difference in sample means, then the resultant test statistic Z or t is large (either positive or negative) and not very close to zero. A large test statistic leads to the inferential conclusion that the two sample means are more different than could have happened due to chance if both samples are taken from the same population. The large difference also leads to the complementary conclusion that the two sample means are most likely taken from two distinct (different) populations, supporting the likelihood of the alternate hypothesis.

Conversely, if the actual difference in sample means (numerator of Equation 10.1) is not much greater than the expected difference in sample means (denominator), then the resultant test statistic is relatively small. The small test statistic leads to the conclusion that the two sample means are taken from the same population and supports the validity of the null hypothesis. It is also possible for the actual difference in sample means to be smaller than the expected difference in sample means, again resulting in the conclusion that the two sample means are taken from the same population.

In most geographic applications of a two-sample difference of means test, the test statistic for Z in Equation 10.1 cannot be used because the variances of the two populations are unknown. In these situations, we must estimate the standard error from the *sample* variances (s^2), and the difference of means test will follow the t distribution. The test statistic for t is

$$t = \frac{\bar{X}_1 - \bar{X}_2}{\sigma_{\bar{X}_1 - \bar{X}_2}} \qquad \textbf{(10.3)}$$

Although Equation 10.3 has the same general appearance as Equation 10.1, the equations differ in the way the standard error (denominator) is estimated. In Equation 10.1, the standard error is derived directly from the known variances of the two populations using Equation 10.2. In Equation 10.3, the standard error can only be estimated from the variances calculated for the samples taken from the two populations. In addition to its usefulness for problems with unknown population variances, the t distribution also provides more accurate results than Z when one or both sample sizes are small (n less than 30).

Two methods exist for using sample data to estimate the standard error in the denominator of Equation 10.3. Selection of the appropriate method depends on the assumed relationship between the two population variances. In the first instance, when the population variances are assumed to be *equal*, $\left(\sigma_1^2 = \sigma_2^2\right)$, a **pooled variance estimate (PVE)** is calculated as the weighted average of the two sample variances:

$$\text{PVE} = \sqrt{\frac{s_1^2(n_1-1) + s_2^2(n_2-1)}{n_1 + n_2 - 2}} \qquad \textbf{(10.4)}$$

The denominator of Equation 10.4 represents the degrees of freedom in the problem. When the two population variances are unknown but assumed equal, the standard error estimate for Equation 10.3 is written as

$$\sigma_{\bar{X}_1 - \bar{X}_2} = \text{PVE}\sqrt{\frac{1}{n_1} + \frac{1}{n_2}} \qquad \textbf{(10.5)}$$

In problems where the population variances are assumed *unequal* (that is, if $\sigma_1 \neq \sigma_2$), the pooled estimate is not appropriate. In such cases, the sample variances are substituted directly into the standard error portion of Equa-

Two-Sample Difference of Means Z or t Test

Primary Objective: Compare two independent random sample means for difference

Requirements and Assumptions:

1. Two independent random samples
2. Each population is normally distributed
3. Variable is measured at interval or ratio scale

Hypotheses:

$H_0 : \mu_1 = \mu_2$
$H_A : \mu_1 \neq \mu_2$ (two-tailed)
$H_A : \mu_1 > \mu_2$ (one-tailed) or
$H_A : \mu_1 < \mu_2$ (one-tailed)

Test Statistic:

$$Z \text{ or } t = \frac{\bar{X}_1 - \bar{X}_2}{\sigma_{\bar{X}_1 - \bar{X}_2}}$$

If sample size ≥ 30 and population variances are known, use Z
If sample size < 30 and population variances are unknown, use t

tion 10.2 as best estimates of the respective population variances. A **separate variance estimate (SVE)** is calculated:

$$SVE = \sigma_{\bar{X}_1 - \bar{X}_2} = \sqrt{\frac{s_1^2}{n_1} + \frac{s_2^2}{n_2}} \quad \textbf{(10.6)}$$

Since geographers rarely know population parameters (like μ or σ), we must rely heavily on sample data to estimate population characteristics. This task supports one of the central goals of inferential statistics discussed earlier-the use of samples to produce unbiased estimates of population parameters. Thus, the method chosen to estimate variance (pooled vs. separate) depends on whether the population *variances* are assumed to be equal. In practice, researchers usually decide equality or inequality of population variances by testing the corresponding sample data. Sample variances are considered the best estimators of population variances.

To determine whether the variances of the two samples are significantly different, a test for equality of variances is needed. A widely used approach is the F statistic for the Levene test for equality of variances, derived by computing a one-way analysis of variance (ANOVA) on the absolute deviations of each observation from its sample mean. Details on the application of analysis of variance and the associated F test are discussed in the next chapter. If the two sample variances are found to be statistically different, the population variances are assumed to be unequal, and the separate variance method is used to test for the difference in the means. However, if the sample variances are not significantly different, the pooled variance estimate is applied. Most computer programs provide the result of an equality of variance test, as well as the t test statistic and p-values for both situations (variances assumed to be equal and not equal).

Wilcoxon Rank Sum *W* (Mann-Whitney *U*) Tests

When testing the difference between two independent samples, the parametric two-sample difference of means tests (Z or t) may not be appropriate for some geographic problems. Sometimes data are available only in ordinal or ranked form, making it impossible to calculate sample means and sample variances. In other situations, you may be working with data measured at the interval-ratio level, but have reason to believe that the samples are from populations that are not normally distributed. When populations exhibit moderate-to-severe deviation from normality, using a difference of means test raises serious questions about the validity of such a parametric procedure. When a parametric test is deemed inappropriate, a nonparametric test for two independent samples often provides a better alternative.

The most widely used nonparametric alternatives for the two-sample difference of means test are the Wilcoxon rank sum W test and the directly related Mann-Whitney U statistic. Like the t test, the Wilcoxon and Mann-Whitney tests examine two independent samples for differences. Rather than using parameters like mean and variance, however, these techniques use the ranks of sample observations to measure the magnitude of the differences in the ranked positions or locations between the two sets of sample data. Although no form of distribution is assumed for the two populations, the procedure requires that the two distributions be similar in shape. This characteristic makes the Wilcoxon and Mann-Whitney tests especially useful for problems with small samples drawn from populations that are not necessarily normal.

To keep the explanation of the nonparametric two-sample difference tests simple, the discussion here focuses on the Wilcoxon test procedure, which uses a "difference test" format similar to the two-sample difference of means Z and t tests. The discussion includes a brief explanation of the direct relationships of the Mann-Whitney and Wilcoxon test statistics.

In the Wilcoxon rank sum W test, the data from the two samples are combined and placed in a single ranked set. When two or more values are tied for a particular rank, the average rank value is assigned to each position. The samples are then considered separately and the sum of ranks (W) calculated for each sample. Suppose the sum of ranks for sample 1 is called W_1, and the sum of ranks for sample 2 is called W_2. If the two samples are drawn from the same population, the ranks should be randomly mixed between the two samples, and the sum of ranks for each sample should be roughly equal when

Wilcoxon Rank Sum Test

Primary Objective: Compare two independent random sample rank sums for difference

Requirements and Assumptions:

1. Two independent random samples
2. Both population distributions have the same shape
3. Variable is measured at ordinal or downgraded from interval/ratio scale to ordinal

Hypotheses:

H_0 : The distribution of measurements for the first population is equal to that of the second population

H_A : The distribution of measurements for the first population is not equal to that of the second population (two-tailed)

H_A : The distribution of measurements for the first population is larger (or smaller) than that for the second population (one-tailed)

Test Statistic:

$$Z_W = \frac{W_i - \overline{W}_i}{s_W}$$

the sample sizes are equal. That is, if $n_1 = n_2$, then W_1 should be roughly equal to W_2. If the sample sizes are not equal, but drawn from the same population, then their respective sum of ranks should be proportionate to their respective sample sizes. These findings would all confirm the null hypothesis of no significant difference between the two samples. However, if the sum of ranks for the first sample (W_1) is quite different from the sum of ranks for the second sample of similar size (W_2), it is more likely that the two samples have been drawn from different populations, making confirmation of the null hypothesis less likely.

The Wilcoxon test uses a variation of the Z test to see if the sum of sample ranks is significantly different from what it should be if the two samples are actually drawn from the same population. The test statistic (Z_W) for the two-sample Wilcoxon procedure is

$$Z_W = \frac{W_i - \bar{W}_i}{s_W} \quad \textbf{(10.7)}$$

where W_i = sum of ranks for sample i

$$\bar{W}_i = \text{mean rank of } W_i = n_i\left(\frac{n_1+n_2+1}{2}\right) \quad \textbf{(10.8)}$$

$$s_W = \text{standard deviation of } W = \sqrt{n_1 n_2\left(\frac{n_1+n_2+1}{12}\right)} \quad \textbf{(10.9)}$$

$\bar{W}_i$ and s_W represent the theoretical mean and standard deviation of W_i, respectively, and are determined totally by the sample sizes. As shown in Equation 10.9, only one standard deviation exists for W. However, because we can determine rank sums (W_1 and W_2) for each sample, two means can also be calculated using Equation 10.8—one for sample 1 $(\bar{W}_1)$ and another for sample 2 $(\bar{W}_2)$.

The Wilcoxon statistic, W, is simply the value or magnitude of the sum of ranks of the group with the *smaller* sample size. The Mann-Whitney statistic, U, which complements the Wilcoxon statistic, is determined by the number of times an observation from the group with the smaller sample size ranks lower than an observation from the group with the larger sample size. Because the Mann-Whitney calculation is often more cumbersome than the Wilcoxon calculation, the Wilcoxon worktable is the only one included in the geographic example. Statistical software packages usually calculate both Wilcoxon and Mann-Whitney test statistics. Both test statistic values are equivalent, in that they always provide the same significance level (p-value) when applied to the same set of data.

Example: Two-Sample Land Wilcoxon Rank Sum—Female Labor Force Participation Rates, More-Developed Countries Versus Less-Developed Countries

Many countries are experiencing profound changes in development as the result of an increasingly integrated global economy. A variety of economic, social, political, and demographic changes are taking place as less-developed countries attempt to advance through economic development and modernization. Measuring changes in modernization and development is a continuing and complicated process, potentially involving literally hundreds of different development indicators, as people across the globe seek to improve their standard of living and quality of life.

This rapidly changing situation offers geographers many opportunities for data analysis and research. An important target indicator involves women's progress towards equality in the global labor market. Have women made any recent advances in labor force participation? If so, what changes have occurred and where have they occurred? Suppose female labor force participation rate is selected as a target indicator for investigation. Does the participation level of women in the labor force vary by the economic level of the country in which they live, or is the relationship more complex? Are past disparities between men and women narrowing or widening with regard to job opportunities and the monetary and social gains of employment?

A 2012 World Development Report from the World Bank, *Gender Equality and Development*, examines a wide variety of problems and issues associated with gender differences and trends in employment. Over the past few decades, women have joined the labor market in increasing numbers; however, the global rate of change has been very gradual. Over the last forty years the global rate of female labor force participation has increased only slightly from 50.2% to 52.1%, indicating that nearly half of all adult women are neither employed nor seeking work.

The current spatial distribution of global female labor force participation shows a moderate to strong regional patterning (Figure 10.1). A great disparity in female labor force participation rates seems to exist across different less-developed regions. Several countries in Sub-Saharan Africa and East Asia have participation rates in excess of 70%. By contrast, various countries in the Middle East, North Africa, and South Asia have female participation rates below 40%.

To explore these spatial patterns of female labor force participation rates, we suggest the following process. Suppose we take two random samples of countries-one sample from more-developed countries (MDCs) and another from less-developed countries (LDCs). The same random samples of MDCs and LDCs are used for each of four years (1989, 1999, 2009, and 2019) so that we can evaluate trends in female labor force participation rates over the last four decades. We propose these two related inferential hypotheses: (1) "For each of the three years, the average female labor force participation rate of MDCs is not

(continued)

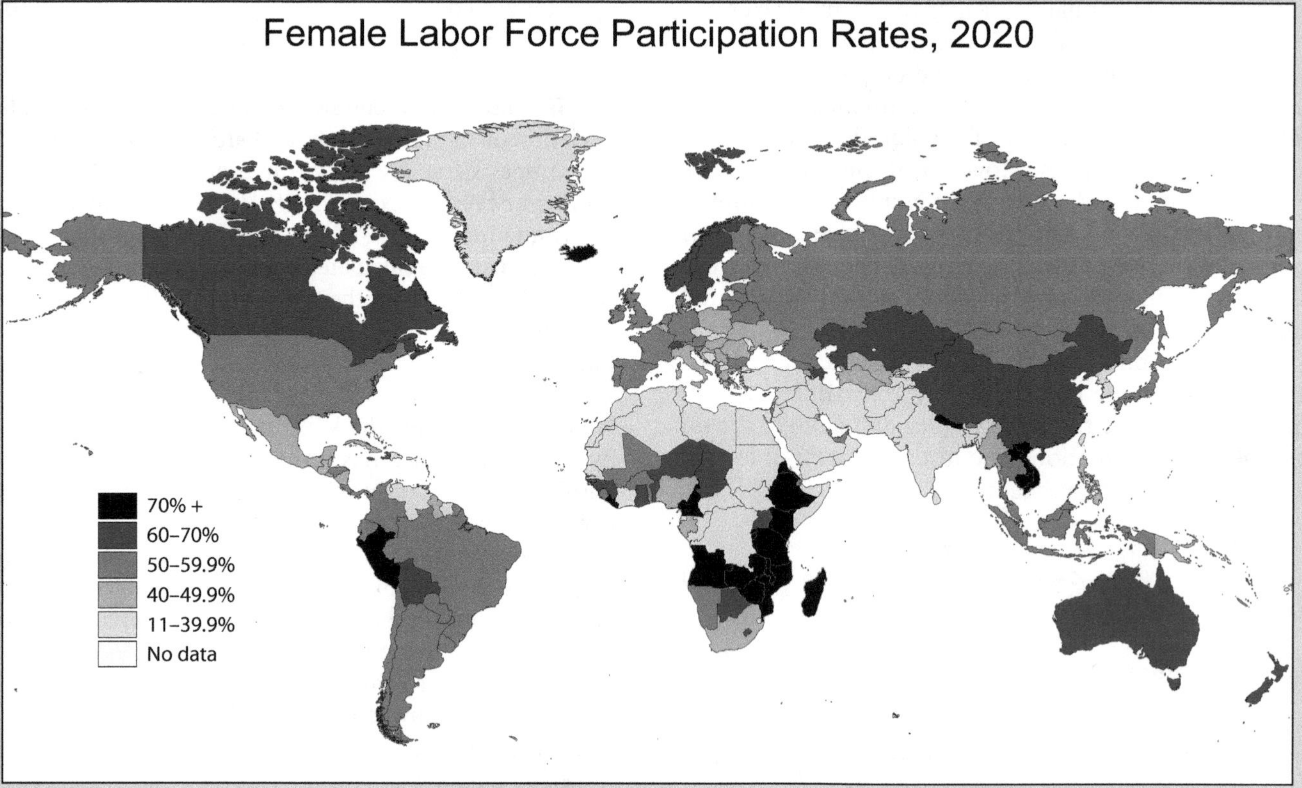

FIGURE 10.1
Female Labor Force Participation Rates, 2020
Source: World Bank, World Development Report 2020: Gender Equality and Development

different than the average female labor force participation rate of LDCs" and (2) "The more recent the year, the smaller the magnitude of difference between MDC and LDC female labor force participation rates." Taken together, these two hypotheses are attempting to conclude that there is no statistically significant difference between the average female labor force participation rates of MDCs and LDCs and that the LDCs are "catching up" or narrowing this already small disparity in female labor force participation as time goes by.

If we decide to run a parametric test, then the appropriate test would be the two-sample difference of means *t* test. (Note: we have more to say later about the appropriateness of selecting a parametric test for this particular problem). Suppose a *t* test is applied to the sample data for each of the four years: 1989, 1999, 2009, and 2019. It is best to use a separate variance estimate (SVE) as there appears to be a much larger variation among LDC female labor force participation rates than among MDC rates. Table 10.1 summarizes all of the relevant sample statistics and basic steps involved in the *t* test calculations.

What information can we glean from the worktable, and what are the basic conclusions that can be drawn from this analysis? First, notice that for all three years the mean female labor force participation rate for the more-developed country sample is slightly larger than the corresponding mean for the less-developed country sample. For example, in 1989 the MDC sample mean is 50.2 and the LDC sample mean is 47.4. In all four years, however, the difference in means is quite small and found to be statistically insignificant. Again, looking at the 1989 results, the calculated *t* test statistic is 0.74 and the associated *p*-value is .462. Furthermore, as time progresses, the magnitude of the calculated *t* test statistic gets smaller and the associated *p*-value becomes larger. Using classical hypothesis testing terminology, the null hypothesis of "no difference" should not be rejected at any conventional significance level (such as .10 or .05) for any of the years. For example, the 1989 *p*-value of .462 provides the exact probability associated with making a Type I error. That is, if we conclude that there is significant difference between the two sample means and reject the null hypothesis, then there is a .462 or 46.2% chance that the wrong or incorrect conclusion was reached. The logical decision in this case is to not reject (that is, accept) the null hypothesis.

The sequence of *p*-values (1989–.462; 1999–.528; 2009–.657; 2019–.703) indicates that the magnitude of difference between the MDC and LDC means is decreasing over time. The hypothesis (educated guess) suggesting that the gap is narrowing between MDC and LDC female labor force participation rates appears correct.

Table 10.1 reveals yet another important finding; for each year of analysis, the standard deviation of the LDC sample is much larger than the standard deviation of the MDC sample. For instance, in 1989 the MDC standard deviation is only 10.0, while the LDC standard deviation is nearly twice as large at 19.0. Furthermore, the difference in magnitude of standard deviation values widens considerably over time.

TABLE 10.1

Worktable for Two-Sample Difference of Means *t* test: Female Labor Force Participation Rates (1989, 1999, 2009, and 2019)

Sample Statistics: 1989

	n	$\bar{X}$	s	
More-developed countries (sample 1)	15	50.2	10.0	t = 0.74
Less-developed countries (sample 2)	45	47.4	19.0	2-tailed p-value = .462

Sample Statistics: 1999

	n	$\bar{X}$	s	
More-developed countries (sample 1)	15	50.87	8.19	t = 0.64
Less-developed countries (sample 2)	45	48.8	17.2	2-tailed p-value = .528

Sample Statistics: 2009

	n	$\bar{X}$	s	
More-developed countries (sample 1)	15	52.60	7.82	t = 0.45
Less-developed countries (sample 2)	45	51.2	16.5	2-tailed p-value = .657

Separate Variance Estimate (SVE) = $\sigma_{\bar{X}_1-\bar{X}_2} = \sqrt{\frac{s_1^2}{n_1}+\frac{s_2^2}{n_2}} = \sqrt{\frac{(7.82)^2}{15}+\frac{(16.5)^2}{45}} = \sqrt{4.077+6.05}$ = 3.18

$$t = \frac{\bar{X}_1 - \bar{X}_2}{\sigma_{\bar{X}_1-\bar{X}_2}} = \frac{52.60 - 51.2}{3.18} = 0.45$$

Sample Statistics: 2019

	n	$\bar{X}$	s	
More-developed countries (sample 1)	15	53.70	6.81	t = 0.54
Less-developed countries (sample 2)	45	52.11	15.97	2-tailed p-value = .703

Separate Variance Estimate (SVE) = $\sigma_{\bar{X}_1-\bar{X}_2} = \sqrt{\frac{s_1^2}{n_1}+\frac{s_2^2}{n_2}} = \sqrt{\frac{(6.81)^2}{15}+\frac{(15.97)^2}{45}} = \sqrt{3.09+5.66}$ = 2.95

$$t = \frac{\bar{X}_1 - \bar{X}_2}{\sigma_{\bar{X}_1-\bar{X}_2}} = \frac{53.70 - 52.11}{2.95} = 0.54$$

(continued)

Let's explore the 2019 data further. Recall from Chapter 3 that an individual value plot is a graph showing the distribution of all observations in a sample. We can display country-level observations in comparative individual value plots with all sampled MDCs plotted in one column and all sampled LDCs in another (Figure 10.2). The 15 MDCs and 45 LDCs have nearly equal mean female labor force participation rates (53.7 and 52.1, respectively). However, the standard deviation of the MDC participation rates is less than half the standard deviation of the LDC rates (6.81 versus 15.97).

These individual value plots clearly show a much larger variation in LDC values as contrasted with MDC values. Several Sub-Saharan African LDCs in the 2019 sample have female labor force participation rates in excess of 70% (Kenya, Zambia, Ethiopia, and Mozambique). At the other extreme, several LDCs in the Middle East and North Africa (Algeria, Morocco, and Tajikistan) have less than 30% of females participating in the labor force.

Clearly, the social and economic status of women varies greatly across the less-developed world. The obvious next question you might ask in the geographic research process is *why*. Why are women in Sub-Saharan Africa and Southeast Asia heavily involved in the economies of their country, while Middle Eastern and Northern African woman are very lightly involved (even excluded) from labor force activities? The reasons are multiple, complex, and deeply rooted in cultural heritage, values, and traditions of each country. Many individual as well as societal factors influence a woman's decision to enter the labor force. It is unknown how many women would like to work but have not done so due to cultural norms. The different rates of increase in women's participation in the work force from one LDC to another are probably linked to improved economic development, rising education among women, declining fertility, and different rates of change in various cultural and religious norms. We need to make one final point that casts severe validity questions on this entire geographic example. The magnitude of certain factors related to female employment is virtually unknown in many countries. These factors include the extent of labor underutilization among females in many countries (both unemployed and working only part time). Also generally unknown is the degree to which women are working in the "informal economy" or in a situation of "vulnerable employment," where wages may not be monetary and women are working as unpaid family members without market economy benefits. The bottom line here is that data regarding female labor force participation rates may be nothing more than very rough guesses (even from reputable organizations such as The World Bank) subject to major differences in data collection from country to country.

The Wilcoxon rank sum W and Mann-Whitney U tests are now applied to the same female labor force participation rate data. "Pairing" these nonparametric tests with the parametric t test is an experiment. For one thing, we do not know whether the two populations from which samples were taken (MDCs and LDCs) are normally distributed. This is an important consideration, since normality of the two populations being tested is one of the assumptions associated with the parametric two-sample difference tests (Z and t). Also, the MDC sample size of 15 is rather small (certainly smaller than 30), so it is not possible to take full advantage of the central

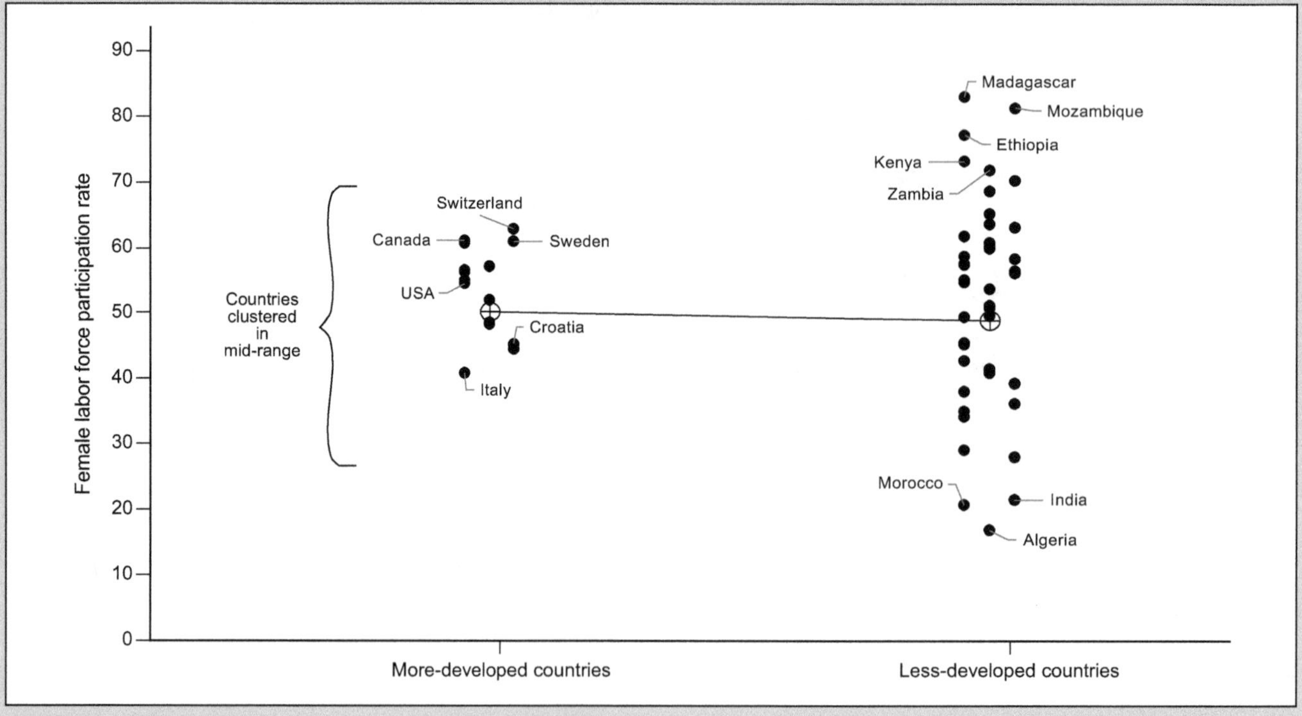

FIGURE 10.2
Individual Value Plots of Female Labor Force Participation Rates, 2019
Source: World Bank, World Development Report 2020: Gender Equality and Development

limit theorem. The worktable for the Wilcoxon rank sum *W* test (Table 10.2) summarizes the application of these tests on the female labor force participation rates for the 2019 data. In that year, the sum of ranks for the smaller MDC sample is 460.5 and for the larger LDC sample it is 1369.5. The Wilcoxon *W* is the sum of ranks of the *smaller* sample (that is, the MDC sample of 460.5). Results are also shown for the standard deviation of *W* and the test statistic Z_W, with a low test statistic value (Z_W = 0.051) and associated *p*-value of .968

The results of the nonparametric two-sample difference tests (Wilcoxon and Mann-Whitney) are about what we should expect in relation to the corresponding two-sample *t* test. The *p*-values associated with the *t* tests are just slightly lower (stronger statistically) than the comparable Wilcoxon (Mann-Whitney) *p*-values. This finding is expected, given the slightly stronger statistical power of the parametric *t* test over its nonparametric equivalents.

For example, a quick comparison of the 2019 *p*-values from Tables 10.1 and 10.2 (.968 for the nonparametric test versus .703 for the parametric *t* test) shows a moderately weaker conclusion for the nonparametric tests (Table 10.3). Clearly, some information about the female labor force participation rate data is lost when this interval-ratio variable is downgraded to ordinal. However, both the parametric and nonparametric results are telling basically the same story-for all three dates, the *p*-values are much larger than conventional cut-off values (.10 or .05) for rejecting the null hypothesis and for concluding there is a significant difference between means. Thus, we must conclude there is no significant difference between MDC and LDC female labor force participation rates.

TABLE 10.3

Comparing *p*-values of Parametric and Nonparametric Tests for Two-Sample Differences: Female Labor Force Participation Rates (1989, 1999, 2009, and 2019)

	t	versus	Z_W
1989	.462		.707
1999	.528		.675
2009	.657		.791
2019	.703		.968

TABLE 10.2

Worktable for Wilcoxon Rank Sum Test (Two Samples): Female Labor Force Participation Rates, 2019

Sample Statistics: 2019

	n	Sum of Ranks	Mean Rank
More-developed countries (sample 1)	15	460.5	30.70
Less-developed countries (sample 2)	45	1369.5	30.43

Wilcoxon W_1 is the sum of ranks of the group with the smaller sample

$W_1 = 460.5$

Mean rank of sample 1 $= \bar{W}_1 = n_1\left(\frac{n_1+n_2+1}{2}\right) = 15\left(\frac{15+45+1}{2}\right) = 457.5$

S_W = standard deviation of W $= \sqrt{n_1 n_2\left(\frac{n_1+n_2+1}{12}\right)} = \sqrt{(15)(45)\left(\frac{61}{12}\right)} = 58.575$

$$Z_W = \frac{W_i - \bar{W}_i}{S_W} = \frac{460.5 - 457.5}{58.575} = \frac{3.0}{58.575} = 0.051$$

2-tailed *p*-value = .968

Note: $W_i - \bar{W}_i$ is the actual difference and S_W is the expected difference.

10.2 Two-Sample Difference of Proportions Test

Geographers and other scientists often work with proportions as well as means, so it should not be surprising that statisticians have developed a two-sample difference of proportions test. Inferences concerning the difference of two population proportions are made by comparing the difference of two sample proportions, using a similar logic (and similar test structure) as the difference of means tests.

For example, a political geographer might want to learn if a difference exists in the President's approval rating among the voting population from two neighboring states or among male and female voters. Random samples of voters in each state could be polled and the proportion of potential voters giving the President a favorable rating tabulated. A two-sample difference of proportions test could then assess the likelihood that a statistically significant difference in approval ratings exists. A recreation planner could ask visitors at a regional park whether they approve of a proposed change in operating hours. If the park visitors were divided into two samples (weekday visitors versus weekend visitors), it is possible to apply a two-sample difference of proportions test to determine if the proportion of those who approve of the new hours differs between weekday and weekend visitors.

In these situations, where a dichotomous (binary) variable has only two possible outcomes or responses, the two-sample difference of proportions test is appropriate. For many problems, the variable analyzed fits this requirement of exactly two categories or classes, such as male-female, approve-disapprove, or yes-no. In situations with more than two possible outcomes or responses, you can sometimes reduce three or more categories to two so that the difference of proportions test is appropriate across the entire data set. For example, a public survey measuring attitudes toward capital punishment might include the following possible responses: strongly favor, mildly favor, mildly oppose, and strongly oppose. If these four possible responses were "collapsed" to two (favor and oppose) then the data would be structured in a way that would allow a single two-sample difference of proportions test to be applied across the entire set of data. Then you could test whether the proportion of elderly (age 65 and above) favoring capital punishment differs from the proportion of young adults (age 20–34) sharing that opinion.

In problems testing for significant differences with a binary variable, one of the two categories is selected as the focus for the analysis. The proportion of the sample in the focus category is termed p_1, while the other proportion is termed p_2. The objective of the difference test is to determine whether the proportion of *population 1* (ρ_1) having the focus attribute differs significantly from the corresponding proportion of *population 2* (ρ_2).

The null hypothesis for a two-sample difference of proportion test is written: H_0: $\rho_1 - \rho_2 = 0$. By expressing the null hypothesis this way, it relates directly to the *difference* of population proportions and it is hypothesized that the difference is zero. It is possible to set the hypothesized difference as something other than zero, but that treatment is not covered here. As in other difference tests, the alternate hypothesis (H_A) can take two forms. If no direction is hypothesized for difference in proportion between the two samples, a two-tailed procedure is applied where H_A: $\rho_1 - \rho_2 \neq 0$. A two-tailed procedure is used if one wishes to test if the proportion of elderly adults in favor of capital punishment differs from the proportion of young adults in favor of capital punishment. On the other hand, if one wishes to test whether one of the sample proportions is larger or smaller than the other (specifying the direction of difference), a one-tailed test is appropriate. A one-tailed test is used for a hypothesis that a greater proportion of elderly favor capital punishment than do young adults. In these instances, the alternate hypothesis is H_A: $\rho_1 - \rho_2 > 0$ if it is assumed that ρ_1 is the larger of the two proportions.

The test statistic for the difference of proportion procedure (Z_p) has a form similar to other two-sample difference tests:

$$Z_p = \frac{p_1 - p_2}{\sigma_{p_1 - p_2}} \quad \textbf{(10.10)}$$

where p_1 = proportion of sample 1 in the category of focus
p_2 = proportion of sample 2 in the category of focus
$\sigma_{p_1 - p_2}$ = standard error of the difference of proportions

Two-Sample Difference of Proportions Test

Primary Objective: Compare two independent random sample proportions for difference

Requirements and Assumptions:

1. Two independent random samples
2. Variable is organized by dichotomous (binary) categories

Hypotheses:

$H_0 : p_1 = p_2$
$H_A : p_1 \neq p_2$ (two-tailed)
$H_A : p_1 > p_2$ (one-tailed) or
$H_A : p_1 < p_2$ (one-tailed)

Test Statistic:

$$Z_p = \frac{p_1 - p_2}{\sigma_{p_1 - p_2}}$$

Just as the sample proportion (p) is the best estimate of the population proportion (ρ), the sample difference of proportions ($p_1 - p_2$) is the best estimate of the population difference of proportions ($\rho_1 - \rho_2$). The denominator of Equation 10.10 represents the standard error of the difference in proportions and is estimated as follows:

$$\sigma_{p_1 - p_2} = \sqrt{\hat{p}(1-\hat{p})\left(\frac{n_1 + n_2}{n_1 n_2}\right)} \quad \textbf{(10.11)}$$

where $\hat{p}$ = pooled estimate of the focus category for the population

The pooled estimate $(\hat{p})$ is the proportion in the focus category if the two samples are combined into one sample. Operationally, the pooled estimate is the weighted proportion from the two samples:

$$\hat{p} = \frac{n_1 p_1 + n_2 p_2}{n_1 + n_2} \quad \textbf{(10.12)}$$

Example: Two-Sample Difference of Proportions-Changes in State-Level Obesity Rates from 2010 to 2020

The two-sample difference of proportions test is illustrated by returning again to the obesity data provided by the Centers for Disease Control and Prevention (CDC). Their excellent sampling processes allow scientists to develop hypotheses from a sequence of independent samples at various spatial levels and over different time periods. One of many such possibilities is to develop hypotheses regarding changes in state-level obesity rates over time. We suggest the following hypotheses:

- "For each state in the United States, the proportion of the adult population classified as obese in 2020 is higher than the proportion of the adult population classified as obese in 2010." This hypothesis is testable inferentially using the two-sample difference of proportions test because the CDC has taken independent samples from the residents of each state for both 2010 and 2020.
- "States with higher obesity levels in 2010 have larger increases in the proportion of their population that is obese from 2010 to 2020 than states with lower obesity levels in 2010." This descriptive statement (from the obesity problem in Chapter 1) suggests that the level of "obesity inequality" by location is increasing and that the "obesity gap" between high and low obesity level states is widening. Some exploratory results are presented, including a scatterplot showing the state-level relationship between the proportion of obese adults in 2010 and the magnitude of change in the proportion of adults classified as obese from 2010 to 2020.

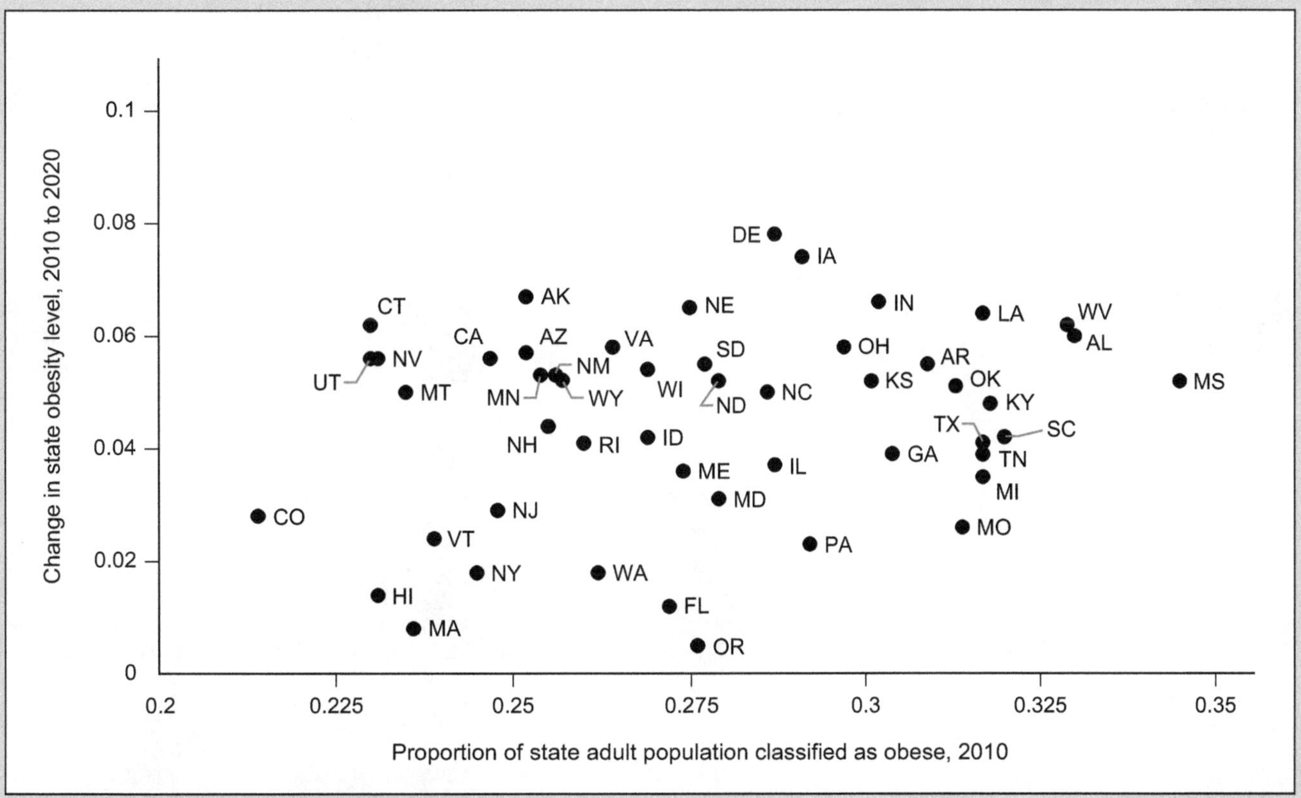

FIGURE 10.3
Scatterplot of Proportion of State Adult Population Classified as Obese (2010) and Change in Obesity Level, 2010 to 2020

(continued)

From a purely descriptive perspective, every state in the United States experienced an increase from 2010 to 2020 in the proportion of adult population classified as obese by the CDC (Table 10.4). The greatest increase in proportion occurred in Delaware, which went from .287 (28.70%) obese in 2010 to .365 (36.50%) in 2020-a fairly significant change in just one decade, considering the increases already included from the previous decades. A number of other states also experienced over a .05 proportional increase (over a 5% increase) in obesity level over the decade, including Alabama, Arkansas, California, Mississippi, Oklahoma, and West Virginia. The states with the least proportional increase in obesity included Oregon .005 (0.50%); Massachusetts .008 (.8%), Florida .012 (1.2%), and Hawaii .014 (1.4%).

TABLE 10.4

Proportion of State Adult Population Classified as Obese (2010 and 2020) and Change in Obesity Level from 2010 to 2020

State	State ID	Obesity Level in 2010 (proportion of adults classified as obese)	Obesity Level in 2020 (proportion of adults classified as obese)	Change in Obesity Level 2010 to 2020
Alabama	AL	0.328	0.390	0.060
Alaska	AK	0.252	0.319	0.067
Arizona	AZ	0.252	0.309	0.057
Arkansas	AR	0.309	0.364	0.055
California	CA	0.247	0.303	0.056
Colorado	CO	0.214	0.242	0.028
Connecticut	CT	0.230	0.292	0.062
Delaware	DE	0.287	0.365	0.078
Florida	FL	0.272	0.284	0.012
Georgia	GA	0.304	0.343	0.039
Hawaii	HI	0.231	0.245	0.014
Idaho	ID	0.269	0.311	0.042
Illinois	IL	0.287	0.324	0.037
Indiana	IN	0.302	0.368	0.066
Iowa	IA	0.291	0.365	0.074
Kansas	KS	0.301	0.353	0.052
Kentucky	KY	0.318	0.366	0.048
Louisiana	LA	0.317	0.381	0.064
Maine	ME	0.274	0.310	0.036
Maryland	MD	0.279	0.310	0.031
Massachusetts	MA	0.236	0.244	0.008
Michigan	MI	0.317	0.352	0.035
Minnesota	MN	0.254	0.307	0.053
Mississippi	MS	0.345	0.397	0.052
Missouri	MO	0.314	0.340	0.026
Montana	MT	0.235	0.285	0.050
Nebraska	NE	0.275	0.340	0.065
Nevada	NV	0.231	0.287	0.056
New Hampshire	NH	0.255	0.299	0.044
New Jersey	NJ	0.248	0.277	0.029
New Mexico	NM	0.256	0.309	0.053
New York	NY	0.245	0.263	0.018
North Carolina	NC	0.286	0.336	0.050
North Dakota	ND	0.279	0.331	0.052
Ohio	OH	0.297	0.355	0.058
Oklahoma	OK	0.313	0.364	0.051
Oregon	OR	0.276	0.281	0.005
Pennsylvania	PA	0.292	0.315	0.023
Rhode Island	RI	0.260	0.301	0.041
South Carolina	SC	0.320	0.362	0.042
South Dakota	SD	0.277	0.332	0.055
Tennessee	TN	0.317	0.356	0.039
Texas	TX	0.317	0.358	0.041
Utah	UT	0.230	0.286	0.056
Vermont	VT	0.239	0.263	0.024
Virginia	VA	0.264	0.322	0.058
Washington	WA	0.262	0.280	0.018
West Virginia	WV	0.329	0.391	0.062
Wisconsin	WI	0.269	0.323	0.054
Wyoming	WY	0.257	0.307	0.052

Source: Centers for Disease Control and Prevention (CDC)

If you are looking for an easily explained spatial pattern in these 2010–2020 obesity level changes, you will probably be disappointed. The descriptive statement we presented earlier suggests that states with higher obesity levels in 2010 increased their obesity levels more dramatically from 2010 to 2020 than did states with lower obesity levels in 2010. Is there tangible evidence of this widening "obesity gap" over the last decade? To explore this contention, a scatterplot is constructed showing the obesity level for each state in 2010 plotted against magnitude of proportional change in obesity from 2010 to 2020 (Figure 10.3, p. 167). There appears to be no strong relationship between these two variables.

In other words, knowing the obesity level of a particular state in 2010 gives little or no clue as to that state's change in obesity level from 2010 to 2020. We calculated the direction and strength of this relationship statistically using correlation (discussed later in Chapter 16), and found no evidence of a significant "widening obesity gap." At this point, it is sufficient to say that only a slight (and statistically insignificant) relationship exists between state obesity level in 2010 and the percent change in state obesity level from 2010 to 2020 (the Pearson's correlation is 0.219 and the associated *p*-value is .126).

The significance of these obesity level changes are generally (but not completely) confirmed statistically. A series of 50 two-sample difference of proportions tests are run, with a separate statistical test for each state. A representative two-sample difference of proportions procedure is shown for the State of Alabama in Table 10.5. The obesity level for adults in Alabama increased from .328 (32.8%) in 2010 to .390 (39.0%) in 2020. Therefore, the observed (or actual) difference of propor-

TABLE 10.5

Worktable for Two-Sample Difference of Proportions: Comparing Alabama Adult Rates (Proportion Obese), 2010 to 2020

$H_0: p_1 - p_2 = 0$

$H_A: p_1 - p_2 > 0$ if it is assumed (hypothesized) that population 1 has the larger of the two proportions

	Year	Total sampled (*N*)	Number classified as obese (*X*)	Sample proportion obese
Sample 1	2010	7,269	2,388	0.33
Sample 2	2020	4,973	1,939	0.39

$p_1 - p_2$ (observed difference of proportions) = 0.06

Pooled proportion $\hat{p} = \frac{n_1 p_1 + n_2 p_2}{n_1 + n_2} = \frac{7269(.33) + 4973(.39)}{7269 + 4973} = .354$ (weighted avg. of the two proportions)

$$p_1 - p_2 \text{ (standard error of the difference of proportions)} = \sqrt{\hat{p}(1-\hat{p})\left(\frac{n_1 + n_2}{n_1 n_2}\right)}$$

$$= \sqrt{0.354(0.646)\left(\frac{7{,}269 + 4{,}973}{7{,}269(4{,}973)}\right)}$$

$$= \sqrt{0.354(0.646)(0.00034)}$$

$$= .008 \text{ (expected difference of proportions, if } H_0 \text{ is true)}$$

$$Z = \frac{p_1 - p_2}{S_W} = \frac{0.06}{0.008} = 7.5 \quad \text{with associated } p\text{-value} = .0000$$

(continued)

tions, $(p_1 - p_2)$, equals .062. Unfortunately for the overall health of Americans, this is a fairly typical state-level increase in the proportion of adults who are obese.

To proceed with the Z test for difference of proportions, $Z_p = (p_1 - p_2)/\sigma_{p_1-p_2}$ the denominator of the test statistic (that is, the standard error of the difference of proportions) must be calculated. To do this, the "pooled" proportion of the two sample proportions is calculated. This "pooled" proportion $\hat{p} = n_1p_1 + n_2p_2 / n_1 + n_2$ is the weighted average of the two proportions (for Alabama, this value is .3081). Using a weighted average is best because the two sample sizes are not the same. For Alabama, the 2020 sample size is quite large. The standard error of the difference of proportions can now be calculated. This standard error is the difference of proportions "expected" if the null hypothesis is true and there really is no difference in the two samples. The value of this expression $\sigma_{p_1-p_2} = \sqrt{\hat{p}(1-\hat{p})(n_1 + n_2 / n_1n_2)}$ for Alabama is .008. The associated test statistic $Z_P = (p_1 - p_2)/\sigma_{p_1-p_2}$ is the ratio of the actual (observed) difference of proportions to the expected difference of proportions if the null hypothesis is true and both samples have been drawn from the same underlying population. The observed difference of proportions is nearly eight times the expected difference of proportions: $Z_P = (p_1 - p_2)/\sigma_{p_1-p_2} = .060/.008 = 7.5$ and the associated p-value is .000. It is clear that the proportion of Alabama's adult population classified as obese has increased significantly from 2010 to 2020, and we can be virtually 100% confident that this is true.

The most significant statistical change in obesity level occurs in Nebraska, which has a Z-value of 12.68 ($p = .000$). The only state not having a p-value of .000 is California. However, by any reasonable standard, California's obesity level, Z-value of 3.14 ($p = .001$), also went up significantly. Some states had higher Z-values than you might expect simply due to the larger 2020 sample sizes taken by the CDC. [Note: remember the general statistical rule—all other things being equal, the larger the sample size, the higher (and more statistically significant) the test statistic value]. This is certainly the case with Florida, which had a huge statewide sample of 33,638 in 2010: nearly double the size of any other state. The state of Washington ranked second in 2010 sample size with 18,571 respondents.

10.3 Dependent-Sample (Matched-Pairs) Difference Tests

As geographers, we sometimes want to test whether two sets of sample values are significantly different, even if they are drawn from only one group of individuals or only one set of locations. For example, we may want to compare the number of migrants moving into a set of counties with the number of migrants leaving the same set of counties. The number of in-migrants and number of out-migrants could be determined for a sample of geographic areas and the differences tested for statistical significance. In another example, residents of an area are sampled to determine the miles traveled per week for purposes other than commuting. Following a major rise in the price of gasoline, which can occur during times of increased political tension in the Middle East, the same individuals are surveyed a second time to determine changes in their discretionary travel behavior. One might ask if the change in gasoline prices led to decreased automobile use for drivers within a particular region. In another example, the same sample of coastal zone residents could be surveyed both before and after a severe winter storm (nor'easter) to determine if attitudes toward coastal zone management change significantly. As a final example, suppose that average male life expectancy *and* average female life expectancy data are collected from a single random sample of villages in Mexico. Are male and female life expectancies in these Mexican villages different from one another?

In each of these examples, the data consist of one set of observations (locations or individuals) collected for two different variables or at two different time periods. At first, the difference of means t or Wilcoxon rank sum W might seem most appropriate to test differences for significance. However, as discussed in Section 10.1, these two difference tests require two **independent samples**. In the examples from the previous paragraph, only *one* sample is drawn. In two of those examples, data are collected for two variables at one time period (in- and out-migrants; male and female life expectancy). In the other two examples, data are collected for one variable at two time periods (miles traveled before and after a price increase for gasoline; attitudes toward managing the coast before and after a winter storm). Thus, neither the two-sample t test nor Wilcoxon W procedure is appropriate for these problems.

When two sets of data are collected for one group of observations, the samples are termed **dependent**, and a **test of matched-pairs** is the proper inferential procedure for examining differences between such samples. As the name implies, each observation or sample member has two values, known as a "matched pair." The differences in the set of matched-pairs are tested for statistical significance and results inferred to the population from which the dependent samples are drawn. In the example of life expectancy in the Mexican villages, two equal-sized samples are drawn-one male and the other female. However, the samples are not independent. In fact, average male and female life expectancies from a particular village are directly related to one another, since men and women at the same location are likely affected by many of the same economic, social, and environmental factors. The male and female life expectancies from each village are a matched pair, and it is appropriate to compare a sample of such matched-pairs of observations to see if they differ significantly.

Two common inferential procedures for testing differences in dependent samples are the matched-pairs *t* test and the Wilcoxon matched-pairs signed-ranks test. The matched-pairs *t* test is a parametric test requiring interval-ratio level data and a normally distributed population. Like the Wilcoxon *W* two-sample difference test, the related matched-pairs signed-ranks test is a distribution-free (that is, the population from which the sample is taken can have any distribution) nonparametric method that uses either ranked data directly or interval-ratio data downgraded to its ordinal equivalent.

Matched-Pairs *t* Test

The matched-pairs *t* test considers the difference between the values for each matched pair. The greater this difference (d), the more dissimilar the results of the two values within the matched pair. The mean of the difference values is $\left(\bar{d}\right)$ determined for the set of all matched pairs in the sample. If the differences within the matched-pair values are small, the average difference value is close to zero. However, in a situation where the matched-pair differences are large, $\bar{d}$ is also large, suggesting a significant difference between the two sets of data.

The null hypothesis in the matched-pair problem states that the mean difference for all pairs in the population (δ) equals zero. The best estimate of the population matched-pair mean difference (δ) is the sample matched-pair mean difference $\left(\bar{d}\right)$. The alternative hypothesis is either nondirectional or directional, depending on *a priori* information about the expected matched-pair differences.

Matched-Pairs *t* Test

Primary Objective: Compare matched pairs from a random sample for difference

Requirements and Assumptions:

1. Random sample
2. Data are collected for two different samples or at two different time periods
3. Population is normally distributed
4. Variable(s) is (are) measured at interval or ratio scale

Hypotheses:

$H_0 : \delta = 0$

$H_A : \delta \neq 0$ (two-tailed)

$H_A : \delta > 0$ (one-tailed) or

$H_A : \delta < 0$ (one-tailed)

Test Statistic:

$$t_{mp} = \frac{\bar{d}}{\sigma_d}$$

The test statistic for the matched-pairs *t* (t_{mp}) is defined as follows:

$$t_{mp} = \frac{\bar{d}}{\sigma_d} \quad \textbf{(10.13)}$$

where $\bar{d}$ = mean of matched-pair differences (d)

σ_d = standard error of the mean difference

The numerator of Equation 10.13 is the mean of matched-pair differences (d):

$$\bar{d} = \frac{\Sigma d_i}{n} \quad \textbf{(10.14)}$$

where d_i = difference for matched-pair i

n = number of matched pairs

The difference (d_i) is found by subtracting the corresponding paired values of the second variable (Y) from those of the first variable (X):

$$di = X_i - Y_i \quad \textbf{(10.15)}$$

The mean of the difference values $\left(\bar{d}\right)$ can also be calculated from the means of the two variables:

$$\bar{d} = \bar{X} - \bar{Y} \quad \textbf{(10.16)}$$

Like other difference tests, the denominator of Equation 10.13 contains a standard error measure. In this case, the standard error refers to the mean difference in matched pairs:

$$\sigma_d = \frac{s_d}{\sqrt{n}} \quad \textbf{(10.17)}$$

where

$$s_d = \sqrt{\frac{\Sigma\left(d_i - \bar{d}\right)^2}{n-1}} \quad \textbf{(10.18)}$$

Standard deviation of the matched-pair differences can also be derived from the computational formula (Table 3.3):

$$s_d = \sqrt{\frac{\Sigma d_i^2 - \frac{\left(\Sigma d_i\right)^2}{n}}{n-1}} \quad \textbf{(10.19)}$$

Wilcoxon Matched-Pairs Signed-Rank Test

In some geographic problems, a matched-pair (dependent-sample) test may be appropriate, but the sample data for analysis are measured at the ordinal level. In other situations, the sample sizes are too small for the reliable use of a parametric technique or the sample data are not drawn from a normally distributed population. In the first instance, the parametric matched-pair t test cannot be applied because the measurement scale is not appropriate. In the other cases, the parametric test may produce biased results. The Wilcoxon signed-ranks test is the nonparametric equivalent for dependent sample or matched-pair problems and is the appropriate procedure in these situations.

The Wilcoxon signed-ranks test uses matched-pair differences ranked from lowest (rank 1) to highest. The matched-pairs data can come either from direct ordinal measurement or from interval-ratio differences downgraded to ranks. The *absolute* difference between the two variables (rather than the positive or negative difference) is used to determine the rank for each matched pair. When the difference for any matched pair is zero, the data are ignored and the sample size reduced accordingly. When differences for matched pairs are tied for a particular rank position, the average rank is assigned to each such pair. The null hypothesis for the signed-ranks test states that the matched-pairs differences (in ranks) for the population from which the sample is drawn equals zero.

Two sums can be calculated from the set of ranked matched pairs: T_p, the sum of ranks for *positive* differences (variable 1 greater than variable 2), and T_n, the sum of ranks for *negative* differences (variable 2 greater than variable 1). If the two variables measured for the single sample show very little difference, T_p should be approximately equal to T_n. However, for a problem in which the differences between the two variables are large, the disparity between T_p and T_n will also be large. In these situations, one of the rank sums (either the positive or negative differences) will be large and the other small.

The Wilcoxon test for dependent samples uses only one of the two possible rank sum values. The decision of which rank sum to test depends on whether the alternative hypothesis (H_A) is directional (one-tailed) or nondirectional (two-tailed). If no direction of difference between the two variables is hypothesized, a two-tailed test is applied, and the *smaller* of T_p and T_n is chosen. In this instance, the hypothesis concerns only a difference between the two variables under study and not which variable is the largest.

The second possibility involves a directional hypothesis and a one-tailed procedure. In this case, the hypothesis states that either the positive or negative differences for the matched pairs are expected to dominate. The value of T corresponding to the *smaller number* of hypothesized differences (either positive or negative) is selected for testing. Thus, if more differences are expected to be positive, T_n, the sum of the negative differences is used.

Wilcoxon Matched-Pairs Signed-Ranks Test

Primary Objective: Compare matched pairs from a random sample for difference

Requirements and Assumptions:

1. Random sample
2. Data are collected for two different variables or at two different time periods
3. Variable(s) is (are) measured at ordinal scale or downgraded from interval/ratio scale to ordinal

Hypotheses:

H_0 : The ranked matched-pair differences of the populations are equal

H_A : The ranked matched-pair differences of the populations are not equal (two-tailed)

H_A : The ranked matched-pair differences of the population are positive or negative (one-tailed)

Test Statistic:

$$Z_W = \frac{T - \frac{n(n+1)}{4}}{\sqrt{\frac{n(n+1)(2n+1)}{24}}}$$

when $n > 10$

When the number of matched pairs exceeds ten, the rank sum (T) can be converted to a Z statistic (Z_W) and tested using the distribution of normal values:

$$Z_W = \frac{T - \frac{n(n+1)}{4}}{\sqrt{\frac{n(n+1)(2n+1)}{24}}} \quad \textbf{(10.20)}$$

where n = number of matched pairs ($n > 10$)

T = rank sum

Example: Matched-Pairs *t* Test—Salmon Growth Rates with (and without) Growth Hormone

A biogeographer is interested in determining if a growth hormone can increase the size of farm-raised salmon faster than the normal (hormone-free) growth rate. Typically, farm-raised salmon require 30 months to reach market size. It is hoped the introduction of a growth hormone will decrease the time to 18 months. Obviously, if the farmer can ship the product to market sooner, profits can increase.

Normally, farm-raised salmon grow an average of .75kg per year. Therefore, even without the hormone, the farmer would expect the salmon to grow by an average of .75kg in a given year. Therefore, the farmer wants to learn if the added growth hormone will increase the size of the salmon significantly more than .75kg. To test this hypothesis, a random sample of 10 salmon are weighed, given the growth hormone, and weighed one year later. Because the farmer is monitoring the same fish before and after the hormone treatment, a matched-pairs procedure is applied. The findings are shown in Table 10.6.

As expected, every salmon gains weight. However, the farmer is interested in whether the growth rate is significantly more than expected (that is, more than .75kg). Therefore, the null hypothesis states that the change in salmon weight before and after the growth hormone is *not significantly greater than* .75kg, making a one-tailed test appropriate (Table 10.7).

As shown in Table 10.7, the calculated matched-pairs *t* value is 2.96, with a corresponding one-tailed *p*-value of .01. It seems clear that we should reject the null hypothesis and conclude that the growth rate of the salmon is significantly greater than .75kg. In fact, we can be over 99% confident that the differences in growth can be attributed to the growth hormone.

TABLE 10.6

Change in Weight of Sample of Salmon After Hormone Treatment

Salmon ID Number	Weight before hormone treatment (kg)	Weight after hormone treatment (kg)	Difference in weight after one year
1	5.2	5.8	0.6
2	5.3	5.8	0.5
3	5.8	7.3	1.5
4	4.1	6.7	2.6
5	4.8	6.7	1.9
6	5.2	7	1.8
7	4.7	7	2.3
8	4.9	5.8	0.9
9	4.9	7.4	2.5
10	4.5	5.1	0.6

TABLE 10.7

Worktable for Dependent-Sample (Matched-Pairs) *t* Test: Salmon Weight Before and After Hormone Treatment

$H_0 : \delta = .75$

$H_A : \delta > .75$

Calculate the average difference $(\bar{d})$:

$\Sigma d = 15.2$ $\Sigma d^2 = 29.18$ $n = 10$ $\bar{X} = 4.94$ $\bar{Y} = 6.46$

Where d = difference in weight before and after hormone treatment

X = salmon weight before hormone treatment

Y = salmon weight after hormone treatment

$$\bar{d} = \frac{\sum d_i}{n} = \frac{15.2}{10} = 1.52$$

Calculate the standard error of *d*:

$$s_d = \sqrt{\frac{\sum d_i^2 - \frac{(\sum d_i)^2}{n}}{n-1}} = \sqrt{\frac{29.18 - \frac{(1.52)^2}{10}}{9}} = 0.82$$

Calculate the test statistic t_{mp} and *p*-value:

$$t_{mp} = \frac{\bar{d}}{\frac{s_d}{\sqrt{n-1}}} = \frac{1.52}{\frac{0.82}{\sqrt{9}}} = 2.96$$

p-value = .01 (one-tailed)

Example: Wilcoxon Matched-Pairs Test— Parkwood Estates Household Water Usage Levels Before (and After) Water Conservation Program

Over the last few decades, the Environmental Protection Agency (EPA) has developed and updated nationwide "Water Conservation Plan Guidelines" for indoor residential water use monitoring and conservation. The members of the Water Conservation Board of Middletown (introduced in Chapter 8) want to select a small sample of households in the Parkwood Estates development and provide them with information about the various water-saving features, strategies, and incentives (such as rebates for the purchase of low-flow appliances) discussed in the EPA "Guidelines". The Conservation Board members hope that households will significantly reduce their water consumption levels.

Twenty randomly selected households initially volunteer to participate in this multi-year study. Baseline indoor water consumption levels are determined before implementing the informational program for each of the following uses (showers, clothes washers, toilets, dishwashers, baths, leaks, faucets, and other domestic uses). The informational program is implemented over a two-year period, to provide enough time for people to adjust their water-use habits and purchase water-saving devices if they wish. At the end of the program, the water consumption levels for all of the uses listed above is measured again.

Only 11 of the 20 households successfully complete the program with useable water consumption data. During the program implementation period, some people change residence, others have a major demographic change such as adding or subtracting a household member, and others decide they don't want to continue with the program. Further complicating the data, 2 of the 11 households actually completing the program are seasonal residents. While it is disappointing to lose almost half of the participants, one often sees large

TABLE 10.8

Worktable for Dependent-Sample Wilcoxon Signed-Ranks Test: Daily Indoor Water Usage by Parkwood Estates Households (in Gallons)

Household ID Number	Daily indoor water usage per household: Before program implementation (X)	After program implementation (Y)	Change in water usage	Rank
1	174.2	146.7	−27.5	11
2	185.0	174.5	−10.5	7
3	57.2	52.3	−4.9	2
4	194.7	177.6	−17.1	9
5	143.4	149.4	+6.0	4
6	116.5	123.8	+7.3	5.5
7	72.1	59.4	−12.7	8
8	206.3	212.2	+5.9	3
9	226.8	202.7	−24.1	10
10	102.9	104.7	+1.8	1
11	201.3	208.6	+7.3	5.5
	$\bar{X} = 152.8$	$\bar{Y} = 146.5$		

T_n = sum of ranks of negative changes (less water used) = 47

T_p = sum of ranks of positive changes (more water used) = 19

Select the T which corresponds to the smaller number of hypothesized changes. Since most changes are expected to be negative (less water used), select T_p.

Calculate the test statistic t_{mp} and p-value:

$$Z_W = \frac{T - \frac{n(n+1)}{4}}{\sqrt{\frac{n(n+1)(2n+1)}{24}}} = \frac{19 - \frac{11(12)}{4}}{\sqrt{\frac{11(12)(23)}{24}}} = \frac{19-33}{\sqrt{126.5}} = -1.245$$

corresponding p-value = .1066 (one-tailed)

attrition rates in real-world sampling. Because the Water Conservation Board is monitoring the same households before and after implementation of the water-saving information program, the appropriate type of statistical test is matched-pairs. However, the very small sample size, coupled with the likelihood that this sample of households has probably not been taken from a normally distributed population, makes the nonparametric Wilcoxon matched-pairs test the only logical choice. The "before-after" household water consumption data are shown in the top portion of Table 10.8.

At first glance, the results seem inconsistent. A majority of the households (6 out of 11) show decreasing water use levels, and the average daily indoor water usage of the 11 households that actually complete the program decreases slightly, going from 152.8 gallons to 146.5 gallons. Are these modest results enough to conclude that less water is being used overall? A one-tailed Wilcoxon matched-pairs test is applied to answer this question (see the bottom portion of Table 10.8).

The calculated test statistic value from Wilcoxon matched-pairs is −1.245. Reading from the normal table, the corresponding p-value is approximately .1066. A p-value of this magnitude indicates there is marginal justification for concluding that there is a decrease in water usage levels. If we conclude the water conservation program is a success in reducing water consumption, there is a 10.66% change we are wrong and are making a Type 1 error. Certainly, when looking at the very slight decrease in average household water usage, very little water is conserved overall. It is probably best to conclude that the water conservation program *does not* significantly reduce water usage levels.

Given the very limited success of the water conservation program, it is now incumbent upon the members of the Water Conservation Board to examine these disappointing results on a household-by-household basis. For example, which households (if any) actually purchased low-flow toilets? Did any households systematically check for drips and leaks in kitchen and bathroom sinks, bathroom showers, and bathroom toilets? Did any households install a low-flow showerhead or aerator/flow restrictor on kitchen and bathroom sinks? Did any households take advantage of Energy Star rebates for purchase of such water-saving devices? If some (or all) of these water conservation strategies were not implemented, what reasons for nonimplementation were given by the household members? Only after these questions are answered can the Water Conservation Board decide what future programs and policies should be tried (if any) in Parkwood Estates.

In several respects, this is a flawed study. If this were an actual water usage survey, our recommendation would be to go back to the drawing board and redesign the entire research process. The sample itself is nonrandom and tainted in several ways: the participants are volunteers, not randomly selected; the overall sample size is very small; and 2 of the 11 households (with only seasonal occupancy and lower water consumption levels) clearly have different water usage patterns than the year-round resident households. You should be able to identify some other weaknesses or characteristics of this research design that could be strengthened.

Note: we present this example as a reminder that as human beings we often have our personal opinions about how we wish an experiment to conclude. Yet, as scientists, we are committed to following the scientific method and honestly reporting our findings. Members of the Water Conservation Board may be genuinely surprised by these results. Hopefully, however, they would recognize that aspects of the research design need strengthening. The goals and objectives of this study are certainly worthwhile and deserving of a better designed research process.

Key Terms

dependent-sample (matched-pairs) difference tests, *170*
independent and dependent samples, *170*
pooled variance estimate (PVE), *159*
separate variance estimate (SVE), *160*
two-sample difference of means Z or t test (parametric), *158*
two-sample difference of proportions test, *166*
Wilcoxon rank sum W (Mann-Whitney U) tests (non-parametric), *160*

Major Goals and Objectives

If you have mastered the material in this chapter, you should now be able to:

1. Recognize those geographic problems or situations for which application of a two-sample difference test is appropriate.
2. Explain the basic difference between independent and dependent samples.
3. Understand the procedure for selecting the proper two-sample difference test based on level of data measurement.
4. Explain the types of geographic problems where a matched-pairs (dependent-sample) difference test is appropriate.

References and Additional Reading

If you are interested in further exploring the geographic examples mentioned in this chapter, the following are good places to start.

- For information regarding gender equality internationally, as well as data dealing with female participation in the labor force, see various World Bank publications (www.worldbank.org), including the 2020 World Development Report, *Gender Equality and Development*, and their extensive World Development Indicators data set. Another excellent source is the International Labour Organization (a specialized agency of the United Nations; www.ilo.org). Especially see the report by Sara Elder, *Women in Labour Markets: Measuring Progress and Identifying Challenges*, 2010.

- For state-level obesity data in the US, the Centers for Disease Control and Prevention (CDC) is once again recommended www.cdc.gov.
- General sources for household energy consumption and expenditures are available from the Energy Information Administration (www.eia.gov). More specifically, see their latest Residential Energy Consumption Survey.
- Detailed water conservation guidelines for households are available through the Environmental Protection Agency (www.epa.gov/watersense/water-conservation-plan-guidelines).

11

Three-or-More-Sample Difference Tests

Analysis of Variance Methods

Many geographic problems involve comparison of three or more independent samples for significant differences. The logic applied to these problems is an extension of the reasoning used with two-sample difference tests. If multiple (three or more) samples are taken from the *same* population, their sample means are expected to vary somewhat from one another just because of sampling variation. However, if multiple sample means vary considerably more than what is expected from ordinary sampling variability, then it can reasonably be concluded that the samples are drawn from *at least two* different populations.

Consider again the example used in Section 10.1 to test for differences between two samples. In that application, a random sample of less-developed countries (LDCs) is compared to a second independent sample of more-developed countries (MDCs) with regard to female labor force participation rates. Suppose that the research design of this two-sample problem is expanded to test for differences between samples of countries from each of the six major world regions identified by the Population Reference Bureau. If this six-group classification scheme is used to test for difference of means, a two-sample difference of means test such as the t test or Wilcoxon rank sum W test is no longer appropriate.

Early in the previous chapter, another possible application of two-sample difference of means tests was suggested. In that example, a geographer is interested in comparing surface water runoff levels in two different environments: a natural setting with little or no landscape modification versus an exurban setting that had been partially modified by development, such as residential subdivisions. If this problem is expanded to three distinct landscapes whose mean surface water runoff levels are compared (e.g., totally natural setting, slightly modified landscape, and urban area substantially modified by human activity), then the t and Wilcoxon W tests are again not applicable.

The statistical procedures used to test three or more (multiple) samples for differences are **analysis of variance (ANOVA)** and the **Kruskal-Wallis test**. ANOVA is a parametric test that requires interval-ratio data drawn from normally distributed populations. In addition, the variances in all groups in the population are assumed equal. The equivalent nonparametric procedure is the Kruskal-Wallis one-way analysis of variance by ranks test. The Kruskal-Wallis test uses ordinal data directly or interval-ratio data downgraded to ordinal if either the assumption of normality or equal variance is badly violated.

11.1 Analysis of Variance (ANOVA)

Recall that the term "variance" is a descriptive statistic that measures the total amount of variability about the mean. Analysis of variance (ANOVA) involves the separation or partitioning of the total variance found in three or more groups or samples into two distinct components: (1) variability *between* the group or category means themselves; and (2) variability of the observations *within* each group around its group mean. Even though the structure of ANOVA uses variation as the key descriptive statistic, the means of the samples are compared for significant differences, making ANOVA a three-or-more-sample difference of means test.

ANOVA is explained this way. If the variability *between* the group means is relatively large as contrasted with a relatively small amount of variability *within* each group around its group mean, then the statistical conclusion is likely that the different groups have been drawn from different populations. As a result, the null hypothesis would be rejected. The null hypothesis for ANOVA takes the form $H_0: \mu_1 = \mu_2 = \ldots = \mu_k$ (where k is the number of independent random samples), so the null hypothesis asserts that all samples are drawn from the same population. Conversely, the alternate hypothesis takes the form $H_A: \mu_1 \neq \mu_2 \neq \ldots \neq \mu_k$, asserting that at least one sample is drawn from a different population than the other samples.

The interpretation of ANOVA becomes more apparent when viewed graphically (Figure 11.1). The null hypothesis states that independent samples have been drawn from the *same population*, whereas the alternate hypothesis asserts that the different samples are from *at least two separate and distinct populations*. In Case 1 of Figure 11.1, only a small difference separates the three sample means $(\bar{X}_1, \bar{X}_2, \text{ and } \bar{X}_3)$, and when ANOVA is applied the likely decision is to not reject the null hypothesis. That is, the three sample means are inferred to be no more different than would be expected with three independent samples drawn from the same population. Conversely, in Case 2, the seemingly large difference between the three sample means leads to the inferential conclusion that the different samples are taken from more than one separate and distinct population. The proper decision in this case is to *reject* the null hypothesis. Note that these decisions are all nondirectional or two-tailed, since the concern is to find a difference between three or more samples. The presence of more than two samples makes it impossible to conduct a directional (one-tailed) test of difference.

Between-group variation focuses on how the sample mean of each group differs from the total or overall mean when all categories are grouped together. Quite simply, the total or overall mean is the weighted average of the individual group or sample means. Therefore, if the individual group or sample means differ significantly from the total or overall mean, the between-group variability is large. Sometimes the between-group variation is called the "explained variation" because variability attributed to differences between group or sample means is a measure of how much variation is explained by (or statistically dependent on) the group structure itself.

Within-group variation measures the variation of observations in each group or sample about the mean of that group or sample. That is, each sample has a mean and variance, and when these estimates of variability are totaled for all groups, the result is the total within-group variability. Internal variation within each of the groups is not "explained" by the grouping or categorization scheme and thus is considered "unexplained," "residual," or "error" variation.

Many different testing structures are available for ANOVA, but only a single-variable model called "one-way analysis of variance" is examined here. This testing model is called "one-way" because observations fall into different groups or samples based on their values for *one* variable. Other, more complicated ANOVA techniques incorporate two or more variables simultaneously in the statistical model. In the problem regarding the percentage of females participating in the labor force, for example, it might be useful to see if another variable such as national literacy rate also varies with female labor force participation. Perhaps literacy should be included with the level of development variable in the analysis.

More complex ANOVA models with multiple variables are often used in agricultural, medical, and psychological research, but less often in geography. In part, this is because geographers infrequently work with controlled experimental research designs containing equal sample sizes and other restrictive assumptions required by the more complex models. ANOVA testing structures having multiple variables are generally discussed only in nonintroductory statistics texts that examine multivariate statistical techniques.

The basic goal of the ANOVA test is to determine which is more dominant or pronounced: between-group variability or within-group variability. The test statistic for ANOVA (F) incorporates these two sources of variation:

$$F = \frac{MS_B}{MS_W} \quad \textbf{(11.1)}$$

where MS_B = between-group mean squares
MS_W = within-group mean squares

Calculating Between-Group Variability

To calculate between-group mean squares (the numerator in Equation 11.1) a three-step procedure is followed:

Step 1: Calculate the total or overall mean $(\bar{X}_T)$, which is defined as the weighted average of the individual group or sample means $(\bar{X}_1, \bar{X}_2, \ldots, \bar{X}_k)$:

$$\bar{X}_T = \frac{\sum_{i=1}^{k} n_i \bar{X}_i}{N} \quad \textbf{(11.2)}$$

where $\bar{X}_i$ = mean of sample i
n_i = number of observations in sample i
k = number of groups or samples
N = total number of observations in all samples

Step 2: Calculate the between-group sum of squares (SS_B):

$$SS_B = \sum_{i=1}^{k} n_i\left(\bar{X}_i - \bar{X}_T\right)^2 \quad \textbf{(11.3)}$$

$$= \left(\sum_{i=1}^{k} n_i\left(\bar{X}_i^2\right)\right) - N\left(\bar{X}_T^2\right) \quad \textbf{(11.4)}$$

Equation 11.3 is the definitional formula of the between-group sum of squares, and Equation 11.4 is the computational formula. Notice what is happening in Equation 11.3. Within the parentheses, the total or overall mean is subtracted from each group mean. If any of these differences between total mean and individual group mean is large, it will make the entire expression (the between-group sum of squares) a larger value. This makes sense, for if an individual group mean is very different from the total or overall mean, it suggests a large amount of variability *between* the group means. In Equation 11.3, each of these differences is squared and multiplied by the number of observations in that group (to give that group its proper weight). Finally, these weighted differences are summed across all k groups.

Step 3: Calculate the between-group mean squares (MS_B):

$$MS_B = \frac{SS_B}{df_B} = \frac{SS_B}{k-1} \quad \textbf{(11.5)}$$

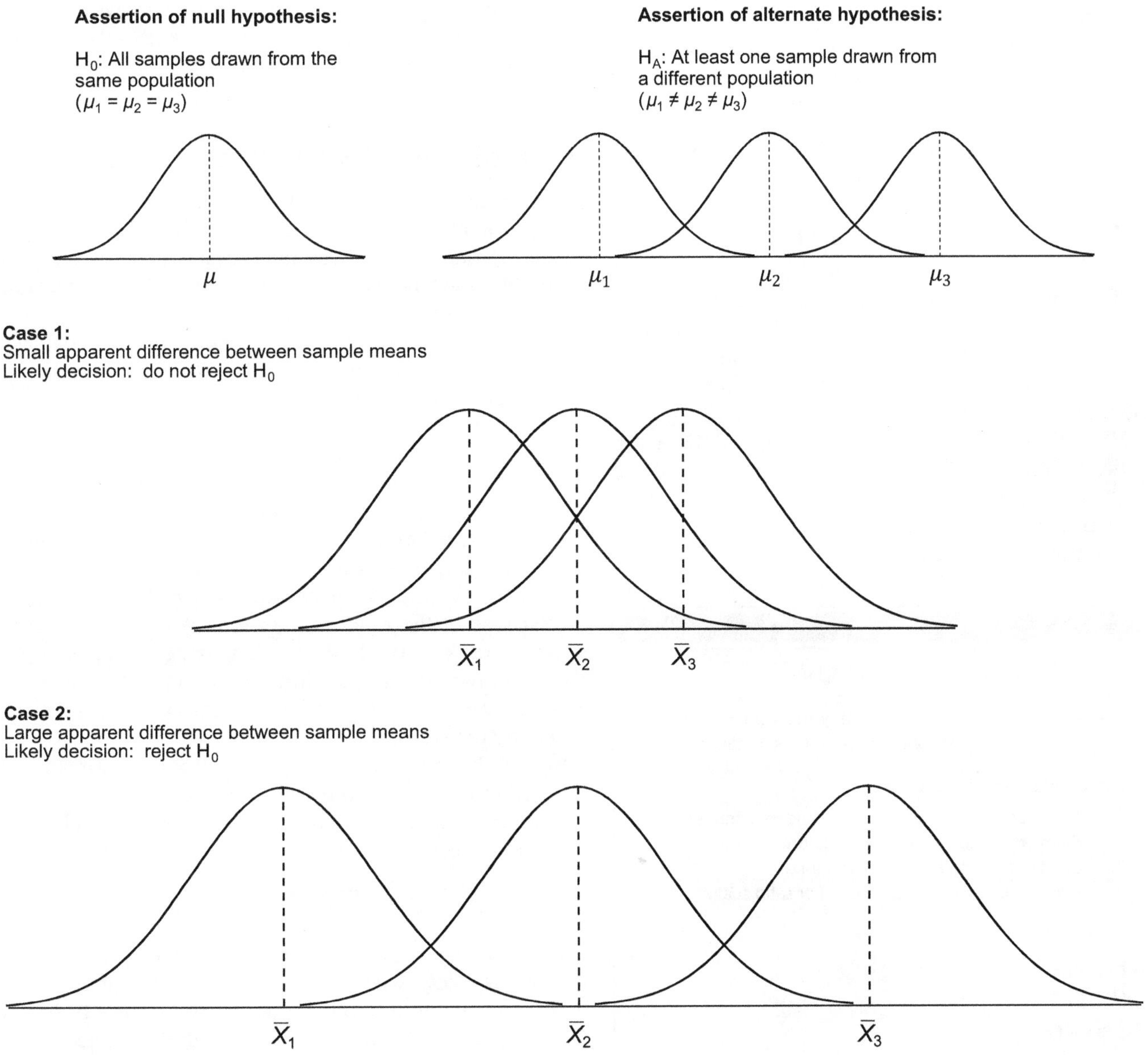

FIGURE 11.1
Null and Alternate Hypotheses in Analysis of Variance (ANOVA)

where df_B = between group degrees of freedom

Note that the between-group mean squares is simply the between-group sum of squares divided by the between-group degrees of freedom (df_B), where df_B depends on the number of groups or samples in the problem.

Calculating Within-Group Variability

The calculation of within-group mean squares (the denominator in Equation 11.1) also involves multiple steps:

Step 1: Calculate the within-group sum of squares (SS_W):

$$SS_W = \sum_{i=1}^{k} (n_i - 1) s_i^2 \qquad \textbf{(11.6)}$$

Interpretation of Equation 11.6 is straightforward. The variance in each group or sample (s_i^2) is multiplied or "weighted" by its sample size ($n_i - 1$). Then all of the weighted variances are summed to obtain the total within-group sum of squares for the groups.

Step 2: Calculate the within-group mean squares (MS_W):

$$MS_W = \frac{SS_W}{df_W} = \frac{SS_W}{N-k} \qquad \textbf{(11.7)}$$

The within-group mean squares is simply the within-group sum of squares divided by the within-group degrees of freedom (df_W), where df_W equals the total number of observations in all samples (N) minus the number of groups or samples (k).

When Equation 11.1 is examined, the ANOVA test statistic, F, defines the ratio of the between-group mean squares (Equation 11.5) to the within-group mean squares (Equation 11.7). How can we interpret the magnitude of this ratio of mean squares?

Analysis of Variance (ANOVA)

Primary Objective: Compare three or more (k) independent random sample means for difference

Requirements and Assumptions:

1. Three or more (k) independent random samples
2. Each population is normally distributed
3. Each population has equal variance
4. Variable is measured at interval or ratio scale

Hypotheses:

$H_0 : \mu_1 = \mu_2 = \ldots = \mu_k$

$H_A : \mu_1 \neq \mu_2 \neq \ldots \neq \mu_k$

Test Statistic:

$$F = \frac{MS_B}{MS_W}$$

Even if multiple independent samples are drawn from the same population, the set of sample means will fluctuate somewhat. This is to be expected because of the nature of sampling. This expected fluctuation is measured by the standard error of the estimate. Now consider what should occur if the null hypothesis is correct; the group or sample means should vary no more than expected if they are a set of independent random samples drawn from a single population. Therefore, when the null hypothesis is correct, the between-group variance would approximately equal the within-group variance. As a result, the F statistic will have a magnitude of about one.

However, if H_0 is incorrect and the population means are significantly different, then the sample means estimating these population means will vary more than expected from simple random fluctuations of multiple samples from the same population. This will cause the between-group variance to be significantly larger than the within-group variance, and the F statistic will be greater than one.

11.2 Kruskal-Wallis Test

The Kruskal-Wallis (K-W) test is the nonparametric equivalent of ANOVA. Kruskal-Wallis may be the most appropriate technique in cases where assumptions required for the use of the parametric ANOVA test (such as normality and equal population variances) are not fully met. Kruskal-Wallis is the nonparametric extension of the Wilcoxon rank sum W test to problems with three or more samples.

In the K-W test, values from all samples are combined into a single overall ranking (as in the Wilcoxon test). The rankings from each sample are summed and the mean rank of each sample (r_i) is then calculated. For example, in sample one the sum of ranks is R_1 and the mean rank is $r_1 = R_1/n_1$. In sample two, the sum of ranks is R_2 and the mean rank is $r_2 = R_2/n_2$, and so on. Because of the random nature of sampling, the sample mean ranks should differ somewhat, even if the samples are drawn from the same population. The K-W test examines whether the mean rank values are significantly different.

If the multiple (k) samples are from the same population, as asserted by the null hypothesis, their mean ranks should be approximately equal. Thus, if the null hypothesis is correct:

$$r_1 = r_2 = \ldots = r_k$$

On the other hand, if the mean ranks differ more than is likely with chance fluctuations it may be concluded that at least one of the samples comes from a different population and the null hypothesis is rejected.

The Kruskal-Wallis H test statistic is

$$H=\left(\frac{12}{N(N+1)}\sum_{i=1}^{k} n_i\bar{r}_i^{\,2}\right)-3(N+1) \qquad \textbf{(11.8)}$$

where N = total number of observations or values in all samples
$= (n_1 + n_2 + \ldots + n_k)$
n_i = number of observations or values in sample i
$\bar{r}_i$ = mean rank in sample i

In some geographic data sets, the ranks for a number of observations across the k samples may be tied. If this affects more than 25% of the values, an adjustment factor is included in the K-W test statistic. As a rule, statistical software packages calculate both the unadjusted and adjusted test statistic automatically. The correction factor for ties is

$$1-\frac{\sum T}{N^3-N} \qquad \textbf{(11.9)}$$

where $T = t^3 - t$
t = number of tied observations in the data
N = total number of observations or values in all k samples

With the correction factor for ties, the H statistic is

$$H=\frac{\left(\frac{12}{N(N+1)}\sum_{i=1}^{k} n_i\bar{r}_i^{\,2}\right)-3(N+1)}{1-\frac{\sum T}{N^3-N}} \qquad \textbf{(11.10)}$$

This modification for ties increases the value of H, reduces the p-value, and improves the likelihood of discovering significant differences between samples.

Kruskal-Wallis Test

Primary Objective: Compare three or more (k) independent random sample mean ranks for difference

Requirements and Assumptions:

1. Three or more (k) independent random samples
2. Each population has an underlying continuous distribution
3. Variable is measured at ordinal scale or downgraded from interval/ratio scale to ordinal

Hypotheses:

H_0 : The populations from which the three or more (k) samples have been drawn are all identical
H_A : The populations from which the three or more (k) samples have been drawn are not all identical

Test Statistic:

$$H=\frac{\left[\frac{12}{N(N+1)}\sum_{i=1}^{k} n_i\bar{r}_i^{2}\right]-3(N+1)}{1-\frac{\sum T}{N^3-N}}$$

The Kruskal-Wallis H distribution approximates the commonly used chi-square (χ^2) distribution, when $k > 3$ and/or at least one sample has a size (n_i) > 5. Thus, to determine a p-value for a calculated Kruskal-Wallis statistic, you would first convert H into chi-square and generate the p-value from this distribution. The sampling distribution of H is chi-square, where degrees of freedom equal $k - 1$. The chi-square distribution will be discussed in Section 12.1.

11.3 Example Applications in Geography

Example: Analysis of Variance (ANOVA)—Differences by Census Region in County-level Obesity Rates

In the first chapter, the spatial pattern of obesity across the United States was one of the four example problems used to illustrate the geographic research process. At that time, we looked at a state-level map showing the percentage of adults classified as obese in 2020 (Figure 1.3). That map reveals that the states with the highest obesity levels are mostly in the mid-South (especially in the Mississippi River Valley) and in Appalachia. Conversely, states with the lowest levels of obesity form two identifiable clusters—one in the West and Rocky Mountains and another in the Northeast, from New Jersey and New York up into New England.

If we disaggregate the obesity data to the county level, inferential hypotheses can be created and tested using a random sample of counties. From those sample county statistics, we can infer to the entire population of all counties in the United States. If a proportional stratified random sample of counties is taken from each of the four official census regions (Midwest, Northeast, South, and West), we can apply ANOVA to test whether there is a statistically significant difference in obesity levels from region to region.

Exactly 200 counties in the conterminous United States are selected using a stratified sample with each census region proportionally represented, resulting in these sample sizes: 65 counties from the Midwest, 14 from the Northeast, 90 from the South, and 31 from the West. Before applying ANOVA, we must determine if the samples are drawn from normally distributed populations and if each population has an equal variance. Testing of the sample data reveals that the samples are normally distributed and that they have equal variances. The Levene test for equality of variance results in a test statistic of 1.53 and a p-value of 0.209. This means that the sample variances are equal, so an equal variance estimate is used in the

(continued)

ANOVA analysis rather than a pooled variance estimate. All the major statistical software packages make this determination (pooled or separate variance estimate) automatically.

Table 11.1 summarizes the basic ANOVA procedure. As you might expect from our Chapter 1 discussion and obesity map, the counties sampled from the Midwest have the highest obesity rates, averaging 35.040% obese, while the West and Northeast have relatively lower obesity rates (under 34%). Even so, the obesity rates for the West and Northeast are much higher than the 2009 values (under 26%) explored in

TABLE 11.1

Worktable for Analysis of Variance (ANOVA): Differences in County-Level Obesity Rates by Census Region

$H_0 : \mu_1 = \mu_2 = \mu_3 = \mu_4$

$H_A : \mu_1 \bullet \mu_2 \bullet \mu_3 \bullet \mu_4$

	Sample statistics		
Census region	n	$\overline{X}$	s
Midwest	65	35.040	4.934
Northeast	14	32.514	4.900
South	90	34.516	6.250
West	31	28.919	5.369

$$N = \sum_n = 200$$

Calculate between group variability:

Step 1: Total or overall mean ($\bar{X}_T$):

$$\bar{X}_T = \frac{\sum_{i=1}^{4} n_i \bar{X}_i}{N} = \frac{65(35.040) + 14(32.514) + 90(34.516) + 31(28.919)}{200} = 33.67$$

Step 2: Between-group sum of squares (SS_B):

$$SS_B = \left(\sum_{i=1}^{4} n_i(\bar{X}_i^2)\right) - N(\bar{X}_T^2)$$

$$= [65(35.040)^2 + 14(32.514)^2 + 90(34.516)^2 + 31(28.919)^2] - 200(33.67)^2 = 1021.02$$

Step 3: Between-group mean squares (MS_B):

$$MS_B = \frac{SS_B}{df_B} = \frac{1021.02}{3} = 340.33$$

Calculate within-group variability:

Step 1: Within-group sum of squares (SS_W):

$$SS_W = \sum_{i=1}^{4} (n_i - 1)s_i^2 = 64(4.934)^2 + 13(4.900)^2 + 89(6.250)^2 + 30(5.369)^2 = 6211.52$$

Step 2: Within-group mean squares (MS_W):

$$MS_W = \frac{SS_W}{df_W} = \frac{SS_W}{N-k} = \frac{6211.52}{200-4} = 31.69$$

Calculate test statistic (F) and associated *p*-value:

$$F = \frac{MS_B}{MS_W} = \frac{340.33}{31.69} = 10.74 \ (p = .000)$$

the previous edition of this textbook. Also, the changes in obesity by census region from 2009 are important to keep in mind as we not only evaluate the statistical significance, but also consider how the continued increase in obesity has changed geographically. The statistical question concerns the likelihood or probability that such observed differences in obesity levels could occur if indeed no real difference exists from one census region to another. If these observed differences in mean obesity level are significantly more than expected from random sampling of a single population, then the result will be a large *F* statistic and a *p*-value close to zero.

The calculated *F* statistic of 10.74 has an associated *p*-value or significance level of .000. The logical conclusion is that *at least one* of the four census regions has an obesity level that differs from the other census regions. We are virtually 100% confident that it is the correct conclusion to reject the null hypothesis and conclude that county obesity levels differ by census region.

To gain further valuable insights (beyond ANOVA) about the precise differences between obesity rates from one census region to another, we follow two exploratory continuations. First, we construct a set of boxplots showing the distribution of county obesity rates by census region (Figure 11.2), and then we conduct a set of six two-sample *t* tests to determine which specific pairs of census regions have significantly different obesity rates and which do not (Table 11.2). The boxplots nicely confirm the ANOVA results. The Midwest and South clearly have the larger mean obesity levels and the South has the largest standard deviation of obesity levels.

The series of two-sample *t* tests (comprising all six possible census region pairs) shows that only four of the six census region pairs have significantly different obesity levels. Therefore, obesity levels previously differ from one region of the country to another.

TABLE 11.2

Two-Sample Difference of Means *t* test Results: County-Level Obesity Rates by Census Region

Census Region Pair	*t*	*df*	*p*-value
Midwest and Northeast	1.75	77	0.086
Midwest and South	.560	153	0.576
Midwest and West	5.52	94	0.000
Northeast and South	−1.14	102	0.256
Northeast and West	2.13	43	0.039
South and West	4.45	119	0.000

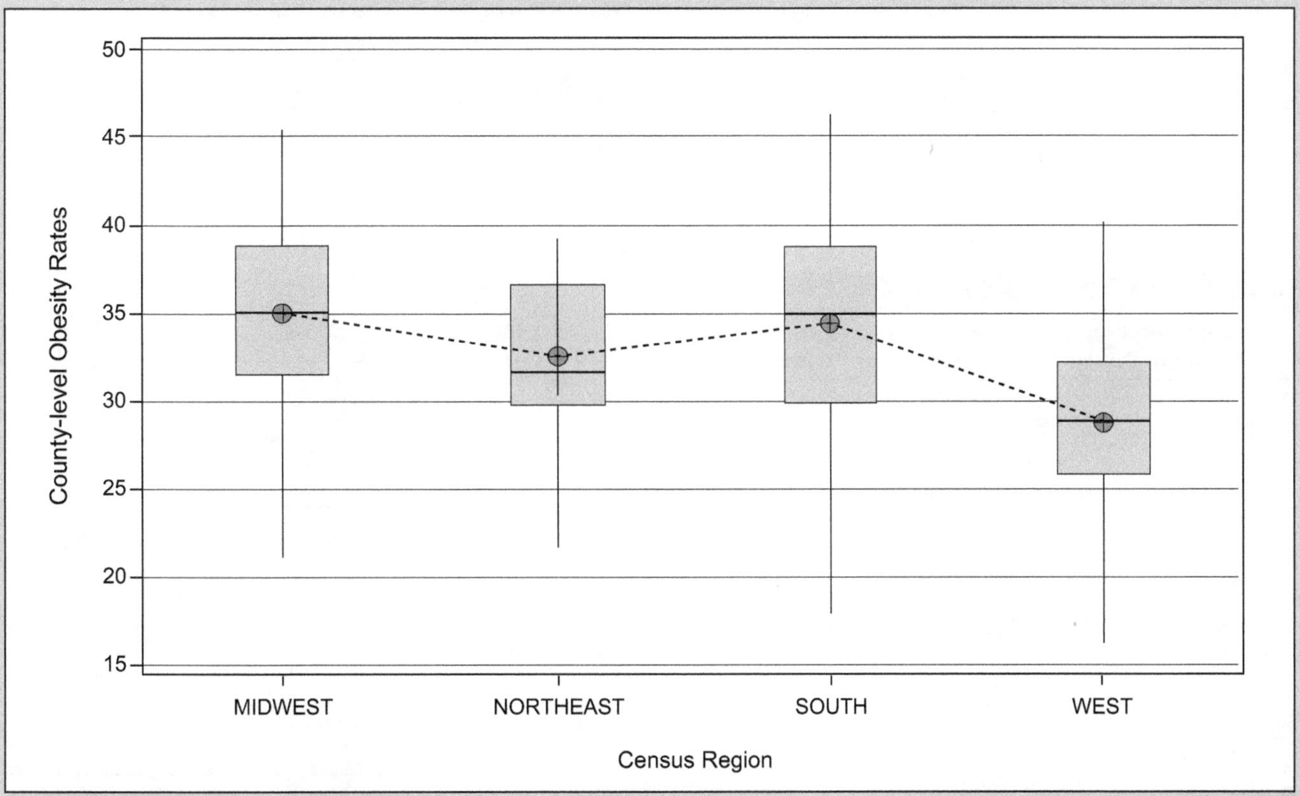

FIGURE 11.2
Boxplots of County-Level Obesity Rates by Census Region, 2020

Example: ANOVA and Kruskal-Wallis Tests—Differences in Home Purchase Price by Middletown District

Suppose that members of the Middletown City Council want to determine whether home purchase prices differ between the four official town districts: Northside, Central, Easton, and Southside. (Note: to refresh your memory about Middletown, you might want to review parts of Chapter 8, which show Middletown-related examples of confidence intervals and a map of Middletown with its four districts).

A random sample of home purchase prices is collected from each district and the multi-sample difference tests of ANOVA and Kruskal-Wallis are applied to answer this question. Table 11.3 shows the home purchase prices of the 34 total homes randomly selected from the four districts.

First, an analysis of variance (ANOVA) test is applied. ANOVA seems appropriate since this is a "difference test" problem with four samples and the data are measured on a ratio scale (with the purchase price of each home expressed in thousands of dollars). Before applying ANOVA, it must be determined if the samples are drawn from normally distributed populations and if each population has an equal variance. Testing of the sample data confirms that the samples are normally distributed and have equal variances.

Table 11.4 summarizes the results of the ANOVA analysis. The average home purchase price ranges from a low of \$186,200 in the Central district to a high of \$242,260 in the Northside district, with a total or city-wide mean purchase price of \$214,971. The statistical question concerns the likelihood or probability that such observed sample differences in purchase price could occur if indeed no real difference exists in the overall Middletown population of home values. If these observed differences in mean purchase price are significantly more than expected from random sampling, then the result will be a large *F* statistic and a *p*-value close to zero.

The calculated *F* statistic of 10.22 has an associated *p*-value or significance level of .000. The logical conclusion is that at least one of the four Middletown districts has home purchase prices that differ from the other districts. We can be nearly 100% confident that rejecting the null hypothesis and concluding that Middletown home purchase prices differ by district is indeed the correct conclusion.

The next logical question is "where do the sample differences occur?" The ANOVA significance level of .000 clearly indicates that significant differences exist *somewhere* in the grouping structure of four districts, but does not indicate where. To determine the location of these differences, a series of two sample difference of means *t* tests is applied. The descriptive statistics from ANOVA show the Central district with the lowest mean home purchase price ($\overline{X}_3$ = \$186,200) and the Northside district with the highest mean purchase price ($\overline{X}_4$ = \$242,260). Therefore, the largest magnitude *t* test result and lowest *p*-value will likely occur when the Central and Northside districts are paired.

The results (Table 11.5) confirm the largest magnitude *t* value exists when these two districts are contrasted ($t_{34} = -4.58$, $p = .001$). However, an equally significant *p*-value occurs when the Easton and Central districts are compared ($t_{23} = 3.97$, $p = .001$). Why is the *p*-value for Central and Northside the same as the *p*-value for Easton and Central when the calculated *t* for Central and Northside is quite a bit larger

TABLE 11.3

Random Sample of Home Purchase Prices by Middletown District (in thousands of dollars)

($n_1 = 9$) Southside	($n_2 = 10$) Easton	($n_3 = 8$) Central	($n_4 = 7$) Northside
218.38	212.28	202.52	212.28
231.80	220.82	154.94	233.02
179.34	235.46	151.28	244.00
180.56	262.30	189.10	256.20
224.48	248.88	173.24	270.84
168.36	223.26	219.60	257.42
204.96	235.46	183.00	222.04
213.50	239.12	215.94	
190.32	184.22		
	250.10		

TABLE 11.4

Summary of Analysis of Variance (ANOVA): Home Purchase Prices by Middletown District (in thousands of dollars)

$H_0: \mu_1 = \mu_2 = \mu_3 = \mu_4$ $\qquad H_A: \mu_1 \neq \mu_2 \neq \mu_3 \neq \mu_4$

Middletown District		s
Southside ($n_1 = 9$)	201.30	22.47
Easton ($n_2 = 10$)	231.19	22.29
Central ($n_3 = 8$)	186.20	25.77
Northside ($n_4 = 7$)	242.26	20.96

$n = 34 \qquad \overline{X}_T = 214.971$

$$F = \frac{MS_B}{MS_W} = \frac{5382}{527} = 10.22 \quad (p = .000)$$

Ninety-five Percent Confidence Intervals by Middletown District *

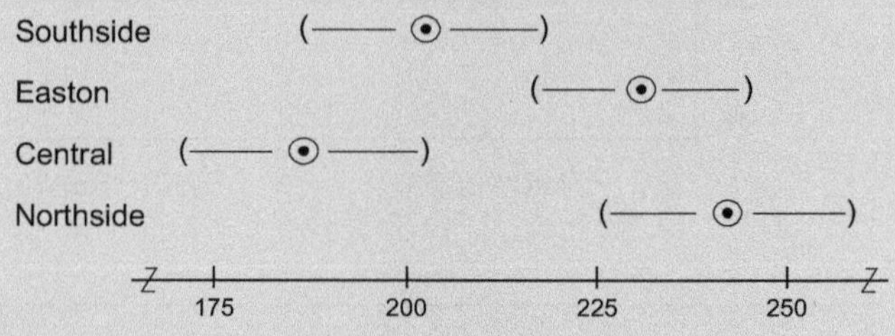

* based on pooled standard deviation = 22.95.

(absolute magnitude 4.58 versus 3.97)? The answer can be found by examining the slightly different sample sizes. Mean home purchase prices differ by the largest magnitude between Central and Northside, but the Northside sample size is somewhat smaller than the Easton sample size (Table 11.3). Quite simply, the smaller sample size from Northside has the effect of keeping the level of statistical confidence from dropping below $p = .001$, even though the magnitude of the test statistic t is higher.

The ANOVA results validate the existence of significant differences in home purchase prices by neighborhood. Furthermore, the normality and equal variance requirements regarding the application of ANOVA have been met. However, in any problem with rather small sample sizes, application of powerful parametric tests like ANOVA must be evaluated cautiously. One conservative strategy is to run the equivalent nonparametric test as well and compare test results. In this case, the equivalent nonparametric test is Kruskal-Wallis.

The Kruskal-Wallis procedure is now applied to the same home purchase price data (Table 11.6). With K-W, the sample values from all Middletown districts are combined into a single overall ranking, the rankings from each sample are summed, and the mean rank of each sample is calculated. The mean ranks of the four districts are quite different [Southside (12.6); Easton (23.1); Central (8.5); and Northside (26.1)]. It is therefore not surprising that the K-W test statistic indicates

TABLE 11.5

Two-Sample Difference of Means *t* Test Results: Home Purchase Prices by Middletown District

District Pair	*t*	*df*	*p*-value
Southside and Easton	−2.91	17	0.010
Southside and Central	1.29	15	0.216
Southside and Northside	−3.72	14	0.002
Easton and Central	3.97	16	0.001
Easton and Northside	−1.03	15	0.318
Central and Northside	−4.58	13	0.001

TABLE 11.6

Worktable for Kruskal-Wallis: Home Purchase Prices by Middletown District

($n_1 = 9$) Southside		($n_2 = 10$) Easton		($n_3 = 8$) Central		($n_4 = 7$) Northside	
Home Purchase Price	Overall Rank	Home Purchase Price	Overall Rank	Home Purchase Price	Overall Rank	Home Purchase Price	Overall Rank
218.38	17	212.28	13.5	202.52	11	212.28	13.5
231.80	23	220.82	19	154.94	2	233.02	24
179.34	5	235.46	25.5	151.28	1	244.00	28
180.56	6	262.30	33	189.10	9	256.20	31
224.48	22	248.88	29	173.24	4	270.84	34
168.36	3	223.26	21	219.60	18	257.42	32
204.96	12	235.46	25.5	183.00	7	222.04	20
213.50	15	239.12	27	215.94	16		
190.32	10	184.22	8				
		250.10	30				

$R_1 = 113$	$R_2 = 231.5$	$R_3 = 68$	$R_4 = 182.5$
$n_1 = 9$	$n_2 = 10$	$n_3 = 8$	$n_4 = 7$
$\bar{r}_1 = 12.6$	$\bar{r}_2 = 23.1$	$\bar{r}_3 = 8.5$	$\bar{r}_4 = 26.1$

$$H = \left(\frac{12}{N(N+1)} \sum_{i=1}^{k} n_i \bar{r}_i^2\right) - 3(N+1)$$

$$= \frac{12}{34(35)} (9(12.6)^2 + 10(23.1)^2 + 8(8.5)^2 + 7(26.1)^2) - 3(35)$$

$$= .010084(12111.41) - 105$$

$$= 17.13145$$ associated p-value = .001 (both adjusted and not adjusted for ties)

(continued)

significant differences in home purchase prices by Middletown district ($H = 17.13$ and $p = .001$). These results clearly confirm the ANOVA findings that somewhere in the four samples, significant differences exist.

Wilcoxon and Mann-Whitney tests are run on each pair of neighborhoods as a follow-up to the K-W testing. The Wilcoxon/Mann-Whitney results identify several significant differences in home purchase prices, findings that closely match the results obtained earlier from corresponding parametric *t* tests (Table 11.7). The most significant difference statistically occurs when home purchase prices in Easton and Central are compared ($p = .0029$), and the next most significant difference is between Central and Northside ($p = .0032$).

Statistically, the ANOVA and Kruskal-Wallis test results are remarkably consistent. Both tests confirm that highly significant differences in home purchase prices exist among the four districts of Middletown. The slightly lower *p*-value associated with ANOVA ($p = .001$).

As a geographer, you should now ask *why* home purchase prices vary significantly among the four Middletown districts. Why do district pairs Easton-Central and Central-Northside differ by the largest magnitudes? These ANOVA K-W results should be viewed as intermediate steps in the research process. You might want to examine other neighborhood characteristics that possibly affect home purchase prices, such as age of home, number of square feet, size of lot, and so on. Here is another instance where the results (output) from statistical analysis provide questions (input) that lead to further analysis. In other words, effective geographic problem solving often involves this cumulative, exploratory process.

TABLE 11.7

Two-Sample Difference Test Results (Wilcoxon and Mann-Whitney): Home Purchase Prices by Middletown District

District specific statistics:

District	Overall mean rank (Wilcoxon)	Median home purchase price (Mann-Whitney)
Southside	12.6	204.96
Easton	23.1	235.46
Central	8.5	186.05
Northside	26.1	244.00

Middletown District Pairing	Wilcoxon and Mann-Whitney *p*-value
Southside-Easton	.0128
Southside-Central	.2685
Southside-Northside	.0081
Easton-Central	.0029
Easton-Northside	.4350
Central-Northside	.0032

Example: ANOVA and Kruskal-Wallis Tests—Differences by Census Region in Immigrant Growth Rates for Large Metro Areas

The Brookings Institution's Metropolitan Policy Program redefines the challenges facing metropolitan America and promotes innovative solutions to help communities grow in more productive, inclusive, and sustainable ways. A key component of the Metro Program is an ongoing examination of immigration patterns. In a report titled "Immigrants in 2010 Metropolitan America: A Decade of Change," Jill H. Wilson and Audrey Singer examined newly released data from the 2010 American Community Survey on the foreign-born population. They focused on immigration flow patterns into the 100 largest metropolitan areas (as ranked by their 2010 census populations) and found that immigration settlement has become less concentrated (more dispersed) during the 2000s as metropolitan areas with relatively small numbers of immigrants in 2000 have been experiencing rapid growth of immigrant population since that time.

Building on their findings, we explore statistically the spatial pattern of immigration growth for these 100 largest metropolitan areas, using percent change in immigrant population from 2000 to 2010 as our measurement parameter. The Brookings policy brief states that "Metro areas experiencing the fastest growth rates were places that had relatively small immigrant populations. A swath of metro areas from Scranton stretching southwest to Indianapolis and Little Rock and sweeping east to encompass most of the Southeast and lower mid-Atlantic . . . saw growth rates on the order of three times that of the 100-largest-metro-areas rate."[1]

Using both ANOVA and Kruskal-Wallis, we apply these tests *in a descriptive fashion only* to explore the existence and strength of a regional concentration of high-percentage change immigrant populations during the last decade. Each of the 100 largest metro areas is assigned to its census region (Northeast, South, Midwest, and West), and both of these difference tests are run. The statistical analysis must be limited to confirming or rejecting descriptive statements. We cannot make direct use of statistical test values and associated *p*-values to make inferential probabilistic statements because our data set is the total population of the 100 largest metro areas, and not a sample of metro areas from some larger population. However, we can use these inferential statistics in an exploratory way and visually show the spatial pattern of metro areas with the most rapidly growing immigrant populations.

[1] The Brookings Institution. "Immigrants in 2010 Metropolitan America: A Decade of Change," Published October 24, 2011. http://www.brookings.edu/research/speeches/2011/10/24-immigration-singer#

The results from both the ANOVA and Kruskal-Wallis tests shown in Table 11.8 describe patterns of change in immigrant growth that seem to vary considerably from one census region to another. Both the ANOVA result ($F = 11.39$ and p-value = .000) and K-W result ($H = 29.04$ and p-value = .000) suggest notable overall regional differences. In addition, the intermediate summary statistics show a separation of the South from the other three census regions. With ANOVA, the mean percentage of immigrant growth in metro areas of the South is 71.19, compared to much lower mean growth rates in the other three census regions [Northeast (40.52), Midwest (43.28), and West (35.28)]. Very similar results are obtained with Kruskal-Wallis, as the average rank of immigrant growth in metro areas of the South is 70.2, much higher than the other census regions [Northeast (37.7), Midwest (43.1), and West (35.3)].

A map showing the percentage change in immigrant population in the largest 100 metro areas contrasts the spatial pattern of the 25 fastest-growing areas with the other 75 slower-growing places (Figure 11.3). This map is a modification of one that appears in the Brookings policy brief cited above. The rapid growth of immigrant population in many southern cities is clearly evident. In the 100 largest metro areas of the United States, 19 of the 25 with the most rapidly growing immigrant populations are located in the South.

Boxplots of each census region are compared in Figure 11.4. The boxplot graphic confirms both the ANOVA and Kruskal-Wallis test results and the map evidence—the large metropolitan areas of the South have large percentage increases in immigrant population from 2000 to 2010.

Scranton, PA, is the lone outlier in this series of boxplots, with an extraordinary immigrant population growth rate of 140.2% between 2000 and 2010. Although the numerical growth in immigrant population is not extraordinary (a 15,907 increase in immigrants), the growth *rate* is much higher than any other large metropolitan area in the Northeast. In fact, it is the fastest immigrant growth rate of all 100 large metro areas in the country!

What is going on in Scranton? We know that most of the immigrant growth is among younger cohorts. For example, according to the 2010 Census, over 80% of the Scranton Wilkes-Barre metropolitan area Hispanic population is less than 44 years old and nearly 40% are under the age of 24. We also know that the city of Scranton faces a total debt of about $300 million and may have to declare bankruptcy (as of July 2012). This includes an unfunded pension obligation of $90 million,

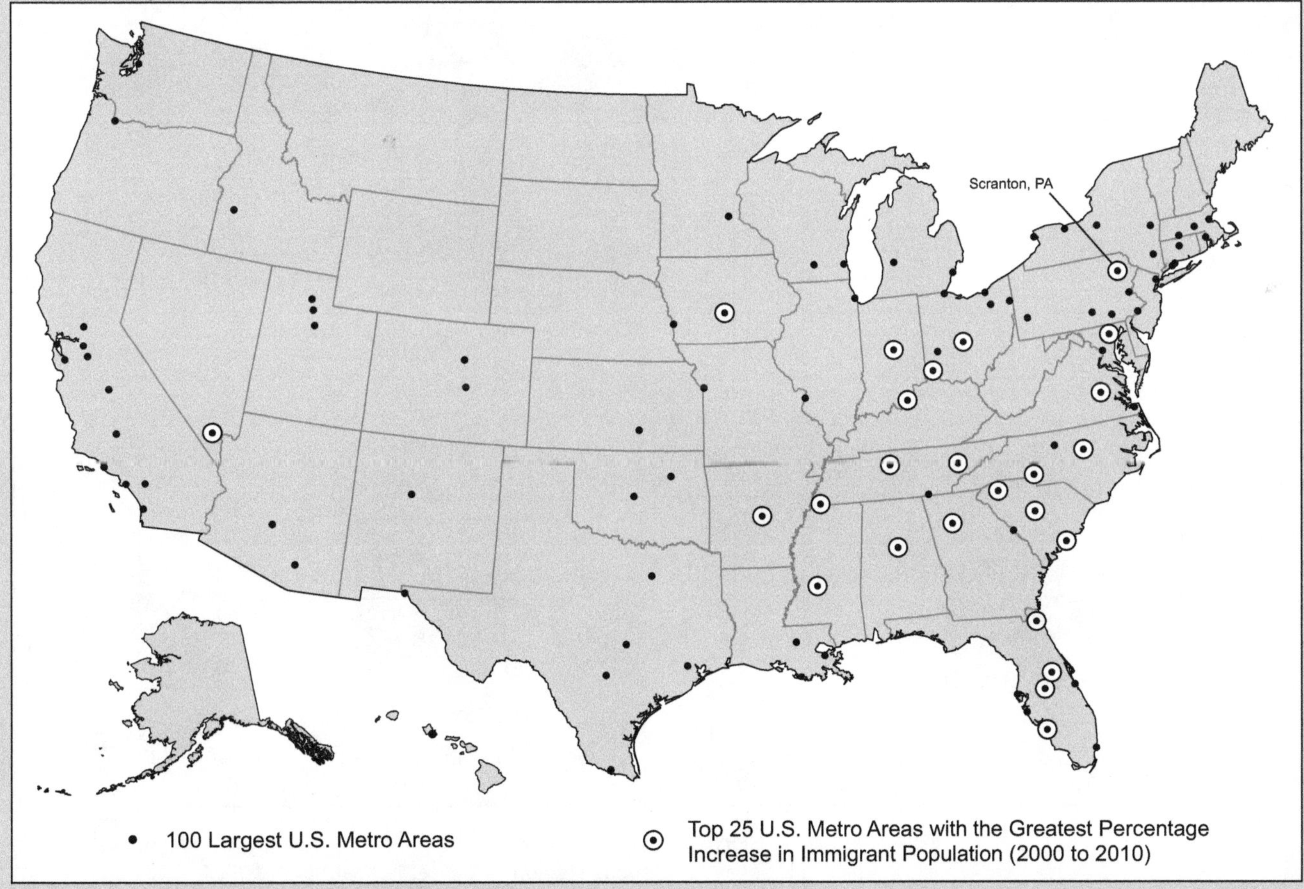

FIGURE 11.3
The Hundred Largest Metro Areas in the United States: Highlighting the Twenty-Five Metro Areas with the Greatest Percentage Increase in Immigrant Population, 2000 to 2010
Source: Brookings Institution, Immigrants in 2010 Metropolitan America: A Decade of Change

(continued)

TABLE 11.8

Hundred Largest Metro Areas in the United States: ANOVA and Kruskal-Wallis Tests for Differences by Census Region in Percent Change in Immigrant Population, 2000 to 2010

Analysis of Variance	Statistics		
Census region	n	$\bar{X}$	s
West	24	35.28	17.09
South	38	71.19	28.63
Midwest	19	43.28	30.99
Northeast	19	40.52	28.70
OVERALL:	100		

Overall Mean = 51.44 F = 11.39 p = .0000

Kruskal-Wallis	Statistics		
Census region	n	*Median*	*Average rank*
West	24	33.35	35.3
South	38	67.50	70.2
Midwest	19	42.80	43.1
Northeast	19	33.00	37.7
OVERALL:	100		50.5

H = 29.04 p = .0000 (both adjusted and not adjusted for ties)

$150 million in debt held directly by the city or the Scranton Parking Authority, and a $15 million mandatory arbitration settlement due to the city's firefighters' union. Why then is this metropolitan area so attractive to immigrants? This is certainly a question worthy of further attention.

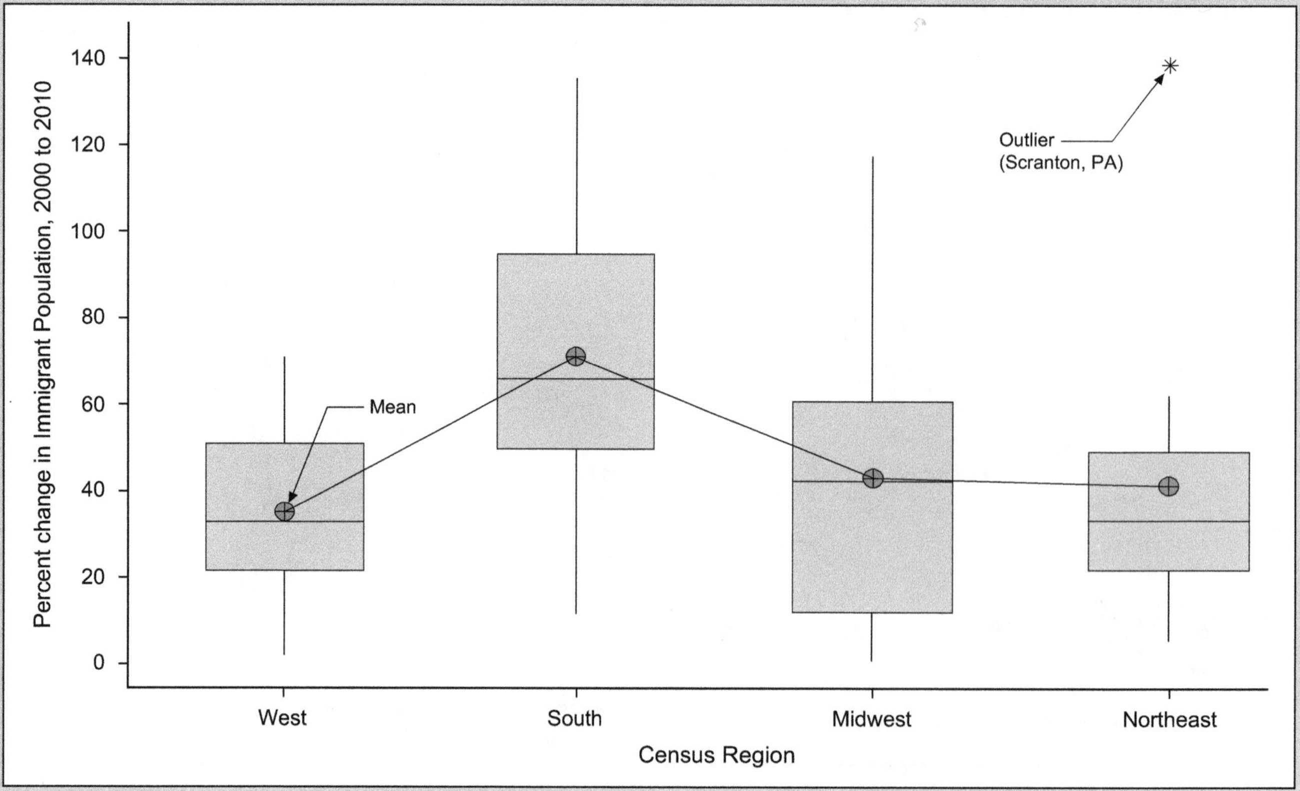

FIGURE 11.4
Boxplots of Percent Change in Immigrant Population, Hundred Largest Metro Areas in the United States, by Census Region, 2000 to 2010

Key Terms

analysis of variance (ANOVA) or *F*-test (parametric), *177*
between-group variation, *178*
Kruskal-Wallis test (non-parametric), *177*
within-group variation, *178*

Major Goals and Objectives

If you have mastered the material in this chapter, you should now be able to:

1. Recognize those geographic problems or situations for which application of a multiple sample difference test is appropriate.
2. Explain the rationale and purposes for conducting a difference test for three or more samples, either analysis of variance (ANOVA) or Kruskal-Wallis.
3. Distinguish between the major components of analysis of variance: between-group variability and within-group variability.

References and Additional Reading

If you are interested in exploring further the geographic examples mentioned in this chapter, the following are good places to start. Once again, see the Centers for Disease Control and Prevention for obesity data (www.cdc.gov). Home purchase price and other housing data are available from many sources. The US Census Bureau has a variety of reports and detailed data related to housing affordability, prices, and starts (www.census.gov/topics/housing.html). Home price data is constantly updated by numerous private agencies, including RealtyNow (www.realtynow.com) and Zillow (www.zillow.com/homes/recently_sold). With regard to immigration data, the US Department of Homeland Security, Office of Immigration Statistics (www.dhs.gov/immigration-statistics) is a fine place to begin. Also see the US Census Bureau. The specific issue of immigration and metropolitan growth is covered by the Brooking Institution, especially in a series of reports dealing with the "State of Metropolitan America" (www.brookings.edu). Of course, many other online organizations and interest groups focus on aspects of the immigration issue.

12

Categorical Difference Tests

In the last few chapters, we have presented a logical progression of difference tests. The progression started in Chapter 9 with *one-sample* difference tests comparing a single sample statistic to a corresponding population parameter. The sequence continued through *two-sample* difference tests that analyze two independent samples for differences and *dependent-sample (matched pairs)* difference tests that analyze one sample twice (Chapter 10). The progression finished with *three-or-more-sample* difference tests that determine whether multiple samples have been drawn from one or more populations (Chapter 11).

The focus now shifts to special types of difference tests that are designed to analyze data organized into categories. Some of these tests, known as **goodness-of-fit tests,** compare actual or observed sample frequency counts of a single variable (that is, the actual or observed number of observations in each category of a single variable) with some expected frequency distribution. In many cases, this type of goodness-of-fit test determines whether the actual data "fits" (is similar to) the distribution of some expected geographic model or theory. In Section 12.1, we look at three goodness-of-fit examples: (1) the application of the chi-square statistic to determine if the actual sample frequency counts of a variable are distributed in a uniform or even manner; (2) the application of the chi-square statistic to determine if the actual frequency counts of a variable are distributed in a proportional manner similar to what is expected in some predictive model; and (3) the application of the Kolmogorov-Smirnov statistic to determine if a set of data is normally distributed.

Sometimes we have a two-way (contingency) table that classifies observations according to two different categorical variables. Our goal is to determine whether any nonrandom relationship exists between these two variables. This type of data structure (**contingency table analysis**) is presented in Section 12.2, with three accompanying geographic examples.

12.1 Goodness-Of-Fit Tests

In some problems, the distribution of frequency counts is expected to be uniform or equal. For example, suppose an environmental geographer wishes to examine nutrient runoff levels from a series of five adjacent tributaries running into the same bay. Water samples measuring nitrogen and phosphorus levels could be taken on a random selection of days simultaneously from each tributary, and the number of sample days that nutrient levels exceed some critical (threshold) value could be counted for each tributary. Although low levels of such nutrients are acceptable, nutrient levels exceeding some threshold value become pollution problems. It might be expected that the nutrient level is exceeded about the same number of sample days in each tributary, if the nutrient levels are similar from tributary to tributary. This expected frequency count information can be compared with the observed frequency counts of high nutrient level days among the bay tributaries to see whether the two frequency counts are statistically different. A goodness-of-fit procedure can determine whether the actual frequency counts are uniform or equal for this set of tributaries.

Goodness-of-fit procedures can also evaluate whether a set of frequency counts fits an expected distribution that is uneven or proportional. Suppose you expect the distribution of high nutrient level days from tributary to tributary to be something other than uniform. For example, a tributary with a greater proportion of its watershed in agricultural land might be expected to have more sample days exceeding critical nutrient levels than another tributary with less of its watershed in farmland. You might hypothesize that the expected number of high pollution days is proportional to the percentage of tributary land area under cultivation.

Proportional goodness-of-fit tests can also be applied to evaluate the validity of a geographic model. For example, suppose a recreation planner is studying attendance patterns at a regional park that serves people in several nearby communities. How many park visitors might be expected from each community? To estimate park usage patterns, a model of spatial interaction might be useful. One simple model predicts that the volume of spatial interaction (that is, the number of park visitors from a community) is directly related to the population size of the community and inversely related to the distance from the community to the park. That is:

$$SI_{ij} = \frac{Pop_i}{D_{ij}} \quad \textbf{(12.1)}$$

where SI_{ij} = "spatial interaction"—that is, the number of park visitors expected (predicted by the model) from community i to park j

Pop_i = population of community i

D_{ij} = distance from community i to park j

The recreation planner can compare the observed number of park visitors from each community with the corresponding number of visitors predicted (expected) by the spatial interaction model. The extent to which the model "fits" (the extent to which the observed and expected visitor counts from the set of communities match) can be tested statistically with a goodness-of-fit procedure.

Goodness-of-fit tests are also used to compare an actual frequency distribution with a theoretical probability distribution such as normal or random. A frequency distribution often needs to be tested for normality—an important assumption for the data to be appropriately applied in a parametric test such as t or ANOVA. Suppose a geomorphologist is studying soil erosion rates of several soil orders (alfisols, aridosols, entisols, etc.) in a large regional study area. To test for significant differences in erosion rate by soil order, a spatial random sample is taken from each type of soil. If the samples are drawn from normally distributed populations, a parametric ANOVA could be applied as the difference test. A goodness-of-fit procedure can be used to determine whether this normality requirement has been met, thereby confirming the application of ANOVA as a valid statistical test.

In some geographic problems, a spatial frequency distribution can be tested for randomness. In these cases, a Poisson probability distribution would be expected. For example, a goodness-of-fit test for randomness could be applied to a spatial pattern of influenza outbreaks in a metropolitan area. This procedure would compare the observed pattern of influenza cases with the pattern expected if the outbreak follows a random spatial process.

Chi-Square (χ^2) Goodness-Of-Fit Test

When applied as a goodness-of-fit test, the chi-square statistic compares the observed frequency counts of a single variable (organized into nominal or ordinal categories) with an expected distribution of frequency counts allocated over the same categories. The chi-square test must use absolute frequency counts and cannot be applied if the observations or sampling units are in relative frequency form, such as percentages, proportions, or rates.

Chi-square is a method to determine if a truly significant difference exists between a set of observed frequency counts and the corresponding expected frequency counts. The focus is on how closely the two frequency counts match, thereby providing a goodness-of-fit measure. The null hypothesis (H_0) states that the population from which the sample has been drawn fits an expected frequency distribution. Thus, H_0 assumes no difference between observed and expected frequency counts. If the magnitude of difference between frequency counts is small across *all* categories, the data are likely to be a random sample drawn from the expected frequency distribution and H_0 is not rejected. Conversely, the alternate hypothesis (H_A) suggests that the magnitude of difference between frequencies is large in *at least one category.* If H_A is true, the data cannot be considered a random sample from the expected frequency distribution and H_0 is rejected.

The formula for the chi-square test statistic is

$$\chi^2 = \sum_{i=1}^{k} \frac{(O_i - E_i)^2}{E_i} \quad \textbf{(12.2)}$$

where O_i = observed or actual frequency count in the ith category

E_i = expected frequency count in the ith category

k = number of nominal or ordinal categories

If the observed and expected frequency counts for each category are similar, then all of the differences $(O_i - E_i)$ will be slight, χ^2 will be small, the goodness-of-fit will be strong, and H_0 will not be rejected. However, if at least one difference between frequency counts is large, then the χ^2 statistic will be large, the observed frequency counts will not necessarily come from the population or model theorized by the expected frequency counts, and H_0 will be rejected. In later example problems, the methodology for calculating expected frequencies will be discussed.

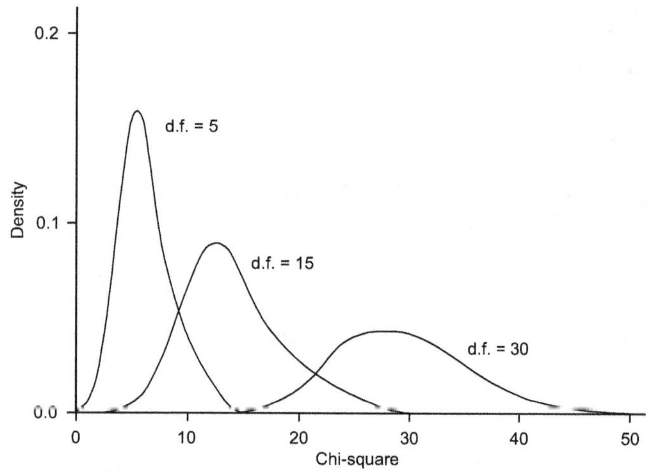

FIGURE 12.1
Chi-square Distributions

The shape of the χ^2 distribution is determined by the degrees of freedom (df), which itself depends on the number of nominal or ordinal categories in the data set (Figure 12.1). The distribution is always positively skewed, but the amount of skew decreases as the degrees of freedom approaches 30. In addition, the total area under the curve is always equal to 1.0. When the df is 30 or more, the χ^2 distribution is closely approximated by the normal distribution.

While chi-square is a very flexible goodness-of-fit test, minimum size restrictions apply under certain circumstances. For example, if the variable has only two categories, the expected frequency in both should be at least 5. For the test to be valid with more than two categories, no more than one-fifth of the expected frequencies should be less than 5, and no expected frequency should be less than 2. Sometimes categories need to be combined to increase the expected frequencies to an appropriate size.

Chi-Square (χ^2) Goodness-Of-Fit Test

Primary Objective: Compare random sample frequency counts of a single variable with expected frequency counts (goodness-of-fit)

Requirements and Assumptions:

1. Single random sample
2. Variables are organized by nominal or ordinal categories; frequency counts by category are input to statistical test
3. If two categories, both expected frequency counts must be at least five, if three or more categories, no more than one-fifth of the expected frequency counts should be less than five, and no expected frequency count should be less than two

Hypotheses:

H_0 : Population from which sample has been drawn fits an expected frequency distribution (uniform or equal, proportional or unequal, random, etc.); no difference between observed and expected frequencies

H_A : Population does not fit an expected frequency distribution; there is a significant difference between observed and expected frequencies

Test Statistic:

$$\chi^2 = \sum_{i=1}^{k} \frac{(O_i - E_i)^2}{E_i}$$

Example: Chi-Square Goodness-Of-Fit (Uniform Distribution)—Comparing High School Student SAT Scores

Suppose a school district superintendent wants to conduct a geographic study that compares the Scholastic Aptitude Test (SAT) scores among juniors from five comparably sized high schools located in different portions of a metropolitan school district. Are students from all five high schools performing equally well on the SATs? More specifically, do about the same number of students in each school earn test scores above the national median? If the educational experiences are similar at all five high schools, it should be expected that about the same number of students from each school will have above-median SAT scores. If performance levels are unequal, the superintendent has indicated she will begin a five-year improvement program, expending particular effort toward improving the SAT scores of the lower-achieving schools, while continuing to improve the performance levels of the better schools.

Suppose an initial random sample of 100 students is taken from each school and the number of sampled students with an SAT score exceeding the national median is recorded (Table 12.1). A chi-square goodness-of-fit test is appropriate since a set of observed frequency counts from nominal categories is being compared with a comparable set of expected frequency counts. Moreover, expected frequency counts are uniform or equal, because each of the five high schools is expected to have an equal number of sampled students with above-median SAT scores.

The chi-square calculation from the "Initial year results" column of Table 12.1 is shown in the top portion of Table 12.2. A combined total of 245 sampled students from all five high schools scored above the national SAT median. If this group of students is uniformly or evenly distributed among the five high schools, the expected frequency per high school would be 49 (245 divided by 5). The statistical question is whether the actual (observed) frequency counts (42, 45, 51, 47, and 60) are significantly different from the expected frequency counts (49, 49, 49, 49, and 49). If the frequency counts are different, then the superintendent has statistical verification to begin her five-year "SAT improvement" program.

The calculated chi-square statistic ($\chi^2 = 3.959$) has an associated p-value or significance level of .412. How should this intermediate-sized p-value be interpreted? If the p-value is very close to 1.0, the actual and expected frequency counts would be very similar, the logical conclusion would be to not reject the null hypothesis, and it would be clear that each of the five high schools has an equal or uniform number of students exceeding the national median SAT score. Alternatively, if the p-value is very close to 0.0, the actual and expected frequency count distributions would be very different, and the decision would be to reject the null hypothesis and conclude that the high schools have different SAT test results.

This intermediate-sized p-value is inconclusive statistically, but should suggest to the superintendent that there seems to

TABLE 12.1

Random Sample of Students with SAT Scores Above National Median, by High School

High School	Number of SAT scores above the national median: Initial year results	Results after fifth year
John Muir H.S.	42	51
Gifford Pinchot H.S.	45	52
Rachel Carson H.S.	51	56
Garrett Hardin H.S.	47	53
Aldo Leopold H.S.	60	64
TOTAL	245	276

TABLE 12.2

Worktable for Goodness-Of-Fit Chi-Square Uniform: Number of Students with SAT Scores Above National Median, by High School

Initial Year Results:

Observed number of students with above-median test scores (from table 12.1)

Muir H.S.	Pinchot H.S.	Carson H.S.	Hardin H.S.	Leopold H.S.	Total
42	45	51	47	60	245

If testing goodness-of-fit uniform or equal distribution, the expected frequency of each category is calculated by dividing the total observed frequency count by the number of categories.

$$E_i = \frac{\sum O_i}{k} \quad \text{for all categories}$$

$$E_i = \frac{245}{5} = 49$$

$$\chi^2 = \sum_{i=1}^{k} \frac{(O_i - E_i)^2}{E_i} = \frac{(42-49)^2}{49} + \frac{(45-49)^2}{49} + \frac{(51-49)^2}{49} + \frac{(47-49)^2}{49} + \frac{(60-49)^2}{49}$$

$$= 3.959 \ (p = .412)$$

Results after fifth year:

Observed number of students with above-median test scores (from table 12.1)

Muir H.S.	Pinchot H.S.	Carson H.S.	Hardin H.S.	Leopold H.S.	Total
51	52	56	53	64	276

$$E_i = \frac{276}{5} = 55.2$$

$$\chi^2 = \sum_{i=1}^{k} \frac{(O_i - E_i)^2}{E_i} = \frac{(51-55.2)^2}{55.2} + \frac{(52-55.2)^2}{55.2} + \frac{(56-55.2)^2}{55.2} + \frac{(53-55.2)^2}{55.2} + \frac{(64-55.2)^2}{55.2}$$

$$= 2.007 \ (p = .734)$$

(continued)

be "room for improvement," particularly if the test scores of Muir (42), Pinchot (45), and Hardin (47) high school students could be substantially improved. The "SAT improvement program" can be deemed a success (at least statistically) if the magnitude of the chi-square test statistic can be reduced and the resultant *p*-value shifted dramatically toward 1.0. A *p*-value of 1.0 will occur only if there is perfect uniformity or equality across all of the frequency count categories—that is, if an equal number of students from each high school were to score above the national median on their SAT test.

Suppose now that the "SAT improvement program" is implemented. Five years later, the effectiveness of the program is evaluated. The number of students scoring above the national median increases at each of the five high schools and the five-school total of students increases from 245 to 276 (Table 12.1, "Results after fifth year"). Looking at the observed frequency counts, there seems to be a shift toward interschool equality. The magnitude of the chi-square statistic declines from 3.959 to 2.007, confirming this shift toward equity among high schools (Table 12.2).

The "SAT improvement program" has apparently been successful. High schools with the lowest frequency counts during the initial year improved their frequency counts the most by the end of the fifth year. Leopold High School had the highest frequency count during the initial year. Their students managed to improve, but not to the same degree as the other high schools. As a result of these changes in student test performance, the *p*-value increases considerably from .412 to .734, indicating stronger confidence that the null hypothesis should not be rejected. If the null hypothesis is rejected and a conclusion drawn that high school test scores are different, a 73.4% chance exists that this conclusion is incorrect.

Example: Chi-Square Goodness-Of-Fit (Proportional Distribution)—Interprovincial Migration to Alberta, Canada

Chi-square goodness-of-fit is now applied to a simple spatial interaction model to see how well the model "fits" an actual set of data. We will also focus on the ways in which the model pattern *does not fit* the actual pattern. Geographers often use spatial interaction models to replicate or predict the volume of movement or amount of activity (interaction) between locations. One simple spatial interaction model assumes that the amount of movement or interaction from a given origin to a destination is directly related to the population size of the origin and inversely related to the distance from the origin to the destination. If the frequency counts of the chi-square goodness-of-fit model closely match the frequency counts that actually exist, then the model is predicting the actual situation effectively.

TABLE 12.3

Interprovincial Migration to Alberta, Canada, 2020

Origin	Actual number of Interprovincial migrants to Alberta, 2020
British Columbia	17,605
Manitoba	4,636
New Brunswick	1,496
Newfoundland and Labrador	2,242
Northwest Territories	715
Nova Scotia	2,211
Nunavut Territory	115
Ontario	14,550
Prince Edward Island	362
Quebec	2,497
Saskatchewan	9,805
Yukon Territory	304
TOTAL	56,538

Source: Statistics Canada

The spatial interaction we are examining is the pattern of interprovincial migration into Alberta, Canada during 2011. During that year, 56,234 out-of-province migrants settled in Alberta (Table 12.3). The recent 2021 Census showed that Alberta experienced a 4.8% growth rate. Much of the population growth is largely due to the continued development of the Athabasca oil sands (or tar sands) located in the northeastern part of the Province. The Athabasca deposit is one of the largest known reservoirs of crude bitumen in the world.

Our question concerns how the actual pattern of Alberta in-migration and population growth compares with the pattern predicted by a simple spatial interaction model. The specific model we are using predicts that the expected number of migrants into Alberta is directly proportional to the population in each origin province *and* inversely proportional to the distance from that origin province to Alberta. That is:

$$E_{ij} = \frac{Pop_i}{D_{ij}} \qquad (12.3)$$

To estimate these distances, we used a mileage guide published by the Canadian Government Travel Bureau. The actual measure used was average road distance in miles from Calgary and Edmonton (the two largest cities in Alberta) to the estimated population centroid of each of the other provinces.

Using the actual number of interprovincial migrants to Alberta (as provided in Statistics Canada), the observed frequencies for the model are a "simulated sample" of 255 in-migrants (maintaining a constant proportion of approximately three-tenths of 1% (.003) of the actual total as closely as possible). For example, the actual number of arrivals from Manitoba is 4,636 (.0819 or 8.19% of the 56,538 total) (Table 12.3). If the "simulated sample" of 255 is used as the total, then Manitoba's in-migration would be .0819(255) = 20.884 ≈ 21 (Table 12.4).

With such a low simulated total of in-migrants (255), we must combine certain low-population provinces and use regional population totals instead of individual province populations. As discussed earlier, a valid application of chi-square requires at least two observations in each cell and no more than 20% (one-fifth) of the cells should have an observed frequency count of 5 or less. Therefore we have chosen to combine Newfoundland and Labrador, Prince Edward Island, Nova Scotia, and New Brunswick into an "Atlantic Canada" region. Also, the three territories of "Northern Canada" (Yukon, Northwest, and Nunavut) are dropped from the analysis due to small simulated sample sizes.

Using the "population/distance" model and assuming a simulated total of 255 in-migrants, we are now able to contrast the observed and expected frequency counts using the chi-square test. The result is a chi-square test statistic of 112.25. By themselves, the magnitude of the chi-square statistic and its associated *p*-value has no inferential value: we are using chi-square in an exploratory and descriptive fashion only. In short, this is because the simulated data are just constant proportions (.003) of the actual total and not true random samples. We cannot make any inferences from the "simulated sample" of 255 in-migrants to the actual total of 56,538 in-migrants to Alberta.

We are applying chi-square as an exploratory and descriptive tool to see where the largest differences between observed and expected frequencies are located. Then we can speculate about the underlying spatial processes causing these differences. The final two columns in Table 12.4 indicate the contribution of each Canadian province and region to the total chi-square value.

One key discovery is that the Atlantic Provinces are sending many more migrants to Alberta than expected using the "population/distance" predictive model. Nearly 82% (81.82%) of the total "contribution to chi-square" is from the Atlantic region. The logical follow-up geographic question is: "*why are many more migrants than expected relocating from the Atlantic region to Alberta*?" We might investigate a variety of economic conditions in these eastern provinces (depressed primary industries such as fishing and forestry, current unemployment rate, and so on) to begin to answer this question.

The other important contributor to the magnitude of the chi-square statistic is Quebec, which accounts for slightly more than 16% of the chi-square value. In this case, *many fewer migrants than expected are moving to Alberta*. As we ask why, we might refer to the "linguistic divide" and other aspects of cultural distinctiveness separating largely French-speaking Quebec from the rest of English-speaking Canada. Other factors may be suggested as well.

TABLE 12.4

Summary Table for Chi-Square Goodness-Of-Fit Proportional: "Simulated" Interprovincial Migration to Alberta, Canada, 2020

Province (Region) of origin included in final simulation	Observed number of interprovincial migrants (simulated)	Expected number of interprovincial migrants (population/distance model)	Contribution to χ^2	Percent of total contribution to χ^2
Atlantic*	29	7	69.14	61.59
British Columbia	81	85	0.18	0.17
Manitoba	21	16	1.56	1.39
Ontario	67	78	1.55	1.38
Quebec	11	43	23.81	21.21
Saskatchewan	45	25	16.00	14.25
TOTAL	254	254	112.25	100.0

* The Atlantic region of Canada comprises the four provinces located on the Atlantic coast, excluding Quebec. That is Newfoundland and Labrador, New Brunswick, Prince Edward Island, and Nova Scotia.

Note: The three territories of Northern Canada (Yukon, Northwest, and Nunavut) were dropped from the analysis due to small simulated sample sizes.

"Population/Distance" model: $E_{ij} = \frac{Pop_i}{D_{ij}}$

$$\chi^2 = \sum_{i=1}^{k} \frac{(O_i - E_i)^2}{E_i} = \frac{(29-7)^2}{7} + \frac{(81-85)^2}{85} + \frac{(21-16)^2}{16} + \frac{(67-78)^2}{78} + \frac{(11-43)^2}{43} + \frac{(45-25)^2}{25}$$

$$= 112.25$$

Kolmogorov-Smirnov Normality Test

The Kolmogorov-Smirnov statistic is an alternative to chi-square for testing similarity between two frequency distributions. Chi-square uses frequencies from either nominal or ordinal classes, but Kolmogorov-Smirnov (K-S) requires data measured at the ordinal level or interval-ratio level downgraded to ordinal. Technically, K-S requires continuously distributed variables, but only very slight errors result when the technique is applied to discrete data. In the most common goodness-of-fit application of K-S—testing a set of sample data to determine if it has been taken from a normally distributed population—the observed distribution of sample data is compared to an expected theoretical normal distribution. The null hypothesis states that no significant difference exists between the two frequency distributions. If it is confirmed that the sample has been taken from a normally distributed population, then it is fully valid to use a parametric statistic such as *t* or ANOVA in the subsequent analysis.

More specifically, the Kolmogorov-Smirnov test for normality compares the **cumulative relative frequencies** of the observed sample data with the cumulative frequencies expected for a perfectly normal distribution. If the two sets of cumulative frequencies closely match, the theoretical and sample distributions can be considered the same and the population considered normal. Any sizable difference between the two cumulative distributions suggests that the sample data are not taken from a normally distributed population.

Kolmogorov-Smirnov Goodness-Of-Fit Test

Primary Objective: Compare random sample frequency counts of a single variable with expected frequency counts (goodness-of-fit)

Requirements and Assumptions:

1. Single random sample
2. Population is continuously distributed (test less valid with discrete distribution)
3. Variable is measured at ordinal scale or downgraded from interval/ratio to ordinal

Hypotheses:

H_0 : Population from which sample has been drawn fits an expected frequency distribution (uniform or equal, proportional or unequal, random, etc.); no difference between observed and expected frequencies

H_A : Population does not fit an expected frequency distribution; there is a significant difference between observed and expected frequencies

Test Statistic:

$$D = \text{maximum } |\text{CRF}_o(X) - \text{CRF}_e(X)|$$

The step-by-step manual calculation of the Kolmogorov-Smirnov is somewhat cumbersome—involving the conversion of the actual ranked data to *Z*-scores, the calculation of cumulative normal values and cumulative relative frequencies from those *Z*-scores, and the subsequent comparison of those with the corresponding cumulative relative frequencies expected with a perfectly normal distribution. Fortunately, we can make use of a statistical software package to do the sequence of calculations for us, so it is not necessary to include a step-by-step worktable of calculations. The resultant K-S statistic, (*D*), is the *maximum absolute difference* between the two sets (observed and expected) of cumulative relative frequencies, that is: $D = \text{maximum } |CRF_o(X) - CRF_e(X)|$. The concept is illustrated in Figure 12.2 with a hypothetical sample of 10 observations (ranked in ascending order) being tested for normality.

where $CRF_o(X)$ = cumulative relative frequencies (observed) for variable X

$CRF_e(X)$ = cumulative relative frequencies (expected if normal) for variable X

Note that the $CRF_e(X)$ is a continuous curve, because its value can be calculated for any magnitude of X. By contrast, $CRF_o(X)$ is a series of points, because its values are determined only at the actual sample points. The K-S test compares the expected and observed relative frequency values only at the sample locations.

When the deviation between the actual and theoretical (normal) distribution is very large, *D* is very large and the null hypothesis of no difference between the two dis-

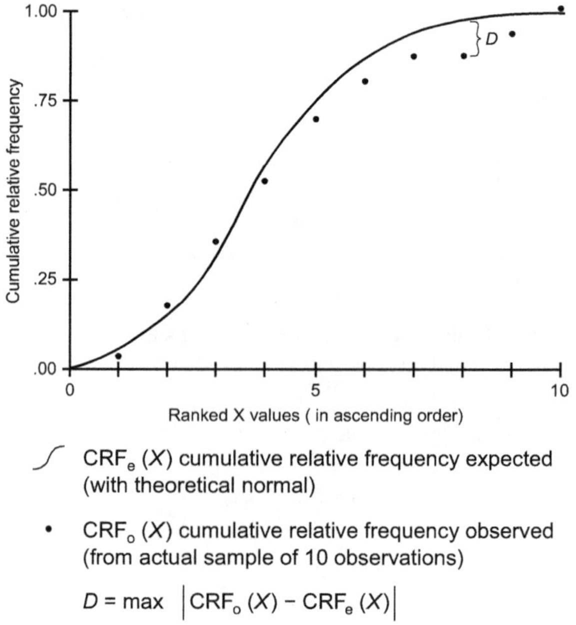

FIGURE 12.2

Kolmogorov-Smirnov Test: Maximum Difference (*D*) between Observed and Expected Cumulative Relative Frequencies tributions is likely to be incorrect. In this instance, if *D* is a large enough value, the actual distribution may be found statistically non-normal. Conversely, if all the differences between the actual and expected distributions are small, then *D* will be small. In this case, the null hypothesis is not rejected, and the observed distribution is considered normal.

In the *p*-value approach to determining the significance of the Kolmogorov-Smirnov test statistic, the approximate value of *p* is estimated by interpolating from a special table of *D* values. Most statistical software packages provide both the K-S (*D*) value and the corresponding *p*-value estimate, along with certain basic statistics such as the mean and standard deviation of the data set.

Example: Kolmogorov-Smirnov Goodness-Of-Fit (Normal Distribution)—Household Annual Energy Expenditures, Parkwood Estates

The Parkwood Estates survey of household water usage change before and after the attempted implementation of a water conservation program was not very successful (see the Wilcoxon matched-pairs test example in Chapter 10). Local planners are not deterred, however, and want to conduct another survey of Parkwood Estates households, this time analyzing annual energy expenditures.

A random sample of 25 households is selected and their total energy expenditures are monitored for an entire year (Table 12.5). Most of the households sustained energy costs in the $1,800 to $2,000 range. This finding is not very surprising to local officials, as the homes in the development were generally built about the same time with similar quality construction materials and insulation levels, and have about the same square footage. For some reason, however, a few of the homes (particularly numbers 2, 5, and 6) registered much lower energy costs.

To explore further, a histogram is constructed (Figure 12.3) and the low-energy cost outliers are easily seen. The shape of the histogram suggests that this sample of households may have been taken from a non-normal population (the "population" being all of the households in Parkwood Estates). If true, significant non-normality would affect which statistical tests could be subsequently used to analyze the energy expenditure data. Recall that many statistical procedures (parametric tests) assume population normality, and using a normality

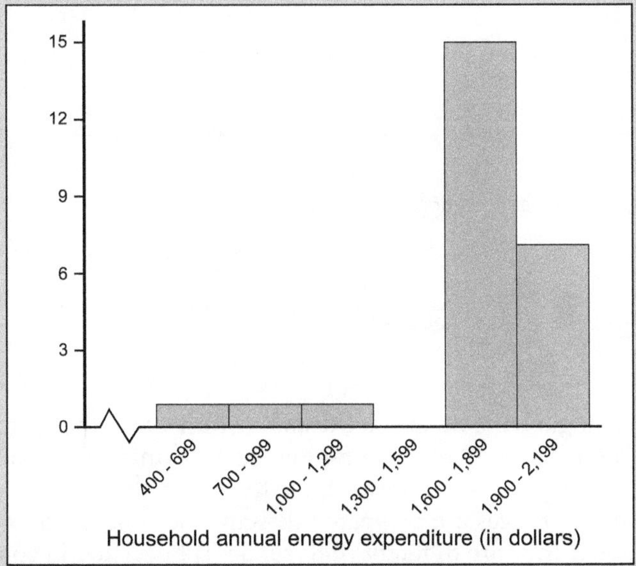

FIGURE 12.3
Histogram of Annual Energy Expenditures for Twenty-Five Parkwood Estates Households

TABLE 12.5

Annual Energy Expenditures for Twenty-Five Parkwood Estates Households

Household number	Annual energy expenditure (dollars)
1	$ 1,915
2	$ 412
3	$ 1,840
4	$ 1,845
5	$ 841
6	$ 1,044
7	$ 1,773
8	$ 1,691
9	$ 1,865
10	$ 1,846
11	$ 1,880
12	$ 1,844
13	$ 1,822
14	$ 1,915
15	$ 2,010
16	$ 1,991
17	$ 1,974
18	$ 1,967
19	$ 1,920
20	$ 1,885
21	$ 1,872
22	$ 1,887
23	$ 1,891
24	$ 1,820
25	$ 1,813

(continued)

test such as K-S is often an important step in the analysis procedure. Kolmogorov-Smirnov must be used as the normality test in this problem because the sample size is less than 30.

We also decided to generate a **probability plot** of the energy expenditure data along with calculation of the K-S statistic (Figure 12.4) to assess population normality visually. This graphic plots the actual ranked data values against the values you would expect to occur if the population from which the sample has been taken is normally distributed.

The sequence of plotted points will form an approximately straight line if the sample has been taken from a normally distributed population: clearly that is *not* the case in the figure. Based on visual evidence, we must conclude that the sample has been taken from a non-normal population. This is confirmed by the Kolmogorov-Smirnov test result. The K-S value of 0.372 indicates the maximum absolute difference in cumulative percentages between the actual and theoretical normal distributions. The likelihood (*p*-value) that a difference as large as 0.372 could occur if the sample is drawn from a normally distributed population is less than 0.010. That is, we are over 99% confident that the sample has been taken from a non-normal population.

Local officials are upset with this result, as they expected household energy costs to be normally distributed. A quick check of the situation revealed that households 2 and 5 were occupied only during the summer months and that a single elderly person was the sole occupant of household 6. The study would have been improved by including only year-round occupants or by using energy expenditure per month of occupancy as an alternative measure.

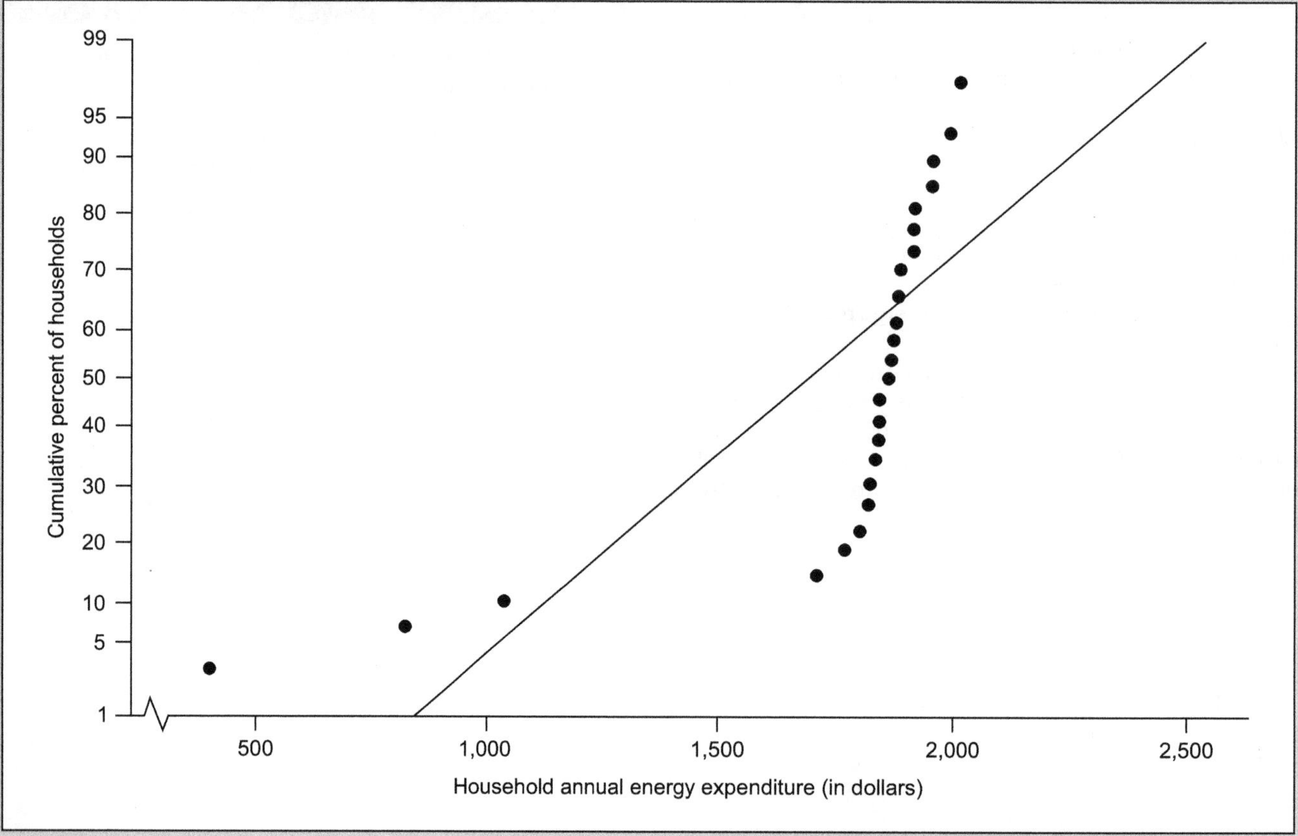

FIGURE 12.4
Probability Plot of Annual Energy Expenditures for Twenty-Five Parkwood Estates Households

12.2 Contingency Table Analysis

The idea of testing an actual frequency distribution against an expected frequency distribution is now expanded to research problems that examine the relationship between two variables. This is a logical extension of the single-sample or single-variable goodness-of-fit tests. In this instance, categorical frequency counts of a single variable are not tested against some theoretical model or expected distribution (such as uniform, proportional, or normal). Instead, the frequency distributions of two variables are compared directly with one another to see if they are different (independent) or related in some way (not independent). The null hypothesis would state that the two categorical variables have no relationship, while the alternate hypothesis would state that they do have a non-random relationship.

For example, suppose a city planning agency is developing a five-year capital improvement program (CIP). This program involves the scheduling of public physical improvements based on estimates of future fiscal resources and project costs. As part of the CIP development process, decision-makers want citizen input in determining the method for financing several proposed cultural facilities. The proposals include:

- improving the grounds of a local historical site
- constructing a new wing on the art museum
- expanding the city aquarium
- building a new science center with science education facilities
- constructing a new branch library

Several methods of financing are available, including general tax revenue, user fees, administrative fees, and revenue bonds. Suppose a random survey of residents is taken and respondents are asked which of the five projects they prefer. For that preferred project, each respondent is then asked which method of funding they prefer. A contingency table could then be constructed showing the number of people who prefer each project/funding source combination. Contingency analysis could then be applied to the frequency counts in this table to determine if the preferred financing methods differ significantly from project to project.

Geographers can also use contingency analysis effectively when comparing maps. Given two spatial pattern (choropleth) maps of the same study area, the extent of areal relationship between the maps can be examined by selecting n identical sets of random point locations from both maps, then allocating each of the sample point combinations to the appropriate cell of a contingency table. For example, suppose a geographer has two maps of the same region. One map shows the suitability of land for general farming (in five ordinal categories: well suited, moderately suited, somewhat poorly suited, poorly suited, and unsuited), and the other map shows soil association (in five nominal categories: alfisols, entisols, histosols, inceptisols, and spodosols). The spatial relationship between these two maps can be measured though contingency analysis of the frequency counts of points assigned to each cell of the table.

When using contingency analysis, data values from both variables are assigned to nominal or ordinal categories, and the frequency count of each category from one variable is cross-tabulated with the frequency count of each category from the second variable. The frequencies are summarized in a two-dimensional **contingency** or **cross-tabulation table**. Each variable must have at least two categories to which observations can be allocated.

The actual frequency count in each cell of the contingency table is compared with the frequency count expected if no relationship exists between the two variables. If *at least one* difference between actual and expected frequency counts is large, then the variables are not likely to be statistically independent, but rather related in some nonrandom (systematic) fashion. The null hypothesis of no relationship is then rejected. Conversely, if the differences between frequency counts in all cells of the table are small, then you would conclude the variables are statistically independent with only a random association, and the null hypothesis is not rejected.

The chi-square statistic evaluates contingency table frequency counts and is a simple extension of the goodness-of-fit test described in Section 12.1. *The entire contingency table is evaluated statistically as a single entity.* That is, the analysis determines whether significantly large differences exist between actual and expected frequency counts of the two variables, but does not indicate which cells in the table contain these differences. When both variables have exactly two categories, this analysis poses no problem. With tables having three or more categories for at least one variable, however, further analysis is sometimes needed to learn exactly where the significantly different frequency counts are located.

Certain restrictions apply to the use of chi-square in contingency analysis. Data must be absolute frequency counts rather than relative frequencies such as percentages or proportions. In addition, no more than one-fifth of the expected frequencies should be less than five and none less than two. As discussed with goodness-of-fit

Contingency Analysis (χ^2)

Primary Objective: Compare random sample frequency counts of a two variables for statistical independence

Requirements and Assumptions:

1. Single random sample
2. Variables are organized by nominal or ordinal categories; frequency counts by category are input to statistical test
3. No more than one-fifth of the expected frequency counts should be less than five, and none of the expected frequencies should be less than two

Hypotheses:

H_0 : There is no relationship between two variables in the population from which sample has been drawn

H_A : There is a relationship between two variables in the population from which sample has been drawn (variables are not statistically independent)

Test Statistic:

$$\chi^2 = \sum_{i=1}^{r} \sum_{i=1}^{j} \frac{\left(O_{ij} - E_{ij}\right)^2}{E_{ij}}$$

procedures, combining categories is sometimes an option if too many expected frequency counts are too low.

The chi-square statistic used in contingency analysis is:

$$\chi^2 = \sum_{i=1}^{r}\sum_{j=1}^{k}\frac{\left(O_{ij}-E_{ij}\right)^2}{E_{ij}} \quad \textbf{(12.4)}$$

where: O_{ij} = observed frequency count in the ith row and j^{th} column

E_{ij} = expected frequency count in the ith row and j^{th} column

r = number of rows in the contingency table

k = number of columns in the contingency table

If the observed and expected frequency counts in each cell of the contingency table are similar, then the differences ($O_{ij} - E_{ij}$) in all of the cells will be small, the statistic will be low, and the null hypothesis will not be rejected. Conversely, if at least one cell in the contingency table has a large difference between observed and expected frequency counts, the test statistic will be large and the null hypothesis will more likely be rejected.

To calculate the expected frequency of cell ij in the contingency table, the sum of all observed frequency counts in row i (R_i) is multiplied by the sum of all observed frequency counts in column j (C_j), then divided by the grand total of all observed frequencies (N):

$$E_{ij} = \frac{R_i C_j}{N} \quad \textbf{(12.5)}$$

Two items regarding these expected frequencies should be noted. First, the frequency count totals of a row or column are often called **marginal totals**. Thus, R_1 is the margin total for row 1, C_3 is the marginal total for column 3, and so on. Second, even though all observed frequency counts are integers, the calculated expected frequency values will generally not be integers. Rounding the expected frequency values to integers is not necessary. Doing so would only result in a less precise test.

Example: Contingency Table Analysis— Comparing Obesity Status, Metropolitan and Nonmetropolitan Counties

A number of interrelated risk factors have been associated with the spatial patterns of obesity in the conterminous United States. Obesity seems to be related to several "meso-environmental" factors such as land use, population density, and level of urbanization. Since obesity data are available at the county level, and each county can be classified as either metropolitan or nonmetropolitan, we propose a research hypothesis testing the relationship between obesity status and type of county (metropolitan or nonmetropolitan).

Suppose we take a random sample of 200 U.S. counties, classify each county as either metropolitan or nonmetropolitan, and classify its obesity status as either "low" (less than the national median of 34%) or "high" (greater than or equal to the national median). We can now formulate the following inferential null hypothesis: "The type of county (metropolitan or non-metropolitan) and the obesity status of the county (low or high obesity level) are not independent." That is, the frequency counts of counties across these two variables are not related. On the other hand, if the alternate or research hypothesis is true, type of county and obesity status of the county will be related (not independent).

Agencies such as the Centers for Disease Control and Prevention (CDC) and the National Center for Health Statistics (NCHS) have made statements that obesity levels are generally higher in rural areas. We should therefore expect that a statistically significant chi-square value will be calculated from the random sample of 200 counties.

The data are entered into a chi-square analysis, with the following results (Table 12.6). The number of nonmetro counties (120) exceeds the number of metro counties (80) by a considerable margin, but this is proportional to the nationwide county counts. The number (frequency count) of "low" obesity counties is almost identical to the number of "high" obesity counties, which is to be expected since we are using the median as the dividing value.

TABLE 12.6

Contingency Table: Difference Between Metro and Nonmetro County-Level Obesity Rates*

	County obesity rates		
County classification**	"Low" obesity level (< 34%)	"High" obesity level (≥ 34%)	TOTAL
Nonmetro	56 (57.6)	64 (62.4)	120
Metro	40 (38.4)	40 (41.6)	80
TOTAL	96	104	200

* Each cell of the table contains the observed frequency count, followed by the expected frequency count, in parentheses.

"Low" obesity level defined as less than the median of 34.
"High" obesity defined as greater than or equal to the median of 34.

All expected frequency counts $E_{ij} = \frac{R_i C_j}{N}$

for example: $E_{11} = \frac{(120)(96)}{200} = 57.6$

$$\chi^2 = \sum_{i=1}^{r}\sum_{j=1}^{k}\frac{(O_{ij}-E_{ij})^2}{E_{ij}} = 0.214 \qquad p = 0.644$$

Source: Centers for Disease Control and Prevention (CDC)

Notice that the magnitudes of the observed and expected frequency counts are very similar in all cells of the table. These small differences result in a chi-square statistic of 0.214, with an associated *p*-value of 0.644. We should therefore conclude that the type of county (metropolitan or nonmetropolitan) and the obesity status of the county (low or high obesity level) are independent. That is, the frequency counts of counties across these two variables appear to be unrelated. This finding seems to contradict the conventional wisdom expressed by the CDC and the NCHS, and it therefore seems logical to explore the sample data further. Perhaps we should look at particular counties in the sample to see if there are any influential anomalies or outliers. In fact, we will return to this set of data when constructing a multiple regression model in Chapter 18.

Example: Contingency Table Analysis—Comparing Congressional Approval Ratings by Census Division

By now, it is common knowledge that most Americans do not have a very favorable view of Congress. In fact, the approval rating of Congress has been quite low for several years: in 2010, it hit an all-time low of 10%. Not surprisingly, Congressional ratings have not improved much given the continuing budget problems facing the country. Paradoxically, while the general approval rating of Congress remains very low, most Americans have a surprisingly strong positive impression about their own representative. The overwhelming majority (90%) of all congressional lawmakers running for reelection were returned to Washington in 2012.

Much less attention is directed toward how congressional approval ratings vary from one part of the country to another. We propose the following inferential hypothesis: "Congressional approval ratings are no different from one census division to another." The latest available survey of approval ratings by census division was done in 2010, as part of the General Social Survey conducted by the National Opinion Research Center at the University of Chicago. The results are shown in Table 12.7, with various levels of confidence in Congress cross-tabulated against census division.

Less than 10% of all respondents (125 out of 1347) had a great deal of confidence in Congress. Most cells of the contingency table show only slight differences between the actual frequency counts and those frequency counts expected if there is no difference in views by census division (the null hypothesis).

The calculated chi-square value of 22.56 has an associated *p*-value of 0.126 (Table 12.8). This result suggests that some minor differences between observed and expected frequency counts exist, but they are relatively insignificant. If we reject the null hypothesis and conclude that there are significant differences from one census division to another in congressional approval levels, we would be wrong 12.6% of the time. If we follow conventional practice, it is probably best to not reject the null hypothesis and conclude that approval ratings do not vary across census divisions; it seems that people all across the country think that Congress is doing a poor job!

TABLE 12.7

Contingency Table: Census Division of Respondent Cross-Tabulated with Level of Confidence in Congress, 2020

	Level of confidence in Congress*			
Census Division	**A great deal**	**Only some**	**Hardly any**	**TOTAL**
New England	4 (4.33)	30 (26.06)	21 (24.61)	55 (55)
Middle Atlantic	11 (9.12)	63 (54.96)	42 (51.92)	116 (116)
E. North Central	9 (12.66)	65 (76.28)	87 (72.06)	161 (161)
W. North Central	6 (7.08)	37 (42.64)	47 (40.28)	90 (90)
South Atlantic	21 (16.04)	91 (96.65)	92 (91.31)	204 (204)
E. South Central	3 (4.25)	22 (25.58)	29 (24.17)	54 (54)
W. South Central	10 (7.79)	45 (46.90)	44 (44.31)	99 (99)
Mountain	6 (9.04)	61 (54.49)	48 (51.47)	115 (115)
Pacific	14 (13.69)	92 (82.44)	68 (77.87)	174 (174)
TOTAL	84 (84)	506 (506)	478 (478)	1,068 (1,068)

* Each cell of the table contains the observed frequency count, followed by the expected frequency count, in parentheses.

All expected frequency counts $E_{ij} = \frac{R_i C_j}{N}$ (for example: $E_{11} = \frac{(55)(84)}{1,068} = 4.33$)

Source: General Social Survey (GSS), conducted by the National Opinion Research Center (NORC)

(continued)

TABLE 12.8

Worktable for Chi-Square Contingency Analysis: Level of Confidence in Congress

H_0: there is no relationship between two variables (variables are statistically independent with only a random association).

H_A: there is a relationship between two variables (variables are not statistically independent, but related to one another in some nonrandom fashion).

$$\chi^2 = \sum_{i=1}^{r}\sum_{j=1}^{k}\frac{(O_{ij} - E_{ij})^2}{E_{ij}}$$

where O_{ij} = the observed frequency count in the ith row and j^{th} column
E_{ij} = the expected frequency count in the ith row and j^{th} column
r = the number of rows in the contingency table
k = the number of columns in the contingency table

The row variable is Census division and the column variable is level of confidence in Congress.

$$E_{ij} = \frac{(R_i)(C_j)}{N}$$

$$E_{11} = \frac{(R_1)(C_1)}{N} = \frac{(55)(84)}{1068} = 4.32$$

$$\chi^2 = \sum_{i=1}^{r}\sum_{i=1}^{k}\frac{(O_{ij} - E_{ij})^2}{E_{ij}} = \frac{(4 - 4.33)^2}{4.33} + \frac{(30 - 26.06)^2}{26.06} + \ldots + \frac{(92 - 82.44)^2}{82.44} + \frac{(68 - 77.87)^2}{77.87} = 21.28$$

$\chi^2 = 21.28$ $\quad$ $p\text{-value} = 0.226$

(1) cell with expected count less than 5

Example: Contingency Table Analysis— Patterns of State-Level Housing Vacancy Rates, Former Owner-Occupied Units and Former Rental Units

As data for the 2010 Census of Population was being collected, the United States was struggling through a major housing crisis. Part of Chapter 8 deals with the problem of housing units being "underwater" (having lost so much value that the amount owed on the mortgage exceeds the current value of the home).

Another unfortunate consequence of this crisis has been the abandonment of many housing units. As geographers, one of many interesting issues we might investigate is the spatial distribution and variation of vacant housing units across the United States. As part of the 2010 Census effort, state-level data were collected concerning housing unit vacancy rates, both for units last occupied by the homeowner and for units last occupied by a renter. At the time these data were collected, the national average homeowner-occupied vacancy rate was 2.4% and the national average rental unit vacancy rate was 9.2%.

To show the spatial patterns of both homeowner-occupied and rental unit vacancy rates, we construct a crosstab map (Figure 12.5). Each state is allocated to its particular combination of vacancy rates, whether above or below the national average.

Certain states were particularly hard-hit by the housing crisis, including Nevada (especially Las Vegas), Arizona (Phoenix), Florida, Michigan (Detroit), and Ohio (Cleveland). As you can see, these are all among the states that have both homeowner and rental vacancy rates above the national average. The Southeast appears to be the region that has suffered the most. For your information, Nevada had the highest homeowner-occupied vacancy rate (5.2%), while South Carolina had the highest rental unit vacancy rate (14.3%).

Conversely, many states in the upper Midwest and New England had vacancy rates below the national average for both former homeowner-occupied units and former rental housing units. Massachusetts had both the lowest former homeowner-occupied vacancy rate (1.5%) and former rental unit vacancy rate (6.5%).

A contingency table analysis confirms a state-level relationship between the owner-occupied and the renter-occupied vacancy rates of housing units (Table 12.9). While only descriptive in nature, a chi-square test statistic of 7.710 (with associated *p*-value of 0.005) strongly suggests that states having above national average owner-occupied housing unit vacancy rates are also likely to have above national average rental unit vacancy rates.

Once again, the basic intriguing question for geographers is *why* these spatial patterns vary in the way they do. Many different factors undoubtedly influence these patterns in complex ways. It is known, for example, that mortgage lending and banking practices vary considerably, even at the local or metropolitan level.

TABLE 12.9

Contingency Table: State-Level Housing Unit Vacancy Rates, Formerly Owner-Occupied Units and Rental Units, 2010

	Vacancy rate of formerly owner-occupied housing units		
Vacancy rate of formerly renter-occupied housing units	**Above national average vacancy rate**	**Below national average vacancy rate**	**TOTAL**
Above national average vacancy rate	13 (8.24)	8 (12.76)	21
Below national average vacancy rate	7 (11.76)	23 (18.24)	30
TOTAL	20	31	51

$\chi^2 = 7.710$ *p*-value = 0.005

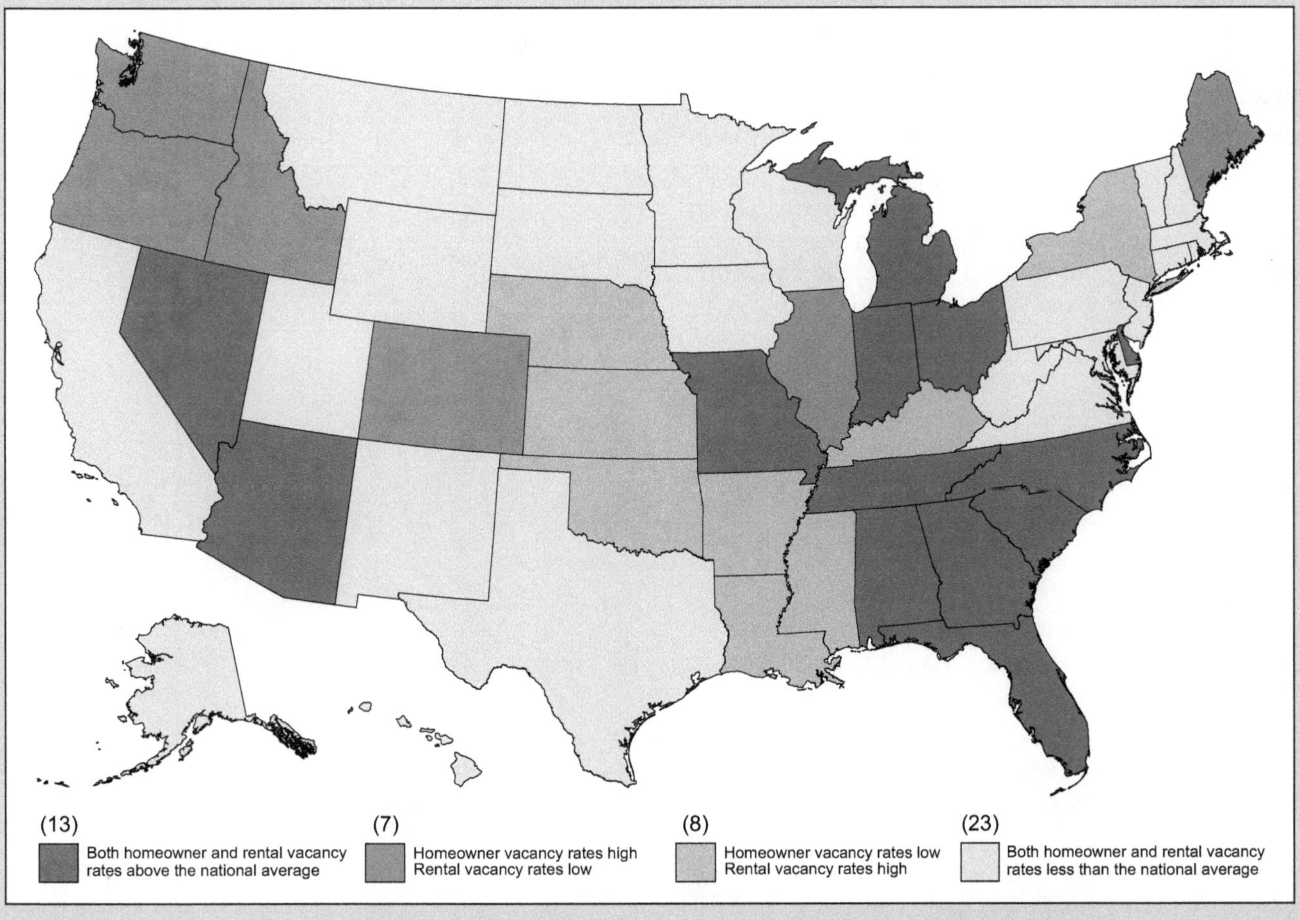

FIGURE 12.5

Crosstab Map: Pattern of State-Level Housing Vacancy Rates, Owner-Occupied and Rental Units (Washington, DC, also included in the analysis)

Source: United States Bureau of the Census, Population and Housing Occupancy Status

Key Terms

chi-square goodness-of-fit (proportional distribution), *194*
chi-square goodness-of-fit (uniform distribution), *192*
contingency (crosstab) map, *202*
contingency table (cross-tabulation table) analysis, *190*
cumulative relative frequency, *196*
Kolmogorov-Smirnov goodness-of-fit (normal distribution), *197*
marginal totals, *200*
probability plot, *198*

Major Goals and Objectives

If you have mastered the material in this chapter, you should now be able to:

1. Recognize the variety of geographic research problems for which goodness-of-fit tests are appropriate.
2. Explain the difference between an actual or observed frequency distribution and an expected frequency distribution in a goodness-of-fit context.
3. Understand the nature of various frequency distributions (uniform or equal, proportional or unequal, and normal) and identify the corresponding goodness-of-fit test.
4. Define the term "cross-tabulation" and identify geographic situations that use cross-tabulation.
5. Identify the types of geographic problems for which contingency analysis is appropriate and be able to create and organize an original "geographic situation" where contingency analysis could be used.

References and Additional Reading

If you are interested in further exploring the geographic examples mentioned in this chapter, the following are good places to start. Additional information about educational testing and the statistical validity, accuracy, and precision of test scores can be found at the National Center for Educational Statistics (NCES) www.nces.ed.gov. Without priority access, however, it will be difficult to obtain location-based test score data (confidentiality restrictions). Migration and immigration pattern data are available from Statistics Canada www.statcan.gc.ca. The Energy Information Administration has the nationwide averages from which the Parkwood Estates example problem data were designed to closely replicate: www.eia.gov/consumption/residential. Attitudes regarding Congressional approval have been collected for decades through Gallup polling www.gallup.com. For a full statistical analysis of congressional approval rating contingency tables by Census Division and Region, see the General Social Survey archives at the Survey Documentation and Analysis (SDA) website: http://sda.berkeley.edu. Finally, housing vacancy rate data (for both owner-occupied and rental units) can be found at www.census.gov/housing/hvs/index.html

PART 5

Inferential Spatial Statistics

13

General Issues in Inferential Spatial Statistics

Geographers often examine spatial patterns on the earth's surface that are produced by physical or cultural processes. These patterns represent the spatial distribution of a variable across a study area. Sometimes geographic variables are displayed as point patterns with dot maps. In Chapter 4, descriptive spatial statistics (or geostatistics), such as the mean center, standard distance, and standard deviational ellipse were introduced to summarize point patterns. In other instances, explicit spatial patterns representing data summarized for a series of subareas within a larger study region can be displayed effectively using choropleth maps.

Whether data are presented as points or areas, geographers often want to describe and explain an existing pattern. In this chapter, we focus on several general issues associated with inferential spatial statistics. A recurring theme is the analysis of a sample point pattern or sample area pattern to determine if the overall or "global" arrangement is random or nonrandom. When testing a point or area pattern for randomness, the question is whether the population pattern from the sample was generated by a spatially random process. A secondary theme has to do with "local" inferential spatial statistics that identify and map clusters ("hot spots" and "cold spots") of points or areas within a larger area that have a concentration of very high (or very low) data values.

13.1 Types of Spatial Pattern

Geographers often want to compare an existing spatial pattern to a particular theoretical pattern. Spatial patterns may appear clustered, dispersed, or random (Figure 13.1). In Case 1, both the point and area patterns have a **clustered**

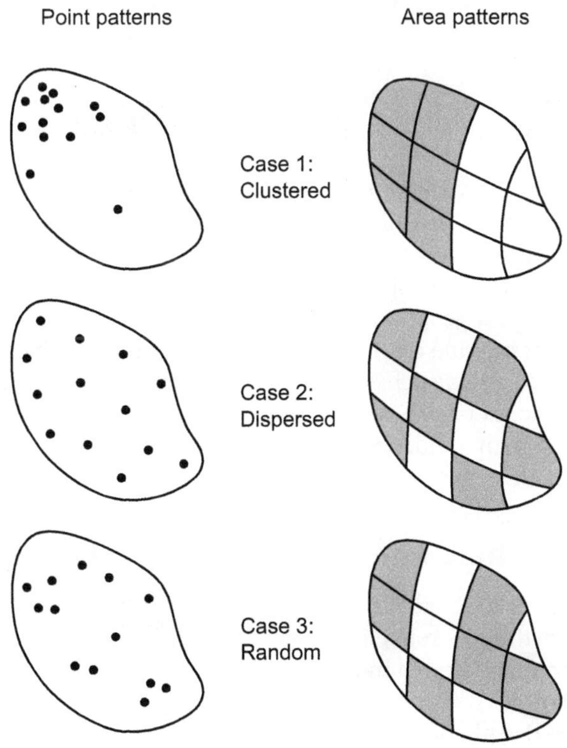

FIGURE 13.1
Types of Point and Area Patterns

appearance. On the point pattern map, the density of points appears to vary significantly from one part of the study area to another, with many points concentrated in the northwest portion of the area. Perhaps the points represent sites of tertiary economic activity (retail and service functions), which often cluster around a location with high accessibility and high profit potential, such as a highway interchange. On the clustered area pattern map, the shaded subareas could represent political precincts where a majority of registered voters are Democrat, with Republican majority precincts not shaded. Such a clustered pattern would likely occur if there is a distinctly nonrandom spatial distribution of voters by income, race, or ethnicity in the region.

Other spatial patterns seem evenly **dispersed** or regular. The set of points in Case 2 (Figure 13.1) appears uniformly distributed across the study area, perhaps suggesting that a systematic spatial process produced the locational pattern. One hypothesis in classical central place theory, for example, is that settlements are uniformly distributed across the landscape to best serve the needs of a dispersed rural population. The area pattern in Case 2 exhibits a regular or alternating type of spatial arrangement. This pattern could represent county populations in the same region where the central place distribution of settlements is hypothesized. Shaded counties could have above average populations while counties not shaded are below average.

The spatial patterns in Case 3 (Figure 13.1) appear **random** in nature, with no dominant trend toward either clustering or dispersion. A random point or area pattern logically suggests that a spatially random (Poisson) process is operating to produce the pattern. In each of these examples we use the term *appear,* based on our subjective opinion of the distribution of the data. However, the New Mexico wildfire distribution example from Chapter 5 illustrates a process that can be statistically evaluated to determine if the wildfire pattern is random or clustered.

In most geographic problems, a point or area pattern will not provide a totally clear indication that the pattern is clustered, dispersed, or random. Rather, many real-world patterns show a combination of these arrangements with tendencies from "purely" random toward either clustered or dispersed.

13.2 The Concept of Autocorrelation

In 1970, geographer Waldo Tobler stated that "Everything is related to everything else, but near things are more related than distant things." This quote became known as Tobler's Law, but, more importantly, it illustrates the phenomenon of **spatial autocorrelation**. Spatial autocorrelation simply measures the degree to which a geographic variable is correlated with itself through space. Like traditional correlation discussed in Chapter 16, spatial correlation is measured as either being positive, negative, or non-existent.

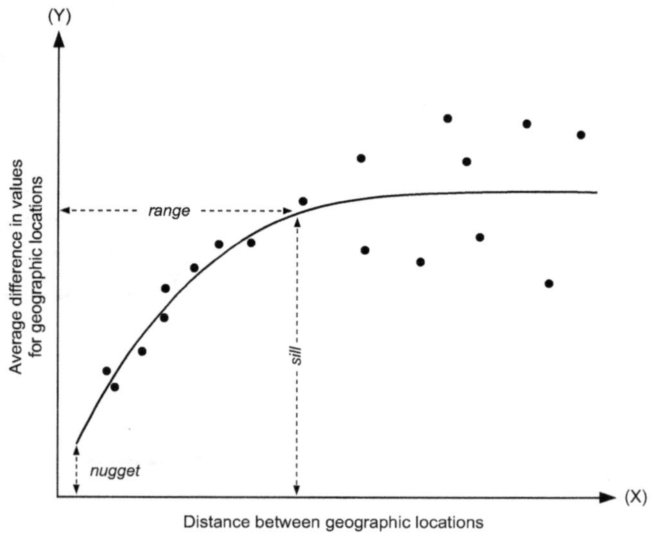

FIGURE 13.2
An Example of a Variogram

Positive spatial autocorrelation means that geographic objects that are near one another tend to be similar. That is, high value features tend to be located near other high value features, medium value features near other medium value features, and so on. Negative spatial autocorrelation means that geographic objects near one another tend to have sharply contrasting values (a high value located adjacent to a low value, for example). Most geographic phenomena have some measure of positive spatial autocorrelation. Nearby cities tend to get similar amounts of rainfall; expensive homes tend to be located near other expensive homes.

One can visualize spatial autocorrelation through use of a **variogram.** A variogram is a type of scatterplot that displays the differences in values between geographic locations against the differences in the distance between the geographic locations. The *Y* axis records the average variance in values among a set of geographic objects, but for reasons beyond our discussion, the actual measure used is *half* the variance (termed *semi-variance*). The *X* axis shows the average distance among the objects, which is termed the **range**. For simplicity, the distances are aggregated into more easily understood groups like 100 miles, 200 miles, etc. By plotting a variogram you can determine the average difference in values among locations that are 200 miles apart, 500 miles apart, and any other logical distance that you would like. A generalized example of a variogram is shown in Figure 13.2.

In this example, the *Y* axis represents the average difference in *values* between geographic locations, while the *X* axis represents the difference in *distances* between geographic objects. You can see that geographic locations near one another tend to have smaller differences, while geographic locations with larger differences are generally further apart. Rather than showing the individual points, a curved line is used to represent the nature of the relationship. While beyond the scope of our discussion, in advanced geostatistics the mathematical properties of the best-fitting curve are used to estimate values at unknown locations. The first part of the variogram (left portion of the X axis) illustrates the change in average differences in data values as the distances become greater, while the flat portion of the variogram (right portion of the X axis) shows there is no change in the average differences in data values when the locations are very far apart. In other words, the flat portion of the graph suggests no relationship between the average difference in values for pairs of geographic objects and their distance from one another, while the earlier portion of the graph shows that the values become less similar as the distances among the locations increase.

The **range** represents the distance at which the differences in values of geographic locations are no longer correlated, while the **sill** represents the average difference in value where there is no relationship between location and value. Typically, if two objects are in the same location, they should have the same value and thus no difference. However, whenever samples are taken, some degree of uncertainty always occurs due to measurement error, the inherent properties of the measuring device, or some other reason. The **nugget** represents the degree of uncertainty when measuring values for geographic locations that are very close to one another.

For our purposes, we do not address the mathematical derivation of the variogram (you can learn more about variograms in some of the more advanced readings listed at the end of this chapter). Rather than going into statistical detail, we illustrate the variogram concept with

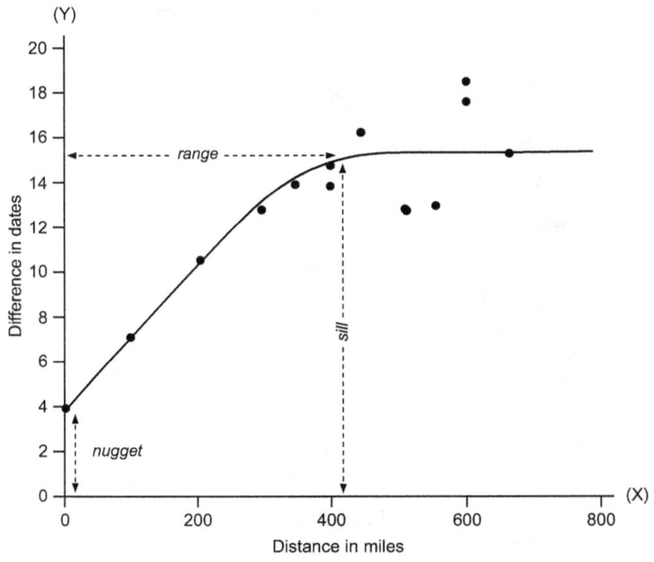

FIGURE 13.3
Variogram of Last Spring Frost (LSF) Date

a short example. In Chapter 1, an example shows the Last Spring Frost (LSF) date for selected weather stations in the Southeast United States. Naturally, we would expect the LSF date for two nearby weather stations to be similar, while stations further separated from one another would have less similarity.

This relationship is illustrated perfectly by the set of observations and curve that "best fits" through those observations (Figure 13.3). Between 0 and 400 miles it appears that if the distance separating a pair of weather stations is large, then their dates of last spring frost are likely to be quite different as well. Furthermore, this relationship between distance and differences in LSF date appears linear. Notice, however, what happens with pairs of weather stations located more than 400 miles apart: the scatterplot of points disperses around the best fit curve and appears random. This means there is no relationship between the LSF dates of weather stations when they are more than 400 miles apart. Therefore, the **range** is said to represent the distance at which the LSF dates are no longer spatially autocorrelated—in this example, about 400 miles.

The concept of spatial autocorrelation is critical in analyzing geographic phenomena for a number of reasons. First, by just pushing a button, GIS software allows users to analyze the relationship between points no matter how far apart they are. However, in our example we have concluded that there is no relationship in LSF values between weather station pairs that are more than 400 miles apart. Therefore, using a GIS or statistical process that assumes a relationship among the data may have limited use if the points are more than 400 miles apart. In short, to make intelligent observations about spatial data, we must understand the spatial structure of the data.

Second, and more importantly, the presence of spatial autocorrelation presents a real challenge when performing statistical analysis. Most inferential statistics assume that observations are independent of one another. Unfortunately, we have just demonstrated that LSF dates are actually spatially correlated (and thus dependent) with one another for weather stations less than 400 miles apart. Why is this important? Simply stated, confidence intervals calculated from spatially autocorrelated data will likely be different than confidence intervals calculated from data that are not spatially autocorrelated.

Recall from Chapter 8 that sample size is an important component in the computation of standard error; the greater the sample size, the better our estimates. In the case of spatially autocorrelated data, imagine a geographer sampling 30 objects, with five of those objects very close to one another. Essentially, selecting five nearby observations (assuming spatial autocorrelation within that distance) is almost like selecting the same observation five times. While you may believe you have 30 independent observations, in reality (since five of them are spatially autocorrelated), it is probably more practical to recognize that you essentially have only 25 independent observations. Therefore, the value of n in the confidence interval equation is too high (you will be erroneously making n equal to 30 rather than 25), thus giving you a smaller standard error than warranted. Also, if the five spatially autocorrelated values are similar, then the standard deviation for the entire dataset will also most likely be lower, yielding an even lower standard error. Therefore, you must take extreme care when interpreting results for inferential analysis when the data exhibits spatial autocorrelation.

A final characteristic of autocorrelation worth mentioning here is that it can be measured either globally or locally. For instance, a social geographer may be interested in determining if the distributions of various racial and ethnic groups in a city have different levels of spatial autocorrelation. Perhaps it is hypothesized that one racial or ethnic group has a higher positive autocorrelation across the entire area of the city than another. That is, spatial autocorrelation is being used as a comparative **global** measure of segregation or integration. It is "global" in the sense that a single spatial autocorrelation value is derived for each group across the entire study area, telling us the degree to which the overall spatial pattern of each group is clustered, random, or dispersed. In Chapter 15, we apply a measure of global autocorrelation to several racial and ethnic groups in Cleveland, Ohio.

By contrast, **local** autocorrelation methods compare each geographic object with its surrounding neighbors, thereby measuring whether the values in the immediate neighborhood are clustered, random, or dispersed. Many global methods for measuring spatial autocorrelation have been modified to provide local measures. Also in Chapter 15, we evaluate the spatial pattern of obesity levels for Pennsylvania counties and identify particular counties having significant local spatial autocorrelation.

For the last half century, geographers and statisticians have continued to study methods to identify and account for the amount of spatial autocorrelation among variables. It is not unlike correlation over time. For instance, during a long recession, a company's sales for one year may be very similar to the sales in a previous year. Economists apply advanced statistical formulas to account for this correlation over time. Geographers are now applying similar principles to account for correlation over space.

13.3 Who is my Neighbor?

In the famous parable of the Good Samaritan, a man asked "who is my neighbor?" Geographers studying spatial autocorrelation have recently been asking the same question. Geographers have different ways of defining the term "neighbor." One approach is to define **neighbor** in a binary way—that is, two geographic objects are either neighbors or they are not. One acceptable criterion for this binary relationship can be represented as a form of geographic association called **adjacency**. For example, Pennsylvania is adjacent to New York because they share a common boundary. However, neither state is adjacent to California. Another approach is to define "neighbor" using a **distance threshold** (or cut-off distance) measure. For example, if we establish a distance threshold of 300 miles, New York City and Boston would be considered neighbors, while Boston and Los Angeles would not.

"Neighbor" may also be defined as a continuous variable, by measuring the strength of "neighborliness" between two objects as a function of the distance separating them. This logic is comparable to spatial interaction models. For example, a simple **inverse-distance** weight (1/distance) could establish a measure of proximity. Using this model, the inverse distance weight of New York City and Boston is 1/189 miles, or .005, while the inverse distance weight of Boston and Los Angeles is 1/2588 miles, or .0004. In this case, the "neighborliness" or interaction measure is 12 times stronger for New York City and Boston than it is for Boston and Los Angeles. Yet another measure of neighborly interaction is **inverse-distance squared** ($1/\text{distance}^2$).

However neighbors are defined, geographers typically represent them in the form of a weight, where w_{ij} is the weight between geographic objects i and j. The weight can be either 0 or 1 for a binary representation or a ratio such as 1/distance for a continuous function.

Key Terms

autocorrelation (spatial autocorrelation), *208*
clustered, random and dispersed patterns, *207, 208*
global and local measures of autocorrelation, *210*
neighbor definitions: adjacency, distance threshold, inverse-distance, inverse-distance squared, *210*
range, sill, and nugget in a variogram, *209*
variogram, *208*

Major Goals and Objectives

If you have mastered the material in this chapter, you should now be able to:

1. Understand the concepts of clustering, dispersion, and randomness in point and area patterns.
2. Recognize situations in which point and area patterns appear clustered, dispersed, or random.
3. Explain why it is important to measure spatial autocorrelation.
4. Explain the basic purpose of a variogram and understand the terminology associated with a variogram.
5. Identify some ways in which the term "neighbor" is defined in spatial statistics.
6. Explain the basic difference between "global" and "local" measures of spatial association.

References and Additional Reading

Included below are a number of references that refer to geostatistics (spatial statistics) in general. Many of these sources also cover the point pattern and area pattern techniques discussed in Chapters 14 and 15.

Anselin, Luc. "Local Indicators of Spatial Association—LISA." *Geographical Analysis,* Vol. 27 No. 2 (1995): 93–115.

Bailey, T. C. and A. C. Gatrell. *Interactive Spatial Data Analysis.* London: Longman, 1995.

Barnes, R. *Variogram Tutorial.* Golden Software Inc. http://www.goldensoftware.com/variogramTutorial.pdf

Cliff, A. and J. Ord. *Spatial Processes: Models and Applications.* London: Pion, 1981.

Ebdon, D. *Statistics in Geography. A Practical Approach.* Oxford: Basil Blackwell, 1985.

Goodchild, M. *Spatial Autocorrelation*, CATMOG No. 47. Norwich, England: Geo Books, 1988.

Griffith, D. A. *Spatial Autocorrelation: A Primer.* Washington, DC: Association of American Geographers, 1988.

Griffith, D. and C. Amrhein. *Statistical Analysis for Geographers.* Englewood Cliffs, NJ: Prentice Hall, 1991.

Isaaks, E. H. and R. M. Srivastava. *An Introduction to Applied Geostatistics.* Oxford: Oxford University Press, 1989.

Lee, J. and D. W. S. Wong. *Statistical Analysis with ArcView GIS.* New York: Wiley, 2001.

Mitchell, A. *The ESRI Guide to GIS Analysis. Vol. 2: Spatial Measurements and Statistics.* Redlands, CA: ESRI Press, 2005.

Odland, J. *Spatial Autocorrelation.* Vol. 9, Scientific Geography Series. Beverly Hills, CA: Sage, 1988.

Ripley, B. *Spatial Statistics.* New York: Wiley, 1981.

14

Point Pattern Analysis

This chapter presents methods for analyzing point patterns that are distributed across an entire study region (that is, they are *global* methods of analysis). Such spatial patterns are characterized by observations appropriately displayed as a series of points or dots on a map and are very common in geographic studies. For example, urban geographers plot the location of settlements as points on a map. Economic geographers study the spatial pattern of retail activities by mapping store locations as dots on the map. Physical geographers show glacial features like drumlins as a series of points, while crime analysts may plot the location of burglaries as a series of dots.

In some problems, the *spacing of individual points* within the overall point pattern is important, in which case a statistical procedure known as "nearest neighbor analysis" is applied. For example, an urban geographer might analyze the existing configuration of fire stations in a city to determine whether the pattern is random or more dispersed than random. Suppose one of the goals in the provision of fire service is to provide an equitable or dispersed distribution of fire stations throughout the city so that no one lives or works extremely far from the nearest fire station. The geographer might suggest the siting of new stations or relocation of existing facilities to meet this goal. The proposed new configuration of fire stations could then be analyzed to determine if it were more dispersed than the existing pattern.

In other instances, the *nature of the overall point pattern* is important, and a statistical test known as "quadrat analysis" is often appropriate. For example, a biogeographer might select a random sample of trees in a national forest to determine which trees are diseased, plot the location of each diseased tree as a point on a map, and analyze the points. The study would determine whether the trees were randomly distributed throughout the study area or clustered in certain portions of the forest. The result of this analysis would provide some guidance concerning the most appropriate type of treatment (e.g., widespread aerial spraying versus concentrated treatment from the ground).

For many geographic studies that begin with analysis of a variable having locations on a dot map, a primary objective is to determine the form of the pattern of points. The nature of the point pattern might reveal information about the underlying processes that produced the distribution. In addition, a series of point patterns of the same variable recorded at different times can help determine temporal changes in the locational process. In short, point pattern analysis offers quantitative tools for examining a spatial arrangement of point locations on the landscape as represented by a conventional dot map.

14.1 Nearest Neighbor Analysis

Nearest neighbor analysis (NNA) is a widely used procedure to determine the spatial arrangement of a pattern of points within a study area. The distance of each point to its "nearest neighbor" is measured and the mean nearest neighbor distance for all points is determined. The spacing within a point pattern can be analyzed by comparing this observed mean distance to some expected mean distance, such as that for a random (Poisson) distribution.

The nearest neighbor technique was originally developed by biologists interested in studying the spacing of plant species within a region. They measured the distance separating each plant from its nearest neighbor of the same species and determined whether this arrangement was organized in some non-random manner or was the result of a random process. Geographers have applied the technique in numerous research problems, including the study of settlements in central place theory, economic functions within an urban region, and the distribution of earthquake epicenters in an active seismic region. In all applications, the objective of nearest neighbor analysis is to describe the pattern of points within a study area and make inferences about the underlying process.

The nearest neighbor methodology is illustrated using the example of seven points shown in Figure 14.1. A coordinate system is created and the horizontal (X) and vertical (Y) positions of the points recorded (Table 14.1). For each of the points, the nearest neighbor (NN) is determined as the point closest in straight-line (Euclidean) distance. The distances to each nearest neighbor (NND) are then calculated from the coordinates or measured from the map. From the set of nearest neighbor distances, the *mean* nearest neighbor distance ($\overline{NND}$) is determined using the basic formula for the mean:

$$\overline{NND} = \frac{\sum NND}{n} \qquad \textbf{(14.1)}$$

where n = number of points.

The mean nearest neighbor distance of the seven-point pattern is 2.67 distance units. This mean nearest neighbor distance provides an index of spacing for the set of points. However, the usefulness of this descriptive index comes from comparing the index value for an observed pattern to the results produced from certain distinct or theoretical point distributions. This objective is analogous to the situation discussed in Section 12.1, where an observed frequency distribution is compared to a perfectly normal frequency distribution.

In NNA, mean nearest neighbor distances can be calculated for three distinct theoretical point arrangements: perfectly random, perfectly dispersed, and perfectly clustered. In each case, the spacing index is determined for an area containing a certain density of points. For example, if points are arranged in a random spatial pattern, the mean nearest neighbor distance ($\overline{NND}_R$) would be determined as follows:

$$\overline{NND}_R = \frac{1}{2\sqrt{Density}} \qquad \textbf{(14.2)}$$

where $\overline{NND}_R$ = mean nearest neighbor distance in a random pattern

$Density$ = number of points (n)/Area

Since the study area displayed in Figure 14.1 has a dimension of 10 by 10 units, the area represented is 100 square units. The corresponding density of points is therefore 7/100 or .07. For an area with this point density, a random arrangement of points within the study area should produce mean nearest neighbor distance:

$$\overline{NND}_R = \frac{1}{2\sqrt{.07}} = 1.89$$

TABLE 14.1

Coordinates and Nearest Neighbor Information for Example*

Point	*X*	Y	NN	NND
A	1.3	0.9	E	3.94
B	3.2	4.4	C	2.00
C	3.3	6.4	B	2.00
D	5.6	3.8	E	1.36
E	4.8	2.7	D	1.36
F	8.1	7.4	G	4.21
G	9.4	3.4	D	3.82
SUM				18.69

where NN = nearest neighbor
NND = nearest neighbor distance

$$\overline{\text{NND}} = \frac{\sum \text{NND}}{n} = \frac{18.69}{7} = 2.67$$

* See Figure 14.1 for graph of point locations.

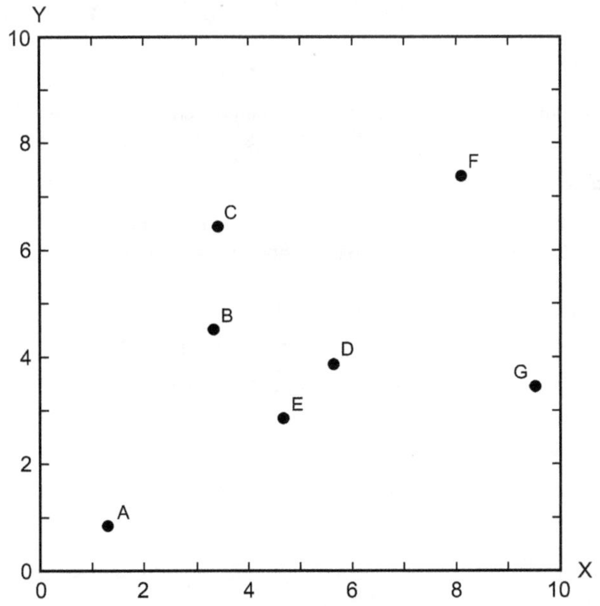

FIGURE 14.1
Location of Points for Nearest Neighbor Problem

If the arrangement of points shows maximum dispersion or a perfectly uniform pattern, the mean nearest neighbor distance ($\overline{NND}_D$) would be determined as:

$$\overline{NND}_D = \frac{1.07453}{\sqrt{Density}} \qquad \textbf{(14.3)}$$

Since this pattern represents the most dispersed or separated arrangement of points, it also serves as the maximum value for the nearest neighbor index. The mean nearest neighbor distance for a dispersed pattern whose point density matches that in Figure 14.1 is:

$$\overline{NND}_D = \frac{1.07453}{\sqrt{.07}} = 4.06$$

The pattern of spacing most distinct from dispersion or uniformity is "clustering." When all points lie at the same position, the pattern shows maximum clustering of points, and each nearest neighbor distance is zero. Therefore, in a perfectly clustered pattern, the mean nearest neighbor distance $\overline{NND}_C$ is also zero, representing the lowest possible value for the index:

$$\overline{NND}_C = 0 \qquad \textbf{(14.4)}$$

Since perfectly clustered and perfectly dispersed patterns provide the extreme spacing arrangements for a set of points, the nearest neighbor index offers a useful method to measure the spacing of locations within an observed point pattern. However, the mean nearest neighbor distance is an absolute (as opposed to relative) index, dependent upon the units used to measure distance. Therefore, direct comparison of results from different problems or different regions is difficult. Although the minimum value of the nearest neighbor index is always zero (a perfectly clustered pattern), the maximum value corresponding to a perfectly dispersed pattern is not constant, but a function of the point density. To overcome this complication, a **standardized nearest neighbor index (R)** is often used. This index is found by dividing the mean nearest neighbor distance ($\overline{NND}$) by the corresponding value for a random distribution with the same point density:

$$R = \frac{\overline{NND}}{\overline{NND}_R} \qquad \textbf{(14.5)}$$

With the standardized index, a perfectly clustered pattern produces an R value of 0.0, a perfectly random distribution 1.0, and a perfectly dispersed arrangement generates the maximum R value of 2.149 (Figure 14.2). Thus, an actual point pattern can be measured for relative spacing along a continuous scale from perfect clustering to perfect dispersal. For the set of points from Figure 14.1, the standardized nearest neighbor index is:

$$R = \frac{2.67}{1.89} = 1.41$$

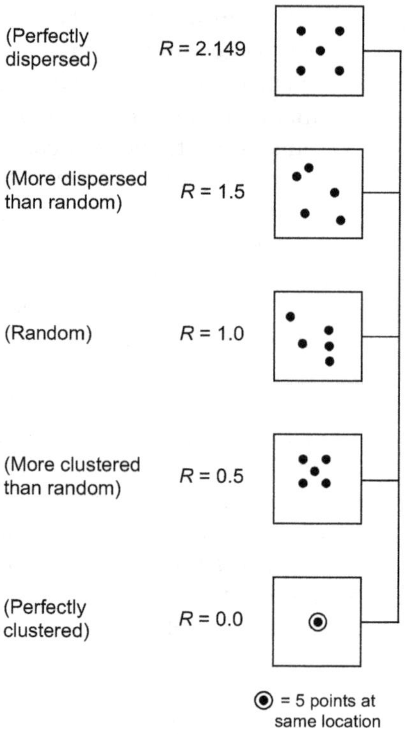

FIGURE 14.2
Continuum of R Values in Nearest Neighbor Analysis

Nearest Neighbor Analysis

Primary Objective: Determine whether a random (Poisson) process has generated a point pattern

Requirements and Assumptions:

1. Random sample of points from a population
2. Sample points are independently selected

Hypotheses:

$H_0 : \overline{NND} = \overline{NND}_R$ (point pattern is random)

$H_A : \overline{NND} \neq \overline{NND}_R$ (point pattern is not random)

$H_A : \overline{NND} > \overline{NND}_R$ (point pattern is more dispersed than random)

$H_A : \overline{NND} < \overline{NND}_R$ (point pattern is more clustered than random)

Test Statistic:

$$Z_n = \frac{\overline{NND} - \overline{NND}_R}{\sigma_{\overline{NND}}}$$

This spacing pattern is moderately dispersed, since it lies between a perfectly dispersed distribution and a random one.

In addition to its use as a descriptive index of point spacing, the nearest neighbor methodology can also be used to infer results from a sample of points to the population from which the sample was drawn. A difference test can be used to determine if the observed nearest neighbor index ($\overline{NND}$) differs significantly from the theoretical norm ($\overline{NND}_R$), which would occur if the points were randomly arranged. The expectation is that a Poisson process operates over space to produce a random pattern of points. The null hypothesis is that no difference exists between observed and random nearest neighbor values. The test statistic (Z_n) follows a format similar to other difference tests discussed in Chapters 9 through 12:

$$Z_n = \frac{\overline{NND} - \overline{NND}_R}{\sigma_{\overline{NND}}} \quad (14.6)$$

where $\sigma_{\overline{NND}}$ = standard error of the mean nearest neighbor distances

The standard error for the nearest neighbor test can be estimated with the following formula:

$$\sigma_{\overline{NND}} = \frac{.26136}{\sqrt{n(Density)}} \quad (14.7)$$

where: n = number of points

Either a directional (one-tailed) or non-directional (two-tailed) approach can be applied, depending on the form of the alternate hypothesis (H_A). If the problem suggests a clear rationale for the actual point pattern being either dispersed or clustered (as opposed to the null hypothesis of randomness), a one-tailed approach is warranted. However, without an underlying reason or theoretical expectation that the null hypothesis should be rejected on one side as opposed to the other, a two-tailed non-directional approach is best.

Example: Nearest Neighbor Analysis—Community Services in Toronto, Canada

The nearest neighbor methodology is now applied to point patterns representing community service sites within Toronto, Ontario. Some public services are best located in a highly dispersed pattern to provide relatively equal service distance for all parts of the region. Such a concern for spatial equity is especially important for emergency services such as police and fire. Other community activities may be sited to offer the region a high degree of efficiency or cost-effectiveness, with less concern for equal spacing of services within the region. Many nonemergency services exhibit more clustered arrangements in their point patterns, possibly reflecting financial rather than spatial constraints.

Four community services in Toronto are selected for a nearest neighbor analysis of their spacing. Two of the services, police and fire, provide emergency protection from sites of their facilities to various locations within the region. The other two services, elementary education and voting locations, provide nonemergency services for persons who travel to these locations. The facility sites for the four services are shown in Figure 14.3.

A nearest neighbor analysis is applied to each pattern to measure the spacing of service sites. The calculation procedure for analyzing the spacing of police facilities is summarized in Table 14.2. The nearest neighbor distances (in kilometers) for each of the 20 sites are shown in Figure 14.4. The mean nearest neighbor distance for police stations in Toronto is 3.63 kilometers. If the stations had been distributed randomly across the city, the mean spacing would have been 2.83 kilometers (Table 14.2). To hold the influence of point density constant and allow more useful comparisons of nearest neighbor results, a standardized nearest neighbor index (R) is calculated. This ratio of $\overline{NND}$ to $\overline{NND}_R$ for police stations in Toronto ($R = 1.28$) suggests that the actual spacing of points is more dispersed than random. Although dispersal of sites is expected for an emergency public service, does the R value of

TABLE 14.2

Worktable for Nearest Neighbor Analysis: Police Service Facilities in Toronto, Ontario

$H_0: \overline{NND} = \overline{NND}_R$ (point pattern is random)
$H_A: \overline{NND} \neq \overline{NND}_R$ (point pattern is not random)

Calculate mean nearest neighbor distance:

$$\overline{NND} = \frac{\sum NND}{n} = \frac{72.68}{20} = 3.63$$

where: n = number of points

Calculate random nearest neighbor distance:

$$\overline{NND}_R = \frac{1}{2\sqrt{Density}}$$

where: $Density = n/Area$

$$\overline{NND}_R = \frac{1}{2\sqrt{\frac{20}{641.45}}} = \frac{1}{2\sqrt{0.0312}} = \frac{1}{0.3532} = 2.83$$

Calculate standardized nearest neighbor distance:

$$R = \frac{\overline{NND}}{\overline{NND}_R} = \frac{3.63}{2.83} = 1.28$$

Calculate test statistic:

$$Z_n = \frac{\overline{NND} - \overline{NND}_R}{\sigma_{\overline{NND}}}$$

where: $Density = n/Area$

$$\sigma_{\overline{NND}} = \frac{0.26136}{\sqrt{20(0.0312)}} = \frac{0.26136}{0.7899} = 0.3309$$

$$Z_n = \frac{3.63 - 2.83}{0.3309} = \frac{0.80}{0.3309} = 2.42 \quad (p = 0.0078)$$

(continued)

1.28 differ significantly from 1.0, the result for a random pattern of points?

Equations 14.6 and 14.7 are used to test for a significant difference between an observed nearest neighbor index and the corresponding index value for a random spacing of points. The null hypothesis is that the observed and random nearest neighbor distances are equal. In this problem, a one-tailed (directional) test is logical, since police facilities are hypothesized to have a dispersed locational pattern. Therefore, the alternate hypothesis is that the observed nearest neighbor index is larger than the corresponding random value. The resulting test statistic for police locations in Toronto ($Z_n = 2.42$; $p = .0078$) indicates that you can reject the null hypothesis with more than 99% certainty and conclude that the spacing of police stations is more dispersed than random.

Inferential interpretations in this problem should be made only with extreme care. It could be argued that the existing pattern of locations is a "natural" sample from the many possible locations where police stations could have been sited. If this argument is not accepted, the inferential assumption of an independently selected random sample of points is not met. If this is the case, you might want to focus primarily on a descriptive comparison of nearest neighbor indices (*R*-values) for the police stations and the other community services.

TABLE 14.3

Nearest Neighbor Values for Selected Public Facilities in Toronto, Ontario

Public Facility	$\overline{NND}$	Density	$\overline{NND}_R$	R	Z_n	p
Fire	2.06	.1278	1.40	1.47	8.21	.0000
Police	3.63	.0312	2.83	1.28	2.42	.0153
Elementary schools	0.59	.7358	0.58	1.02	0.95	.3439
Voting locations	0.27	2.4351	0.32	0.84	−12.23	.0000

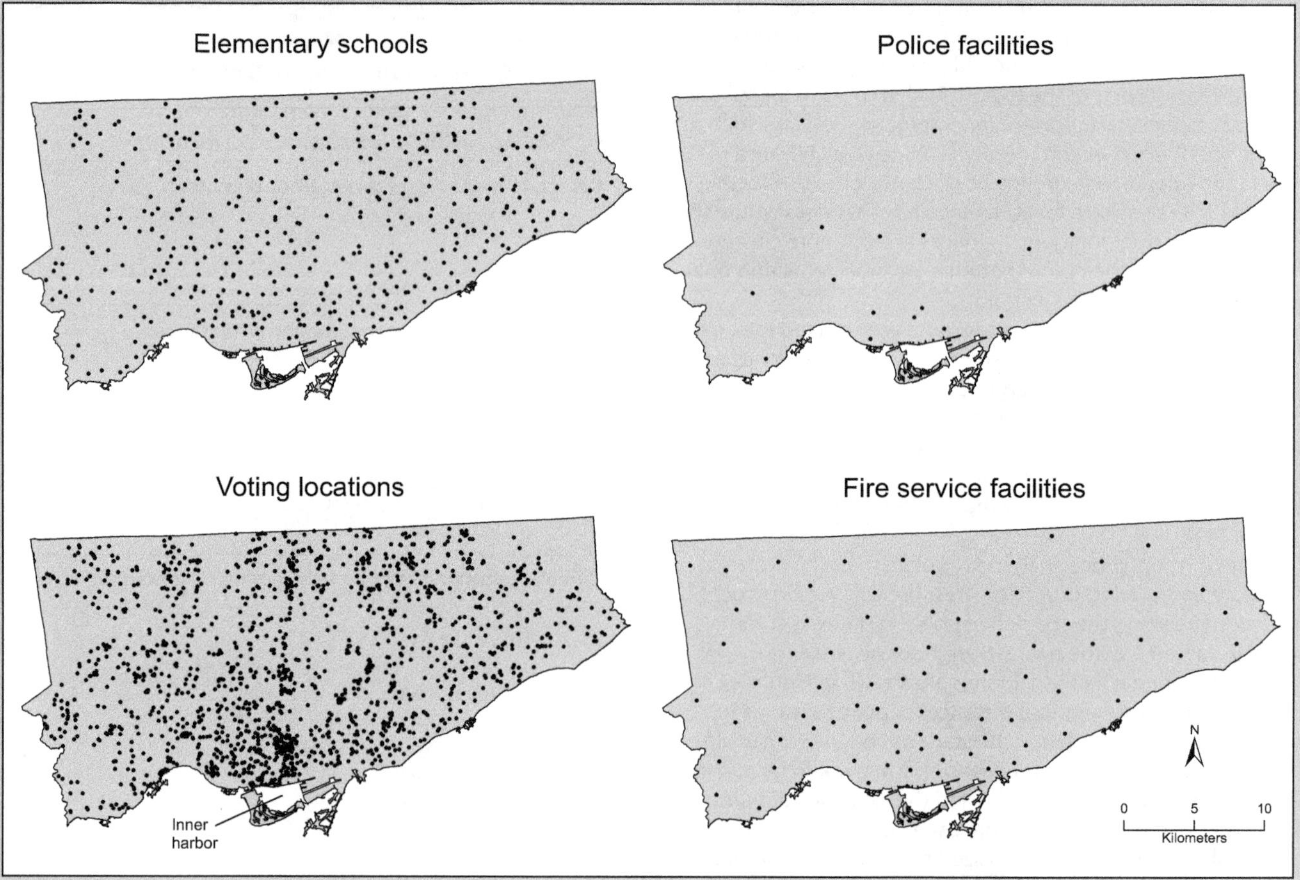

FIGURE 14.3
Location of Selected Public Facilities in Toronto, Ontario
Source: City of Toronto Planning Division

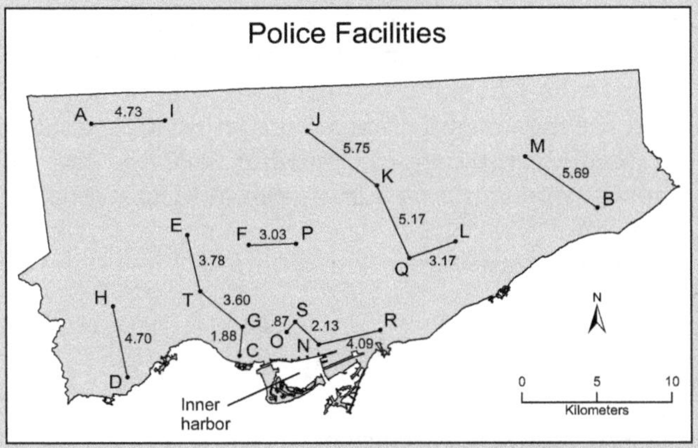

FIGURE 14.4
Nearest Neighbor Distances for Police Service Facilities in Toronto, Ontario

The corresponding nearest neighbor values and test statistic results for each of the four service patterns are shown in Table 14.3. The public services are ranked in decreasing size of the R index (from more dispersed to more clustered). The pattern of fire stations in Toronto is the most dispersed ($R = 1.47$) followed by police stations ($R = 1.28$), thus supporting the assumption of an even distribution of emergency services to meet an equity requirement. The third most dispersed service is elementary education facilities ($R = 1.02$). However, statistical evidence ($p = 0.34$) indicates that this dispersion is not significantly different from a random pattern.

The most clustered spacing pattern among the four services is voting locations ($R = 0.84$), the only service where the random nearest neighbor distance exceeds the observed average nearest neighbor distance. Unlike emergency services which are located to provide relatively equal response times for all parts of a region, voters must utilize a specific voting location based on their place of residence. This suggests that areas with higher population densities would require additional voting locations producing a more clustered spatial pattern. This clustered pattern is reflected in a lower R index and is evident in the areas just north of Toronto's inner harbor.

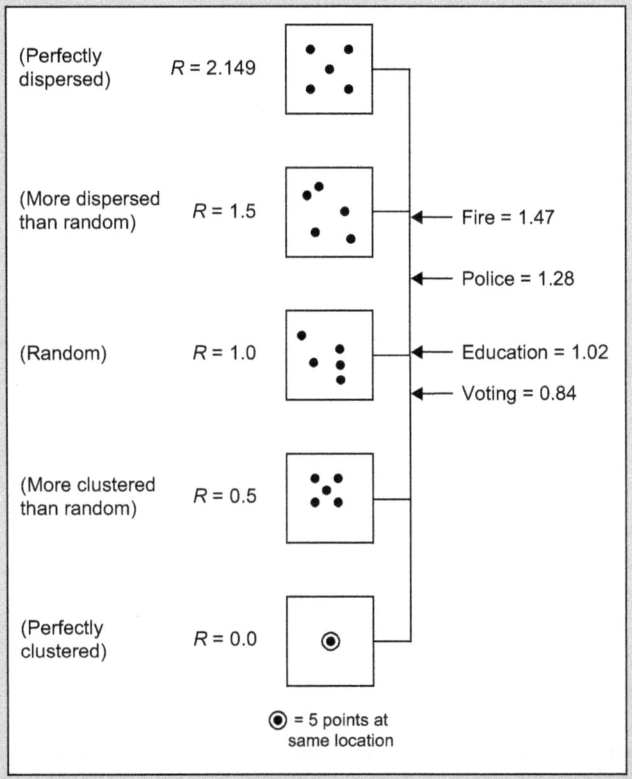

FIGURE 14.5
Continuum of R Values in Nearest Neighbor Analysis: Toronto, Ontario

Spacing of facilities for the four selected services in Toronto confirms the expected arrangements of emergency and non-emergency services (Figure 14.5). As hypothesized, the pattern of emergency fire services exhibits the most dispersed spacing, followed closely by the pattern of police facilities. Also as expected, voting locations in Toronto are more clustered than random. Although education services visually appear more dispersed than random, statistical evidence suggests that the spacing does not differ significantly from a random pattern.

These results are clearly influenced by the way in which the problem is structured. A critical issue when using the nearest neighbor procedure is the delimitation of the study area boundary. In many research problems (like the analysis of public services in Toronto), a political boundary defines the study area. In other problems, where no formal or logical boundary exists for enclosing the point locations, you must designate an operational boundary or territorial limit for the region under study. The position of the boundary does not directly affect the distance between nearest neighbors or the mean nearest neighbor index. However, boundary position does affect both the area of the study region and the point density, factors that determine the random nearest neighbor distance. Therefore, specification of the study area boundary influences the outcome of a point pattern analysis. Ideally, a functional study area boundary should be defined that is consistent with the type of points being analyzed. Some researchers suggest that the boundary should be defined just beyond the outermost points within the study area.

Another issue related to study area boundary is the specification of nearest neighbors at the edge of a region. In some actual situations, the nearest neighbor for a point located near the boundary may actually lie outside the study area. Perhaps the nearest fire station or school lies across the border at a shorter distance than the nearest neighbor lying inside the study area. Therefore, results are influenced by how nearest neighbor distances are handled near the study area boundary. Administrative boundaries do not necessarily offer the best approach to delimiting the study area in point pattern problems.

To adapt to various situations, geographers have expanded the NNA concept to include not just the single nearest neighbor, but *n* nearest neighbors (where *n* represents all nearby neighbors, and the definition of "nearby" varies depending on the specific situation). If the complete distribution of all point-pair distances in a spatial pattern is considered, then we have a classical *k* nearest neighbor (often called a *Ripley's K* function). In evaluating more than one nearest neighbor, you simply determine the mean distance from each point to the *n* closest points in the distribution. For example, we can determine the average distance of the closest four police stations to any other police station in Toronto.

The *k* function is a true second order statistic that evaluates the distribution of points over the full set of distances in the point set. That is, for each point the *k* function determines how many points are within distance *d*. Because distance is a continuous variable, the *k* function typically breaks the data up into distance intervals. For example, we might determine how many police stations are within 1 km, 2 km, 3 km, . . . and 10 km of each other and compare that to an expected number of police stations from a theoretical distribution of a completely random set of points. The purpose is to determine if a set of points (services or facilities) is clustered at multiple distances.

14.2 Quadrat Analysis

An alternate methodology for studying the spatial arrangement of point locations is **quadrat analysis**. Rather than focusing on the spacing of points within a study area, quadrat analysis examines the frequency of points occurring in various parts of the area. A set of quadrats (usually square cells) is superimposed on the study area, and the number of points in each cell is determined. By analyzing the distribution of cell frequencies, the point pattern arrangement within the study area can be described.

Whereas nearest neighbor analysis uses the average spacing of the closest points, quadrat analysis considers the variability in the number of points per cell (Figure 14.6). If each of the quadrats contains the same number of points (Case 1), the pattern would show no variability in frequencies from cell to cell and would be perfectly dispersed. By contrast, if a wide disparity exists in the number of points per cell for the set of quadrats examined (Case 2), the variability of the cell frequencies would be large, and the pattern would display a clustered arrangement. In a third alternative (Case 3), the variability of cell frequencies is moderate, and the pattern of points would reflect a random or near random spatial arrangement.

However, the absolute variability of the cell frequencies cannot be used as an effective descriptive measure of the point pattern because it is influenced by the density of points—that is, the mean number of points per cell. This relationship is directly analogous to the influence of the mean on the standard deviation of a variable. Recall from Chapter 3 that the coefficient of variation is used for meaningful comparisons of relative variability between distributions. In quadrat analysis, an index known as the **variance-mean ratio (VMR)** standardizes the degree of variability in cell frequencies relative to the mean cell frequency:

$$VMR = \frac{VAR}{MEAN} \tag{14.8}$$

where: VMR = variance-mean ratio
VAR = variance of the cell frequencies
$MEAN$ = mean cell frequency

The mean and variance are the basic descriptive statistics that summarize the central tendency and variability of a

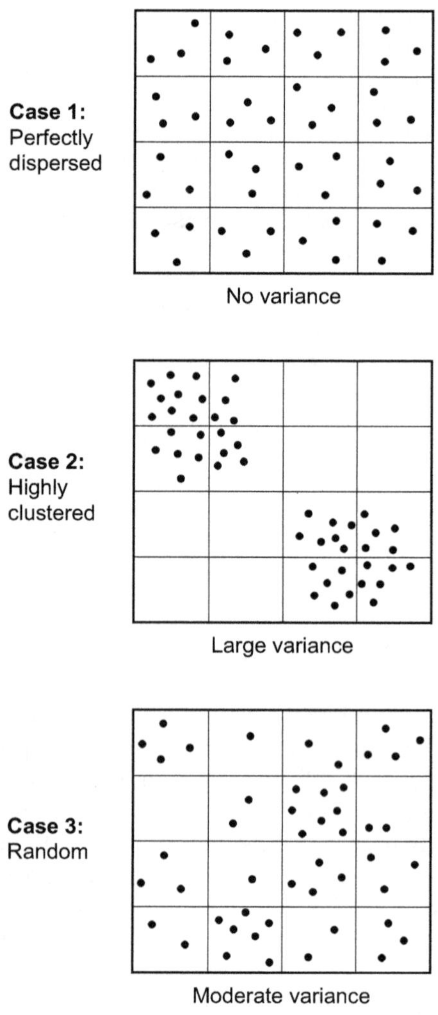

FIGURE 14.6
Quadrat Analysis: The Relationship of Cell Frequency Variability and Point Pattern

variable. Since the data from quadrat analysis are usually summarized as frequency counts (number of points in each quadrat), formulas for the weighted mean and weighted variance (weighted standard deviation squared) are usually applied (see Tables 3.4 and 3.8).

Interpretation of the variance-mean ratio offers insight into the spatial pattern of points within the study area. In a dispersed distribution of points, for example, the cell frequencies will be similar, and variance of cell frequency counts will be very low. In an extreme case (such as Case 1 in Figure 14.6), if each quadrat contains exactly the same number of points, the variance will be zero, making the variance-mean ratio zero. Conversely, if a point pattern is highly clustered with most cells containing no points and a few cells having many points, the variance of cell frequencies will be large relative to the mean cell frequency. This type of spatial pattern will produce a large value for the variance-mean ratio.

If the set of points is randomly arranged across the cells of the study area, an intermediate value of variance

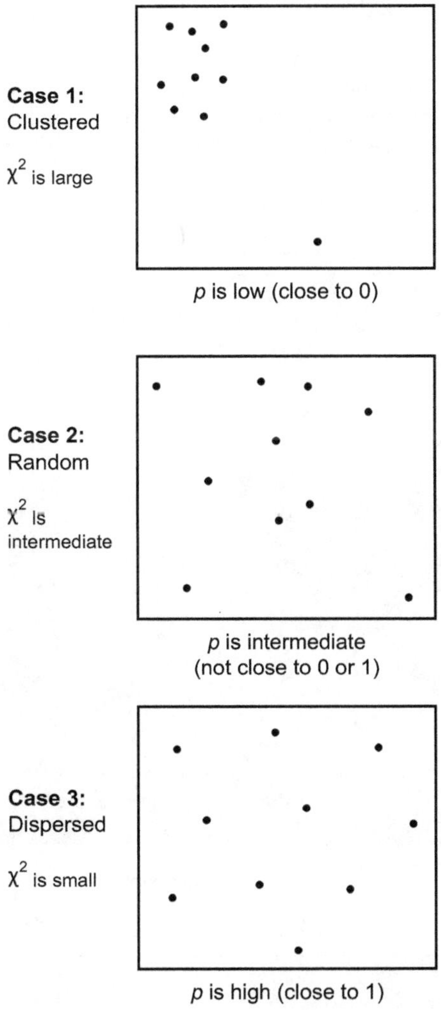

FIGURE 14.8
Quadrat Analysis: The Relationship of Chi-square Value and p

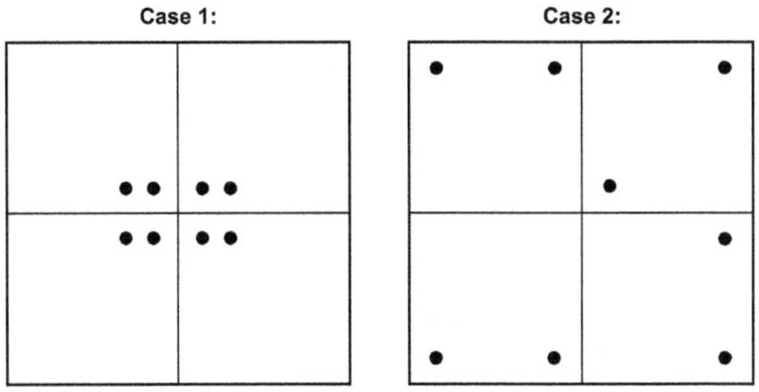

FIGURE 14.7
Two Point Patterns with the Same Variance-mean Ratio, but Different Visual Impressions

will occur. For a perfectly random point pattern, the variance of the cell frequency is equal to the mean cell frequency. To understand this result, recall from Section 5.4 that the Poisson distribution is used to describe the frequency of values for a randomly generated spatial or temporal pattern. In the Poisson distribution, the mean frequency equals the variance of the frequencies. Therefore, when using the variance-mean ratio to describe spatial point patterns, a result close to one (variance equals mean) suggests that the distribution has a random arrangement.

Since quadrat analysis measures dispersion related to the frequencies of points in the cell boundaries and not the explicit location of points in relationship to one another, you must be careful when interpreting the VMR. For example, both Case 1 and Case 2 in Figure 14.7 have the same VMR. There are eight points and four quadrats. However, the actual distribution of the points in relationship to one another in the large study area (ignoring the four internal quadrats) would suggest that Case 1 has a

Quadrat Analysis

Primary Objective: Determine whether a random (Poisson) process has generated a point pattern

Requirements and Assumptions:

1. Random sample of points from a population
2. Sample points are independently selected

Hypotheses:

$H_0 : VMR = 1$ (point pattern is random)

$H_A : VMR \neq 1$ (point pattern is not random)

$H_A : VMR < 1$ (point pattern is more dispersed than random)

$H_A : VMR > 1$ (point pattern is more clustered than random)

Test Statistic:

$$\chi^2 = VMR\,(m-1)$$

clustered pattern of points while the distribution of points in Case 2 is dispersed.

In addition to its use as a descriptive index, the variance-mean ratio can also be applied inferentially to test a distribution for randomness. The test statistic is chi-square, defined as a function of both the *VMR* and number of cells (m):

$$\chi^2 = VMR(m - 1) \quad \textbf{(14.9)}$$

The null hypothesis for this test is expressed as no difference between the observed distribution of points and a distribution of points resulting from a random process (that is, $VMR = 1$). Either a directional (one-tailed) or non-directional (two-tailed) approach can be used, depending on the test objective of the alternate hypothesis.

Rejection of the null hypothesis can occur if a point pattern is more clustered than random ($VMR > 1$) or more dispersed than random ($VMR < 1$). The difference test determines whether an observed *VMR* value differs significantly from a theoretical random value ($VMR = 1.0$). The interpretation of the chi-square test statistic and corresponding p-value reflects these two possibilities. A large *VMR* generally produces a larger chi-square value (with a small p-value), suggesting greater variability of the cell frequencies and a clustered arrangement of points. Conversely, a small *VMR* produces a smaller chi-square value, indicating lower variability of cell frequencies and a more dispersed distribution of points. In such instances, the corresponding p-value is larger. Intermediate chi-square values result from variance-mean ratios closer to one, suggesting a lower probability of rejecting the null hypothesis and therefore more likelihood that the distribution of points is actually random. Thus, a continuum of p-values from 0 to 1 indicates transition from clustered to random to dispersed (Figure 14.8). Either very high or very low p-values suggest rejection of the null hypothesis, while intermediate values do not support rejection.

Example: Quadrat Analysis—Distribution of Wildfires in New Mexico

The example of wildfire distribution (Section 5.4) is re-examined here using quadrat analysis. The objective now is to describe the pattern of points within the study area by analyzing the distribution of cell frequencies. The resulting frequency pattern indicates whether the wildfire starting points have a dispersed, random, or clustered arrangement. The chi-square difference test is then applied to determine whether the observed pattern differs significantly from a theoretical random pattern.

In this analysis, wildfire locations are assumed to be independent random events. This assumption is justified if the sample is considered a "natural" sample over one time period. If a natural sampling process is assumed, this particular pattern of wildfire starting points represents a random sample from an infinite number of sample patterns that could have occurred during a given time period.

As discussed in Section 5.4, 122 cells covering New Mexico have at least half of their area in the state. This set of 122 quadrats excludes only 116 of the over 16,000 recorded wildfire starting points for the state over the entire 30-year time period. One of our objectives is to determine if the distributions of wildfires in New Mexico are different over three decades: 1980s, 1990s, and 2000s.

In addition to an analysis of wildfire distributions for all three decades, we also focus attention on a random set of 2,500 points. Using GIS processing, a random set of 2,500 points is generated, and yields a mean frequency of 20.9 wildfires per cell—2,500 wildfires in 120 cells (Figure 14.9). The variability in cell frequencies around these means largely determines the nature of the point pattern. For this set of quadrats, the variance of the random points is 18.9 per cell. The two summary statistics produce a variance-mean ratio of .904, suggesting a random distribution (Table 14.4).

A difference test is applied to determine whether the *VMR* of .904 of our computer-generated random set of points does not differ significantly from that of a theoretical random pattern ($VMR = 1$). The null hypothesis assumes no difference between the observed and expected *VMR*s. Selection of a directional (one-tailed) or non-directional (two-tailed) test depends on whether *a priori* evidence exists to suggest a point pattern that is more clustered than random or more

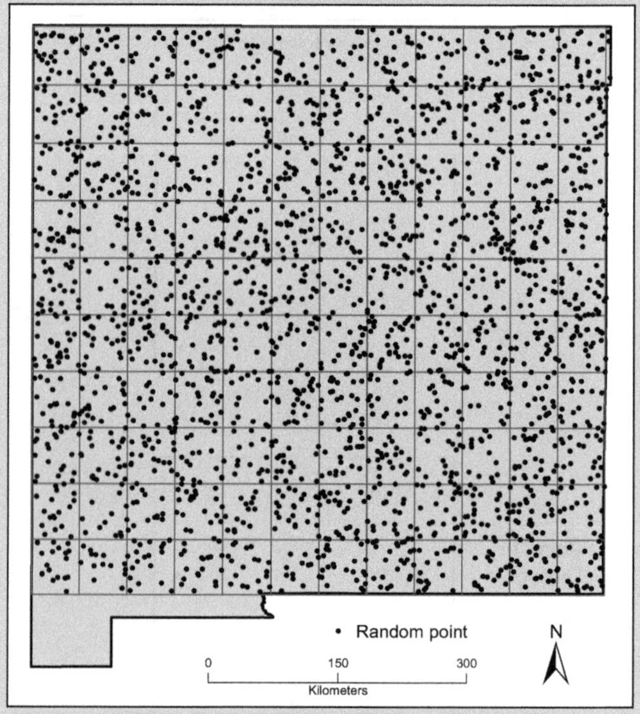

FIGURE 14.9
GIS-generated Random Set of 2,500 Points Superimposed on 120 Cells in New Mexico

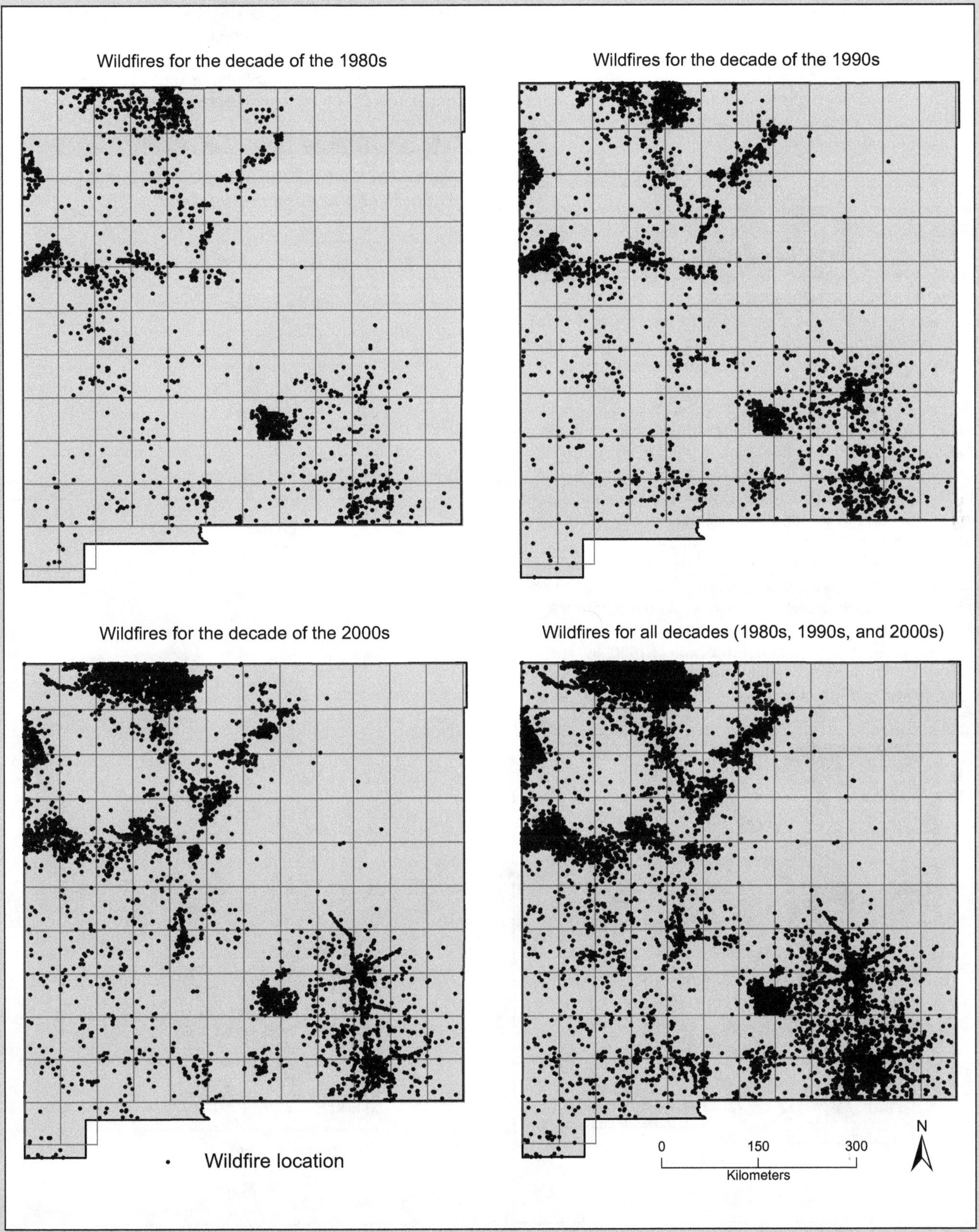

FIGURE 14.10
Point Patterns for Wildfires in New Mexico: 1980s, 1990s, and 2000s
Source: U.S. Geological Survey, Federal Wildland Fire Occurrence Website

(continued)

dispersed than random. Because we generated a fictitious set of random points, there is no *a priori* evidence suggesting the pattern should be clustered (or dispersed), therefore a nondirectional, two-tailed difference test is appropriate. Conversely, an alternate hypothesis that assumes, *a priori*, a clustered or dispersed point pattern would require a directional, one-tailed difference test.

As could be expected, the resulting chi-square test statistic of 107 produces a rather large *p*-value of 0.77 (Table 14.4). Thus, we have very strong confidence that the null hypothesis should be accepted and that our computer generated pattern is, in fact, random. Figure 14.10 shows the point pattern distribution for wildfires in New Mexico for the decades of 1980–1989, 1990–1999, and 2000–2009.

A cursory view of the maps seems to show that the distribution of wildfire starting points is very highly clustered over all three decades. This makes intuitive sense, as many areas in New Mexico are desert, and only a few areas are heavily forested. In cases like this, a *VMR* will almost certainly describe a highly clustered distribution. However, even though the "natural pattern" of forested areas would logically result in a clustered distribution of wildfires, there is still merit in applying the *VMR* technique to compare multiple decades (and detect any changes in the overall clustering pattern of wildfires). Table 14.5 shows the *VMR* results for each decade.

You can see that each decade exhibits a highly clustered distribution, but as we said, that is not surprising given the distribution of forests, deserts, and human settlements in the state. However, the decades of 1990–99 and 2000–09 are dramatically more clustered than the decade of 1980–89. The increasing clustering of wildfire starting points could possibly be explained by the fact that expansion of "the human footprint" across New Mexico has relegated forested areas to an ever-smaller portion of the state. Quite simply, there are fewer places where a wildfire could start, and these places are increasingly clustered on public lands such as national parks, national forests, and Bureau of Land Management lands.

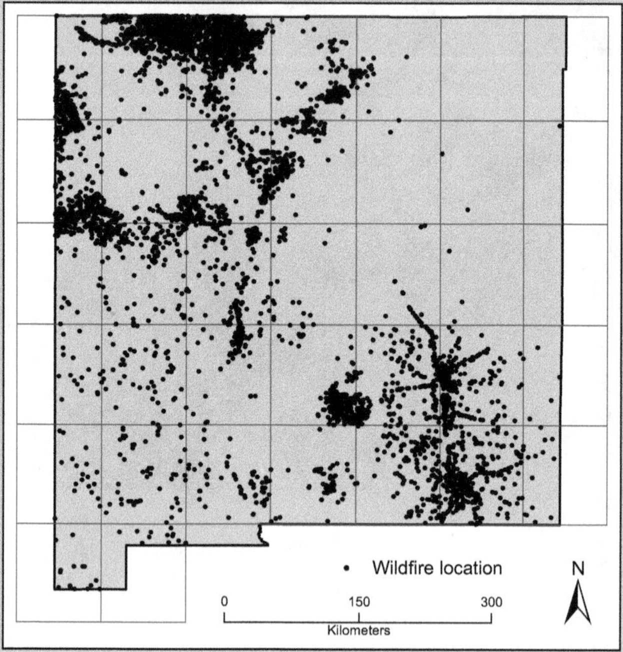

FIGURE 14.11
New Mexico Wildfires from 2000 to 2009 Overlain with a Larger Quadrat Size

TABLE 14.4

Worktable for Quadrat Analysis: New Mexico Wildfires (2,500 Random Points)

$H_0: VMR = 1$ (point pattern is random)
$H_A: VMR \neq 1$ (point pattern is not random)

Calculate mean cell frequency:

Mean cell frequency = *n*/*m*

where: *n* = number of points
m = number of cells

$$\text{MEAN} = \frac{2{,}509}{120} = 20.9$$

Calculate variance of cell frequencies:

Number of points per cell (*x*)	Number of cells (*f*)	Total points	fx^2
11	1	11	121
12	2	24	288
13	2	26	338
14	4	56	784
15	3	45	675
16	4	64	1,024
17	9	153	2,601
18	8	144	2,592
19	15	285	5,415
20	10	200	4,000
21	10	210	4,410
22	11	242	5,324
23	9	207	4,761
24	13	312	7,488
25	3	75	1,875
26	2	52	1,352
27	3	81	2,187
28	8	224	6,272
32	1	32	1,024
33	2	66	2,178
SUM	120	2,509	54,709

$$VAR = \frac{\sum f_i X_i^2 - \left[\frac{(\sum f_i X_i)^2}{m}\right]}{m-1}$$

where: f_i = frequency of cells with *i* wildfires
X_i = number of wildfires per cell

$$VAR = \frac{54{,}709 - \left(\frac{2{,}509^2}{120}\right)}{120-1} = \frac{54{,}709 - 52{,}459}{119} = 18.9$$

Calculate variance-mean ratio:

$$VMR = \frac{VAR}{MEAN} = \frac{18.9}{20.9} = 0.904$$

Calculate test statistic:

$$\chi^2 = VMR(m-1) = 0.9(119) = 107 \quad (p = 0.77)$$

TABLE 14.5

Variance-Mean Ratios (VMRs) for Wildfire Distribution in New Mexico

Distribution	Variance-Mean Ratio
Computer generated random sample	.904
1980–1989	20,381
1990–1999	32,645
2000–2009	30,818

Does the size of quadrats used to cover the study area influence results from quadrat analysis? To examine this question, the 2000–09 wildfire distribution is evaluated using a cell design with quadrat areas four times as large. This cell structure produces 37 larger quadrats covering 7,177 wildfires for the decade (Figure 14.11). To be consistent, boundary cells whose area is mostly (more than half) outside New Mexico are eliminated.

The larger cell network produces fewer quadrats with higher frequencies per quadrat and a larger mean cell frequency of 211 wildfires per quadrat. The variance of the cell frequencies about this mean is also higher at 81,203; since the variance remains much larger than the mean, the pattern once again displays a highly clustered arrangement. Standardizing this variance to the mean produces a variance-mean ratio of 15,070, much lower than the value found with the smaller quadrat size ($VMR = 30{,}818$), although still highly significant. You might want to speculate about why the *VMR* for the larger quadrats is considerably smaller than the *VMR* for the smaller quadrats.

Although quadrat analysis offers a useful approach to studying the spatial arrangement of points, several procedural issues must be considered. Different cell sizes produce varying levels of mean point frequency and variance per cell. Moreover, studies have shown that if the point pattern is held constant but the quadrat size is decreased (i.e., the number of quadrats is increased), the variance of the point frequencies usually declines faster than the mean. However, this is not always the case as we can see from our example. Thus, different cell sizes in a single problem can generally be expected to produce different values of the variance-mean ratio.

Although more research needs to be conducted on optimal quadrat size, two guidelines can be offered. When visual examination of a point pattern reveals distinct clusters, the size of quadrats should match that of the clusters. This cell size allows the results from quadrat analysis to reflect more accurately the visual impression of the pattern. Other researchers have suggested a general "rule of thumb" that cell size should be equal to twice the average area per point. In other words, the mean frequency of points should be close to 2.0 per quadrat. This guideline approximates the cell structure used with the 122 cells covering the New Mexico forest fire example. Taylor (1977) presents a more detailed discussion of cell size in quadrat problems.

Another issue in the use of quadrat analysis also requires some attention. Like nearest neighbor analysis, delimitation of the study area can influence the results of a study. In the use of quadrat analysis, a way to handle cells around the study area boundary needs to be determined. Cells on the boundary will inevitably contain area both inside and outside the study region. In the New Mexico wildfire problem, cells with more than half of their area outside the study region were eliminated. Alternative decision rules can be used, and these may change the results obtained from the analysis. No matter what decision rule is applied in a particular problem, the geographer needs to be consistent in handling the boundary issue.

Key Terms

nearest neighbor analysis, *212*
point pattern analysis, *212*
quadrat analysis, *218*
standardized nearest neighbor index, *214*
variance-mean ratio (VMR), *218*

Major Goals and Objectives

If you have mastered the material in this chapter, you should now be able to:

1. Identify the types of geographic problems or situations for which nearest neighbor analysis is appropriate.
2. Identify the types of geographic problems or situations for which quadrat analysis is appropriate.
3. Explain the logic and procedures used in nearest neighbor analysis and explain how they differ from those used in quadrat analysis.

References and Additional Reading

Boots, B. N. and A. Getis. *Point Pattern Analysis.* Newbury Park, CA: Sage, 1988.

Getis, A. "Temporal Analysis of Land Use Patterns with Nearest Neighbor and Quadrat Methods." *Annals, Association of American Geographers.* Vol. 54 (1964): 391–399.

Taylor, P. J. *Quantitative Methods in Geography.* Houghton Mifflin: Boston, 1977.

Thomas, R. *An Introduction to Quadrat Analysis.* CATMOG No. 12. Norwich, England: Geo Books, 1977.

For those interested in looking further at the distribution of public services in Toronto, Canada, see the City of Toronto Planning Division website: www.toronto.ca/planning. For those interested in the spatial distribution of wildfire occurrences, see the Department of the Interior, U.S. Geological Survey, Federal Fire Occurrence Website: http://wildfire.cr.usgs.gov/firehistory/data.html

15

Area Pattern Analysis

This chapter presents several methods for analyzing area patterns. Some of these methods are global in nature, encompassing an entire study area, while one method (a variation of Moran's I) analyzes a local region. Two statistical procedures, the join count statistic and Moran's I Index, are discussed in more detail. The join count statistic basically examines the structure of all subarea connections (joins) within the larger study area. Each particular subarea is either "joined" or "not joined" to a number of adjacent subareas, and the overall pattern of this subarea joining structure is analyzed. The Moran's Index analyzes ordinal or interval/ratio area patterns to determine if they are significantly autocorrelated.

For example, an economic development planner might be interested in evaluating changing spatial patterns of poverty in Appalachia. Through area pattern analysis of a chronological series of county-level choropleth maps showing Appalachian poverty, the spatial distribution of poverty can be evaluated over time to determine whether it is becoming more dispersed or clustered. In another example, an urban geographer could analyze a map pattern depicting the number of existing home sales by census tract to determine the degree to which sales are spatially concentrated in certain portions of the city. If a nonrandom clustering of high turnover rates is found in certain segments of the city, attention could be focused on why the sales exhibit such spatial patterns.

The goals and objectives in **area pattern analysis** are similar to those in point pattern analysis. Just as nearest neighbor and quadrat analysis examine the random or nonrandom nature of point patterns, descriptive and inferential procedures are available to analyze area patterns. A choropleth or area pattern map is considered clustered if adjacent or nearby areas tend to have highly similar values or scores. Alternatively, if the values of adjacent or nearby areas tend to be dissimilar, then the spatial pattern is considered more dispersed than random (refer back to Figure 13.1).

Area pattern analysis is appropriate for studying many practical problems in geography. It is particularly valuable when examining the way in which area patterns of a specific variable change over time. For example, suppose a medical geographer is concerned with the diffusion of influenza in a metropolitan area. If the number of cases is reported by census tract for several time periods, the influenza morbidity rate could be displayed on a series of choropleth maps and the degree of non-randomness in the patterns analyzed. If the incidence of the disease becomes more dispersed over successive time periods, researchers would be very concerned because the dispersion indicates that the influenza is spreading more widely throughout the metro area. It could also mean a significant number of cases were appearing in other areas not close to existing areas of high incidence. Conversely, if the morbidity rate patterns were becoming more clustered over successive time periods, then a closer examination of those areas with higher incidence of influenza would be appropriate. Such knowledge could lead to the implementation of effective strategies for disease control.

Suppose a political geographer wants to study the spatial pattern of voter registration rates across a number of precincts within a community. If registration rates have been mapped by precinct before each of the last several elections, a useful comparative analysis of area patterns is possible. If the pattern is becoming more clustered over successive time periods, a different registration strategy may be necessary. An analysis of area pattern trends might also provide insights into demographic or economic variables that seem to be related to the voter registration rate.

Join count analysis is used when areas on a map are represented in binary form, such as yes/no, presence/absence, or above/below a particular threshold. For example, each census tract in an influenza morbidity rate map could be classified as having either an above average or below average incidence of influenza. Moran's I is used when areas on a map are represented in ordinal or interval/ratio form, with the actual values or ranks of values. For example, the census tracts might indicate the actual influenza morbidity rate, or a ranking of census tracts from the highest to lowest morbidity rate.

15.1 Join Count Analysis

The **join count** is the basic organizational statistic for the analysis of area patterns. A "join" is operationally defined as two areas sharing a common edge or boundary. The procedures involved in calculating a join count statistic are relatively straightforward. The fundamental building block is the number of joins in the pattern and the nature of the join structure in the study area. Figure 15.1 illustrates the simplest situation: a binary classification of data and a two-category choropleth map. Many geographic variables are binary in nature, or can be downgraded effectively to binary without sacrificing important locational information. This figure shows the join structure for the subareas in the study region. For example, area A is joined to two other areas, B and D. The number of joins associated with each area is listed adjacent to the join structure map.

To comply with accepted practice, each area can be identified as either "black" or "white." It is then possible to refer to the *observed number* of "black-white," "black-black," and "white-white" joins in the pattern. In Case 1 of Figure 15.1, the area pattern appears clustered, with relatively few black-white (dissimilar) joins and a rather large number of black-black and white-white (similar) joins. In Case 2, a dispersed (checkerboard-like) pattern has more black-white joins and fewer joins of similar-category areas. With random patterning (Case 3), the result is an intermediate number of both similar and dissimilar joins. These situations are all reflected in Table 15.1

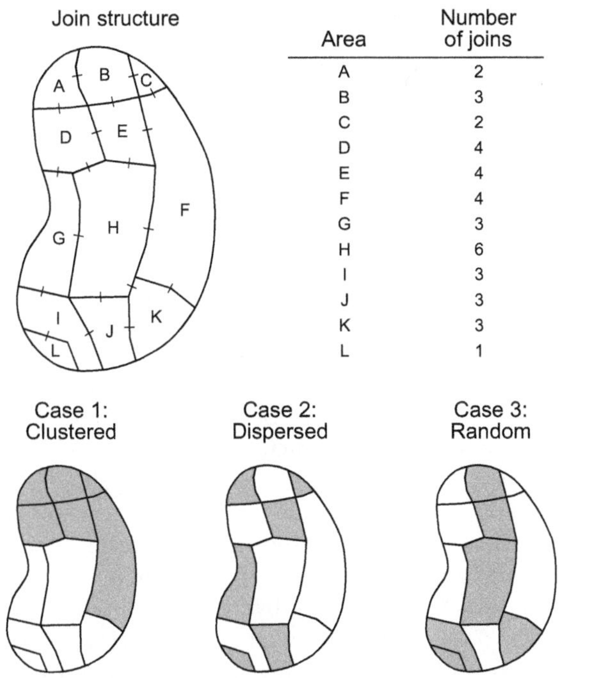

FIGURE 15.1
Join Structure and Examples of Area Patterns Appearing Clustered, Dispersed, and Random

To determine if a join count distribution has been generated by a random process, the number of black-white joins that would be expected from a theoretical random arrangement of binary observations must be calculated. That is, the expected number of black-white joins in a purely random pattern is compared with the one actually observed. If the observed join count is significantly different from the expected join count, the observed pattern can be described as nonrandom.

How do you determine the expected number of black-white joins? The locational context in which the overall study area pattern is being evaluated suggests two

TABLE 15.1

Join Counts Associated with Area Patterns

		Dissimilar areas joined	Similar areas joined		
Case	**Total joins ***	**Black-white joins**	**Black-black joins**	**White-white joins**	**Total**
1 (clustered)	19	5	7	7	14
2 (dispersed)	19	15	0	4	4
3 (random)	19	12	4	3	7

* See Figure 15.1 for map of join structure and area patterns.

possible approaches for generating expected join counts. Some area pattern analyses may be conducted under the hypothesis of **free sampling**. Free sampling should be used if the researcher can determine the probability of an area being either black or white based on some theoretical idea or by referring to a larger study area. Once the probability of an area being black or white has been determined, the expected number of black-white joins in a random pattern having those probabilities can also be determined. For example, in the influenza morbidity rate problem, suppose each census tract throughout the metropolitan area is classified as having either above average incidence of the disease (black) or below average incidence (white). If the pattern of disease is being analyzed in only one portion of the metropolitan area, logic would suggest that the probability of above- or below-average incidence of influenza should be typical or representative of that found throughout the metro area, and the free sampling hypothesis would be appropriate.

The **non-free sampling** hypothesis is used when no reference to a larger study area or general theory is appropriate. In this case, only the study region itself is considered in the analysis, and the expected number of black-white joins is estimated from a random patterning of only those subareas in the study region. For instance, a political geographer studying the spatial pattern of community voter registration rates by precinct may not have any reason to believe this pattern is typical or representative of any larger area (such as the state in which the community is located). Therefore, if it is hypothesized that the area pattern has resulted from the unique characteristics found within the study area, then a non-free sampling hypothesis should be applied.

In many geographic problems, you will find it difficult to conclude with any degree of confidence that a subarea pattern is typical of a larger area. If this is the case, use of the expected number of black-white joins from a non-free sampling hypothesis frees you from having to make any restrictive assumption. For the sake of brevity, we will only show the slightly simpler calculation procedure for area pattern analysis in a non-free sampling context.

In a non-free sampling context the observed number of black-white joins is compared with the number of such joins expected in a purely random arrangement. This will help you determine if the area pattern has been generated by a random process. The null hypothesis is that the observed number of black-white joins is equal to the expected number of black-white joins in a purely random area pattern. If reason exists to hypothesize that the observed area pattern is non-random in a certain direction, then a one-tailed alternate hypothesis is appropriate. For instance, an area pattern could be hypothesized either as more clustered than random or more dispersed than random. Without an *a priori* reason to hypothesize a clustered or dispersed pattern, a two-tailed alternate hypothesis is selected, and the conclusion will be that an area pattern is random or non-random.

The test statistic (Z_b) for non-free sampling is:

$$Z_b = \frac{O_{BW} - E_{BW}}{\sigma_{BW}} \tag{15.1}$$

where: O_{BW} = observed number of black-white joins
E_{BW} = expected number of black-white joins
σ_{BW} = standard error of the expected number of black-white joins

In non-free sampling, the expected number of black-white joins is calculated by incorporating the observed number of black areas and white areas directly into the equation:

$$E_{BW} = \frac{2JBW}{N(N-1)} \tag{15.2}$$

where: E_{BW} = expected number of black-white joins
J = total number of joins
B = number of black areas
W = number of white areas
N = total number of areas (black plus white) $= B + W$

The expected number of black-white joins in the study region is determined by asking how many black-white joins would occur from a theoretical random pattern containing the same number of black and white areas as the observed area pattern. That is, the question becomes how many black-white joins would be generated from a random arrangement of the observed number of black and white areas in the area pattern.

Certain generalizations can be made concerning the expected number of black-white joins. Given a particular total number of areas ($N = B + W$), the value of the denominator in Equation 15.2 is constant. If most of the areas in the observed pattern are either black or white, then their product (BW) in the numerator will be a smaller number, producing a smaller expected number of black-white joins. Conversely, if about the same number of black and white areas is found in the observed pattern, then their product will be a larger number, producing a larger expected number of black-white joins. These results seem logical, for a pattern with a large number of black areas and very few white areas (or vice versa) would not be expected to have a large number of black-white joins. The expected number of black-white joins is maximized when an equal number of black and white areas occur in the overall pattern.

Not all randomly generated patterns will contain exactly the same number of black-white joins, so a measure of the amount of variability expected as a result of sampling must be included. The standard error of expected black-white joins is:

$$\sigma_{BW} = \sqrt{E_{BW} + \frac{\sum L(L-1)BW}{N(N-1)} + \frac{4\left[J(J-1) - \sum L(L-1)\right]B(B-1)W(W-1)}{N(N-1)(N-2)(N-3)} - E_{BW}^2} \tag{15.3}$$

where: σ_{BW} = standard error of the expected number of black-white joins

J = total number of joins

$\sum L$ = total number of links = $2J$

B = number of black areas

W = number of white areas

N = total number of areas (black plus white) = $B + W$

Note: $\sum L = 2J$. If areas A and B are joined, then A is "linked" to B and B is "linked" to A, making the sum of all links twice the number of joins.

The area pattern may now be analyzed with the non-free sampling hypothesis (Equation 15.1 with supplementary Equations 15.2 and 15.3). If the observed number of black-white joins is larger than the expected number of black-white joins, then Z_b will be a positive value, indicating a pattern more dispersed than random. A relatively large number of black-white (dissimilar) joins will occur in a dispersed pattern (Figure 15.1). Conversely, if the observed number of black-white joins is smaller than the expected number of such joins, then Z_b will be negative, and the pattern will appear more clustered than random. The greater the magnitude of Z_b (either positive or negative), the greater the likelihood that the area pattern being analyzed is not random. Just like other Z-score difference tests, this Z_b test statistic for area pattern analysis indicates the number of standard deviations separating the expected random black-white join count from the observed value.

Join Count Analysis (Binary Categories)

Primary Objective: Determine whether a random process has generated a binary (two-category) area pattern

Requirements and Assumptions:

1. Each area is assigned to one of two categories
2. Each pair of areas must be defined as either adjacent ("joined") or nonadjacent (not "joined") in a consistent manner

Hypotheses:

$H_0: O_{BW} = E_{BW}$ (area pattern is random)

$H_A: O_{BW} \neq E_{BW}$ (area pattern is not random)

$H_A: O_{BW} > E_{BW}$ (area pattern is more dispersed than random)

$H_A: O_{BW} < E_{BW}$ (area pattern is more clustered than random)

Test Statistic:

$$Z_b = \frac{O_{BW} - E_{BW}}{\sigma_{BW}}$$

Example: Join Count Analysis—Obesity Distribution Patterns across the United States: 1995, 2002, 2009, 2020

Returning to the discussion of obesity, we have shown previously that the obesity rate in the United States has increased each year. However, while the nation suffers a high obesity rate, the geographic distribution of highly obese states may not be evenly distributed. That is, only certain portions of the country may exhibit high levels of obesity. A geographer might be interested in determining if the areas of high obesity in the United States are clustered, random, or dispersed in order to tailor a specific approach for combating the epidemic. You also might want to know if the overall spatial pattern of obesity is becoming more dispersed, random, or clustered over time. Figure 15.2 shows binary maps of obesity in the conterminous United States (excluding Alaska and Hawaii, since they are obviously not "joined" to any other state) for four different years: 1995, 2002, 2009, and 2020. Black areas represent the states that are above the national mean obesity level and white areas represent states below this level. As shown in Table 15.2 the number of states with obesity levels above the national mean is relatively similar for each year. Across the conterminous United States there are 212 links and 106 joins (recall from Equation 15.3 that $\sum L = 2J$); Table 15.3 shows the state linkage pattern needed for the area pattern analysis. This general state linkage pattern serves all four dates of analysis (1995, 2002, 2009, and 2020) although the detailed calculation procedure which follows is shown only for the 2020 pattern.

TABLE 15.2

Number of States with Obesity Levels Above the National Average

Year	States with obesity above the national average	Number of black-white joins
1995	26	21
2002	27	32
2009	25	25
2020	25	22

(continued)

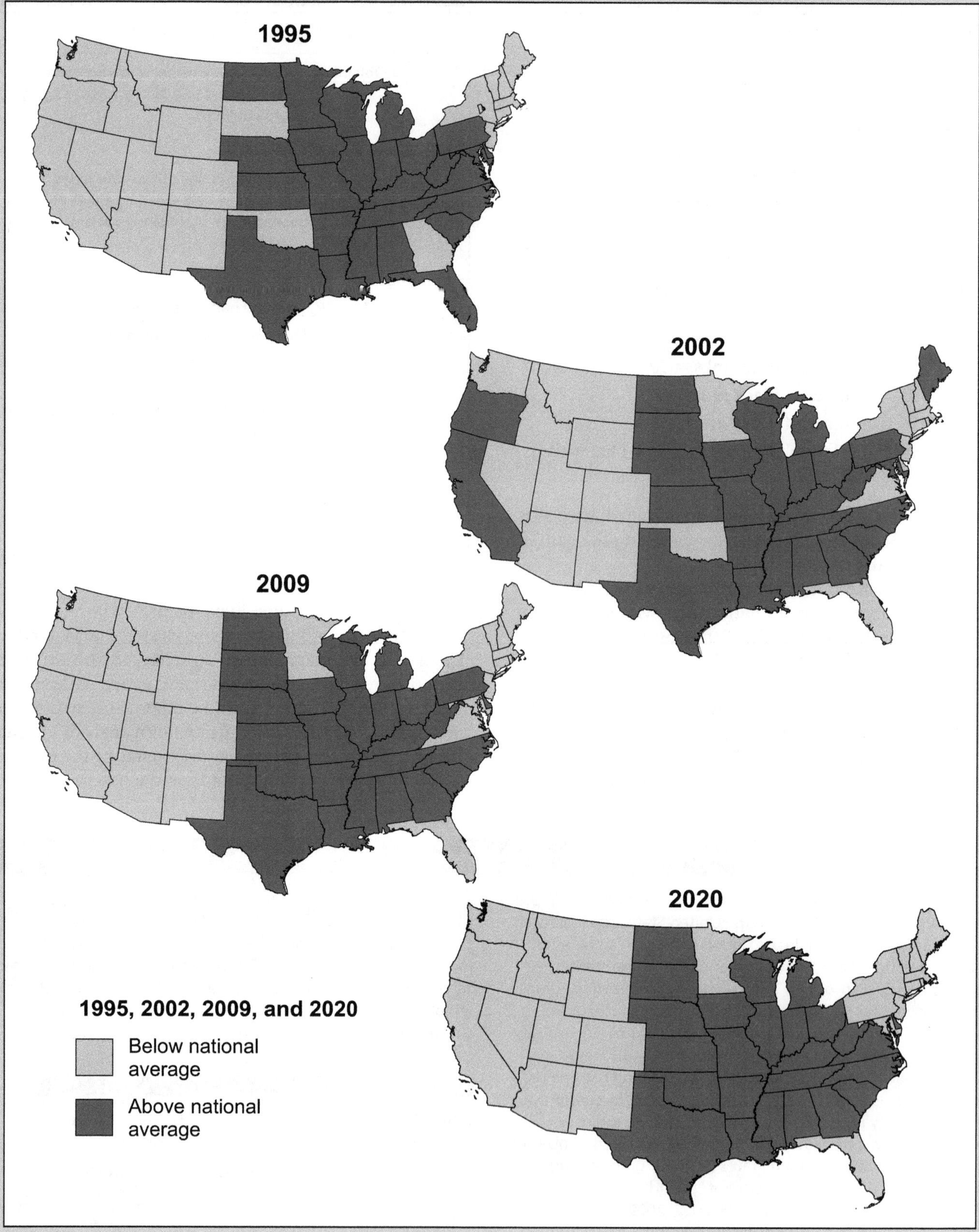

FIGURE 15.2
State Obesity Levels in the United States: 1995, 2002, 2009, and 2020
Source: Centers for Disease Control and Prevention (CDC)

In this example, a "black-white" or dissimilar join occurs when a state with an obesity rate above the national mean shares a common boundary with a state having an obesity rate below the national mean. Conversely, a similar join occurs when either two states with obesity rates above (or below) the national mean share a common boundary. On the 2020 map, 22 of the 106 joins are dissimilar, while the other 85 joins are similar. Our research task is to determine if the observed number of dissimilar joins is significantly different from the number of dissimilar joins expected in a random obesity level pattern.

TABLE 15.3

State Linkage Pattern

State Name	Number of links (L)	$L-1$	$L(L-1)$
Alabama	4	3	12
Arizona	5	4	20
Arkansas	6	5	30
California	3	2	6
Colorado	7	6	42
Connecticut	3	2	6
Delaware	3	2	6
Florida	2	1	2
Georgia	5	4	20
Idaho	6	5	30
Illinois	5	4	20
Indiana	4	3	12
Iowa	6	5	30
Kansas	4	3	12
Kentucky	7	6	42
Louisiana	3	2	6
Maine	1	0	0
Maryland	4	3	12
Massachusetts	5	4	20
Michigan	2	1	2
Minnesota	4	3	12
Mississippi	4	3	12
Missouri	8	7	56
Montana	4	3	12
Nebraska	6	5	30
Nevada	5	4	20
New Hampshire	3	2	6
New Jersey	3	2	6
New Mexico	5	4	20
New York	5	4	20
North Carolina	4	3	12
North Dakota	3	2	6
Ohio	5	4	20
Oklahoma	6	5	30
Oregon	4	3	12
Pennsylvania	6	5	30
Rhode Island	2	1	2
South Carolina	2	1	2
South Dakota	6	5	30
Tennessee	8	7	56
Texas	4	3	12
Utah	6	5	30
Vermont	3	2	6
Virginia	5	4	20
Washington	2	1	2
West Virginia	5	4	20
Wisconsin	4	3	12
Wyoming	6	5	30
TOTAL	**212**		**856**

The worktable for the 2020 join count analysis is presented in Table 15.4. The expected number of dissimilar joins in a purely random pattern is calculated by incorporating the observed number of states with obesity levels above and below the national mean directly into the equation. The result indicates that nearly 54 (54.03) dissimilar joins are expected in a random pattern.

To determine whether this contrast between the number of observed dissimilar joins (22) and the number of expected dissimilar joins (54.03) is statistically significant, the standard error of expected dissimilar joins (σ_{BW}) must be incorporated into the test statistic. The standard error of expected dissimilar joins is 4.96 (Table 15.4). The resultant test statistic value (Z_b) is –6.40, with an associated p-value of .000. These statistics indicate a highly significant tendency toward clustering of states having obesity rates above the national mean.

Using a similar statistical test procedure, analyses of the 1995, 2002, 2009, and 2020 obesity patterns (Figure 15.2) have also been generated. The results of these analyses are summarized in Table 15.5, and an interesting trend emerges across the United States. In all four years, the distribution of states with obesity levels above the national mean is very highly clustered. However, 1995 appeared to have the most clustering over the 25-year period.

The knowledgeable geographer should be able to offer some tentative hypotheses about why the obesity rate in the United States was initially most clustered in 1995, became a bit less clustered in 2002, became more clustered again in 2009, and finally returned to a clustering level similar to 1995 in 2020. First, the overall patterns are all highly clustered (examine Figure 15.2 and Table 15.5). Generally speaking, for all four dates, similar observable regional patterns are evident. Most states in the New England, Rocky Mountain, and Pacific census divisions have below national mean obesity rates most of the time. Conversely, most states in the South and Midwest census regions have above national mean obesity rates most of the time. These dominant regional patterns would explain the continuing high levels of obesity clustering. Perhaps we are over-emphasizing the importance of these changes. After all, the obesity patterns are all extremely clustered.

When there are exceptions to this overall pattern, measurable differences in the level of clustering seem to emerge. The most clustered pattern is that of 1995 ($Z_p = -6.63$) with only 21 black-white or dissimilar joins. All of the New England, Rocky Mountain, and Pacific states have below national average obesity rates. Georgia is one of only a few notable outliers on the 1995 pattern. If Georgia's obesity rate was just a little higher, it would have been above the national mean, and the national pattern would have been even more clustered. Perhaps some additional insights might emerge if we took a closer look at obesity pattern changes within Georgia. Maybe the state obesity level was below the national mean back in 1995 because of the rapid growth in the Atlanta area (where

(continued)

obesity levels are generally lower than in the rest of Georgia). As the rapid growth in Atlanta has slowed a bit in recent years, perhaps this demographic change allowed Georgia to slip back above the national obesity level.

The least clustered of the three patterns (although still very highly clustered) is 2002 ($Z_p = -4.33$). On the 2002 map, there are more notable outliers (Florida, Maine, Virginia, Minnesota, California, and Oregon) which produces more (32) dissimilar joins (Table 15.2).

Are these "cluster shifts" significant in any practical policy context? That is, do these country-wide join count analyses direct us to any particular new obesity reduction plan? The answer is probably no, at least not directly. But the newly emerging outlier states that don't fit the general regional pattern are worth exploring further. In addition to the Georgia situation just discussed, we might hypothesize why the entire Rocky Mountain and Pacific census divisions are now less obese than the national average. What (if anything) distinguishes the people of the West from much of the rest of the country? Will these regional differences become stronger as time goes by?

TABLE 15.5

Results from the Join Count Analysis—Trends in Obesity Level Clustering: 1995, 2002, 2009, and 2020

Year	*Z*	*p*-value
1995	−6.66	0.000
2002	−4.33	0.000
2009	−5.89	0.000
2020	−6.40	0.000

TABLE 15.4

Worktable for Join Count Analysis: Obesity in the Conterminous United States, 2020

$H_0: O_{BW} = E_{BW}$ (area pattern is random)

$H_A: O_{BW} \cdot E_{BW}$ (area pattern is not random)

number of "black" states = 25
number of "white" states = 23

$$O_{BW} = 21$$

$$E_{BW} = \frac{2JBW}{N(N-1)} = \frac{2(106)(25)(23)}{48(47)} = \frac{121{,}900}{2{,}256} = 54.03$$

$$\sigma_{BW} = \sqrt{E_{BW} + \frac{\sum L(L-1)BW}{N(N-1)} + \frac{4[J(J-1) - \sum L(L-1)]\,B(B-1)\,W(W-1)}{N(N-1)(N-2)(N-3)} - E_{BW}^2}$$

$$= \sqrt{54.03 + \frac{(856)(25)(23)}{48(47)} + \frac{4[(106)(105) - 856]\,25(24)(23)(22)}{48(47)(46)(45)} - 54.03^2}$$

$$= \sqrt{54.03 + \frac{492{,}200}{2{,}256} + \frac{4[11{,}130 - 856]\,(600)(506)}{4{,}669{,}920} - 2{,}919.24}$$

$$= \sqrt{54.03 + 218.17 + 2{,}671.72 - 2{,}919.24}$$

$$= \sqrt{24.68}$$

$$= 4.96$$

$$Z_D = \frac{O_{BW} - E_{BW}}{\sigma_{BW}} = \frac{22 - 53.75}{4.96} = \frac{-31.75}{4.96} = -6.40 \quad (p = 0.000)$$

15.2 Moran's *I* Index (Global)

The Moran's Index (I), developed by Australian statistician Patrick Moran in 1948, is a popular technique for quantifying the level of spatial autocorrelation in a set of geographic areas, with a value assigned to each area. In contrast to nearest neighbor analysis which examines patterns in the *location* of points, or the join count analysis that only evaluates a binary areal representation of the data, Moran's Index takes into account the geographic locations (points or areas) in addition to the actual *attribute values* to determine if areas are clustered, randomly located, or dispersed across the overall study area. The index is positive when nearby locations have similar attribute values (clustered), negative when nearby geographic locations have dissimilar attribute values (dispersed) and approximately zero when attribute values are randomly dispersed throughout the study area.

Geographers have applied the Moran's Index in numerous research problems, such as exploring the spatial distribution of income, rates of disease, and the racial and ethnic composition of urban neighborhoods. In all applications, the objective of Moran's technique is to identify significant spatial patterns within a study area. This technique is appropriate for both point and polygon (area) features, and attribute values must be measured on an ordinal or interval/ratio scale.

Moran's Index proceeds by identifying each unique pair of neighboring locations in the study area. The relationship between neighboring locations is shown for a simplified study region comprising four areas as shown in Figure 15.3. Each area is identified by a letter and the magnitude of the area's attribute value appears in parentheses.

The general form of Moran's Index for areas is shown in Equation 15.4, and the mathematical equivalent, as presented in Ebdon (1988), appears in Equation 15.5:

$$I = \frac{\begin{pmatrix}\text{number}\\ \text{of areas}\end{pmatrix}\begin{pmatrix}\text{sum of cross-products}\\ \text{for all contiguous pairs } (i,j)\end{pmatrix}}{\begin{pmatrix}\text{number}\\ \text{of joins}\end{pmatrix}\begin{pmatrix}\text{difference in area}\\ \text{value from the mean}\end{pmatrix}} \quad \textbf{(15.4)}$$

$$I = \frac{n\sum(x_i - \bar{x})(x_j - \bar{x})}{J\sum(x - \bar{x})^2} \quad \textbf{(15.5)}$$

where:

n = the number of areas
x = an area attribute value
$\bar{x}$ = the mean of all area attribute values
x_i and x_j = the values of contiguous pairs
$\sum(x_i - \bar{x})(x_j - \bar{x})$ = the sum of all contiguous pairs
J = the number of joins
$\sum(x - \bar{x})^2$ = difference in area value from the mean.

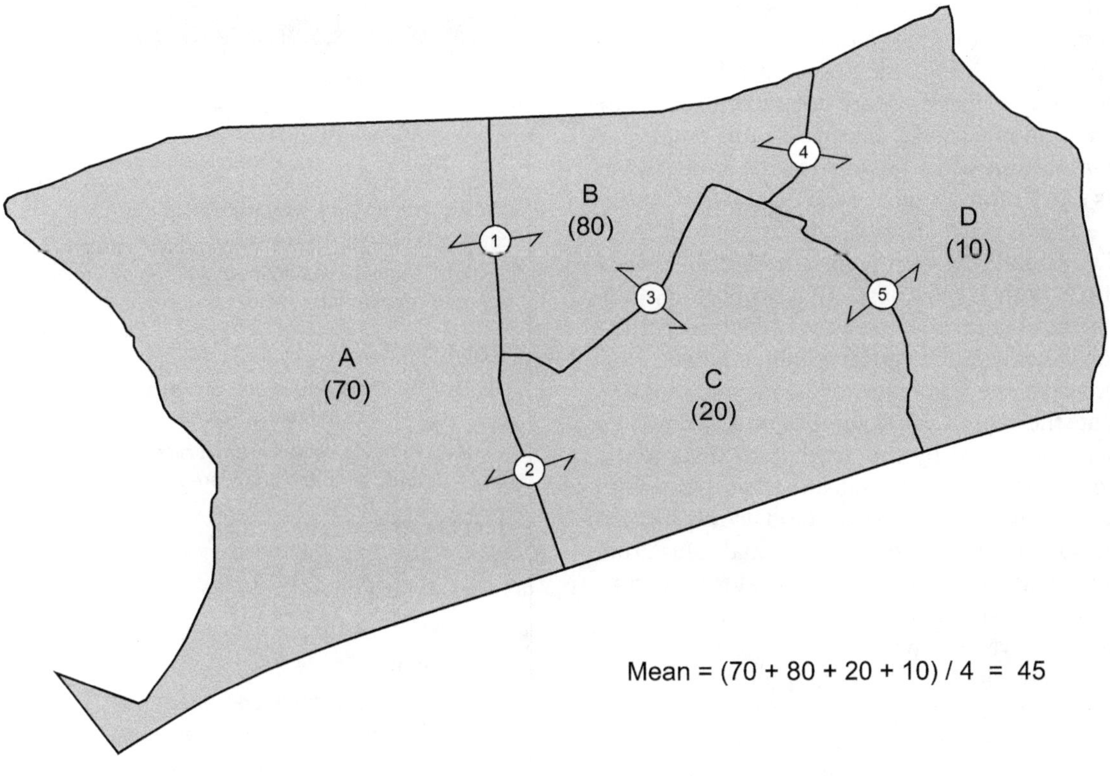

FIGURE 15.3
Areas and Attribute Values for Moran's *I*

The deviation values for each pair of neighboring locations or contiguous pairs (i and j) are multiplied together and summed $\Sigma(x_i - \bar{x})(x_j - \bar{x})$ to create a weighted cross-product. When a neighboring pair of areas both have larger values than the study area mean, their cross-product will be positive (this is the case, for example, with pair AB in Figure 15.3). The cross-product of a neighboring pair will also be positive if both have smaller values than the study area mean (which is the case with pair CD). Positive cross-products therefore indicate that areas with similar attribute values (either high or low) are clustered. In addition, the larger the deviations from the study area mean, the larger the absolute magnitude of the cross-product.

Conversely, when one area of a neighboring pair is larger than the study area mean, while the other area is smaller, the cross-product will be negative (as with pair BD). A negative cross-product therefore indicates dispersion. Once again, the larger the deviations from the study area mean, the larger the absolute magnitude of the cross-product.

It is important to note that the Moran's Index of autocorrelation explicitly takes into account the locations of areas relative to one another. In the procedure as discussed here, the key factor is whether two areas are joined (making them a neighboring pair) or not. We are using *simple adjacency weighting or contiguity* to determine the degree of autocorrelation-only those areas sharing a common boundary are included in the calculations. For instance, areas A and C share a boundary and are included in the calculations, while areas A and D do not share a boundary and are not included.

Other spatial weighting schemes are available to calculate Moran's Index, but these are not shown here. One alternative method is *inverse distance.* With this option, all pairs of geographic areas are included in the analysis, but nearby areas carry more weight in the calculation. If this method were used in our analysis, the cross-product for areas B and C would heavily influence the Moran's Index value (by virtue of their closeness or proximity). By contrast, since areas A and D are relatively far apart, their cross-product value would be given a lighter weight.

To summarize, the numerator of Equations 15.3 and 15.4 represents the sum of all contiguous or adjacent area cross-products multiplied by the number of areas in the study region. A positive sum indicates that areas with similar values (either high or low) tend to be located nearby each other, which suggests spatial clustering among the subareas. A negative sum indicates that areas with high attribute values tend to be located near features with low attribute values, which indicates a dispersed pattern. Sometimes the positive cross-products are balanced by the negative cross-products and their overall sum will be close to zero, which indicates a spatially random pattern. The weighted cross-product numerator is divided by the variance and the sum of the spatial weights (represented as either "joins" or a distance-based weight). This rescales Moran's Index so that it ranges from –1.00 to 1.00. A Moran's Index value less than zero indicates a dispersed pattern of area attribute values, an Index near zero indicates a random pattern, and an Index greater than zero indicates a clustered pattern. Mathematically, the Moran's Index is similar to a correlation coefficient, as we shall see in the next chapter.

While Moran's technique provides an index for the level of spatial autocorrelation for a set of areas, the question becomes how different from zero (randomness) must the Index be in order to indicate significant dispersion or clustering? To answer this question, the Index is usually converted to a Z-score using Equation 15.6. A p-value is then calculated which allows you to determine the significance of the difference between the observed Moran's Index and the corresponding expected Index value for a purely random distribution of values across the same number of areal units.

The null hypothesis is that the area values are arranged in a completely random spatial pattern, and that spatial autocorrelation is not present within the study area. The Z score is computed as:

$$Z = \frac{I - E_I}{\sqrt{Var_I}} \tag{15.6}$$

where: $E_I = \dfrac{-1}{(n-1)}$

Moran's Index

Primary Objective: Identify significant spatial patterns within a study area

Requirements and Assumptions:

1. Minimum of (30) geographic features
2. Attribute values measured on an ordinal or interval/ratio scale

Hypotheses:

H_0 : Attribute values are randomly distributed across features in the study area

H_A : Attribute values are not randomly distributed across features in the study area

Test Statistic:

$$I = \frac{n\Sigma(x_i - \bar{x})(x_j - \bar{x})}{J\Sigma(x - \bar{x})^2}$$

Interpretation:

Assuming a significant p-value:

$I < 0$ (observed pattern is dispersed)

$I = 0$ (observed pattern is random)

$I > 0$ (observed pattern is clustered)

Similar to the free and non-free sampling techniques mentioned with join count analysis and binary data, the variance for Moran's I may be calculated under the assumption of normality or randomization. *Normality* is similar to free sampling in that the observed values of a variable are taken randomly from a normally distributed population, whereas *randomization* is similar to non-free sampling in that the values are taken directly from the data set. The variance under the *normality* assumption is computed as:

$$Var_I = \frac{n^2 J + 3J^2 - n\sum L^2}{J^2(n^2 - 1)}$$

where: n = the number of objects

J = the number of joins, and

$\sum L$ = the sum of the number of joins for each individual area

Variance under the *randomization* assumption is computed as:

$$Var_I = \frac{n[J(n^2+3-3n)+3J^2-n\sum L^2]-k[J(n^2-n)+6J^2-2n\sum L^2]}{J^2(n-1)(n-2)(n-3)}$$

where: k represents the kurtosis for the variable x

If the p-value is not significant, then you should not reject the null hypothesis, as we cannot confidently say that the observed pattern is different from complete spatial randomness. This conclusion is equivalent to a statement that the study area does not exhibit significant spatial autocorrelation. However, if the p-value is significant and the Z-score is positive, we can confidently reject the null hypothesis and say that similar area attribute values (either high or low) are clustered in the study area. If the p-value is significant and the Z-score is negative, we

TABLE 15.6

Worktable for Moran's *I*: Five-area Example Pattern

H_0: no spatial autocorrelation in the data

H_A: spatial autocorrelation exists

	Joins		Area values		
Area	L	L^2	x	$(x-\bar{x})$	$(x-\bar{x})^2$
A	2	4	70	25	625
B	3	9	80	35	1,225
C	3	9	20	-25	625
D	2	4	10	-35	1,225
Joins(J) = 5		$\sum L^2 = 26$	$\bar{x} = \frac{180}{4} = 45$	$\sum(x-\bar{x})^2 = 3{,}700$	

Join number	x_i	$(x_i-\bar{x})$	x_j	$(x_j-\bar{x})$	$(x_i-\bar{x})(x_j-\bar{x})$
1	70	25	80	35	875
2	70	25	20	-25	-625
3	20	-25	80	35	-875
4	80	35	10	-35	-1,225
5	10	-35	20	-25	875
SUM					-975

$$E_I = \frac{-1}{(n-1)} = -\frac{1}{4-1} = -0.33$$

$$I = \frac{n\sum(x_i-\bar{x})(x_j-\bar{x})}{J\sum(x-\bar{x})^2} = \frac{4(-975)}{5(3{,}700)} = -0.21$$

$$Var_I = \frac{n^2J + 3J^2 - n\sum L^2}{J^2(n^2-1)} = \frac{(4^2 \times 5) + (3 \times 5^2) - (4 \times 26)}{5^2(4^2-1)} = \frac{51}{375} = 0.136$$

$$Z = \frac{I - E_I}{\sqrt{Var_I}} = \frac{-0.21 - (-0.33)}{\sqrt{0.136}} = 0.326$$

$p = 0.74$

can confidently reject the null hypothesis and say that the area attribute values are more dispersed across the study area than what we would expect by random chance.

The Moran's I computation procedure is presented in Table 15.6 using the simplified study area having four subareas and five joins, as shown in Figure 15.3. Calculations are only shown using the normalization assumption. In this example, a Moran's I of $-.21$ and a p value of .74 indicates that we do not have enough evidence to reject the null hypothesis and should therefore conclude that the study area does not exhibit significant spatial autocorrelation.

Example: Moran's *I* Index (Global)—Distribution of Racial and Ethnic Groups in Cleveland, Ohio Census Block Groups

The Moran's Index is applied at the census block group level to explore the spatial distribution of racial and ethnic groups within Cleveland, Ohio (Figure 15.4). The objective is to determine the extent to which various racial and ethnic populations are dispersed, randomly located or clustered. Such a concern is important because while progress has been made toward providing equal access to housing and other important public services for all racial and ethnic groups, historical patterns of segregation still exist, particularly in older, industrial American cities. Among other concerns, urban geographers have demonstrated a strong link between patterns of segregation and higher rates of crime and poverty.

Three racial groups and one ethnic group in Cleveland, Ohio are selected to assess the magnitude of spatial autocorre-

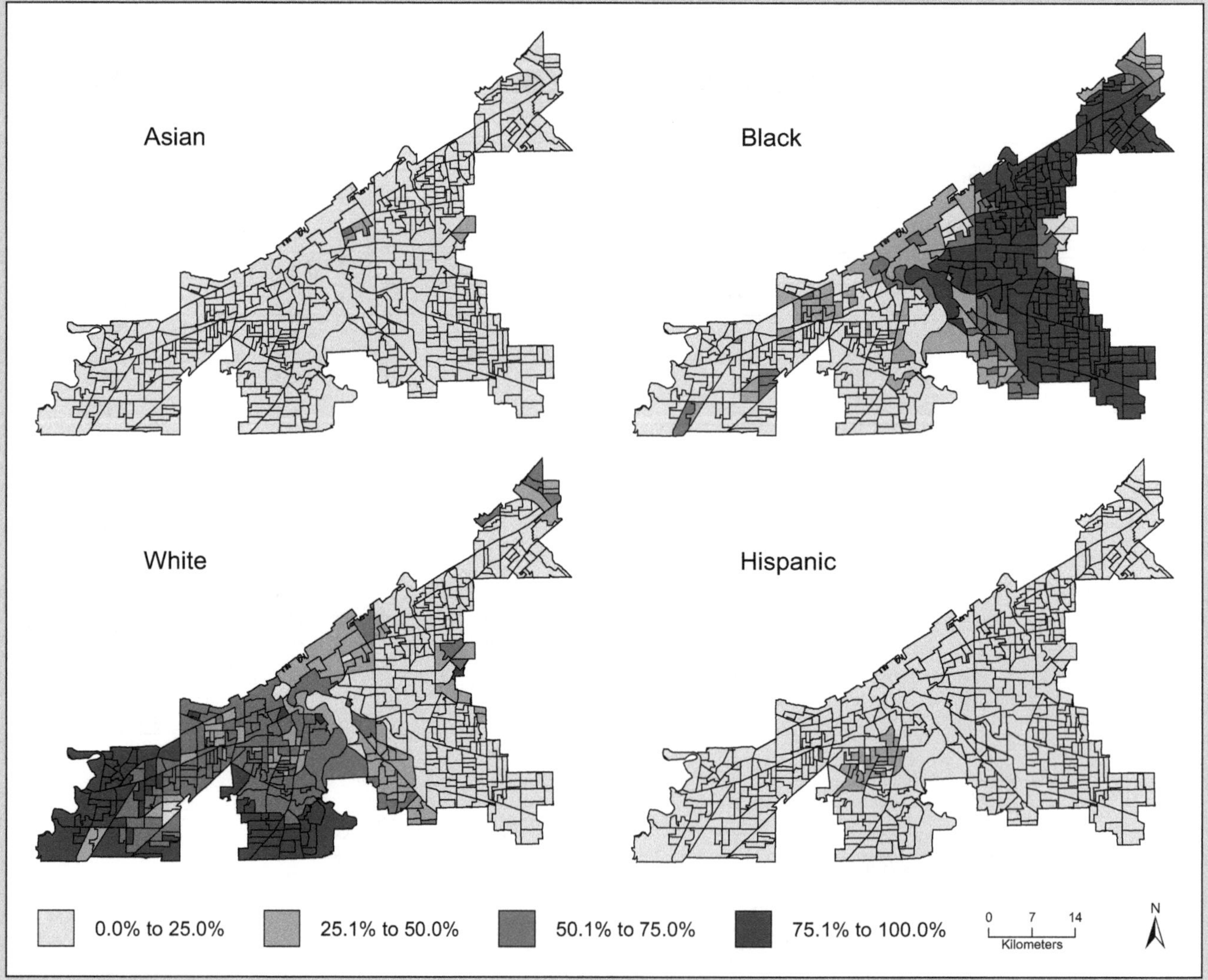

FIGURE 15.4
Distribution of Various Racial and Ethnic Groups by Census Block Group in Cleveland, Ohio
Source: United States Bureau of the Census

TABLE 15.7

Moran's *I* Values and Test Statistics for Selected Racial and Ethnic Groups in Cleveland, Ohio Census Block Groups

Group	Moran's Index	Z	*p*-value
Asian	0.35	18.39	0.000
Black	0.58	64.6	0.000
White	0.64	65.4	0.000
Hispanic	0.81	65.7	0.000

lation using Moran's Index. As reported in the 2010 Census, Cleveland is racially diverse with 53.3% of the population reporting their race as black, 33.4% white (non-Hispanic), 1.8% Asian, and 10.0% reporting their ethnicity as Hispanic/Latino.

A global Moran's analysis is applied to each group to determine the degree to which areas with similar (either high or low) racial/ethnic composition cluster together across the city. Simple adjacency or contiguity is the method used to determine the magnitude of spatial autocorrelation. The null hypothesis in each case is that the values representing the percentage of each racial/ethnic group are arranged in a completely random spatial pattern. The Moran's indices and test statistic results for each of the four racial/ethnic groups are shown in Table 15.7.

In this table, the groups are ranked in increasing magnitude on the Moran's index (that is, increasing levels of clustering since all of the *Z* values are positive). The distribution of Asians is the least clustered. With an index value of 0.35, we can ask whether that value differs significantly from the expected value of –0.002 for a purely random distribution across the same number of census block groups. The resulting test statistics for percent Asian population ($Z = 18.39, p = .000$) indicate that you can reject the null hypothesis with near 100% certainty and conclude that the spatial pattern of Asians in Cleveland is non-random. Indeed, with large positive test statistics and *p*-values of .000, we can conclude that all groups are significantly more clustered than random. In summary, we can conclude that a high level of residential segregation (and spatial autocorrelation) currently exists in Cleveland.

Strongly stated inferential conclusions in this type of problem should be made with extreme care. The underlying null hypothesis of complete spatial randomness is very restrictive. Arguably, almost any spatial distribution that would be of interest to a geographer will show some measurable level of spatial autocorrelation. A more reasonable approach to the interpretation of these results is to consider them descriptive summaries of the patterns within the study area. Rather than a strict inferential interpretation, it might be better to focus on a relative comparison of the magnitude of the Moran's indices across the four racial/ethnic groups. Such a comparison of Moran's values across groups is valid in this instance because all analyses are conducted for the same study area, using similar variables (all related to racial/ethnic composition), and making the same assumptions regarding adjacency and contiguity when creating hypotheses.

15.3 Moran's *I* Index (Local)

In addition to looking at global measures of spatial autocorrelation, geographers may also be interested in assessing whether spatial autocorrelation exists at a local level. As a global indicator of spatial autocorrelation, Moran's *I* may miss certain subtleties within a geographic dataset. That is, while the overall spatial arrangement might indicate that no global spatial autocorrelation exists, there may in fact be pockets (typically called hotspots) of spatial autocorrelation not picked up by the global measure. To resolve this issue, geographers have adapted global methods of spatial autocorrelation like Moran's *I* to include a local component.

This concept, described by economist Luc Anselin (1995) and others, is termed ***L****ocal* ***I****ndicators of* ***S****patial* ***A****ssociation* (LISA). LISA statistics quantify the similarity of each geographic observation with an identified group of geographic neighbors, based on one of the geographic weighting schemes discussed earlier. It is often used to identify local clusters (geographic locations where adjacent or nearby areas have similar values) or spatial outliers (geographic locations that are different from adjacent or nearby areas). Therefore, rather than a single measure of spatial autocorrelation, the LISA methods determine individual measures for each geographic entity as follows:

$$I = \frac{(x_i - \bar{X})}{S_i^2} \sum_{j=1, j \neq i} w_{ij}(x_j - \bar{X})$$
$$S_i^2 = \sum_{j=1, j \neq i} \frac{(x_j - \bar{X})^2}{n-1} - \bar{X}^2 \qquad (15.7)$$

where: x_i = the value for a particular geographic entity
x_j = the value for the neighboring geographic entity
$\bar{X}$ = the average of all attributes
w_{ij} = the spatial weights.

The overall computation of the local Moran index is beyond the scope of this book, but a short example will help illustrate its usefulness.

Example: Moran's *I* Index (Local)—Distribution of Obesity in Pennsylvania Counties

Consider the pattern of obesity levels in Pennsylvania counties shown in Figure 15.5. You can definitely see a pattern of high obesity in the southwest corner of the Commonwealth, while the southeastern corner has lower obesity values. However, due to the overall configuration, the Moran's *I* statistic for the Commonwealth is a single measure, computed as .69 with a *p*-value of .25, indicating no spatial autocorrelation at the global (statewide) level. However, computation of the local Moran's *I* value provides a *p*-value for each county.

Figure 15.6 shows the counties with a *Z*-score above 2.0 or less than –2.0 ($p = 0.00$). Those counties with high positive *Z*-scores indicate places whose obesity level is very similar to neighboring counties, suggesting local clustering. It is not all that surprising to see the metropolitan Philadelphia counties highlighted as an urban region with a cluster of low obesity level counties. A pair of adjacent counties east of Pittsburgh (Cambria County and Blair County) with smaller urban areas (Johnstown and Altoona, respectively) is also identified as very similar to neighboring counties in a region of high obesity levels. Finally, Fayette County in southwest Pennsylvania has a large negative *Z*-score indicating that it is an outlier in obesity level when compared to its neighboring counties. This pattern of local spatial autocorrelation values should raise other basic geographic questions about why the pattern has the specific distribution and variation characteristics shown on the two maps.

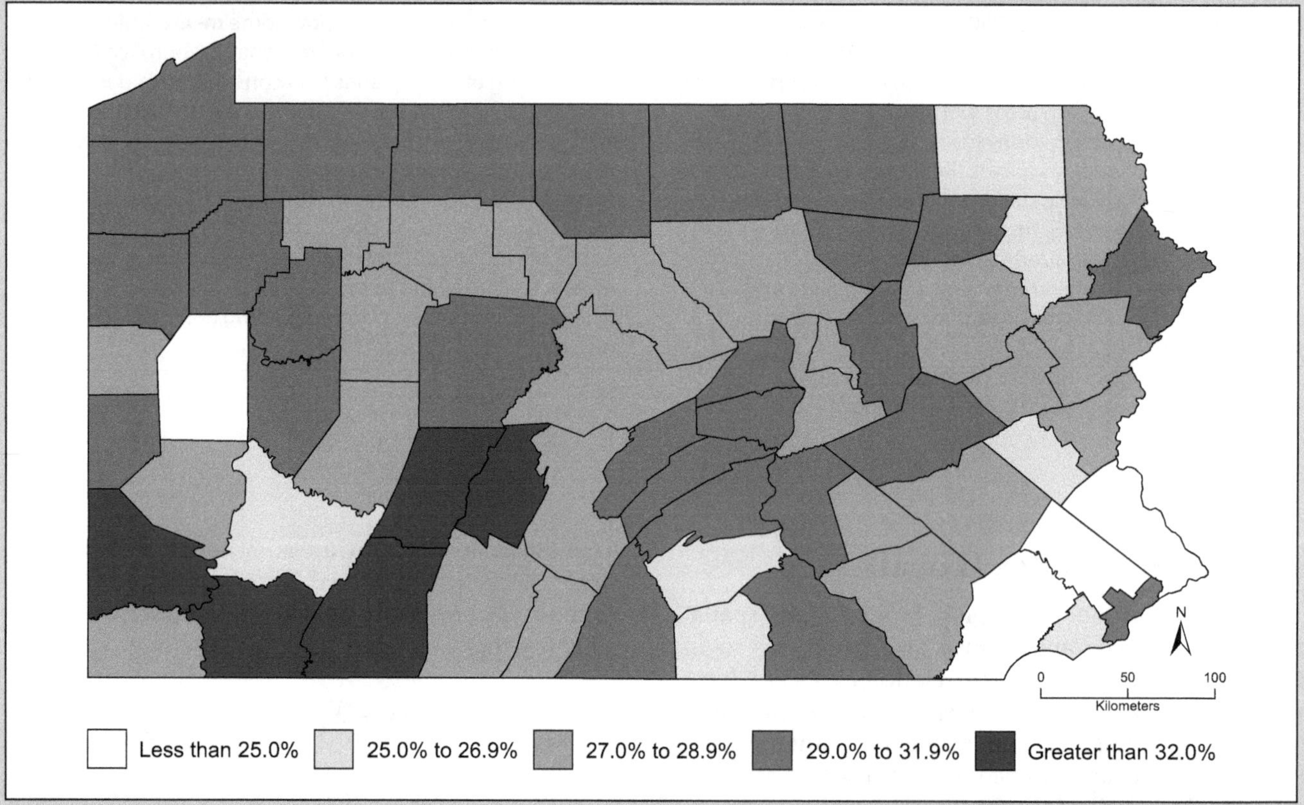

FIGURE 15.5
Distribution of Obesity Levels in Pennsylvania Counties, 2010
Source: Centers for Disease Control and Prevention (CDC)

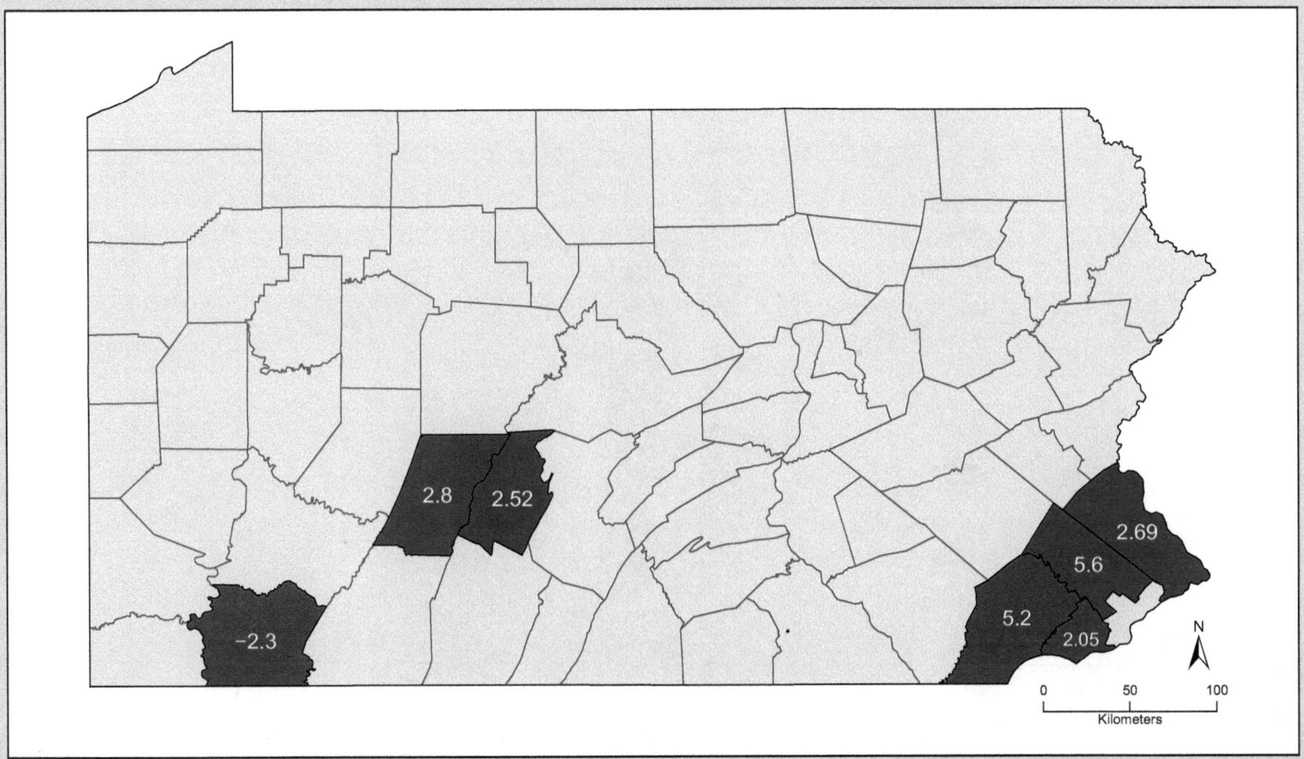

FIGURE 15.6
Pennsylvania Counties with Significant Local Spatial Autocorrelation

Key Terms

area pattern analysis, *224*
free and non-free sampling, *226*
join count analysis, *225*
Moran's *I* index (global and local), *231, 235*

Major Goals and Objectives

If you have mastered the material in this chapter, you should now be able to:

1. Understand the purposes of area pattern analysis.
2. Define and explain join structure and the joint count statistic in area pattern analysis.
3. Distinguish between the contexts or problem settings in which the hypotheses of free sampling and non-free sampling are best used.
4. Explain the basic objectives of Moran's index of autocorrelation.
5. Understand the different goals of global and local autocorrelation.

References and Additional Reading

Anselin, L. "Local Indicators of Spatial Association—LISA." *Geographical Analysis*, Vol. 27, No. 2 (1995): 93–115.

Bailey, T. C. and A. C. Gatrell. *Interactive Spatial Data Analysis.* London: Longman, 1995.

Ebdon, D. *Statistics in Geography. A Practical Approach.* 2nd ed. New York: Wiley, 1991.

Griffith, D. and C. Amrhein. *Statistical Analysis for Geographers.* Englewood Cliffs, NJ: Prentice Hall, 1991.

Lee, J. and D. W. S. Wong. *Statistical Analysis with ArcView GIS.* New York: Wiley, 2001.

Mitchell, A. *The ESRI Guide to GIS Analysis. Vol. 2: Spatial Measurements and Statistics.* Redlands, CA: ESRI Press, 2005.

Ripley, B. *Spatial Statistics.* New York: Wiley, 1981.

For state-level obesity data and county-level obesity data (for Pennsylvania or any other state), see the Centers for Disease Control and Prevention (www.cdc.gov). For spatial distributions of racial and ethnic groups by census block group or by census tract for Cleveland, Ohio or any other metropolitan area, see the U.S. Census Bureau (www.census.gov).

PART 6

Statistical Relationships between Variables

16

Correlation

One of the more important concerns in geographic analysis is the exploration and analysis of relationships between variables that have spatial patterns and variability. Many geographic studies get started by noting the similarities and differences between two mapped variables and then measuring the statistical degree of relationship between the sets of data from which the map patterns were drawn. However, just making a visual comparison of maps to estimate their association is subjective because only a general impression of similarity or contrast between variables is gained. Two people can view the same maps or examine two sets of data and interpret their association very differently.

Correlation analysis provides a precise, quantitative set of statistical methods to measure both the direction and strength of association between a pair of spatial variables. In Section 16.1, the nature of correlation analysis is discussed. The scatterplot is highlighted as a graphic tool used to examine both the direction and strength of association. We have already used scatterplots to gain insights about the nature of relationship between variables in different geographic problems. For example, in Chapter 1 we looked at the association between life expectancy and total health care expenditures per capita for many of the world's countries. The scatterplot nicely illustrated that the United States spends much more per capita on health care, but without any associated lengthening of life expectancy. Just by viewing the scatterplot, we were also able to gain insights about a possible "life expectancy transition" that seems to occur as a country goes through the economic development and modernization process. In Chapter 10 we again used a scatterplot to look at the relationship between state obesity level in 2000 and change in obesity level from 2000 to 2010. Although the relationship did not seem very strong (the points were rather "scattered" in the scatterplot), many of the states in the South and Appalachia that had high obesity levels in 2000 continued to increase their obesity levels from 2000 to 2010. In Section 16.1, the concepts of direction of relationship and strength of association are both further clarified using simple scatterplot illustrations.

The most widely used index of correlation, Pearson's correlation coefficient, is applied only to interval-ratio data and discussed in Section 16.2. To illustrate the calculation process, we return to the example first presented in Chapter 1 hypothesizing a relationship between the date of last spring frost and latitude for a random sample of weather stations in the southeastern United States. A second geographic example of Pearson's index measures the direction and strength of relationship between immigrant growth rates and cost of living index (COLI) values in the largest metro areas of the United States.

Other geographic studies involve the use of ordinal or rank-order data. Section 16.3 presents Spearman's correlation index—the most frequently used coefficient for measuring the direction and strength of association between two ordinal variables. In a "real-data" example illustrating the Spearman's calculation procedure, we explore the relationships among various factors used to rank "America's top states for business" and look for nonrandom patterns among those factors.

16.1 The Nature of Correlation

We have repeatedly seen that geographic investigation often begins with a graphic display of the data, such as a scatterplot. We have also learned that both direction and strength of association between variables can be viewed and roughly estimated by looking at a scatterplot. To clarify further, if the general trend of points is from lower left to upper right (Figure 16.1, Case 1), the direction of association is **positive**. In a positive or direct relationship, a larger value in one variable generally corresponds to a larger value in the second variable. Alternatively, a smaller value in the first variable usually coincides with a smaller value in the second variable. Such a correspondence will result in a positive correlation.

In geography, a positive correlation is found with many variables related to population size. For example, the association between population size and number of retail functions in a sample of settlements is very likely to exhibit a positive or direct relationship. As demonstrated in central place theory, settlements with more people at higher levels of the urban hierarchy generally contain more retail establishments.

The direction of relationship is **negative** if the general trend of points in the scatterplot is from upper left to lower right (Figure 16.1, Case 2). When two variables have a negative or inverse correlation, a larger value in the first variable is generally associated with a smaller value in the second variable and a smaller value in the first variable usually corresponds with a larger value in the second variable.

A negative correlation between two variables can be illustrated by applying the general principle of "distance decay." In such relationships, some phenomenon or idea declines with increasing distance from a source or origin. For example, using pollution data from a sample of monitoring sites, the level of air pollution and distance downwind from the pollution source would probably exhibit a negative or inverse relationship. The further downwind from a pollution point, the more likely the level of exposure to the pollutant will be lower.

Some variables may not exhibit either a positive or negative relationship. If the pattern of points in the scatterplot is **random** (Figure 16.1, Case 3), little or no association exists between the two variables. In such examples, the values of one variable are not associated in any systematic way with the values of the other. This is sometimes referred to as a **neutral** relationship.

In geographic research, variables may show no relationship or pattern on a scatterplot. Suppose you are studying housing characteristics in a city and want to determine if an association exists between house size (square feet of living area) and the age of the home. It is possible that no significant relationship exists between these two variables. Depending on continuing complex changes in the housing market and constantly changing demographics in the population, there may be no overriding rationale for expecting a correlation between house size and age of unit.

The **strength of association** between two variables is roughly estimated by examining the amount of point spread in a scatterplot (Figure 16.2). If the point pattern is tightly packed, as in a "pencil-shaped" pattern, the relationship is said to be strong (Case 2). However, if the points are more widely spread or dispersed, as in a "football-shaped" pattern, the association is moderate to weak (Case 3). The strongest relationship, or perfect correlation, occurs when the points are in a straight line or linear pattern (Case 1) and the weakest association occurs when the points are distributed with very little or no pattern, as in a circular arrangement (Case 4).

Any two variables can be correlated, and the strength and direction of relationship calculated. How-

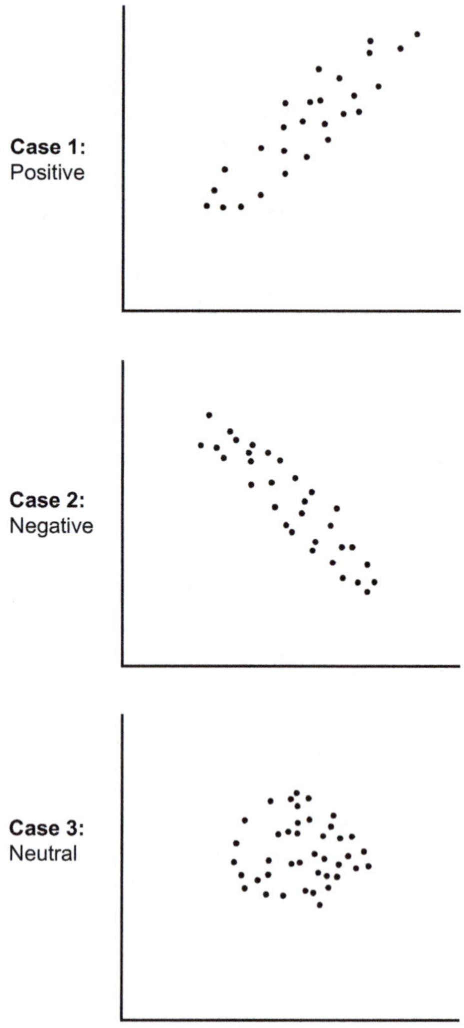

FIGURE 16.1
Generalized Scatterplots Showing Directional Relationships

ever, extreme caution must be used when evaluating or interpreting correlations. The existence of a relationship or association between variables does not necessarily imply that one variable is the "cause" and the other variable is the "effect" or "result" of that cause. One of the most famous statistical phrases you will hear is "correlation does not imply causation." Oftentimes, researchers erroneously assume that if two variables have a high correlation, then they must have a functional relationship that makes them dependent upon one another.

However, in many examples of correlation (and in all cases when applying the companion technique of regression, which follows in the next chapter) the two variables play different roles. The variable of interest that you hope to explain or predict should be placed on the *Y* axis. This *Y* variable is sometimes called the **response** (or **dependent**) **variable**. The **explanatory, predictor** (or **independent**) **variable** should be placed on the *X* axis. The *X* variable is used to account for, "explain," or "predict" variation in the *Y* variable.

When we constructed the scatterplot showing the relationship between life expectancy and total health care expenditures per capita for selected countries (Figure 1.5), we made certain that life expectancy values were plotted on the *Y* axis, as this was the response or dependent variable whose variation we were trying to account for or explain. Total health care expenditures per capita for each country were plotted on the *X* axis, as this was the explanatory, predictor, or independent variable being used to explain that life expectancy value.

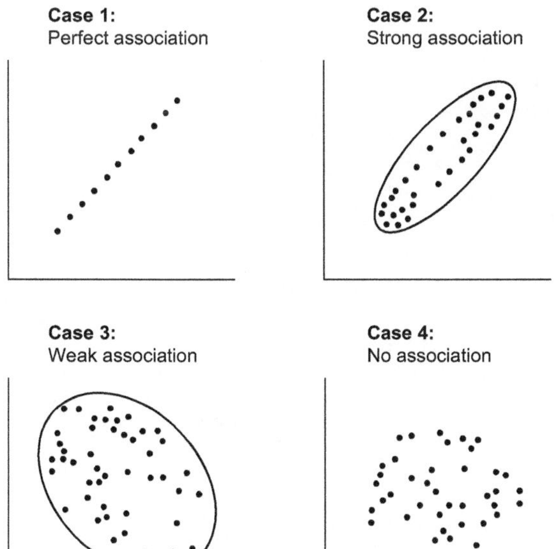

FIGURE 16.2
Generalized Scatterplots Showing Strength of Association

Correlation treats the *X* and *Y* variables symmetrically. That is, the magnitude of correlation of *X* with *Y* is the same as the magnitude of correlation of *Y* with *X*. However, the companion technique with correlation, regression, takes the statistical analysis further, making predictions or estimates about the magnitude of the response variable given the value of the explanatory or predictor variable. This extra step necessitates distinguishing between the roles of the *X* (causal) and *Y* (effect) variables.

As a humorous "cause-effect" example, a correlation analysis may indicate there is a strong *negative* correlation between the number of ice cream cones eaten in a community and the number of flu cases. That is, when a large quantity of ice cream cones is eaten, the area seems to have fewer cases of flu. If one assumes that correlation *does* imply causation, we might conclude that the best way to fight the flu is to eat more ice cream cones! Obviously, another confounding factor is the weather. In the summer, when the seasonal effects of flu are minimal, more people eat ice cream because it is a refreshing treat on a hot day. In the winter, when more people catch the flu, it is often too cold to enjoy ice cream, so less ice cream is consumed. Therefore, it is erroneous to conclude that ice cream consumption causes a reduction of flu cases.

While our ice cream and flu example is obvious, other correlations may not be, such as a correlation of toxic release inventory locations and minority housing, or the application of a single fertilizer on an agricultural plot and the volume of crop production. We must be careful to assess properly the relationship between two variables before attributing any influence of one variable on another.

Statisticians have defined various indices, called "correlation coefficients," to measure the direction and strength of relationships. Most of these coefficients are constructed to have a maximum value of +1.0, which indicates perfect positive or direct correlation between variables. A minimum value of –1.0 represents a perfect negative or inverse correlation, and a value of 0.0 denotes the total lack of correlation between variables. More precisely than a scatterplot, a correlation coefficient can indicate both the direction of the relationship (positive or negative) and the strength of association.

As in other areas of inferential statistical analysis, the level of measurement (nominal, ordinal, or interval-ratio) largely determines which index of correlation is applied to a problem. The most commonly used correlation coefficients are Pearson's product-moment for interval or ratio data and Spearman's rank-order for ordinal or ranked data. As we discussed in Chapter 12, indices that measure association of nominal (categorical) data, such as measuring the strength of association or "cross-tabulation" in a contingency table, are also important.

16.2 Association of Interval-Ratio Variables

The most powerful and widely used index to measure the association or correlation between two variables is **Pearson's product-moment correlation coefficient**. In fact, when the term "correlation" appears in the literature, it is often assumed that the writer is referring to Pearson's correlation index. To use this statistical measure, data must be of interval or ratio scale. It is also assumed that the variables have a linear relationship. In addition, if the index is used in an inferential rather than descriptive manner, both variables should be samples derived from normally distributed populations.

Pearson's correlation coefficient relates closely to the statistical concept of **covariation:** the degree to which two variables "covary" (vary together or jointly). If the values of the two variables covary in a similar manner, the data contain a large covariation, and the two variables will show strong correlation. Alternatively, if the paired values of the variables show little consistency in how they covary, the correlation will be weak.

The concept of covariation and its relationship to correlation can be better understood by comparing a set of scatterplots, each having four quadrants produced from the mean values of the X (horizontal) and Y (vertical) variables (Figure 16.3). In each diagram, the mean values and total variation in the two variables are held constant. This allows a direct comparison of the relative covariation, which differs on each plot.

To understand covariation, we should first look at the deviations of X $(X-\bar{X})$ and Y $(Y-\bar{Y})$ from their respective means. These deviations are the basic building blocks of standard deviation and standardized scores. The X and Y deviations of each data value (matched pair) are multiplied together and summed for the set of values to produce

$$CV_{XY} = \Sigma(X-\bar{X})(Y-\bar{Y}) \qquad \textbf{(16.1)}$$

where CV_{XY} = covariation between X and Y
$(X-\bar{X})$ = deviation of X from its mean $(\bar{X})$
$(Y-\bar{Y})$ = deviation of Y from its mean $(\bar{Y})$

Mathematically, covariation is analogous to the important concept of total variation, the sum of the squared deviations from the mean, an integral component of analysis of variance and regression:

$$TV_X = \Sigma(X-\bar{X})(X-\bar{X}) = \Sigma(X-\bar{X})^2 \qquad \textbf{(16.2)}$$

$$TV_Y = \Sigma(Y-\bar{Y})(Y-\bar{Y}) = \Sigma(Y-\bar{Y})^2 \qquad \textbf{(16.3)}$$

where TV_X = total variation in X
TV_Y = total variation in Y

The point patterns in Figure 16.3 represent different examples of covariation. In Case 1, virtually all points are in quadrants I and III. In quadrant I, the deviations of X $(X-\bar{X})$ and Y $(Y-\bar{Y})$ are both positive since each X and Y value is greater than its respective mean. In quadrant III, where most other points in Case 1 are located, the X and Y deviations are both negative since all X and Y values are less than their respective means. Because the X and Y values for each point in quadrants I and III covary in the same direction from their means, the product of the two deviations will produce a *positive* result for each point. When these individual products are summed according to Equation 16.1, the resultant value will be large and positive. Thus, Case 1 represents an example of a large positive covariation. As seen earlier, this scatterplot also corresponds to a positive or direct correlation between the two variables.

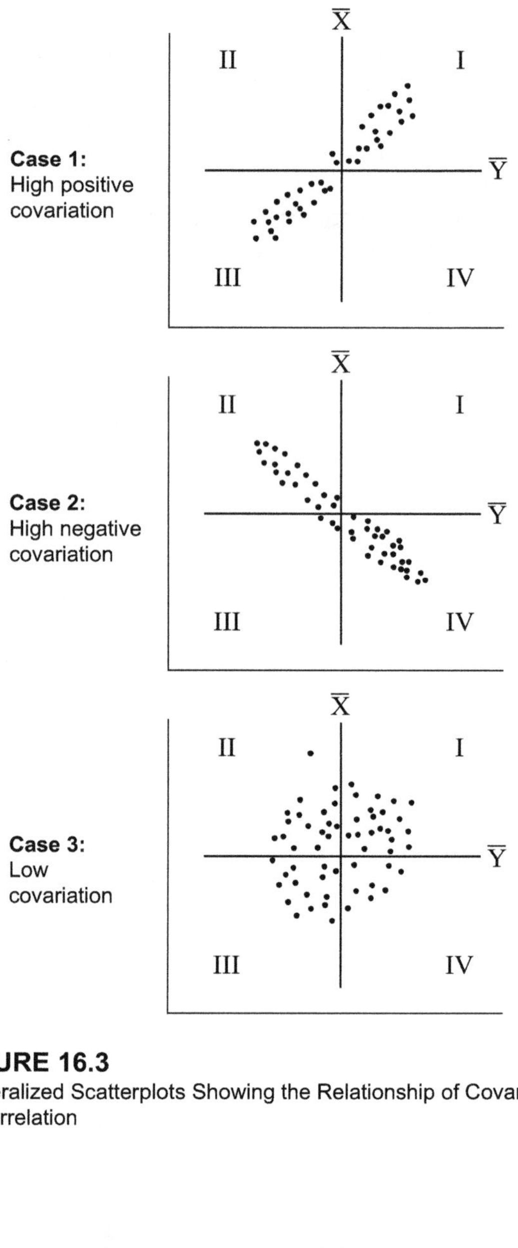

FIGURE 16.3
Generalized Scatterplots Showing the Relationship of Covariation to Correlation

Pearson Correlation Analysis

Primary Objective: Determine if an association exists between two variables

Requirements and Assumptions:

1. Random sample of paired variables
2. Variables have a linear association
3. Variables are measured at interval or ratio scale
4. Variables are bivariate normally distributed

Hypotheses:

$H_0 : \rho = 0$

$H_A : \rho \neq 0$ (two-tailed)

$H_A : \rho > 0$ (one-tailed) or

$H_A : \rho < 0$ (one-tailed)

Test Statistic:

$$t = \frac{r\sqrt{n-2}}{\sqrt{1-r^2}}$$

The situation is different for the plot in Figure 16.3, Case 2, where most points lie in quadrants II and IV. In quadrant IV, deviations in X are positive (X values are greater than $\bar{X}$). However, deviations in Y are negative because values of Y are less than $\bar{Y}$. A reversed situation occurs in quadrant II, where deviations in X are negative (X values are less than $\bar{X}$) and deviations in Y are positive. In Case 2, the product of the deviations will be *negative* because the values of the X and Y variables in both quadrants II and IV covary in opposite directions from their means. When the products from the X and Y deviations are summed, the resultant covariation will be large and negative. The example of covariation in Case 2 represents two variables that have a large inverse correlation.

In the third diagram (Figure 16.3, Case 3) points are scattered in a random pattern, with nearly equal dispersal of points in each quadrant. Some of the X and Y deviations will be positive and others negative. When the deviations are multiplied together for each unit of data, both positive and negative products will occur. When these products are summed for all points, the values generally cancel each other out and produce a covariation close to zero. This low covariation corresponds to a very small correlation, suggesting little or no relationship between the two variables.

Pearson's correlation coefficient (r) can be expressed mathematically in several different ways:

1. with deviations from the mean and standard deviations (Equation 16.4)
2. with X and Y values transformed to Z-scores (Equation 16.5)
3. with the original values of the X and Y variables (Equation 16.6)

All of these formulas produce an equivalent result. Of course we usually calculate correlations using a spreadsheet or computer software program; however, these alternatives are shown to give you further insights into the calculation procedures associated with correlation.

Based on the conceptual definition, Pearson's correlation is expressed as the ratio of the covariance in X and Y to the product of the standard deviations of the two variables:

$$r = \frac{\left[\Sigma(X-\bar{X})(Y-\bar{Y})\right]/N}{S_X S_Y} \qquad \textbf{(16.4)}$$

where r = Pearson's correlation coefficient
N = number of paired data values
S_X, S_Y = standard deviation of X and Y, respectively

If the data are standardized or converted to Z-score form (Section 6.1), an alternative formula exists to calculate the Pearson's correlation coefficient:

$$r = \frac{\Sigma\left[Z_X Z_Y\right]}{N} \qquad \textbf{(16.5)}$$

where: Z_X = X variable transformed to Z-score = $\dfrac{(X-\bar{X})}{S_X}$

Z_Y = Y variable transformed to Z-score = $\dfrac{(Y-\bar{Y})}{S_Y}$

N = number of paired data values

Using Equation 16.5, the correlation between two variables is equal to the sum of the product of Z-scores for each data value divided by the number of paired values. This formula is valid because Z-scores take into account deviations from the mean and the standard deviation.

If one wishes to use the original values of the X and Y variables directly, a computational formula is used to derive Pearson's correlation coefficient:

$$r = \frac{\Sigma XY - ((\Sigma X)(\Sigma Y)/N)}{\sqrt{\left[\Sigma X^2 - \left((\Sigma X)^2/N\right)\right]}\sqrt{\left[\Sigma Y^2 - \left((\Sigma Y)^2/N\right)\right]}} \qquad \textbf{(16.6)}$$

Although this formula appears more complex than the previous equations, it uses only the original data, and no prior calculation of means, standard deviations, or deviations from the mean is needed.

In addition to its use as a descriptive index of the strength and direction of association, correlation can also be used to infer results from a sample to a population. The sample correlation coefficient (r) is the best estimator of the population correlation coefficient (ρ). In this application, the null hypothesis states that no correlation exists in the populations of the two variables (H_0: $\rho = 0$). Since the populations of the variables are assumed to have a linear association (see boxed insert), the null hypothesis of no correlation is equivalent to stating that *X* and *Y* are independent.

As in most other inferential tests, the alternate hypothesis (H_A) can be stated as directional (one-tailed) or non-directional (two-tailed). If you use a one-tailed approach, you should have a logical basis for expecting the correlation to be either positive or negative. When no rationale exists for the direction of correlation between two variables, a two-tailed alternate hypothesis should be used.

The most common test statistic is

$$t = \frac{r}{s_r} \quad \textbf{(16.7)}$$

where:

$$s_r = \text{standard error of the correlation estimate}$$
$$= \sqrt{\frac{1-r^2}{n-2}} \quad \textbf{(16.8)}$$

The test statistic for Pearson's *r* can therefore be rewritten as

$$t = \frac{r\sqrt{n-2}}{\sqrt{1-r^2}} \quad \textbf{(16.9)}$$

**Example: Pearson's Correlation Coefficient—
Strength of Relationship between Date of Last Spring Frost and Latitude in Southeastern United States**

Pearson's product moment correlation coefficient is now applied to one of the research hypotheses presented in Chapter 1. After looking at the spatial pattern of average spring frost dates for a sample of seventy-six weather stations distributed across the southeastern United States (Figure 1.6), we suggested the following hypothesis: "Weather stations with more northerly latitudes (further from the equator) have average last spring frost dates later in the spring than weather stations with less northerly latitudes (closer to the equator)." In general, the pattern shown in Figure 1.6 seems to suggest that latitude is associated with average last spring frost date, but it is impossible to tell how strong that association might be just by looking at the map. Pearson's correlation provides us with the direction of association (positive or negative), the strength of association (from +1.0 to −1.0), and the associated *p*-value indicating the probability that this association is statistically significant.

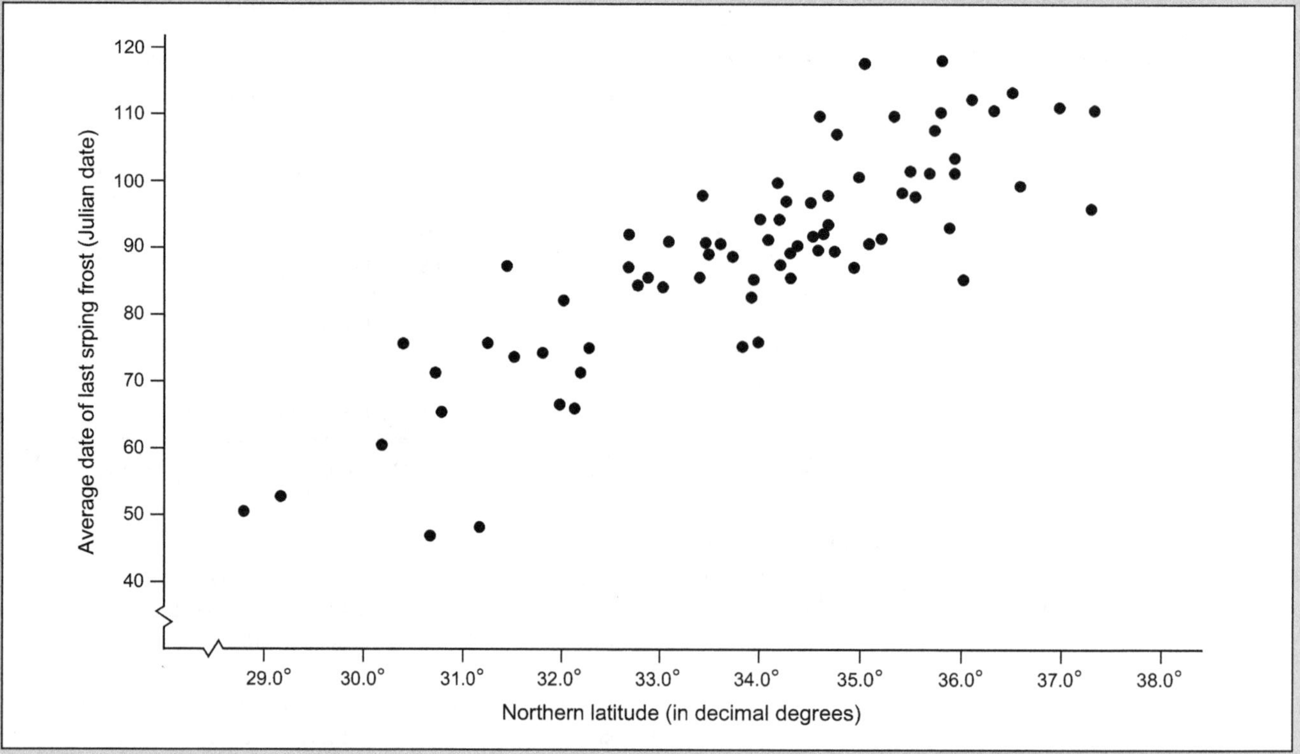

FIGURE 16.4
Scatterplot Showing Relationship Between Average Date of Last Spring Frost and Latitude, Selected Weather Stations in the Southeastern U.S.

If the data meet the requirements and assumptions for Pearson's correlation, we are now in a position to test this hypothesis statistically. Both variables, latitude and average date of last spring frost, are measured on an interval-ratio scale and we have a random sample of weather stations from this region. We tested samples of both variables (using the Kolmogorov-Smirnov normality test) and concluded statistically that they had been taken from normally distributed populations. Finally, we looked at a scatterplot of latitude and average date of last spring frost for these seventy-six weather stations, and concluded that the form or nature of the relationship appears to be linear (Figure 16.4). It seems that we have met all the requirements and assumptions for running a Pearson's correlation.

Notice that the scatterplot displays latitude on the *X* axis (as the explanatory or independent variable) and average date of last spring frost on the *Y* axis (as the response or dependent variable). The date of average last spring frost is expressed using a Julian date calendar (for non-leap years) rather than a regular calendar. In a Julian date calendar, January 1 is day 1, February 1 is day 32, March 1 is day 60, and so on. This conversion to Julian date is done to allow the computation of summary statistics such as mean and standard deviation.

The calculation procedure for Pearson's correlation is summarized in Table 16.1. (Not all 76 weather stations are listed in the table, but a complete listing is available for download on this book's webpage). We are showing the computational formula that uses the original data directly, so you can best see how the correlation coefficient is derived. Of course, any of the mathematical expressions for Pearson's could be used, and the resulting correlation coefficient would be the same.

The resulting *r* value of 0.849 indicates that a strong positive association exists between latitude and average date of last spring frost and that the association it is statistically significant ($p = 0.000$). However, we often want to know more about the form or nature of this association. For example, how useful (precise) is latitude in predicting the average date of last spring frost? Given the latitude of a weather station in the southeastern United States (but not among the 76 in this random sample), can we predict the date of average last spring frost at that weather station, and estimate the precision of that prediction? These questions will be answered in the next chapter, which deals with simple linear regression.

TABLE 16.1

Worktable for Correlation: Latitude and Average Date of Last Spring Frost *

H_0: no association exists between latitude and average date of last spring frost

H_A: a positive association exists between latitude and average date of last spring frost

		X_i	Y_i			
	Weather station	Latitude	Average date of last spring frost	X_i^2	Y_i^2	X_iY_i
1	Troy, AL	31.80°	74.41	1,011.24	5,536.85	2,366.24
2	Union Springs, AL	32.02°	82.27	1,025.28	6,768.35	2,634.29
3	Talledega, AL	33.42°	97.83	1,116.90	9,570.71	3,269.48
...	...	...	...	...	...	...
76	Farmville, VA	37.33°	110.62	1,393.53	12,236.78	4,129.44

$N = 76 \quad \sum X_i = 2{,}578.31 \quad \sum Y_i = 6{,}797.59 \quad \sum X_i^2 = 87{,}736 \quad \sum Y_i^2 = 626{,}720 \quad \sum X_iY_i = 232{,}506$

$$r = \frac{\sum X_iY_i - \left(\frac{(\sum X_i)(\sum Y_i)}{N}\right)}{\sqrt{\frac{\sum X_i^2 - (\sum X_i^2)}{n}}\sqrt{\frac{\sum Y_i^2 - (\sum Y_i^2)}{n}}} = \frac{232{,}506 - \left(\frac{(2{,}578.31)(6{,}797.59)}{76}\right)}{\sqrt{87{,}736 - \frac{(2{,}578.31)^2}{76}}} = \frac{1{,}896.87}{\sqrt{266.495}\sqrt{18{,}730.15}} = 0.849 \quad (p = 0.000)$$

$$t = \frac{r\sqrt{n-2}}{\sqrt{1-r^2}} = \frac{0.849\sqrt{74}}{\sqrt{1-0.849^2}} = 13.82 \quad (p = 0.000)$$

* Latitude expressed in hundredths of a degree (for Troy, AL, 31° 48' is the same as 31.80°)

* Average date of last spring frost expressed in Julian calendar days (non-leap year) (for Troy, AL, 74.41 ≈ 74, which is the same as March 15)

Example: Pearson's Correlation Coefficient— Measuring the Strength of Relationship between Percent Growth of Foreign-Born Population (2000–2010) and the Cost of Living Index (COLI) in America's Largest Metropolitan Areas

Recall that in Chapter 11 we used both ANOVA and the Kruskal-Wallis test to explore descriptively the differences from one census region to another in the rate of immigration growth in large metropolitan areas from 2000 to 2010. We discovered that rates of immigrant growth in metro areas vary considerably by census region. Confirming the results from several studies by The Brookings Institution, large cities in the South have experienced much higher percentages of foreign-born population growth than large cities in the other three census regions. These descriptive findings were shown statistically (through ANOVA and Kruskal-Wallis), in mapped form (Figure 11.3), and with census region boxplots (Figure 11.4)

A logical question for us to ask is why immigrants have been particularly attracted to metro areas of the South over the last decade. Naturally, the answer is complex and multifaceted. One possible explanation is that a large number of immigrants are now attracted to places where the cost of living is fairly low, coupled with the perception by potential immigrants that realistic employment opportunities exist in these cities.

For the 100 largest metropolitan areas in the United States, data are available for both the percent growth in foreign-born population and cost of living index. These data are graphed on a scatterplot (Figure 16.5), with the 2010 Cost of Living Index (COLI) placed on the *X* axis as a possible explanatory (independent) variable and the 2000–2010 percent growth of foreign-born population on the *Y* axis as the response (dependent) variable. Notice the "funnel-like" or "cone-shaped" distribution shown in the scatterplot. The variation of the points clearly changes dramatically given the magnitude of the COLI. This violates the requirement that there is a bivariate normal distribution (see the Pearson correlation analysis boxed inset). There appears to be much more variability of percent growth in foreign-born between metro areas having a COLI near 100 and much less variability when metro area COLI values are considerably larger than 100. There are now two valid reasons for applying only a descriptive correlation analysis (with no inferential analysis): the data are not random samples and the distribution of variables is not bivariate normal.

When a Pearson's correlation analysis is applied to this data set, the result is a correlation coefficient of −.356, indicating a relatively weak negative relationship between the COLI in 2000 and percent growth in foreign-born population from 2000 to 2010 for the 100 largest metropolitan areas of the United States.

From a geographic perspective, additional exploratory analysis seems warranted. The scatterplot depicts a fairly weak negative relationship between the two variables. If we look

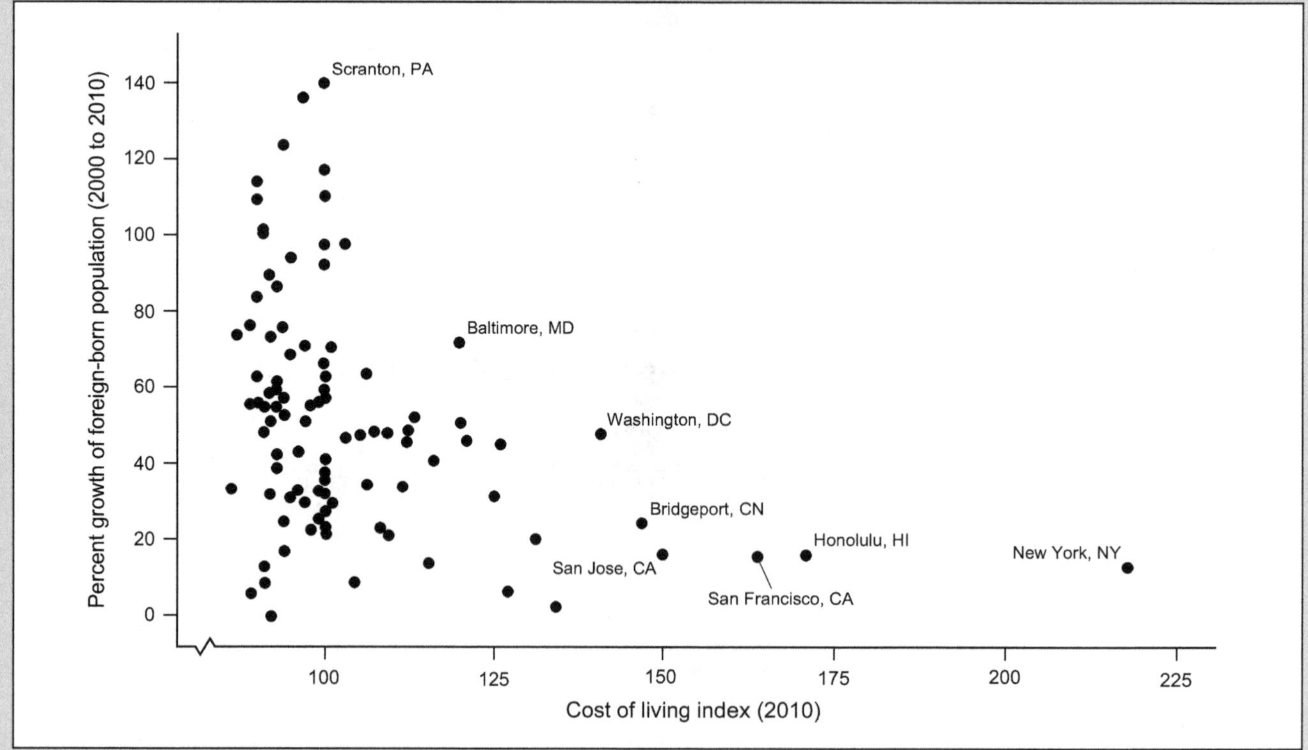

FIGURE 16.5
Scatterplot Showing Relationship between Percent Growth of Foreign-Born Population (2000 to 2010) and Cost of Living Index (2010), America's 100 Largest Metropolitan Areas
Source: Brookings Institution, Immigrants in 2010 Metropolitan America: A Decade of Change

closer, however, a more specific descriptive statement can be directed toward the minority of the 20 or so metro areas with COLI values above 110: "Those large metro areas that currently have a higher-than-national-average COLI (say 110 and above) have generally experienced a low percent growth in foreign-born population (less than 55%) over the last decade."

The scatterplot nicely confirms this descriptive statement. In fact, of the 20 metro areas with COLI values above 110, only Baltimore, MD, has a foreign-born growth rate above 55%. Despite having a relatively high COLI of 120, Baltimore has a 2000–2010 foreign-born growth rate of nearly 72%. Clearly, metro areas with high cost of living indices do not show high rates of immigrant growth.

As we look to the future, we might ask whether this immigrant dispersal to less-expensive metro areas (such as Scranton, PA) will continue. Will we see other "new frontiers" for immigrant settlement in the future? In what ways might possible changes in immigration law (such as altering drastically the conditions for gaining citizenship) affect the future spatial pattern of foreign-born residency choice?

16.3 Association of Ordinal Variables

In geographic problems with data in ranked form, **Spearman's rank correlation coefficient (r_s)** is the measure of choice to determine the strength of association between two variables. It is appropriate when variables are measured on an ordinal (ranked) scale or interval-ratio data have been converted to ranks. Spearman's correlation coefficient is also sometimes appropriate when an assumption of Pearson's correlation is clearly not met. For example, Spearman's is likely the better choice if samples are drawn from highly skewed or severely non-normal populations. The statistical power ("power efficiency") of Spearman's correlation is nearly equal to that of Pearson's r.

Similar to Pearson's correlation, r_s ranges from a maximum of +1.0 for perfect positive or direct correlation to a minimum of –1.0 for perfect negative or inverse correlation. When no association exists between variables, r_s equals 0.0. Spearman's correlation coefficient measures the degree of association between two sets of ranks using the following equation:

$$r_s = 1 - \frac{6(\Sigma d^2)}{N^3 - N} \tag{16.10}$$

where d = difference in ranks of variables X and Y for each paired data value

Σd^2 = sum of the squared differences in ranks

N = number of paired data values

Similar to other statistical tests using ordinal data, the presence of tied rankings influences the Spearman's coefficient. However, the effect of ties on the resultant correlation index will be significant only when the proportion of tied rankings to total number of values sampled is very large. In these instances, a correction factor for ties needs to be applied. As a general rule, the correction factor is not necessary if the number of tied rankings is less than 25% of the total number of pairs.

Spearman's correlation coefficient is sometimes used as a descriptive index of association between two ordinal variables. However, when the paired variables represent a sample drawn at random from a population of bivariate data values, an r_s value can be tested for statistically significant difference from 0. The sample correlation coefficient (r_s) is the best estimator of the population correlation coefficient (ρ_s). In these applications, the null hypothesis states that no relationship exists between the two variables in the population (H_0: $\rho_s = 0$). Confirmation of the null hypothesis is equivalent to affirming independence between the X and Y variables.

Either a one- or two-tailed test is used, depending on the form of the alternate hypothesis. When the Z distribution is used, the test statistic is determined by the Spearman correlation and the sample size:

$$Z_{r_s} = r_s\sqrt{n-1} \tag{16.11}$$

Spearman Rank Correlation Analysis

Primary Objective: Determine if an association exists between two variables

Requirements and Assumptions:

1. Random sample of paired variables
2. Variables have a monotonically increasing or decreasing association (the magnitude of one variable increases as the other variables also increases, or where the magnitude of one variable increases as the other variable also decreases)
3. Variables are measured at ordinal scale or downgraded from interval/ratio to ordinal

Hypotheses:

$H_0 : \rho_s = 0$

$H_A : \rho_s \neq 0$ (two-tailed)

$H_A : \rho_s > 0$ (one-tailed) or

$H_A : \rho_s < 0$ (one-tailed)

Test Statistic:

$$Z_{r_s} = r_s\sqrt{n-1}$$

Example: Spearman's Rank Correlation Coefficient—Exploring Relationships among Various Factors Used to Rank "America's Top States for Business"

Each year CNBC has conducted a study to determine "America's Top States for Business." An annual study from 2011 ranks all 50 states using 43 specific measures allocated into 10 broad categories of competitiveness. These measures are developed with input from business groups such as the National Association of Manufacturers and the Council on Competitiveness. The idea is to compare states by their own standard—the selling points they use to attract business. Each of the 10 categories is weighted by how frequently they are cited in state economic development marketing materials.

The 10 categories and their weightings are: cost of doing business (350); workforce (350); quality of life (350); infrastructure and transportation (325); economy (300); education (225); technology and innovation (225); business friendliness (200); access to capital (100); and cost of living (50).

As part of the analysis, each of the 50 states is ranked on each of the 10 general weighted categories of competitiveness. CNBC then totaled up the points each state received to create an overall ranking. The overall ranking of each state was the headline feature CNBC emphasized on their television programs and internet publications. These overall rankings are summarized in Figure 16.6. A quick glance at the map doesn't seem to reveal a strong spatial pattern, although many of the states in the South and Midwest census regions seem to have above average rankings. For your information, the top 10 states for overall business competitiveness in 2011 are: Virginia, Texas, North Carolina, Georgia, Colorado, Massachusetts, Minnesota, Utah, Iowa, and Nebraska.

If the state rankings for each pair of categories are correlated using Spearman's correlation coefficient, the resulting matrix of correlations allows direct comparison of the relative direction and strength of any two factors of business competitiveness (Table 16.2). The Spearman's correlations have a wide range, from strong negative (−.517 for Quality of Life with Cost of Living) to very strong positive (+.874 for Technology and Innovation with Access to Capital). To illustrate the Spearman's calculation procedure, the Infrastructure and Transportation factor is correlated with the Technology and Innovation factor, resulting in $r_s = .670$ (Table 16.3).

From a spatial perspective, we may want to ask which state-level factors of business competitiveness vary the most by census region. To answer this question statistically, the appropriate test is Kruskal-Wallis (the non-parametric analysis of variance). Kruskal-Wallis is the test of choice because the various factors of business competitiveness are calibrated by ordinal rank, and more than two groups are being compared for differences (four census regions). Table 16.4 summarizes the Kruskal-Wallis results. We are limited to a descriptive presentation of these results: no inferential (probabilistic) analysis is valid because we do not have a sample of states taken from each region (rather,

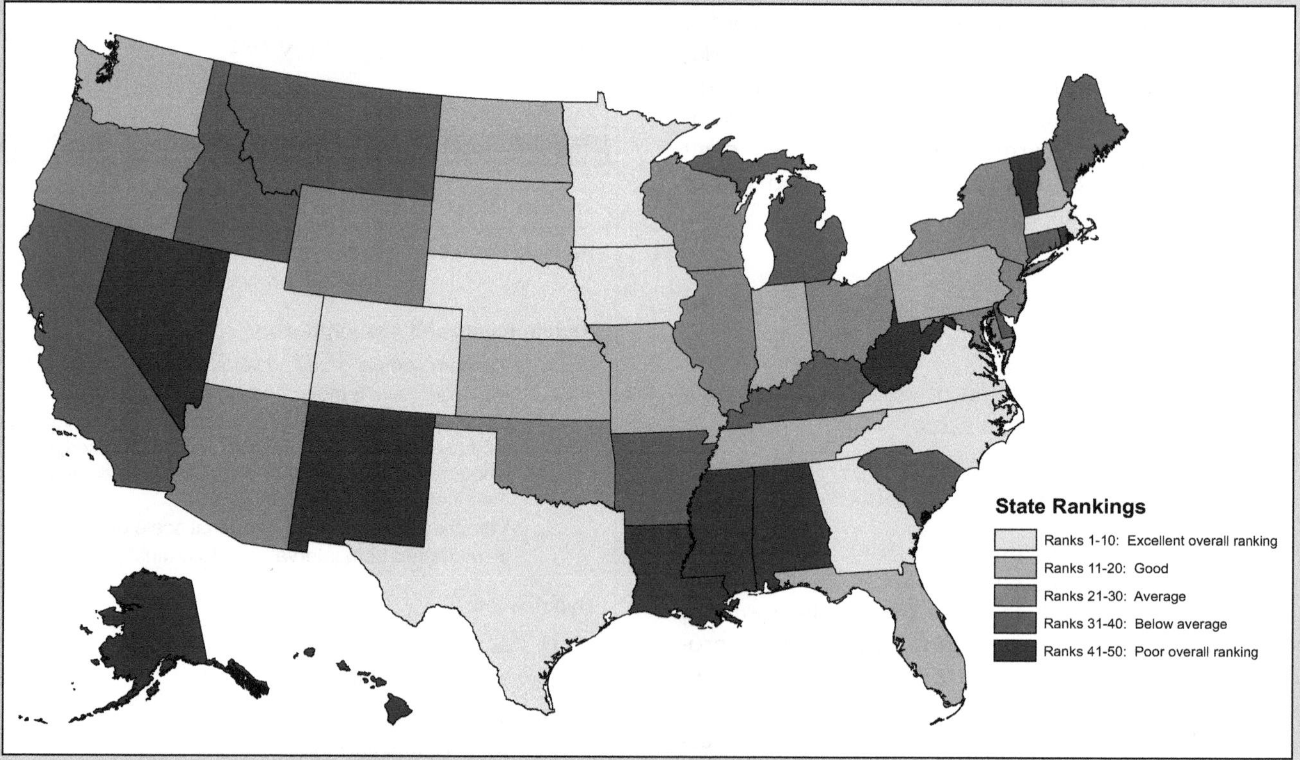

FIGURE 16.6
CNBC's 2011 Overall Ranking of States for Business Competitiveness
Source: CNBC, America's Top States for Business

we have the entire population of all 50 states). Therefore, the computer-calculated p-value of 0.091 has no valid inferential interpretation. However, a glance at the overall census region rankings in Table 16.4 clearly shows that the Midwest Census Region emerges as distinctly better than the other three regions with regard to business competitiveness. As a geographer, you should now ask *why* the Midwestern states are showing better business competitiveness rankings.

With regard to the 10 different factors used in the business competitiveness rankings, some factors show large differences from one census region to another, while other factors show only small differences across the regions of the country. The top portion of Table 16.5 indicates that education, cost of living, and quality of life vary greatly from one region of the U.S. to another. The single most dramatic regional difference is in education. The Northeastern states

TABLE 16.2

Matrix of Correlation Coefficients: Factors Used to Rank Top States for Business

	Cost of doing business	Workforce	Quality of life	Economy	Infrastructure & transportation	Technology & innovation	Education	Business friendliness	Access to capital
Workforce	0.235	—	—	—	—	—	—	—	—
Quality of life	−0.346	−0.222	—	—	—	—	—	—	—
Economy	0.077	−0.187	0.365	—	—	—	—	—	—
Infrastructure & transportation	0.188	0.191	−0.414	−0.379	—	—	—	—	—
Technology & innovation	−0.245	−0.109	−0.068	−0.304	0.670	—	—	—	—
Education	−0.141	−0.418	0.426	0.272	−0.029	0.352	—	—	—
Business friendliness	0.274	0.464	0.098	0.191	0.107	0.040	0.096	—	—
Access to capital	−0.299	−0.145	0.090	−0.197	0.592	0.874	0.405	0.060	—
Cost of living	0.756	0.410	−0.517	0.075	0.336	−0.211	−0.313	0.211	−0.260

TABLE 16.3

Spearman's Correlation Example: State Ranks on Infrastructure–Transportation and Technology–Innovation, 2011

State *	Ranked infrastructure–transportation	Ranked technology–innovation	Difference (d)	d^2
Alabama	24.0	33.0	−9	81
Alaska	47.0	43.0	4	16
Arizona	11.0	18.0	−7	49
Arkansas	40.5	44.0	−3.5	12.25
California	7.0	1.0	6	36
Colorado	26.5	14.0	12.5	156.25
. . .	. . .	. . .	. . .	. . .
Vermont	49.5	40.5	9	81
Virginia	11.0	11.0	0	0
Washington	18.0	5.0	13	169
West Virginia	44.0	47.0	−3	9
Wisconsin	22.0	21.0	1	1
Wyoming	38.0	50.0	−12	144
SUM				6,867.5

* Data listed is for only 12 states

$$r_s = 1 - \frac{6(\sum d^2)}{N^3 - N} = 1 - \frac{6(6{,}867.5)}{50^3 - 50} = 1 - \frac{41{,}205}{124{,}950} = 1 - 0.3298 = .670$$

TABLE 16.4

Kruskal–Wallis: Analysis of Differences Between States by Census Region for Overall Rankings of Business Competitiveness

Census region	Number of states in census region	Median rank	Average rank	Z
Northeast	9	30	29.3	0.87
South	16	30.5	26.3	0.26
Midwest	12	14	16.6	−2.43
West	13	31	30.1	1.33
Overall	50	—	25.5	—

H = 6.46 (p = 0.091)

TABLE 16.5

Kruskal–Wallis: Largest and Smallest Differences by Census Region, Factors Related to Business Competitiveness

Factors of business competitiveness with largest differences by census region

	Median rank		
Census region	Education	Cost of living	Quality of life
Northeast	4.0	42.0	11.0
South	32.5	13.5	39.5
Midwest	18.5	16.5	22.5
West	39.0	33.0	16.0
	H = 26.42 (p = 0.000)	H = 25.71 (p = 0.000)	H = 24.38 (p = 0.000)

Factors of business competitiveness with smallest differences by census region

	Median rank		
Census region	Technology & innovation	Economy	Access to capital
Northeast	19.0	21.0	17.0
South	29.0	31.5	30.0
Midwest	22.5	23.0	27.5
West	31.0	26.0	25.0
	H = 2.15 (p = 0.543)	H = 2.21 (p = 0.531)	H = 3.32 (p = 0.345)

are all evaluated as very strong in education (the median ranking of the Northeast Census Region is 4.0, and the top five states nationwide are Massachusetts, New Jersey, New York, Pennsylvania, and Vermont). By contrast, states in the South and West generally rank low in education (with median ranks of 32.5 and 39.0, respectively).

The cost-of-living and quality-of-life factors also show substantial differences in rankings from one census region to another. However, the spatial patterns are quite different. With regard to cost of living, the South does best (a least-expensive median rank of 13.5) while the Northeast does worst (a most-expensive median rank of 42.0). Just the opposite occurs with quality of life, as states in the South are poorly ranked (median rank of 39.5) and states of the Northeast are highly ranked (median rank of 11.0).

The bottom portion of Table 16.5 shows the three most-evenly distributed factors of business competitiveness (with relatively small differences in ranking between census regions) are technology and innovation, economy, and access to capital. With respect to these dimensions of business competitiveness, it does not seem to matter much which part of the U.S. is being examined; the opportunities for business success are about the same. Once again, as geographers we should now ask *why* these factors are rather evenly distributed around the country.

Key Terms

correlation analysis, *241*
covariation, *244*
direction of correlation: positive, negative, random or neutral, *242*
predictor, explanatory (or independent) variable, *243*
Pearson's product-moment correlation coefficient, *244*
response (or dependent) variable, *243*
Spearman's rank correlation coefficient, *249*
strength of association, *242*

Major Goals and Objectives

If you have mastered the material in this chapter, you should now be able to:

1. Explain the nature and purposes of correlation analysis.
2. Distinguish the various directional relationships between variables (positive, negative, and neutral) and among different strengths of association between variables (perfect, strong, weak, and none).
3. Explain the concept of covariation, and understand how scattergrams can depict the relationship between covariation and correlation.
4. Create a geographic research problem or situation for which Pearson's correlation coefficient would be appropriate.
5. Create a geographic research problem or situation for which Spearman's rank correlation coefficient would be appropriate.

References and Additional Reading

Abler, R., J. Adams, and P. Gould. *Spatial Analysis.* Englewood Cliffs, NJ: Prentice-Hall, 1971.

Burt, J. E., G. M. Barber, and D. L. Rigby. *Elementary Statistics for Geographers.* 3rd ed. New York: Guilford Press, 2009.

Ebdon, D. *Statistics in Geography: A Practical Approach.* 2nd ed. Oxford: Basil Blackwell, 1991.

Falk, R. and A. D. Well. "Many Faces of the Correlation Coefficient," *Journal of Statistics Education*, 1997. Accessed February 2023 from www.tandfonline.com/doi/full/10.1080/10691898.1997.11910597

Rogers, J. L. and W. A. Nicewander. "Thirteen Ways to Look at the Correlation Coefficient." *The American Statistician* 42 (1988): 59–66.

If you are interested in further exploring the geographic examples mentioned in this chapter, the following are good places to start. The most complete source for last spring frost data is the National Centers for Environmental Information (NCEI) at www.ncei.noaa.gov. The immigration growth data for the largest metropolitan areas comes from the Brookings Institution www.brookings.edu. If you want to learn more about the Cost of Living Index (COLI), see various publications provided by the Bureau of Labor Statistics www.bls.gov. Detailed discussions of issues and problems related to the COLI and related indices are found at The Council for Community and Economic Research website www.coli.org.

17

Simple Linear Regression

In the previous chapter Pearson's correlation coefficient was discussed as the index for computing the degree of association between variables measured on an interval/ratio scale. In correlation analysis, you should *not* automatically assume there is a functional or causal relationship between the two variables. A correlation can be computed for any two variables, as long as the correlation index is consistent with the level of measurement for the data.

Geographers often work in research areas where associations between variables need to be explored in more detail. The assumption or hypothesis may be that one variable influences or affects another or that a functional relationship ties one variable to another. For these geographical problems, regression analysis is a useful statistical procedure that supplements correlation.

A classic problem studied by geographers is the spatial relationship between level of precipitation in an agricultural region and the population density the area can support. It is hypothesized that the amount of moisture available at locations within a region influences or affects the density pattern of farm population in the region. Data could be collected for the two variables at various sites in the region, and regression used to answer questions about how population density relates to precipitation level. The nature or form of this relationship can be explored and the strength of the relationship determined. Assuming that the relationship is not exact or perfect, regression allows sources of error in the relationship to be examined and helps uncover additional variables that may influence the geographic pattern. The problem differs from simple correlation analysis because a functional relationship is expected between the variables, and the nature of that relationship needs to be explored more fully. Simply stated, regression should not be used unless a clear rationale or model links one variable to another, as is the case with population density and rainfall.

Regression is applied successfully in all areas of geography. For example, a medical geographer could examine the relationship between the number of physicians located in the counties of a state and the income level of persons residing in these areas. Regression analysis could be used to predict or estimate the number of county physicians based on a county's income profile. An environmental geographer may wish to examine the form of relationship between the acidity level in various locations in a chain of lakes and distance from a point pollution source. Regression analysis could predict acidity level in a lake based on distance from the pollution source. A political geographer might want to compare the strength of votes for a political party and the educational, financial, or racial composition of voters in the wards of a city. Based on the socioeconomic composition of a city ward, regression could then be applied to predict political party voting strength. A cultural geographer surveying current attitudes at various locations could apply regression to predict the level of support for a controversial issue like "women's right to choose" as related to such socioeconomic characteristics as occupation, income, religious belief, or education at those locations.

Simple linear regression (also known as **bivariate regression**) examines the influence of one variable on another and is discussed in Sections 17.1 through 17.4. Just as recommended in correlation, the variable supplying the influence or affect is called the **explanatory, predictor,** or **independent variable** and is placed on the *X* axis. The variable receiving the influence or affect is termed the **response** or **dependent variable** and is placed on the *Y* axis. In regression terminology, the response variable is hypothesized to be affected or influenced by the explanatory or predictor variable. The regression sections include a detailed examination of the form or nature of relationship between latitude and average date of last spring frost, so we are continuing the discussion of this association from the correlation chapter.

The most common application of regression is the identification of linear relationships between variables. In linear form, changes in values of the variables are constant across the range of the data. In these instances, the pattern of points from a scatterplot approximates a straight line. Real-world relationships, however, sometimes show variables that are related in curvilinear ways where variable changes are not constant across all values. In this chapter regression is applied only to a linear relationship.

17.1 Form of Relationship in Simple Linear Regression

Like correlation, simple linear regression measures how one variable relates to another. The basic question posed by regression is: "What is the *form* or *nature* of the relationship between the variables under study?" As with correlation, the association between two variables is easily visualized by constructing a scatterplot, but now the scatterplot is supplemented with a "best-fitting" regression line (and equation) that best replicates or models the relationship.

We continue now with the example in which latitude is associated with the average date of last spring frost in the southeastern United States. Recall that data for these two variables are plotted for 76 weather stations in the study area to produce a scatterplot (Figure 16.4). We always recommend that you look at a scatterplot before proceeding with any regression analysis. If the scatterplot does not appear reasonably linear, a simple linear regression model may not be valid. Pearson's correlation analysis results in a strong positive association of 0.849 (Table 16.1). Generally, the greater the latitude north of the equator for a weather station, the later the average date of the last spring frost for that station.

By calculating a "best-fitting" line and plotting it over a scatterplot of points, we can describe the pattern of points more objectively. Although an infinite number of lines could be drawn to summarize the points in a scatterplot, the "best-fitting" line is always defined as the unique line that minimizes the sum of squared vertical distances between each data point and the line. The best-fitting line is therefore often referred to as the **least-squares regression line** (Figure 17.1). As the name implies, the line minimizes the sum of squared vertical distances between each data point and the line.

No other line can be generated where the sum of the squared distances between the points and the line (measured vertically) is smaller than that calculated for the least-squares line. This line defines the best estimate of the relationship between the explanatory variable (*X*) and the response variable (*Y*). It also serves as a predictive model by generating estimates of the response variable using both the values of the explanatory variable and knowledge of the relationship which connects the two variables.

In a simple linear regression with explanatory variable (*X*) and response variable (*Y*), the least-squares regression line is denoted using the general definition of a line:

$$Y = a + bX \qquad \textbf{(17.1)}$$

In addition to the two variables, the equation contains two constants or parameters (*a* is the ***Y*-intercept** and *b* is the **slope**), which are calculated from the actual set of data. These values uniquely define the equation and establish the position of the best-fitting line on the scatterplot. The equation of the least-squares line for the average date of last spring frost example is:

$$Y = -152.01 + 7.1171X$$

where: X = latitude (expressed in hundredths of a degree)
Y = average date of last spring frost (expressed by Julian date)
a = −152.01 (the *Y*-intercept)
b = 7.1171 (the slope)

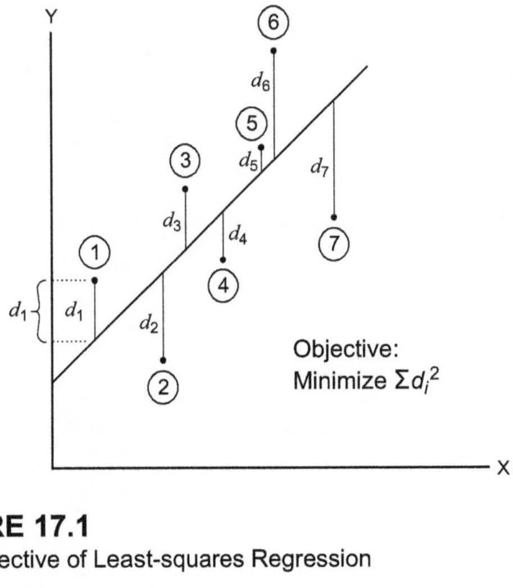

FIGURE 17.1
The Objective of Least-squares Regression

The best-fitting regression line is added to the 76-point scatterplot in Figure 17.2. The constant a represents the expected value of Y when the value of X is zero—that is, the Y-intercept indicates where the regression line crosses the Y axis. In this example, the value of a is –152.01, and is not shown on Figure 17.2. At first glance, this result doesn't seem to make any sense. How can the Julian date be a negative number? The answer is—it can't! So how did we get this seemingly impossible number? It should be noted that a is the predicted best estimate of the Y value (average date of last spring frost) when the X value (latitude) is *zero*. Obviously, we are not interested in knowing the last spring frost date on the equator! We are interested in predicting the last spring frost only in the southeastern United States, roughly between the latitudes of 29 and 37 degrees north. This range of latitude from 29 to 37 degrees north is formally termed the **domain of X**. Zero degrees latitude (the equator) is clearly far outside the domain of our study area, so the regression model is not valid for that value of X.

Whether the X variable represents a study area (like range of latitude) or is a non-spatial explanatory variable (such as per capita total expenditures on health care or percent change in obesity level) it is *always* true that a regression model is not fully valid outside the domain of X. If you attempt to extrapolate beyond this domain, you are venturing into dangerous territory because you are assuming that the linear relationship between X and Y will continue beyond even the most extreme observations of the predictor or explanatory variable, X. A similar problem exists if the explanatory variable is temporal. Extrapolating predictions into the unknown future is a risky business (as any investor in the stock market will readily admit)!

The other constant in the regression equation, b, represents the **slope** (tangent) of the best-fit line. This value, also called the **regression coefficient**, shows the *absolute* change of the line in the Y (vertical) direction associated with an increase of 1 in the X (horizontal) direction. The slope reveals how the response variable will change given a unit increase in the explanatory or predictor variable.

Both the sign and magnitude of b offer useful information about the simple linear relationship. The sign (+ or –) of the slope determines the direction of relationship between the two variables (Figure 17.3). If the slope is positive ($b > 0$), the line trends upward from low values of X to high values of X (Case 1). On the other hand, if the slope is negative ($b < 0$), the line trends downward (Case 2). This interpretation of direction is equivalent to that for the correlation coefficient. When the relationship between the explanatory and response variables is direct, b will be positive and the line trends up from left to right. However, with an inverse relationship, the value of b is negative, and the line trends down. When no relationship exists, the value of b is zero, and the line parallels the X axis (Case 3).

In absolute terms, the magnitude of the parameter b indicates the flatness or steepness of the regression line when moving from lower to higher values of X. When b is large (regardless of sign), the change in Y is large relative to a unit increase in X. In this situation, the slope of the

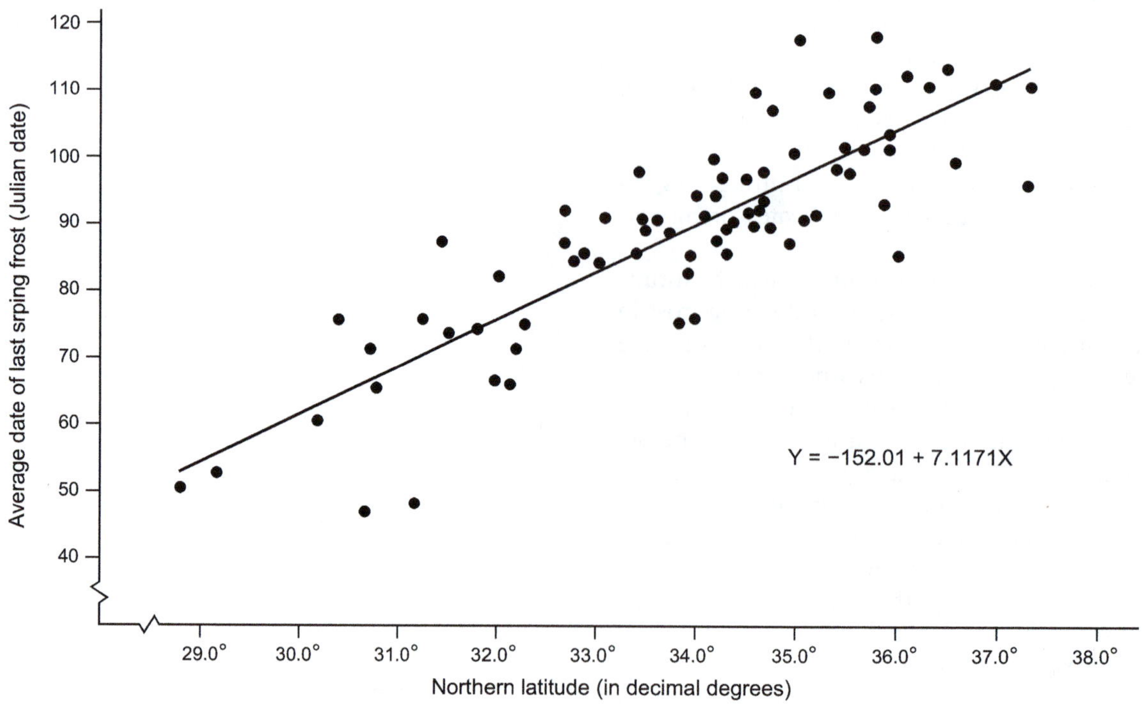

FIGURE 17.2
Best-fitting Regression Line Placed on Scatterplot: Latitude and Average Date of Last Spring Frost

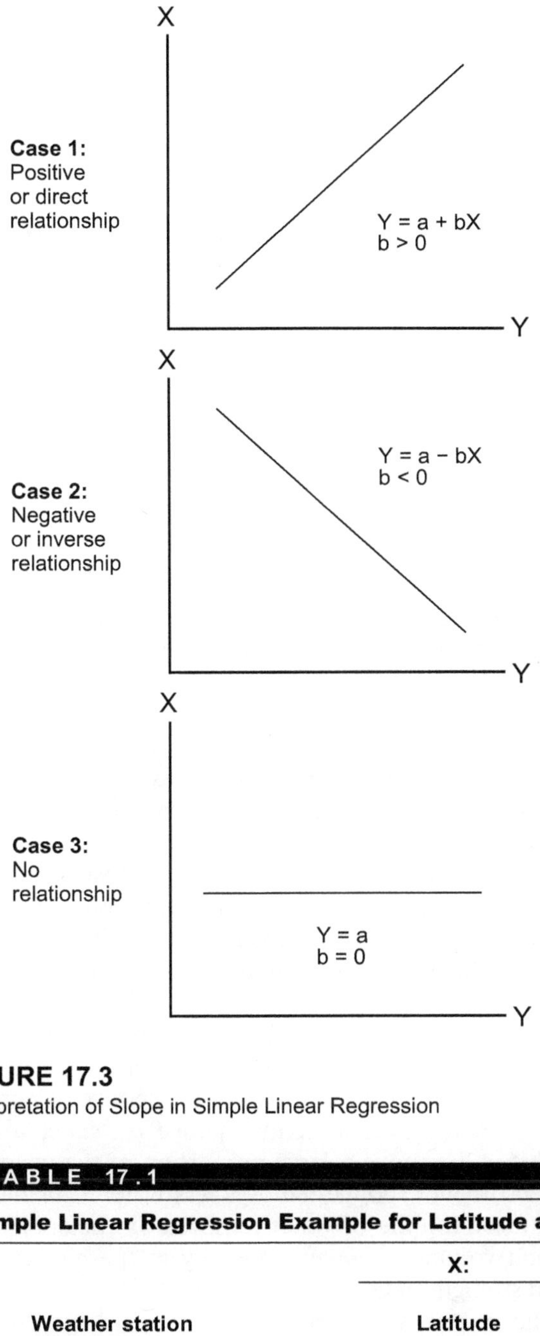

FIGURE 17.3
Interpretation of Slope in Simple Linear Regression

regression line tends to be steep. When b is small, the opposite interpretation occurs. The change in the response variable is small when compared to a unit increase in the explanatory variable, and the line has a flatter slope.

In the last spring frost example, the calculated slope or b value is 7.1171. Since the regression coefficient is positive, the least-squares line slopes upward from left to right on the scatterplot, indicating the last spring frost date will be later (larger or higher Julian date) if the latitude of the weather station is further north (as shown in Figure 17.2). We can now be very specific in our predictions. For each unit increase in X (one degree further north latitude), the average date of last spring frost will increase by 7.1171 Julian days. This means our simple linear regression model predicts that a weather station located one degree further north latitude than another weather station will have a last spring frost date approximately one week later in the spring.

The a and b parameters for the best-fitting regression line are calculated:

$$b = \frac{n\sum XY - (\sum X)(\sum Y)}{n\sum X^2 - (\sum X)^2} \tag{17.2}$$

$$a = \frac{\sum Y - b\sum X}{n} \tag{17.3}$$

where: ΣX = sum of the values for variable X
ΣY = sum of the values for variable Y
ΣX^2 = sum of the squared values for variable X
ΣXY = sum of the product of corresponding X and Y values
n = number of observations

TABLE 17.1

Simple Linear Regression Example for Latitude and Average Date of Last Spring Frost

	Weather station	X: Latitude	Y: Average date of last spring frost	X_i^2	X_iY_i
1	Troy, AL	31.80°	74.41	1,011.24	2,366.24
2	Union Springs, AL	32.02°	82.27	1,025.28	2,634.29
3	Talledega, AL	33.42°	97.83	1,116.90	3,269.48
. . .	. . .	. . .	. . .	. . .	. . .
76	Farmville, VA	37.33°	110.62	1,393.53	4,129.44
		$\sum$ 2,578.31	$\sum$ 6,797.59	$\sum$ 87,736	$\sum$ 232,506

$$b = \frac{n\sum XY - (\sum X)(\sum Y)}{n\sum X^2 - (\sum X)^2} = \frac{76(232{,}506) - (2{,}578.31)(6{,}797.59)}{76(87{,}736) - (2{,}578.31)^2} = 7.1171$$

$$a = \frac{\sum Y - b\sum X}{n} = \frac{6{,}797.59 - 7.1171(2{,}578.31)}{76} = -152.01$$

If you ever have to do calculations manually (hopefully not), the slope or regression coefficient is calculated first, then used to compute the *Y*-intercept or *a* value. Table 17.1 shows the intermediate steps used to calculate *a* and *b* for the average date of last spring frost example.

The two parameters uniquely define the least-squares regression line that best summarizes the relationship between the explanatory and response variables. This line can be placed on the scatterplot by calculating and plotting two points that lie on the line and then connecting them with a straight line. The two points can be determined by selecting any two values of *X*, substituting them into the regression equation, and calculating the corresponding values of *Y*. Fortunately, most statistical software packages will plot the regression line for you.

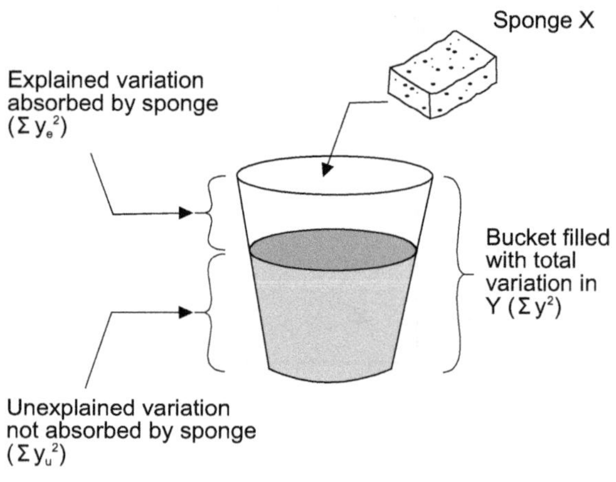

FIGURE 17.4
Bucket and Sponge Analogy in Simple Linear Regression
Source: Abler R., Adams, J., and Gould, P. 1971

17.2 Strength of Relationship in Simple Linear Regression

For any realistic application, the scatterplot of points will not be depicted perfectly by the least-squares line of regression. All points will not lie exactly on the line, thus indicating that the predictor variable cannot fully explain the response variable. This "error" in regression analysis can be traced to several sources. Since geographers study patterns on the earth's surface that are frequently produced by complex processes, it is unreasonable to expect a single predictor (independent) variable in a regression model to account fully for the variation in the response variable. Even when multiple influences are considered, some portion of most real-world patterns is attributable to additional known variables not included in the regression model, other unknown variables, or unpredictable, random occurrences. The inability to measure or operationalize variables in an accurate or valid fashion may be another source of error. Despite these problems, we must determine the strength of relationship between variables if we want to evaluate the explanatory ability of the regression model.

The issue of strength in regression analysis can be viewed in both conceptual and practical terms. In simple linear regression, a single explanatory variable (*X*) is used to explain or account for variation in the response variable (*Y*). The ability of this explanatory variable to account for the variation in *Y* provides a measure of strength or level of explanation. To understand this process from another perspective, the strength of relationship in regression is determined by the amount of deviation between the points on the scatterplot and the position of the best-fit line. In general, the closer the set of points to the regression line, the stronger the linear relationship between the variables. Determining the strength of relationship between two variables in regression is the same as measuring the relative ability of the explanatory variable to account for variation in the response variable.

Abler, Adams, and Gould (1971) created a useful analogy involving a bucket and sponge to illustrate the simple linear regression process (Figure 17.4). A bucket full of water represents the total variation in the response variable, *Y*. A sponge denotes the explanatory or predictor variable (*X*) which will be used to explain variation in *Y*. When the sponge is dipped into the bucket and removed, some of the water (symbolizing variation in *Y*) is absorbed. This represents the amount of variation in *Y* that can be explained by *X*. Although some water is absorbed by the sponge, some of it remains in the bucket. This residual amount is the portion of variation in *Y* that cannot be explained by *X*.

The sponge analogy can be developed further. The central issue in measuring the strength of relationship is to determine the relative ability of the sponge to remove water from the bucket. If the sponge is "super absorbent," it will remove a large proportion of the water. A "less absorbent" sponge removes less water. Calculating the ratio of volume of water removed by the sponge to the total volume of water originally in the bucket provides a strength index for the sponge.

The bucket and sponge analogy translates directly into regression terminology. The **total variation in *Y***, the response variable, is represented by the original volume of water in the bucket. This provides a measure of the variation available for explanation by the predictor variable:

$$\Sigma y^2 = \Sigma\left(Y - \bar{Y}\right)^2 \quad \textbf{(17.4)}$$

where: Σy^2 = total variation in *Y*

Note that *Y* ("large *y*") symbolizes values of the response variable, whereas *y* ("small *y*") denotes the deviation of each value of the response variable from its mean: $(Y - \bar{Y})$. Because the calculation of total variation involves summing the squared deviations from the mean, the term is often referred to as the **total sum of squares.** An alternative formula for calculating the total sum of squares is:

$$TSS = \Sigma y^2 = \Sigma Y^2 - \frac{(\Sigma Y)^2}{n} \quad \textbf{(17.5)}$$

where: TSS = total sum of squares

As illustrated in the bucket and sponge analogy, the total variation in the response variable can be broken into two parts, the **explained variation** and the **unexplained variation**:

$$\Sigma y^2 = \Sigma y_e^2 + \Sigma y_u^2 \quad \textbf{(17.6)}$$

where: Σy_e^2 = explained variation

Σy_u^2 = unexplained variation

The explained variation, also called the "explained sum of squares," is the amount of variation that can be accounted for by the explanatory or predictor variable X. In the analogy, it is the amount of water absorbed by the sponge. The unexplained or "residual" variation is the portion of total variation in the response (dependent) variable that cannot be accounted for by the explanatory variable. It is analogous to the water not removed by the sponge. Because unexplained variation relates directly to analysis of residuals or error in regression, this concept will be discussed more fully in the next section.

The explained variation is the ratio of the square of the covariation between X and Y to the variation in X:

$$\Sigma y_e^2 = \frac{(\Sigma xy)^2}{\Sigma x^2} \quad \textbf{(17.7)}$$

where: Σxy = covariation of X and Y

Σx^2 = total variation of X

As discussed in Section 16.2, the covariation between X and Y indicates how the two variables vary together (covary) and is used to help interpret the correlation coefficient. If two variables tend to covary consistently in the same direction, covariation (and the correlation coefficient) is high and positive (see Figure 16.4, Case 1). If they vary systematically, but in opposite directions, the covariation and resulting correlation is high and negative (see Figure 16.4, Case 2).

Although the direction of association represented by the sign of the covariation term is important in correlation analysis, it can be ignored when using covariation to

TABLE 17.2

Worktable for Calculating Strength of Relationship for Latitude and Last Spring Frost Example

Total variation:

$$\text{TSS} = \Sigma y^2 = \Sigma(Y - \hat{Y})^2 = \Sigma Y^2 - \frac{(\Sigma Y)^2}{N}$$

$$= 626{,}720 - \frac{(46{,}207{,}229)}{76}$$

$$= 18{,}730.15$$

Explained variation:

$$\sum y_e^2 = \frac{(\Sigma xy)^2}{\Sigma x^2}$$

$$\sum xy = \sum XY - \frac{\Sigma X \Sigma Y}{N} = 232{,}506 - \frac{(2{,}578.31)(6{,}797.59)}{76} = 1{,}896.87$$

$$\sum x^2 = \sum X^2 - \frac{(\Sigma X)^2}{N} = 87{,}736 - \frac{(2{,}578.31)^2}{76} = 266.495$$

$$\sum y_e^2 = \frac{(\Sigma xy)^2}{\Sigma x^2} = \frac{(1{,}896.87)^2}{266.495} = 13{,}501.625$$

Coefficient of determination:

$$r^2 = \frac{\Sigma y_e^2}{\Sigma y^2} = \frac{13{,}501.625}{18{,}730.15} = 0.721 \qquad (p = 0.000)$$

determine explained variation. Remember that regression is concerned with the ability of the explanatory variable to account for variation in the response variable. By squaring the covariation in Equation 17.7, the influence of a negative covariation is eliminated.

When the squared covariation (numerator of Equation 17.7) is large relative to the total variation in X (denominator in Equation 17.7) the explained variation is also large. In such cases, the points of the scatterplot will tend to lie close to the regression line and the explanatory variable X will account for more of the variation in the response variable Y. This is equivalent to selecting an absorbent sponge.

On the other hand, if the explanatory and response variables do not covary systematically in a scatterplot, the resulting covariation measure (and correlation coefficient) will be low (See Figure 16.4, Case 3). In this situation, the ratio of the squared covariation to the total variation in X will also be low, producing a smaller amount of explained variation. When the level of covariation of the X and Y variables is weak, the points tend to scatter more widely about the regression line and the general strength of relationship is weak.

The amount of explained variation is an absolute measure calculated in units of the response variable and is comparable to the volume of water removed by the sponge in the analogy. A more useful way to express the strength of a regression relationship is with a relative index that is not tied to the units of measurement. Termed the **coefficient of determination** or r^2, the relative strength index for regression is simply the ratio of the explained variation to the total variation in Y, and measures the *proportion* of variation in the response variable that is accounted for by the explanatory variable:

$$r^2 = \frac{\sum y_e^2}{\sum y^2} \tag{17.8}$$

The index is often multiplied by 100 for ease of interpretation. In this form, the strength index ranges from 0 to 100 and can be interpreted as the *percentage* of variation in the response variable accounted for by the explanatory variable (or in the bucket-sponge analogy, the percentage of water removed by the sponge).

The worktable concerning the strength of relationship for the average date of last spring frost example is shown in Table 17.2. The coefficient of determination (r^2 = .721) indicates that latitude accounts for 72.1% of the variation in average date of last spring frost across this portion of the southeast. The remaining 27.9% of last spring frost variation is not explained by the latitude variable in this regression model.

Although the coefficient of determination (r^2) is closely related to the correlation coefficient (r), the two indices have different purposes and interpretations. The correlation coefficient shows the direction and level of association between any two variables and does not imply a functional or causal relationship. The coefficient of determination, on the other hand, is used as a regression index to measure the degree of fit of the points to the regression line or the ability of the explanatory variable to account for variation in the response variable. As a result, the use of r^2 requires a logical rationale for the existing relationship and the specification of explanatory and response variables.

17.3 Residual or Error Analysis in Simple Linear Regression

Residual analysis provides additional information about the variation in the response variable that cannot be explained by the explanatory variable. Because geographic relationships seldom allow perfect explanations of the response variable, points will only rarely lie on the regression line in the scatterplot. The amount of deviation of each point from the regression line is termed the **absolute residual**. It represents the vertical difference between the actual and predicted values of Y, expressed in units of the variable itself:

$$RES = Y - \hat{Y} \tag{17.9}$$

where: RES = residual
Y = actual value of the response variable
$\hat{Y}$ = predicted regression line value of Y

The predicted values of Y are generated from the regression line. Each $\hat{Y}$ value represents a best estimate of the response variable produced from the a and b parameters that define the regression line. By substituting any value of X (for example, X_1) into the regression equation, the corresponding predicted value ($\hat{Y}_1$) is calculated:

$$\hat{Y}_1 = a + bX_1 \tag{17.10}$$

For example, Farmville, VA (weather station #76) is located at 37° 20′ north latitude (37.33), therefore:

$$\hat{Y}_{76} = -152.01 + 7.1171(37.33) = 113.67$$

This predicted value is quite close to the actual average date of last spring frost calculated directly from the data (110.62). The regression model predicts Farmville will have its last spring frost on about April 24 (Julian date 114), whereas the actual date of last spring frost (averaging about 111) is April 21. The model slightly overestimates the actual last spring frost date by about 3 days and Farmville has an absolute residual of –3.055 (as taken from the computer output).

By using the parameters of the line and the values of the explanatory and response variables, absolute residuals can also be calculated without having to compute the predicted Y values directly:

$$RES = Y - (a + bX) \quad \textbf{(17.11)}$$

Why is so much attention given to calculation of the absolute residuals from regression? The reason is simple: we can gain two general insights when examining residual values associated with a matched pair of data values. First, the size or magnitude of each residual provides the absolute amount of error associated with that data value. For some values the residual or error is small and the explanatory variable accurately predicts the value of the response variable. In other cases the residual is large, indicating poor prediction of the response variable. The smaller the absolute magnitude of a residual, the smaller the vertical distance on the scatterplot separating the point from the regression line and the less error associated with the data value.

Second, the direction of residuals from the line is important. For some data values the actual magnitude of Y exceeds the predicted magnitude and the residual is positive ($RES > 0$). In these instances, the point lies above the regression line on the graph and the model underestimates the actual value of the response variable. For other units of data the predicted magnitude of Y exceeds the corresponding actual magnitude, the residuals are negative ($RES < 0$), and the points lie below the line. In these cases, the regression model has overestimated the actual value of Y.

The insights that can be gained from analysis of absolute residuals are illustrated by examining two weather stations from the average date of last spring frost example: Hawkinsville, GA (weather station #20) and Talladega, AL (weather station #3). Using Equation 17.11 and the data values listed in Tables 16.1 and 17.1, residuals for these weather stations are calculated as follows:

Hawkinsville, GA:

$$\begin{aligned} RES_{20} &= Y_{20} - (a + bX_{20}) \\ &= 75.22 - [-152.01 + (7.1171)(33.28)] \\ &= 75.22 - 84.85 = -9.63 \end{aligned}$$

Talladega, AL:

$$\begin{aligned} RES_{3} &= Y_{3} - (a + bX_{3}) \\ &= 97.83 - [-152.01 + (7.1171)(33.42)] \\ &= 97.83 - 85.84 = 11.99 \end{aligned}$$

Hawkinsville, GA (located about 40 miles southeast of Macon, GA and 120 miles southeast of Atlanta, GA) is a weather station having an absolute residual that is fairly large and negative (–9.63). Positioned below the regression line on the scatterplot with a predicted Y value of 84.85, the simple linear regression model overestimates the average date of last spring frost by nearly 10 days. The date of average last spring frost (derived from the data) is about Julian day 75 (March 16), while the model predicts that the average date of last spring frost should be Julian day 85 (March 26) at that latitude.

Located about 45 miles east of Birmingham, AL, the weather station at Talladega has a fairly large positive residual (11.99). With an actual average last spring frost date of 97.83, the point representing Talladega is well above the predicted regression line value of 85.84. That is, while the model predicts the date of last spring frost should be about March 27 (Julian day 86), the average date derived from the data is April 8 (Julian date 98).

These residual analyses lead to some obvious questions. Why is the last spring frost in Hawkinsville occurring about 10 days earlier than predicted, and why is the last spring frost in Talladega occurring about 12 days later than predicted? Are residuals of this magnitude typical of the 76 weather stations distributed around the region?

Latitude is certainly not a perfect predictor of when the last spring frost will occur at a weather station. We already know this to be the case since latitude accounts for only 72.1% of the variation in average last spring frost dates among the 76 sample weather stations. The simple regression model using only latitude to predict last spring frost date contains a set of residuals (bits of unexplained variation or errors in the model) that leaves 27.9% of the total variation in the response variable unexplained.

To explore further, we should take the next logical geographic step and **map the residuals**. This allows us to examine the spatial pattern of unexplained variation in last spring frost dates. If the spatial pattern of residuals is random (if there is no discernible non-random pattern to the residuals) then it may be difficult to suggest another explanatory variable beyond latitude to add to the model. Hopefully, however, there is a non-random pattern to the residuals that suggests another explanatory variable which could be added to our model. If so, then we can perhaps build a multiple regression model (see Chapter 18) with more than one predictor variable. The generalized pattern of absolute (and standardized) residuals is shown in Figure 17.5.

Of the 76 weather stations included in this problem, only 22 have a residual whose absolute value is greater than 8.41 days, and these are highlighted in Figure 17.5. We mention the specific value of 8.41 because that is the **standard error of the estimate** (an index of relative error) which defines the typical distance separating a point from the best-fitting regression line on the scatterplot. The standard error of the estimate is in fact the standard deviation of the residual values. We will see shortly how this value is calculated. The simple linear regression model using latitude to predict the average date of last spring frost has a standard error of 8.41 days. With some weather stations the regression model predicts the last spring frost date quite a bit earlier than it actually occurs (like Talladega, AL). With other stations (such as Hawkinsville, GA) the model predicts the last spring frost date much later than it actually occurs. With many other stations (such as Farmville, VA) the predicted date of last spring frost is much closer to the date that the frost occurs. If we express the absolute residual of a weather station in terms of the number of standard errors it is from the value predicted by the regression model, then it is the **standardized residual**.

The last three columns in Table 17.3 show the predicted date of last spring frost, the absolute residual, and the standardized residual for some of the weather stations (with the sign of the residual indicating whether the point is above or below the best-fitting regression line).

The standard error of the estimate can be calculated in several ways, using the total error, the residuals, or the actual and predicted values of Y:

$$SE = \sqrt{\frac{\Sigma y_u^2}{n-2}} = \sqrt{\frac{\Sigma RES}{n-2}} = \sqrt{\frac{\Sigma\left(Y-\hat{Y}\right)^2}{n-2}} \qquad \textbf{(17.12)}$$

where: SE = standard error of the estimate
$n-2$ = degrees of freedom

Calculation of total and relative error for the last spring frost example illustrates the different uses of these regression indices. The unexplained variation or residual sum of squares can be calculated from the residuals or from the actual and predicted Y values:

$$\Sigma y_u^2 = \Sigma (RES)^2 = \Sigma\left(Y-\hat{Y}\right)^2 = 5233$$

It can also be calculated from the total and explained variation:

$$\Sigma y_u^2 = \Sigma y^2 - \Sigma y_e^2 = 18730 - 13497 = 5233$$

These calculations show that of the 18730 units of total variation in the response or dependent variable, 5233 units of variation (average date of last spring frost) are left unaccounted for by the single explanatory (independent)

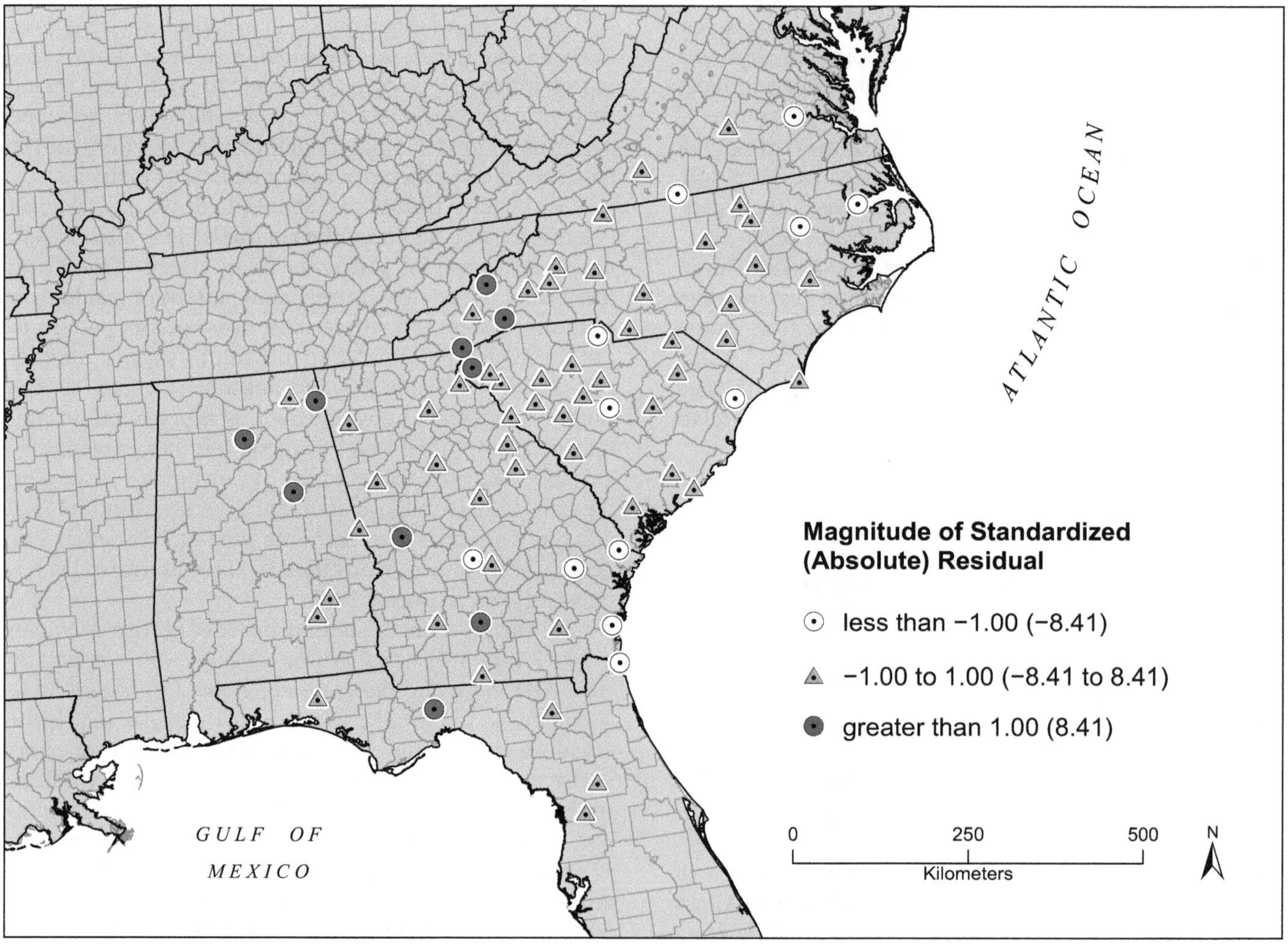

FIGURE 17.5
Pattern of Residuals: Regression Model Using Latitude to Predict Average Date of Last Spring Frost

variable of latitude. To see how much of this total error is associated with a typical data weather station in the southeast, the standard error of the estimate that we were discussing earlier is now calculated:

$$SE = \sqrt{\frac{\Sigma y_u^2}{n-2}} = \sqrt{\frac{5233}{74}} = 8.41$$

The standard error of the estimate produces **standardized residual values** by converting absolute residuals into relative residuals. This simple procedure is analogous to generating *Z*-scores for a distribution:

$$SRES = \frac{RES}{SE} \qquad \textbf{(17.13)}$$

where: *SRES* = standardized residual

In this form, standardized residuals relate the magnitude of each residual to the size of the typical residual, represented by the standard error. They can be interpreted as the typical amount of error associated with a value, measured in standard error units.

Look again at the map of residuals (Figure17.5). The map displays the spatial pattern of error in the regression model when the single explanatory variable of latitude is used to predict the date of last spring frost across the region. All weather stations with a standardized residual more than one standard error from the predicted value (either positive or negative) are highlighted. Do you see any non-random pattern in the very large positive or very large negative residuals?

Many of the largest negative residuals are located in weather stations relatively close to the Atlantic coast. That is, many coastal communities seem to have last spring frost dates that occur more than a week earlier in the spring than predicted by the regression model. Conversely, almost all of the largest positive residuals are weather stations located well inland, often in mountainous areas where the upper piedmont merges with the foothills of the Blue Ridge Mountains (places like western North Carolina and northern Georgia).

Clearly, the spatial patterns of the largest residuals are not random. As we consider making improvements to the regression model (hopefully increasing the magnitude of the coefficient of determination (r^2) above .721), distance from the coast and elevation both seem to be excellent candidates for additional explanatory or predictor variables. In fact, we will statistically examine these two variables (both individually and collectively) in the next chapter.

17.4 Inferential Use of Regression

Sometimes the results from simple linear regression can be tested for significance and results from a sample inferred to the population from which the sample was drawn. This can be done in several ways. For example, inferences can be made concerning the two parameters that define the regression line, the slope and the *Y*-intercept.

TABLE 17.3

Regression Model Parameters for Selected Weather Stations: Latitude and Last Spring Frost Example

		X:	Y:	$\hat{Y}$		
	Weather station	**Latitude**	**Average date of last spring frost**	**Predicted date of last spring frost (from regression model)**	**Absolute residual (RES)**	**Standardized residual (SRES)**
				(may differ slightly from computer output due to rounding)		
1	Troy, AL	31.80°	74.41	74.31	0.10	0.01
2	Union Springs, AL	32.02°	82.27	75.88	6.39	0.76
3	Talledega, AL	33.42°	97.83	85.84	11.99	1.43
...	...	...	...	...	...	...
76	Farmville, VA	37.33°	110.62	113.66	−3.04	−0.37

a = −152.01

b = 7.1171

Analysis of variance (ANOVA) (provided by computer output):

Total variation in the response or dependent variable = 18,730

Explained variation (regression model sum of squares) = 13,497

Unexplained variation (residual sum of squares) = 5,233

R-squared (proportion (percent) of total variation in the response or dependent variable accounted for by the explanatory or predictor variable) = 13,497/18,730 = 0.721 or (72.1%)

Inferential testing can also be applied to the coefficient of determination, measuring the statistical significance of the strength of the regression model. However, no matter which inferential procedure is chosen, a stringent set of assumptions apply.

Regression has more assumptions than the other inferential procedures discussed in the text. All of the assumptions that apply to Pearson's correlation also apply to regression. One basic requirement is that **both variables must be measured on an interval or ratio scale** (neither ordinal nor categorical variables can be used). Both latitude and date of last spring frost meet this condition. Also, **random samples are required**, using one of two modeling schemes. In a fixed-X model, the investigator preselects certain values of the explanatory or independent variable (X), perhaps as part of a controlled experiment. Sample values for the dependent variable (Y) are then derived using a component of randomness. In the random-X model, sample values for both the explanatory and response (dependent) variables are chosen at random. For the last spring frost problem, the weather stations are chosen randomly from throughout the southeastern U.S. study region.

Since the simple linear regression model places a best-fitting line through a scatterplot, **the variables are assumed to have a linear relationship**. If the association between variables is not reasonably straight, then the simple linear model should not be used. In some cases one or both of the variables can be transformed to create a reasonably straight scatterplot, but discussion of transformations is beyond the scope of this book. With regard to the latitude and last spring frost example problem, we have already confirmed from the scatterplot in the correlation chapter (Figure 16.4) that the relationship between variables is linear.

Another important requirement is the **equal variance assumption**. Probably the simplest way to check whether this assumption is met is to plot either the absolute or standardized residual values against the predicted values of Y. There should be a similar amount of scatter about the line whatever the magnitude of the predicted value. That is, this scatterplot should not have any notable features. There should be no significant direction, bend, or shape to the residual plot. Also, there should be no extreme or extraordinary points (no outliers or outlier groups) on the residual plot. The scatterplot of residuals against predicted date of last spring frost (Figure 17.6) appears very nondescript and featureless, suggesting the equal variance assumption has been met. Meeting these requirements also makes it virtually certain that **variables X and Y are themselves normally distributed**. Fortunately, most statistical packages provide the option to plot the residuals against the predicted Y-values.

Both the Y-intercept and the slope of the sample regression model can be tested for statistical significance. *Inferential* testing of the regression parameters considers the Y-intercept (α) and slope (β), which define the population regression line relating the independent and dependent variables:

$$Y = \alpha + \beta X \tag{17.14}$$

The sample Y-intercept (α) and sample slope (β) are the best estimators of their respective population parameters.

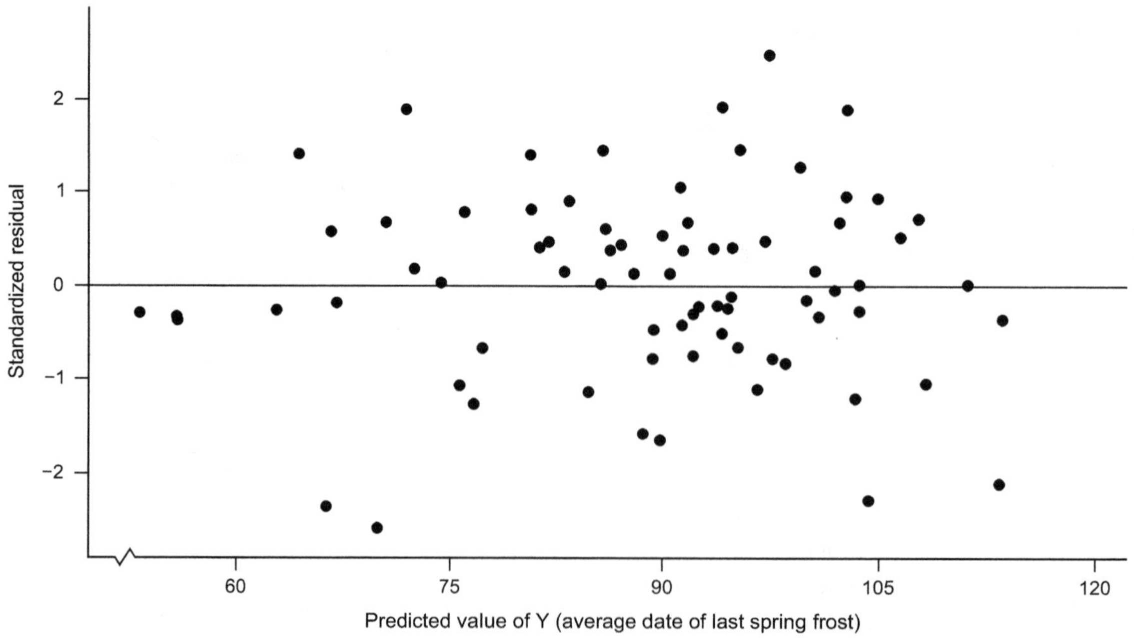

FIGURE 17.6
Scatterplot of Residuals Against Predicted Values of Y (Average Date of Last Spring Frost)

Inferential testing of the sample regression line parameters tells us the likelihood that the Y-intercept and slope are statistically significant. Even if the Y-intercept lies well outside the domain of X, testing its significance is still worthwhile. In the latitude and last spring frost example, the Y-intercept is –152.01, the associated t-value is –8.68, and the p-value is 0.000. The slope is 7.1171, with associated t-value of 13.81and p-value of 0.000. These figures are found in the "analysis of variance" portion of the computer output. We conclude that the regression model contains parameters that are statistically significant. That is, if we collected data from all weather stations in the southeastern U.S., we would find a significant nonrandom slope and Y-intercept between latitude and date of last spring frost.

Inferential significance tests can also be applied to the population coefficient of determination (ρ^2), using the sample coefficient of determination (r^2) as the best estimator. Analysis of variance (or F statistic) is used to evaluate the significance of r^2. In this context, the null hypothesis is that the population coefficient of determination is not significantly greater than zero (H_0: ρ^2 not > 0), and the alternate hypothesis is the converse (H_A: $\rho^2 > 0$).

Recall that the total variation in the dependent variable is equal to the sum of two components, the explained variation and the unexplained variation (Equation 17.6). It is also true that the coefficient of determination is the ratio of explained variation to total variation in Y (Equation 17.8). Thus, returning to the bucket and sponge analogy, testing the significance of r^2 to determine if it is significantly greater than zero is equivalent to testing whether the sponge (explained variation) removes a significant amount of water (total variation) from the bucket.

These components of variation in Y may be expressed in terms of the "sum of squares," illustrating the direct equivalence between regression and analysis of variance:

$\Sigma y^2 = \Sigma y_e^2 + \Sigma y_u^2$ (from Equation 17.8)

$\Sigma y^2 = TSS$ (total sum of squares)

$\Sigma y_e^2 = RSS$ (regression or explained sum of squares)

$\Sigma y_u^2 = ESS$ (error, residual, or unexplained sum of squares)

Therefore: $TSS = RSS + ESS$ and

$$r^2 = \frac{\Sigma y_e^2}{\Sigma y^2} = \frac{RSS}{TSS}$$

Simple Linear Regression Analysis

Primary Objective: Determine if an independent (predictor) variable (X) accounts for a significant portion of the total variation in a dependent (response) variable (Y)

Requirements and Assumptions:

1. Variables are measured on interval or ratio scale
2. Fixed-X model: Values of independent variable (X) chosen by the investigator, and values of dependent variable (Y) randomly selected for each X
   Random-X model: Values of both X and Y randomly selected
3. Variables have a linear association
4. For every value of X_i, the distribution of residuals $(Y - \hat{Y})$ should be normal, and the mean of the residuals should equal zero
5. For every value of X, the variance of residual error is equal (homoscedastic)
6. The value of each residual is independent of all other residual values (no autocorrelation)

Hypotheses:

$H_0: \rho^2 = 0$

$H_A: \rho^2 \neq 0$

Test Statistic:

$$F = \frac{r^2(n-2)}{1-r^2}$$

TABLE 17.4

Summary Results of Simple Linear Regression: Latitude and Average Date of Last Spring Frost

Dependent (response) variable is average date of last spring frost.

Independent (predictor) variable is latitude.

The regression equation is:

average date of last spring frost = −152 + 7.12 (latitude)

Variable	Coefficient	SE (coefficient)	t-ratio	p-value
Intercept	−152.01	17.50	−8.68	0.000
Latitude	7.1171	0.5152	13.81	0.000

SE = 8.41 r^2 = 0.721 (72.1%) (coefficient of determination) r = 0.849 (84.9%) (simple correlation coefficient)

Analysis of Variance:

Source	DF	Sum of squares	Mean square	F-ratio	p-value
Regression model	1	13,502	13,502	191.23	0.000
Residual error	74	5,233	71		
TOTAL	75	18,735			

The F statistic from ANOVA is expressed in terms of the coefficient of determination:

$$F = \frac{r^2(n-2)}{1-r^2} \tag{17.15}$$

This F statistic is the square of the t statistic used in the previous chapter to test the significance of r, the simple correlation coefficient (Equation 16.9). In the last spring frost example the significance testing results in the following:

$$F = \frac{r^2(n-2)}{1-r^2} = \frac{.721(76-2)}{1-.721} = 191.23$$

$$p = 0.000$$

It can be concluded that the explanatory or independent variable (latitude) accounts for a significant (nonzero) amount of the total variation in the response or dependent variable (average date of last spring frost). In conclusion, the summary results of this simple linear regression of latitude as a predictor of last spring frost data are presented in Table 17.4.

17.5 Example: Simple Linear Regression—Lake Effect Snow in Northeastern Ohio

Example: Simple Linear Regression—Lake Effect Snow in Northeastern Ohio

One of the mechanisms influencing precipitation on land areas is the presence of nearby large bodies of water. These water bodies provide an important source of moisture that produces sizable amounts of precipitation when brought over a land area by prevailing winds. However, as distance from the body of water increases, the water's influence decreases, and precipitation levels generally decline. Thus, areas adjacent to water tend to have higher amounts of precipitation than do areas farther away.

An interesting example of this land-water relationship occurs in regions adjacent to the southern and eastern shores of the Great Lakes in New York, Pennsylvania, Ohio, and Indiana. The presence of the Great Lakes and a strong northwesterly wind flow during the winter combine with other

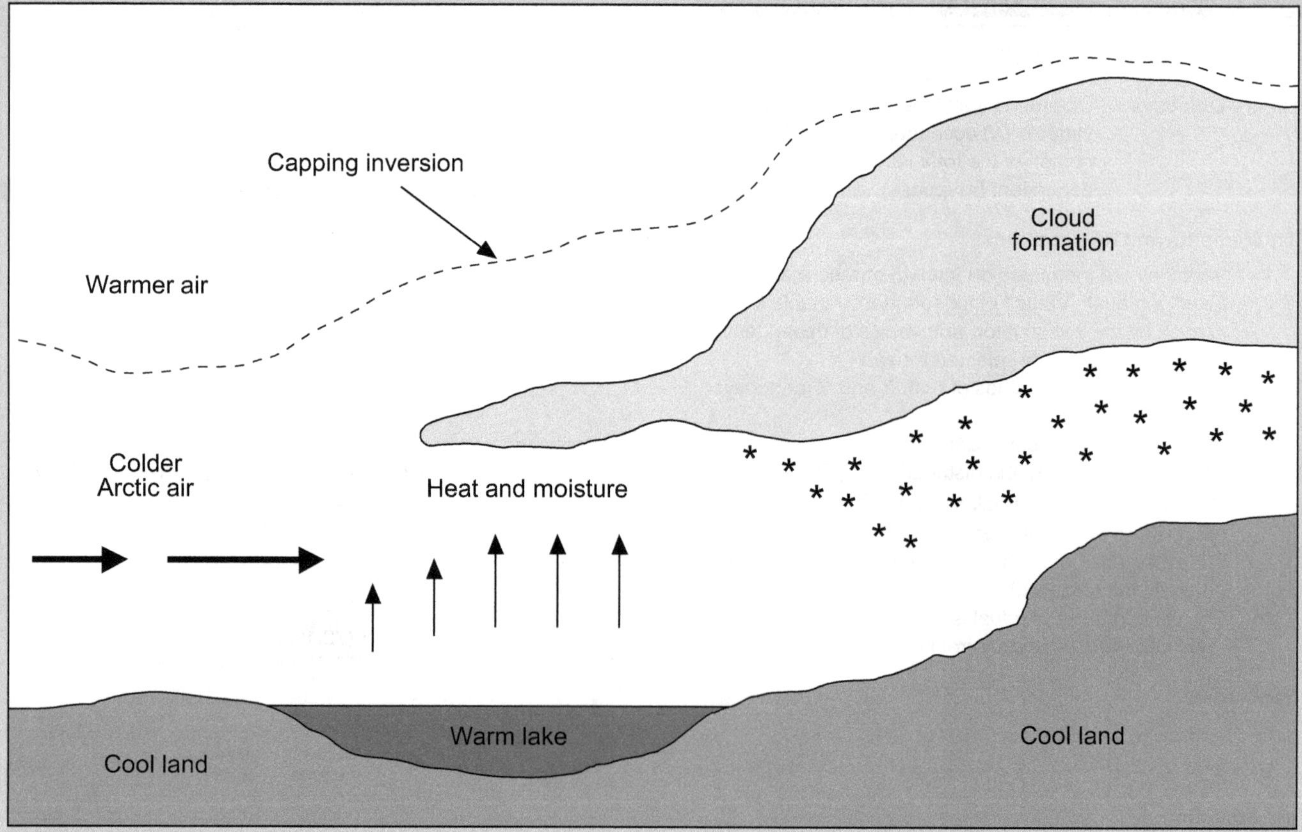

FIGURE 17.7
Formation of Lake Effect Snow

influences to create an important regional climatic phenomenon called "lake effect snow" (Figure 17.7). Using correlation-regression analysis, the spatial pattern of snowfall in extreme northeastern Ohio along the shore of Lake Erie is examined.

Snowfall data from this region seems to indicate that distance from Lake Erie strongly affects the levels of average annual snowfall. To test this relationship, a systematic sample of 38 locations is taken from an isoline map showing average annual snowfall amounts for a set of counties in extreme northeastern Ohio. For each of the 38 sample points, the magnitude of two variables is recorded: (1) average snowfall, interpolated from the map of isohyets, and (2) straight-line distance from Lake Erie. When these data are graphed, the following scatterplot emerges (Figure 17.8). Note that distance from Lake Erie is the independent (predictor) variable (placed on the *X* axis) and average annual snowfall is the dependent (response) variable (placed on the *Y* axis). In general, these two variables seem to indicate a negative correlation—as distance from the Lake increases, average annual snowfall seems to decrease.

However, the overall pattern of scatterplot points is clearly not linear. You can see that average annual snowfall amounts for the sample points closest to the Lake are slightly lower than snowfall levels a bit farther from the Lake. A more complicated curvilinear relationship seems to exist. Perhaps a simple linear regression model should not be applied as we might be violating an important statistical test assumption. Before deciding what to do, we need to take a closer look at the climatological processes which seem to be operating in this lake effect snow region.

During the early winter, the water temperature of Lake Erie is warmer than adjacent land areas. As cold air currents from the northwest pass over this relatively warmer water, the air is heated and can hold more moisture. As the moisture-laden air continues moving onto cooler land surfaces south and east of the Lake, it is cooled, the moisture often condenses, and precipitation occurs. Immediately adjacent to the Lake, the air temperature is often warm enough for the precipitation to fall as rain. Somewhat further from the Lake (about 7–10 miles) the air is frequently cold enough to create snowfall.

Strictly speaking, the straightforward application of Pearson's correlation and simple linear regression may not be the optimal methodology for this complete curvilinear set of data. Some alternate approaches could be used to continue the investigation. One alternative is to apply a nonlinear correlation model matching the curvilinear pattern on the scatterplot. While fully valid, this methodology is beyond the introductory level of this text. Another possibility would be to convert the nonlinear data into linear form by transforming one or both of the coordinate axes and then apply Pearson's correlation to this transformed data. Data transformation has been used effectively in many geographic studies, but also has a level of statistical sophistication beyond the introductory level. Also, a transformation does not seem fully appropriate in this problem setting because the nonlinear portion of

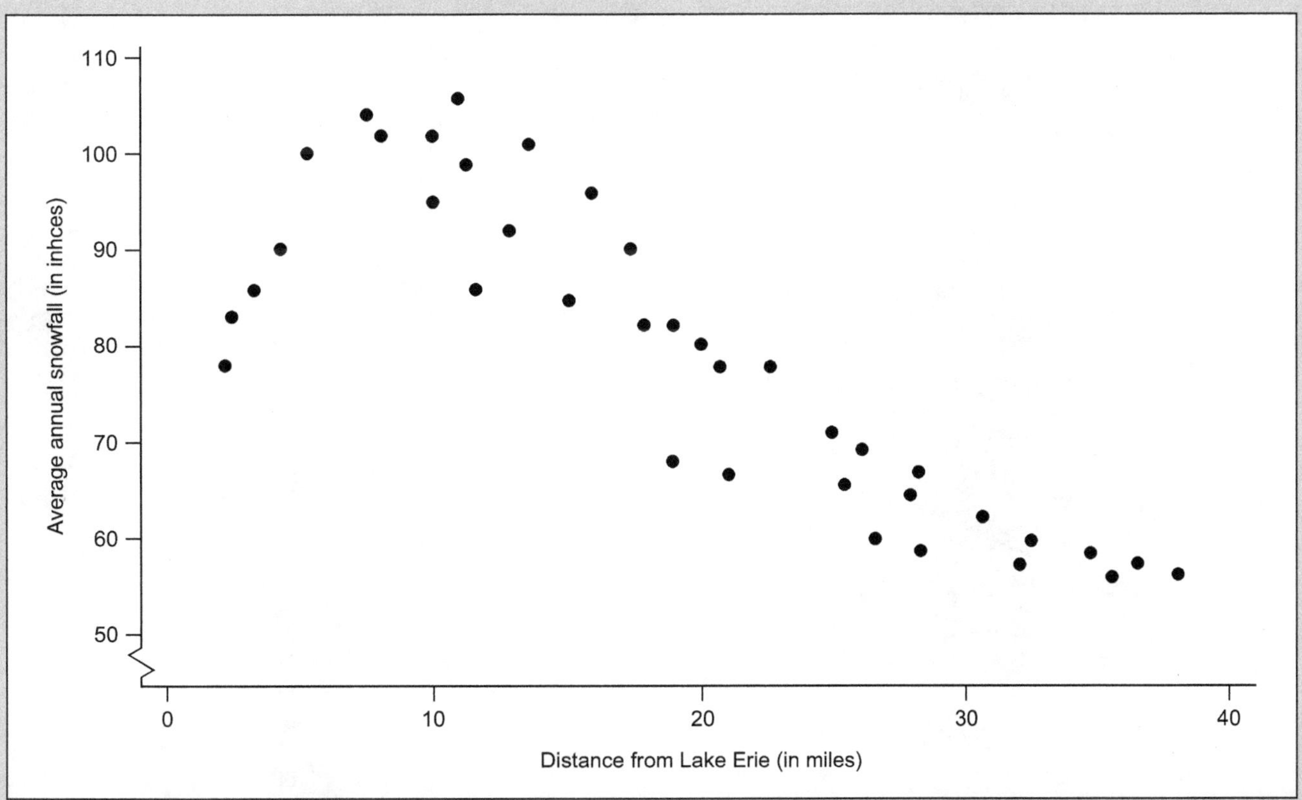

FIGURE 17.8
Original Study Area: Scatterplot Showing Relationship Between Average Annual Snowfall and Distance from Lake Erie

(continued)

the scatterplot is located only in that narrow band of distances from zero to about six miles from Lake Erie.

Yet another alternative is to eliminate that portion of the study area less than about six miles from the Lakeshore. If we select this alternative, the remaining 33 sample points in the revised study region generate a scatterplot that shows a linear relationship between distance from the Lake and amount of snowfall. This strategy allows us to not only meet the linearity assumption of Pearson's correlation and simple linear regression, but also can be justified on a climatological basis, *when the domain is restricted to a range from six to forty miles from the Lake.*

We are now ready to apply correlation-regression analysis to evaluate the relationship of distance from Lake Erie and average annual snowfall. The location of 33 sample points in the revised study area is shown in Figure 17.9, along with the latitude and longitude values (expressed both in degrees-seconds and in degrees-hundredths). Superimposed on this sample point grid is the isohyet pattern of average annual snowfall. Table 17.5 provides all pertinent data for each of the 33 sample points needed to run the correlation and regression analyses. Finally, Figure 17.10 shows the 33 sample points in the revised study region (notice their linear alignment in the domain range of six to forty miles), along with the best-fitting regression line and its formula.

The summary results of the correlation-regression analysis are provided in Table 17.6. The regression equation ($Y = a + bX$) is: [snowfall = 114.6 – 1.719 (distance from Lake Erie)]. The Y-intercept is 114.6 miles (correct for the simple linear regression model but not appropriate for direct model interpretation since zero miles from Lake Erie is beyond the domain of the study area). The regression model slope is –1.719, predicting that for each additional mile of distance from Lake Erie, average annual snowfall will decrease by 1.719 inches.

Both the Y-intercept and the slope are significantly different from zero at all levels of confidence. The t-ratio value for the Y-intercept is 43.26 with an associated p-value of 0.000 and the t-ratio value for the slope is –15.13 with an associated p-value of 0.000.

The standard error of the regression model is 5.80. This indicates that the typical difference between actual average annual snowfall and the predicted average annual snowfall (that is, the typical residual) is 5.80 inches of snowfall. This standard error can be interpreted as how precisely the simple linear regression model predicts average annual snowfall in this region.

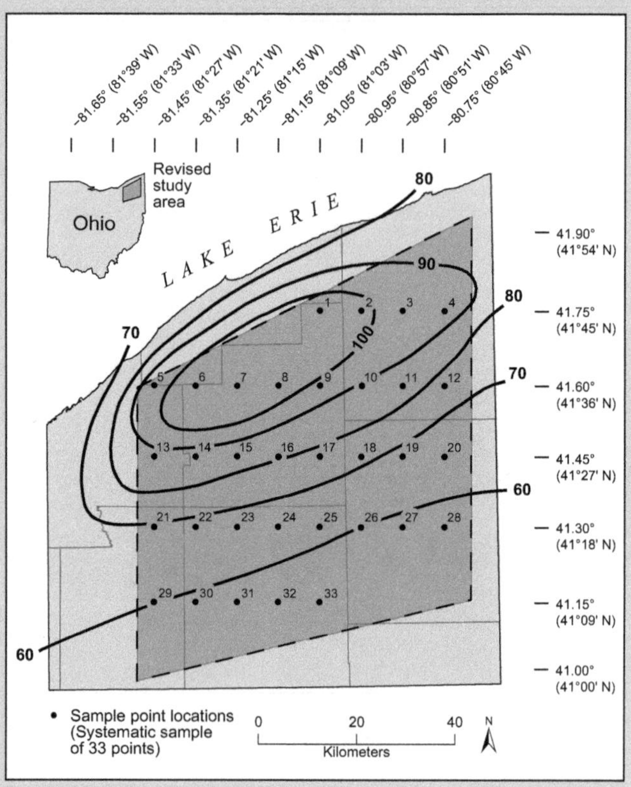

FIGURE 17.9

Revised Northeastern Ohio Study Area: Average Annual Snowfall, in Inches

TABLE 17.5

Revised Study Area: Northeastern Ohio Lake Effect Snow Data

	X	Y		
Systematic Sample of Observations	Distance from Lake Erie (in miles)	Average Annual Snowfall (in inches)	Latitude	Longitude
1	7.7	104	41.75	–81.05
2	9.9	102	41.75	–80.95
3	11.2	99	41.75	–80.85
4	12.9	92	41.75	–80.75
5	9.9	95	41.60	–81.45
6	8.3	102	41.60	–81.35
7	10.9	105	41.60	–81.25
8	13.6	101	41.60	–81.15
9	15.8	96	41.60	–81.05
10	17.1	90	41.60	–80.95
11	18.8	82	41.60	–80.85
12	20.6	78	41.60	–80.75
13	11.4	86	41.45	–81.45
14	14.9	85	41.45	–81.35
15	17.5	82	41.45	–81.25
16	20.1	80	41.45	–81.15
17	22.8	78	41.45	–81.05
18	25.0	71	41.45	–80.95
19	26.3	69	41.45	–80.85
20	28.5	67	41.45	–80.75
21	18.8	68	41.30	–81.45
22	21.0	67	41.30	–81.35
23	25.4	66	41.30	–81.25
24	28.0	64	41.30	–81.15
25	30.7	62	41.30	–81.05
26	32.4	60	41.30	–80.95
27	34.6	59	41.30	–80.85
28	36.4	58	41.30	–80.75
29	26.7	60	41.15	–81.45
30	28.5	59	41.15	–81.35
31	32.0	58	41.15	–81.25
32	35.5	57	41.15	–81.15
33	37.7	57	41.15	–81.05

Source: Kent, R. B. (editor), 1992

TABLE 17.6

Summary Results of Simple Linear Regression: Distance from Lake Erie and Average Annual Snowfall

Dependent (response) variable is average annual snowfall.

Independent (predictor) variable is distance from Lake Erie.

The regression equation is: snowfall = 114.6 − 1.719 (distance from Lake Erie)

Variable	Coefficient	SE (coefficient)	*t*-ratio	*p*-value
Intercept	114.582	2.648	43.26	0.000
Slope	−1.7192	0.1137	−15.13	0.000

SE = 5.80 $r^2 = 0.881$ (88.1%) (coefficient of determination) $r = -0.9385$ (simple correlation coefficient)

Analysis of Variance:

Source	DF	Sum of squares	Mean square	*F*-ratio	*p*-value
Regression model	1	7,699.1	7,699.1	228.82	0.000
Residual error	31	1,043.1	33.6		
TOTAL	32	8,742.2			

The coefficient of determination (r^2) is .881 and the correlation coefficient (r) is −.9386. The strongly negative correlation coefficient and coefficient of determination make it clear that a very high percentage (88.1%) of the variation in average annual snowfall across this study area can be accounted for by distance from Lake Erie.

In the ANOVA portion of Table 17.6 we can see the actual amounts of variance explained and not explained by the regression model. From a total variation of 8742.2, 88.1% (7699.1) is accounted for by the model, while 1043.1 is not explained and is therefore considered residual error. This leads to an extremely large F-ratio value of 228.82, with an associated p-value of 0.000. We can virtually be 100% confident that the simple linear regression model is statistically significant.

Suppose you own an undeveloped piece of property in this northeastern Ohio region, and are planning to build a retirement home at this location. Before proceeding with construction, you want to know how much snowfall to expect each winter. With no official weather station nearby, the best estimate available is from this regression model. If your property is 20 miles (as the crow flies) from Lake Erie, what is the predicted average annual snowfall?

The best-fitting regression line equation is: $Y = 114.582 - 1.7192X$. Since your property is 20 miles from the Lake, $X = 20$. Substituting into the equation gives: $Y = 114.582 - 1.7192(20) = 80.20$. You can expect about 80 inches of snowfall each winter on your property.

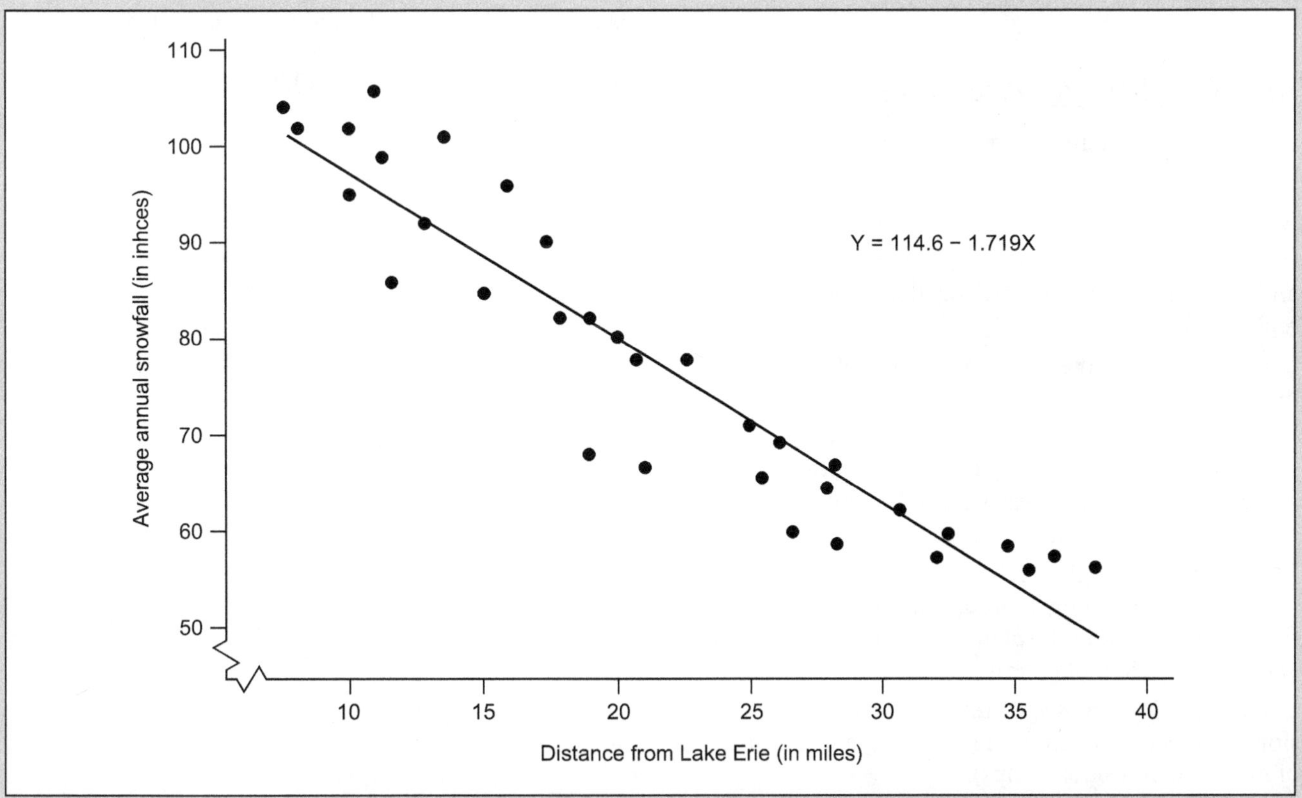

FIGURE 17.10
Best-fitting Regression Line Placed on Scatterplot: Distance from Lake Erie and Average Annual Snowfall

(continued)

One important issue in this geographic research problem is the lack of detailed site-specific snowfall data. This multi-county region has very few official weather stations that are monitoring snowfall. All that is available is a generalized isoline map taken from an atlas. It would certainly be preferable to measure snowfall amounts directly at monitoring stations throughout the study region, but this strategy is not practical for us. Because point-specific data are not available, the precision of the snowfall figures (and hence the quality of the regression model) is only as good as the procedure used to place the isolines on the map in the atlas, which we are not able to assess.

Key Terms

absolute and relative residual, *260*
coefficient of determination, *260*
domain of *X*, *256*
equal variance assumption, *264*
explained and unexplained variation in *Y*, *259*
least-squares regression line, *255*
linear relationship assumption, *264*
normally-distributed variables assumption, *264*
predictor, explanatory (or independent) variable, *255*
residual mapping, *261*
response (or dependent) variable, *255*
simple linear regression (bivariate regression), *255*
slope (regression coefficient), *256*
standard error of the estimate, *261*
standardized residual, *262*
total sum of squares, *259*
total variation in *Y*, *258*
Y-intercept, *255*

Major Goals and Objectives

If you have mastered the material in this chapter, you should now be able to:

1. Explain the nature and purposes of regression analysis.
2. Recognize geographic problems for which regression analysis is a more useful statistical procedure than correlation.
3. Understand how the form or nature of relationship between variables is measured with a least-square regression line.
4. Distinguish between positive (direct), negative (inverse), and no relationship between variables and identify the corresponding slope of regression line.
5. Explain how strength of relationship between variables is measured, including an understanding of total variation, explained and unexplained variation, covariation, and the coefficient of determination.
6. Explain the importance of residual analysis in regression for geographic research (including an explanation of the value of residual maps).
7. List (and briefly explain) the assumptions that must be met when using regression for inferential problem solving.

References and Additional Reading

Abler, R. F., J. S. Adams, and P. R. Gould. *Spatial Organization: The Geographer's View of the World.* Englewood Cliffs, NJ: Prentice Hall, 1971.

Burt, J. E., G. M. Barber and D. L. Rigby. *Elementary Statistics for Geographers.* 3rd ed. New York: Guilford Press, 2009.

Draper, N. R. and H. Smith. *Applied Regression Analysis.* 3rd ed. New York: Wiley, 1998.

Ebdon, D. *Statistics in Geography. A Practical Approach.* 2nd ed. Oxford: Basil Blackwell, 1991.

Ferguson. R. *Linear Regression in Geography.* CATMOG No. 15. Norwich, England: Geo Abstracts, 1976.

Montgomery, D. C., E. A. Peck and G. G. Vining. *Introduction to Linear Regression Analysis.* 6th ed. New York: Wiley, 2021.

The two examples from this chapter deal with last spring frost data and annual snowfall data as related to lake-effect snow. Data for both of these climatic variables are found at the National Centers for Environmental Information (NCEI), www.ncei.noaa.gov. Both of these climatological topics are widely discussed in physical geography and climatology textbooks. The source for lake effect snow in northeastern Ohio is: Kent, R. B., ed. *Region in Transition: An Economic and Social Atlas of Northeastern Ohio.* Akron: The University of Akron Press, 1992.

Part 7

Multivariate Problem-Solving In Geography

18

Examples of Multivariate Problem-Solving In Geography

18.1 The Basics of Multiple Regression

Recall at the start of the text we identified geography as an integrative science that attempts to explain and predict the spatial distribution and variation of human activity and physical features on the earth's surface. The research process in geography entails asking a number of *where, why,* and *what-to-do* questions about spatial patterns, their underlying spatial process, and possible spatial policy and planning strategies that might result. Multiple regression is certainly one of the most popular statistical tools that geographers use to answer these research questions, because most realistic problems in geography are complex and involve multiple causes.

Our primary intentions here are to introduce the basic objectives and principles of multiple regression and demonstrate how geographers develop and apply multiple regression models. Think of this as an initial exposure to multiple regression from a geographer's perspective. We highlight the geographic problem-solving capabilities of multiple regression and emphasize how a well-designed multiple regression model can help better explain (and predict) why a particular spatial pattern exists. Quite simply, how can you use multiple regression effectively to solve geographic problems and make spatial policy decisions?

Please note that we have given less attention to the statistical procedures and technical details of multiple regression itself. We are not suggesting that these procedures and details of multiple regression are unimportant. Rather, we are more concerned that you get a basic feel for how multiple regression can help you make spatial predictions and solve problems. Many intermediate and advanced textbooks provide additional statistical details on multiple regression: some of these are referenced at the end of this chapter.

Fortunately, the basic logic and principles of multiple regression are very similar to those found in simple linear regression. The underlying objective is the same: find a least-squares solution whose coefficients minimize the sum of the squared residuals. That is, find a regression equation that minimizes the amount of unexplained variation. The difference is that we have more coefficients in the multiple regression model. More specifically, the model contains one or more additional predictor (independent) variables, which can account for more variation of the dependent variable. Think back to the bucket and sponge analogy from the last chapter. Multiple regression is equivalent to using additional sponges to absorb a greater volume of water from the bucket.

As in simple linear regression, the selection of independent variables for multiple regression is an important step in applying this statistical method properly. Since use of regression suggests a functional relationship between variables, each independent variable needs to be evaluated carefully to ensure that it shows a logical relationship to the dependent variable.

Multivariate regression is similar in structure to simple linear regression analysis. The statistical technique measures the direction and strength of the functional relationships linking the independent variables to the dependent variable and the remaining error is analyzed. Multiple regression results also provide information on the absolute and relative ability of each predictor variable to explain the dependent (response) variable, while holding the effect of other predictor variables constant.

The assumptions and requirements for multiple regression are very similar to those which apply in simple linear regression. These conditions are discussed in the previous chapter (Section 17.4). The assumptions include: linearity between predictor variables, normality of variables, equal variance of residuals for different values of the predictor variables, independence of errors in the underlying regression model, and random samples drawn from identifiable populations. It is important to check that these various assumptions and conditions are met, at least to a reasonable extent. However, to save space in presenting the following multiple regression examples, many details and various checks of model validity have been completed, but are not shown.

Multicollinearity is a problem that often arises when using multiple regression. This term applies to situations when different predictor or explanatory variables are not independent, but correlated with one another. When multicollinearity is present, the regression coefficients may be both imprecise and inaccurate, and it becomes difficult to assign change in the response (dependent) variable to any one of the predictor variables. It is always a good idea to calculate all correlation coefficients between variables (that is, calculate the correlation matrix) before proceeding with the multiple regression analysis. As a general guideline, whenever the correlation between two predictor variables is above 0.8, you should recognize that collinearity could be an issue. If a particular correlation coefficient is above 0.95, collinearity becomes a serious problem, and one of the two highly correlated predictors should probably be removed from the analysis.

Reviewing quickly, in simple linear regression a best-fitting line is generated to represent the prevailing trend of points on a two-dimensional scatterplot:

$$Y = a + bX \qquad \textbf{(18.1)}$$

In those multivariate problems having two predictor variables, data values are located in three-dimensional space and a best-fitting plane can be derived:

$$Y = a + b_1X_1 + b_2X_2 \qquad \textbf{(18.2)}$$

For problems having three or more independent variables, the same procedure is applied algebraically in multi-dimensional space:

$$Y = a + b_1X_1 + b_2X_2 + \ldots + b_nX_n \qquad \textbf{(18.3)}$$

where: Y = dependent (response) variable
$X_1 \ldots X_n$ = independent (predictor) variables
a = Y-intercept or constant
$b_1 \ldots b_n$ = regression coefficients

In both the simple linear and multivariate equations, the constant or a value shows the value of Y when the values of all X variables are zero. The major difference between the equations lies in the regression coefficients. In the two-variable problem, one regression coefficient or b-value is produced to show the influence of a single independent (predictor) variable on the dependent (response) variable. In the multivariate case, however, a regression coefficient (b_i) is calculated for each independent variable, (X_i). Each coefficient indicates the absolute influence of an independent variable on the dependent variable. Like the simple regression model, each regression coefficient shows the change in Y associated with a unit change in the given X variable.

However, because multiple regression uses more than one predictor variable to explain a response variable, determining the relative importance of each variable is a major concern. At first, it may appear that the regression coefficients could be used directly to measure the importance of independent variables. However, because the magnitude of the measurement units influences these parameters, they cannot be used directly.

This situation is analogous to the problems encountered when using standard deviation to compare relative levels of variation between two or more variables. As discussed in Chapter 3, when variables are measured on different scales, the coefficient of variation provides a better measure for comparing the relative variation between variables. In multiple regression, a valid comparison of the *relative* ability of each independent variable to explain the variation in the dependent variable is accomplished by computing **standardized regression coefficients**. These coefficients serve as relative indices of strength, allowing a direct comparison of the influence of each independent variable in accounting for the variation in the dependent variable. Among the set of independent variables, the one with the largest standardized regression coefficient produces the strongest relationship to the dependent variable.

Multiple Regression Analysis

Primary Objective: Determine if multiple (2 or more) independent (predictor) variables (X_1, X_2, . . ., X_n) account for a significant portion of the total variation in a dependent (response) variable (Y)

Requirements and Assumptions:

1. Variables are all measured on interval or ratio scale
2. All independent (predictor) variables have a linear association with the dependent (response) variable and with one another
3. Samples of all variables taken from normally-distributed populations
4. For all independent (predictor) variables, the variance of residuals is equal for all values of the dependent (response) variable (homoscedasticity)
5. Residual errors are independent of one another in the underlying regression model (no autocorrelation)

Hypotheses:

$H_0 : p^2 = 0$

$H_A : p^2 \neq 0$

In our earlier discussion of simple linear regression, we discussed two different indices that provide absolute and relative measures of strength. Explained variation measures the total level of variation in the dependent variable that is accounted for by the single independent variable. This explained variation is then converted to a relative index called the "coefficient of determination" or r^2.

Similarly, in multivariate regression, a relative index called the **coefficient of multiple determination (R^2)** is calculated:

$$R^2 = \frac{\sum y_e^2}{\sum y^2} \tag{18.4}$$

where: R^2 = coefficient of multiple determination

$\sum y_e^2$ = explained variation

$\sum y^2$ = total variation in Y

This index measures the ratio of variation explained by the set of independent variables relative to the total variation in the dependent variable. When the R^2 index is multiplied by 100, the result is interpreted as the percentage of total variation explained.

Example: Predicting the Date of Last Spring Frost in the Southeastern United States

In the previous chapter we developed a simple regression model using latitude as the only predictor (independent) variable of the average date of last spring frost in the southeastern United States. The model is quite effective. As summarized in Table 17.4, the regression equation is: (predicted average date of last spring frost) = −152 + 7.12 (latitude).

The coefficient of determination (r^2) is .721, indicating that latitude accounts for 72.1% of the total variation in (Y), the average date of last spring frost.

At that time, based on the characteristics of the non-random map pattern of residuals, we suggested that both distance from the coast and elevation seem to be excellent candidates for additional explanatory or predictor variables. What follows is a summary table showing a multiple regression model predicting the average date of last spring frost as a function of all three predictor (independent) variables: latitude, distance from the coast, and elevation (Table 18.1).

The regression equation is:

(predicted average date of last spring frost) =
−115 + 5.77 (latitude) + 0.00415 (elevation)
+ 0.0461 (distance from coast)

This multiple regression model meets the least-squares criterion in the same way as the simple regression model: the sum of the squared residuals is minimized. However, interpretation of the regression coefficients is somewhat different than in simple regression.

- Each additional degree of latitude is associated with about a 5.77 increase in days (later in the spring) for the last spring frost (keeping elevation and distance from coast constant).
- Each additional foot of elevation is associated with about a 0.00415 increase in days (later in the spring) for the last spring frost (keeping latitude and distance from coast constant. By extension, a 1000-foot elevation increase is associated with about a 4.15 increase in days (later in the spring) keeping the other predictor variables constant.
- Each additional mile of distance from the coast is associated with about a 0.0461 increase in days (later in the spring) for the last spring frost (keeping latitude and elevation constant). By extension, a 100-mile increase in distance from coast is associated with about a 4.61 increase in days (later in the spring) keeping the other predictor variables constant.

The coefficient of multiple determination (R^2) is 0.829, indicating that 82.9% of the total variation in average date of last spring frost is explained or accounted for by the three predictor (independent) variables. The adjusted R^2 is 0.822, just slightly less than the unadjusted coefficient. This adjusted formulation takes into account the number of independent variables being used, and can decrease if additional variables are not statistically significant. The adjusted R^2 will increase (over unadjusted R^2) only when the new variable improves the model more than would be expected by chance. Therefore, the adjusted R^2 is the best unbiased estimator of the contribution of a set of independent (predictor) variables to the expla-

(continued)

nation of the dependent (response) variable. For this multiple regression model seeking to explain variation in the average date of last spring frost the results are substantially better than the 72.1% variation explained by latitude alone in the simple regression model.

The residuals (components of the unexplained variation) have a standard deviation of 6.67 days (SE = 6.67). This index (also called standard error of the residuals) provides us with an indication of how precise the prediction of average date of last spring frost can be with this regression model. Stated a bit differently, the typical prediction of last spring frost date is off target by almost a week (6.67 days). The standard error in the simple linear regression was 8.41 (Table 17.4). Using the multiple regression model, we have reduced the standard error in prediction by nearly two days.

The standard errors for each of the regression model coefficients are relatively small compared to the absolute values of the coefficients themselves. For example, the standard error of the intercept is 14.95 while the intercept value is −114.52, the standard error of the latitude slope is 0.4546 while the latitude coefficient is 5.7666, and so on. This suggests that the coefficient estimates are quite precise, and this is reflected by the relatively large *t*-ratio values and the statistically significant *p*-values (all below 0.05).

The *F*-ratio in the ANOVA portion of Table 18.1 is a measurement of the likelihood that all of the regression model coefficients are zero. That is, the *F*-ratio is measuring the statistical significance of the overall regression model itself. If the regression model parameters are zero, then the null hypothesis would not be rejected, and the conclusion would be that there is no relationship between average date of last spring frost and the three predictors. Also, if the null hypothesis was not rejected, the magnitude of the *F*-ratio would be about one. Since *F* is very large (116.48), we can easily reject the null hypothesis (*p*-value = 0.000). Clearly, a statistically significant relationship exists between last spring frost date and all three of the predictor variables.

Do we now have the best multiple regression model? It is impossible to fully answer this question, as there is no single way to define "best." One model might be more suitable for some purposes, while another alternative seems better or more useful for other tasks. Keep in mind that all realistic mod-

TABLE 18.1

Predicting Average Date of Last Spring Frost: Three-predictor Multiple Regression Model

Dependent (response) variable is average date of last spring frost.

Independent (predictor) variables are latitude, elevation, and distance from coast.

The regression equation is:
avg. date of last spring frost = −115 + 5.77 (latitude) + 0.0415 (elevation) + 0.0461 (dist from coast)

Variable	Coefficient	SE (coefficient)	*t*-ratio	*p*-value
Intercept	−114.52	14.95	−7.66	0.000
Latitude	5.7666	0.4546	12.68	0.000
Elevation	0.004145	0.001910	2.17	0.033
Distance from coast	0.04608	0.01450	3.18	0.002

SE = 6.67 R^2 = 0.829 (82.9%) R^2 (adjusted) = 0.822 (82.2%)

Analysis of Variance:

Source	DF	Sum of squares	Mean square	*F*-ratio	*p*-value
Regression model	3	15,530.0	5,167.7	116.48	0.000
Residual error	72	3,199.9	44.4		
TOTAL	75	18,729.9			

Source of regression model	Sequential SS
Latitude	13,496.6
Elevation	1,584.7
Distance to coast	448.7
TOTAL	15,530.1

TABLE 18.2

Predicting Average Date of Last Spring Frost: Two-predictor Multiple Regression Model

Dependent (response) variable is average date of last spring frost.

Independent (predictor) variables are latitude and elevation

The regression equation is:
avg. date of last spring frost = −122 + 6.08 (latitude) + 0.0083 (elevation)

Variable	Coefficient	SE (coefficient)	*t*-ratio	*p*-value
Intercept	−121.66	15.67	−7.76	0.000
Latitude	6.0832	0.4704	12.93	0.000
Elevation	0.008304	0.001475	5.63	0.000

SE = 7.07 R^2 = 0.805 (80.5%) R^2 (adjusted) = 0.800 (80.0%)

Analysis of Variance:

Source	DF	Sum of squares	Mean square	*F*-ratio	*p*-value
Regression model	2	15,081.4	7,540.7	150.87	0.000
Residual error	73	3,648.6	50.0		
TOTAL	75	18,729.9			

Source of regression model	Sequential SS
Latitude	13,496.6
Elevation	1,584.7
TOTAL	15,081.4

els are imperfect (wrong) to some extent, because every regression model has some unexplained variation.

We would like to build a *parsimonious* model that includes as few predictor variables as possible, while also trying not to leave out any independent variables that are potentially important. This is the delicate balance or tradeoff that is placed at the very center of how to construct a good regression model.

The key question in this example seems to be whether to include both elevation and distance to coast in the model or whether to drop one of these independent variables. From the correlation matrix that we ran prior to the regression analysis, we found that all three independent variables correlate significantly ($p = 0.000$) with average date of last spring frost:

(latitude 0.849; elevation 0.599; and distance to coast 0.6430).

A separate correlation analysis shows that elevation and distance to coast are highly correlated (0.737) with each other, creating a potential multicollinearity problem. Since these two predictors are redundant, then one of them should be removed to improve the model's parsimony and efficiency.

Returning to the three-predictor regression model summary (Table 18.1), the bottom portion of the table provides the sequential sum of squares showing the contribution of each independent variable in accounting for variation in the dependent variable. Since the contribution of elevation (1,584.7) is much larger than the contribution of distance to coast (448.7), the results strongly suggest that we should include only latitude and elevation in the revised model (Table 18.2).

The two independent variables account for 80.0% of the total variation in average date of last spring frost. This compares very favorably with the other regression models we have examined:

- Latitude only: $R^2 = 0.721$ (72.1% of total variation explained by model)
- Latitude and elevation: adjusted $R^2 = 0.800$ (80.0% of total variation explained by model)
- Latitude, elevation, and distance from coast: adjusted $R^2 =$ 0.822 (82.2% of total variation explained by model)

If we eliminate distance from the coast from the model, very little of the explained variation is lost (a reduction from 0.822 very slightly down to 0.800), the collinearity problem between elevation and distance from the coast is eliminated, and the model becomes more parsimonious and efficient. In short, this model looks like the best alternative.

Example: Predicting Obesity Levels for Counties of the U.S.

Back in Chapter 11, we examined the spatial pattern of county obesity levels using analysis of variance. Selecting a stratified random sample of two hundred counties, proportionally represented across the four major census regions of the United States (excluding Alaska and Hawaii), we discovered highly significant differences in obesity rates between census regions ($F = 22.54$ and $p = 0.000$).

Using this same sample of two hundred counties, we now present a multiple regression model that can account for the spatial variation in these county obesity rates. Then, because the sample contains a randomization component, we will be able to test the regression coefficients inferentially and use the model to predict the obesity level in any county in the conterminous U.S. with a known degree of precision.

From several different sources, we identified a large number of potential independent (predictor) variables that have been associated with obesity in past research. From that set of possible predictors, we selected seven variables that seem to provide a representative mix for input into an initial exploratory regression model. For the sample of 200 counties, we calculated the correlation matrix of these seven predictors (Table 18.3).

Notice that all seven predictor variables are significantly correlated with obesity level (see the first column of Table 18.3). However, many of these predictors are significantly correlated themselves, suggesting that multicollinearity will likely have a substantial effect on the regression model.

The summary results of the multiple regression model are shown in Table 18.4. The combined set of seven independent variables accounts for 65.0% of the total variation in obesity. The model is moderately successful in predicting obesity rates, but there seems to be some potential for reducing the number of independent variables without sacrificing too much explained variation. Two important related questions are: which variables should be removed and in what order? If we look at the bottom segment of Table 18.4, we get some excellent guidance on how to proceed. The "source of regression model—sequential sum of squares" list tells how much each of the seven variables contributes to the total amount of explained variance in the model. The comparative importance of the variables is quite clear: (1) by a large margin, the percentage of county adult population that is physically inactive is the largest "explainer" of variance in obesity, with a sum of squares of 1,341.03 (out of a total of 1,658.62); (2) a strong second is percentage of county population aged 20 years or older having diabetes (205.32 out of 1,658.62; (3) a clear third place goes to percentage of county population of Hispanic origin (95.03 out of 1,658.62); (4) the other four predictor variables all contribute very small amounts to the total explained variation in the model.

If we were to show the entire multiple regression process, we would remove or eliminate only one predictor at a time from the model. This is because each predictor affects how all the others contribute to the model. This can be accomplished easily by having the computer software conduct what is known as a **stepwise regression,** which (in this case) removes one independent variable at a time from the model.

(continued)

For brevity, we don't present these six successive steps of variable removal. However, after running stepwise regression, the "best" model that resulted contains the three predictors just mentioned (Table 18.5).

This looks like a very good model. Even though four of the seven predictors have been eliminated, 63.4% of the total variation in county obesity rate is still explained. In addition, very little explained variation is lost while the complexity of the model is sharply reduced, allowing us to focus attention on those few factors that seem most related to obesity.

As a closing comment, perhaps you noticed something unusual in this analysis: the percentage of county population that is Hispanic *negatively correlates* with obesity, although the absolute magnitude of the correlation is smaller than any other predictor. This contradicts most research that looks at factors related to obesity, which often cites higher than national average obesity rates for Hispanics. In addition, why is percentage Hispanic staying in the regression model, when its correlation to obesity is so low?

The multiple regression output (not shown in Table 18.5) provides a list entitled "unusual observations." The computer software results identify seven out of the nearly 200 counties that have "**large leverage.**" "Leverage" is a measure of an observation's ability to alter the regression model with just a slight change in the magnitude of the dependent (response) variable—in this case, percentage of adults classified as obese. In other words, observations (counties) with large leverage can have an inordinate amount of influence on the

TABLE 18.3

Matrix of Correlation Coefficients: Factors Used in County Obesity Regression Model

	%Obese08	% PhIn08	%Diab08	%NC1MI	%LI1MI	PovRate08	%Black08
%PhIn08	0.732 0.000	—	—	—	—	—	—
%Diab08	0.744 0.000	0.765 0.000	—	—	—	—	—
%NC1MI	0.526 0.000	0.582 0.000	0.696 0.000	—	—	—	—
%LI1MI	0.485 0.000	0.598 0.000	0.616 0.000	0.813 0.000	—	—	—
%PovRate08	0.512 0.000	0.561 0.000	0.590 0.000	0.655 0.000	0.663 0.000	—	—
%Black08	0.397 0.000	0.333 0.000	0.458 0.000	0.379 0.000	0.099 0.171	0.416 0.000	—
%Hisp08	−0.367 0.000	−0.259 0.000	−0.280 0.000	−0.153 0.033	−0.067 0.356	0.199 0.006	−0.112 0.120

<u>**Cell contents**</u>: Pearson's correlation
p-value

%Hisp08	Percentage of county resident population that is of Hispanic origin, 2008.
%PhIn08	Percentage of county adult population that reported no leisure time physical activity, 2008. [physical inactivity]
%Diab08	Percentage of county population aged 20 years or older having diabetes, 2008.
%NC1MI	Percentage of housing units in a county that have no car and are more than one mile from a supermarket or large grocery store. [no car, 1 mile plus]
%LI1MI	Percentage of county total population that is low income and lives more than one mile from a supermarket or large grocery store. [low income, 1 mile plus]
PovRate08	Percentage of county residents with household income below the poverty threshold, 2008.
%Black08	Percentage of county resident population that is non-Hispanic, black, or African American origin, 2008.
%Obese08	Percentage of age-adjusted persons aged 20 years or older classified as obese, where obese is defined as a Body Mass Index (BMI) ≥ 30, 2008. (see written narrative in chapter 1 for more details)

overall regression model result, if the value of their dependent variable is even slightly altered. Six of the seven counties with very high leverage are also counties with a very high percentage of the total population being Hispanic: Zavala, TX (93.9%); Dimmit, TX (86.2%); Santa Cruz, AZ (82.8%); El Paso, TX (82.2%); Valencia, NM (58.3); and Taos, NM (55.8%). In short, these few "Hispanic-majority" counties have a great deal of influence on results of the overall regression model.

TABLE 18.4

Predicting County Obesity Levels: Seven-predictor Multiple Regression Model

Dependent (response) variable is percent of adult county population classified as obese.

Independent (predictor) variables are %PhIn08, %Diab08, %NC1MI, %LI1MI, PovRate08, %Black08, and %Hisp08.

The regression equation is:

%Obese08 = 16.2 + 0.215 (%PhIn08) + 0.573 (%Diab08) − 0.123 (%NC1MI) − 0.0061 (%LI1MI) + 0.120 (PovRate08) + 0.0113 (%Black08) − 0.0599 (%Hisp08)

Variable	Coefficient	SE (coefficient)	*t*-ratio	*p*-value
Intercept	16.191	1.005	16.10	0.000
%PhIn08	0.21521	0.04679	4.60	0.000
%Diab08	0.5729	0.1295	4.42	0.000
%NC1MI	−0.1235	0.1173	−1.05	0.294
%LI1MI	−0.00612	0.02808	−0.22	0.828
PovRate08	0.11982	0.04024	2.98	0.003
%Black08	0.01132	0.01473	0.77	0.443
%Hisp08	−0.05994	0.01314	−4.56	0.000

SE = 2.14 R^2 = 0.663 (66.3%) R^2 (adjusted) = 0.650 (65.0%)

Analysis of Variance:

Source	DF	Sum of squares	Mean square	*F*-ratio	*p*-value
Regression model	7	1,658.62	236.95	51.90	0.000
Residual error	185	844.62	4.57		
TOTAL	192	2,503.24			

Source of regression model	Sequential SS
%PhIn08	1,341.03
%Diab08	205.32
%NC1MI	0.67
%LI1MI	1.69
PovRate08	7.95
%Black08	6.93
%Hisp08	95.03
TOTAL	1,658.62

TABLE 18.5

Predicting County Obesity Levels: Three-predictor Multiple Regression Model

Dependent (response) variable is percentage of adult county population classified as obese.

Independent (predictor) variables are %PhIn08, %Diab08, and %Hisp08

The regression equation is:

%Obese08 = 15.6 + 0.248 (%PhIn08) + 0.673 (Diab08) − 0.0386 (%Hisp08)

Variable	Coefficient	SE (coefficient)	*t*-ratio	*p*-value
Intercept	15.6350	0.8843	17.68	0.000
%PhIn08	0.24777	0.04495	5.51	0.000
%Diab08	0.6727	0.1110	6.06	0.000
%Hisp08	−0.03860	0.01142	−3.38	0.001

SE = 2.185 R^2 = 0.640 (64.0%) R^2 (adjusted) = 0.634 (63.4%)

Analysis of Variance:

Source	DF	Sum of squares	Mean square	*F*-ratio	*p*-value
Regression model	3	1,600.94	533.65	111.78	0.000
Residual error	189	902.29	4.77		
TOTAL	192	2,503.24			

Source of regression model	Sequential SS
%PhIn08	1,341.03
%Diab08	205.32
%Hisp08	54.59
TOTAL	2,503.24

Example: Predicting Life Expectancy for Countries of the World

One of the four illustrations of the geographic research process introduced in Chapter 1 dealt with life expectancy at the country and world region levels. After looking at the global pattern of life expectancy (Figure 1.4), a series of descriptive statements and inferential hypotheses were suggested for possible future analysis. In addition, a possible "life expectancy transition" was discussed. Of course, many different factors are directly or indirectly related to life expectancy. In *World Health Statistics, 2010,* the World Health Organization lists well over 100 "Global Health Indicators" organized into nine major categories.

From this complexity, we select different combinations of possible independent (predictor) variables as candidates for a series of multiple regression models. Our exploratory analysis results in a variety of different models. A full discussion of these models is well beyond the scope of this text, but we briefly present some model results that influence the exploratory research process.

One of the five-predictor models does quite well in accounting for variability in life expectancy, with a coefficient of determination (R^2) of 0.833 (83.3%). In this model, three independent variables focus on health risk and causes of morbidity (percentage of population using improved drinking water sources, percentage of population using improved sanitation, and percentage prevalence of HIV among adults aged 15–49). The other two independent variables represent measures of health expenditure (total expenditure on health as a percentage of gross domestic product and per capita total expenditure on health in international dollars). Of the 119 countries with complete data (many less-developed countries did not have HIV estimates), the analysis identifies only 9 countries as "unusual observations" having either a very high absolute residual or very high leverage. Eight of these countries are in Africa, and the ninth country is the United States. The eight African countries all have high health-risk numbers (combinations of bad water, poor sanitation, or HIV). The United States has very great leverage in this model because of the extremely high per capita health-care expenditures (look again at Figure 1.5).

With this model result, we decide to try another model containing only the three risk and morbidity factors. Our thinking is that countries in Africa seem to be rather distinctive when compared to the rest of the world. This judgment is confirmed, as a three-predictor model focusing just on risk and morbidity factors accounts for 79.9% of the total variation in life expectancy globally (Table 18.6). As health standards like clean water, proper sanitation, and control of HIV improve in African countries, global life expectancies will likely converge dramatically, and other predictive regression models for life expectancy will no doubt emerge.

At the beginning of this chapter we reference the "what to do" questions that might inform policy decisions. The regression results in Table 18.6 indicate that access to clean water (GoodWater%) is the largest contributor (SS = 9,052.3) to the overall model. Additionally, the coefficient for GoodWater% indicates that a nearly 4% increase nationwide in access to clean water accounts for an additional nationwide year of life expectancy. Similarly, a 1% reduction in HIV% cases results in an additional year of life expectancy. With this information in mind, policy makers can evaluate different strategies for improving life expectancy, such as whether to invest in HIV education or preventive care.

TABLE 18.6

Predicting Country Life Expectancy: Three-predictor Multiple Regression Model

Dependent (response) variable is life expectancy in years.

Independent (predictor) variables are GoodWater%, GoodSanit%, and HIV%.

The regression equation is:

$$\text{LifeExp} = 37.0 + 0.275\,(\text{GoodWater\%}) + 0.127\,(\text{GoodSanit\%}) - 0.887\,(\text{HIV\%})$$

Variable	Coefficient	SE (coefficient)	*t*-ratio	*p*-value
Intercept	36.995	2.771	13.35	0.000
GoodWater%	0.27486	0.04636	5.93	0.000
GoodSanit%	0.12654	0.02631	4.81	0.000
HIV%	−0.8868	0.1079	−8.22	0.000

$SE = 5.007$ $R^2 = 0.804$ (80.4%) R^2 (adjusted) = 0.799 (79.9%)

Analysis of Variance:

Source	DF	Sum of squares	Mean square	*F*-ratio	*p*-value
Regression model	3	11,823.5	3,941.2	157.22	0.000
Residual error	115	2,882.8	25.1		
TOTAL	118	14,706.4			

Source of regression model	Sequential SS
GoodWater%	9,052.3
GoodSanit%	1,078.0
HIV%	1,693.2
TOTAL	11,823.5

18.2 The Basics of Cluster Analysis

The term **cluster analysis** refers to a set of multivariate methods that group or classify a set of observations so that similar observations are placed in the same group and dissimilar observations are placed in different groups. Synonymous with *numerical taxonomy* or *typological analysis,* clustering is a common technique of statistical data analysis in many different fields.

Cluster analysis is not one specific algorithm or procedure, but rather a general goal to be accomplished (assign similar observations to the same group while assigning dissimilar observations to different groups). Procedures vary depending on what constitutes a cluster and how best to identify those clusters. The possibilities are extensive: well over 100 different "clustering models" or algorithms are in active use. Some of the most basic considerations in cluster analysis include:

- Approach—the underlying cluster model approach can be either agglomerative or divisive. This parallels the two conceptual strategies of classification (agglomeration and subdivision) discussed in Chapter 2.
- Level of measurement—different cluster analysis procedures can be used on interval/ratio, ordinal, and nominal (categorical) data.
- Measure of distance—options include Euclidean, squared Euclidean, Manhattan, Pearson's, and squared Pearson's. Euclidean and Manhattan distances have already been defined. Pearson's methods calculate the square root of the sum of square distances divided by variances. This method is used when observations need standardizing.
- Linkage method—linkage alternatives include single, average, median, centroid, complete, Ward's, and McQuitty's. The linkage method selected determines how distance between clusters is defined. Basically, a linkage rule is needed when calculating inter-cluster distances for multiple observations in a cluster.

In geographic applications, the "observations" being allocated to clusters are often locations (countries, states, provinces, etc.) for which there are data on multiple variables. Very often our task is to reduce the complexity of the data through creation of multi-variable clusters of observations. Hence, cluster analysis comprises a set of data-reduction techniques. Just as factor analysis (another important multivariate statistical procedure) combines a large number of individual variables into a smaller number of "variable-groups," cluster analysis reduces a large number of locations to a smaller number of "location-groups" or "location-types."

Since cluster analysis is a complex multivariate technique, it is useful to summarize the basic goal and procedures of this statistical method in written narrative form, without using any specific formulas or statistical formulations. Following the explanation, we will look at some geographic examples. We start with a set of locations (for example, states of the United States or counties in Alabama), for which we have multiple variables of data (e.g., measures of health such as level of obesity, level of diabetes, and level of physical inactivity). Our basic goal is to group (cluster) together those locations that are most similar (that is, close to one another in multi-dimensional space: with regard to level of obesity, level of diabetes, *and* level of physical inactivity) into the same category, while placing locations that are most different (dissimilar) from one another into different categories. Furthermore, the process of grouping or clustering locations *considers all variables simultaneously* (making this a "multivariate" technique).

Let us suppose that we want to group the states of the United States into five categories or classes based on three health indicators (obesity, diabetes, and physical inactivity). With three variables, we are working in three-dimensional space, with each dimension representing a variable. Using this technique, states will be combined with other states (or groups of states) having similar values of the three health indicators so that ultimately, we can generate a choropleth map that shows each state allocated to one of these five categories of "health level." This will allow us to see the overall spatial pattern of the most healthy and least healthy states. Let us further suppose we want to use an agglomerative approach to clustering, starting with each state in its own separate group, and iteratively (step-by-step) group states that are "closest together" (most similar) in this three-dimensional space. We can illustrate this iterative procedure graphically by using a dendrogram and specifying a "cut-off" line to allocate states to their proper category (cluster).

The cluster analysis procedure also calculates some useful summary descriptive statistics. For each cluster, we can determine its mean value on each variable and calculate a **cluster centroid** that indicates the center of the cluster in three-dimensional space. This allows us to graph the centroids in three-dimensional space and see the distance separating cluster centroids. In effect, the closer the centroids in the three-dimensional space, the more similar the clusters (groups) are based on the variables used for the clustering.

Cluster analysis is always descriptive in nature; no inferential hypotheses are being tested. In other words, we are not trying to infer population parameters from sample statistics. In addition, we are not attempting here to provide a comprehensive coverage of clustering procedures. Our main objective is to illustrate how cluster analysis is useful in the investigative exploration of spatial data sets. Those interested in studying these procedures in more detail should examine some of the chapter references.

Example: Grouping African Countries Based on Factors Influencing Life Expectancy

As we developed multiple regression models related to the global distribution and variation of life expectancy, the distinctiveness of African countries quickly emerged. We discovered that access to safe drinking water, access to improved sanitation, and HIV prevalence among adults were all important life expectancy predictors. We further discovered that nearly all of the "unusual observations" (that is, large residuals and observations with lots of leverage) were countries in Africa.

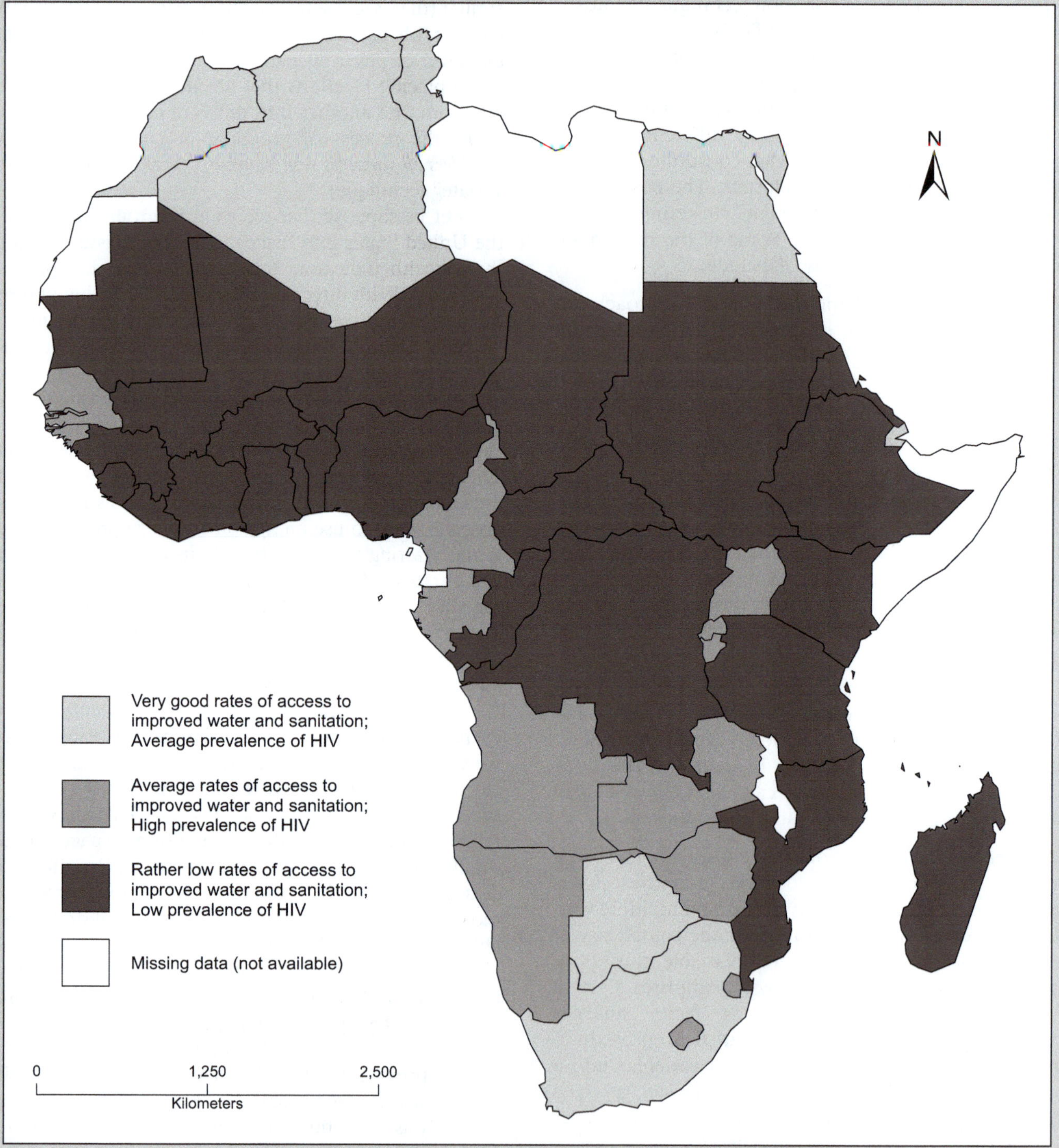

FIGURE 18.1
Cluster Analysis Map of Africa: Based on Three Predictors of Life Expectancy—Access to Good Water and Sanitation and Prevalence of HIV
Source: World Health Organization (WHO), World Health Statistics, 2010

Therefore, a logical next step in the exploratory geographic research process is to look at the spatial distribution and variation of African countries themselves with regard to these three independent variables. Since we want to group or classify observations over multiple variables simultaneously, a cluster analysis procedure is appropriate.

We selected the three-cluster solution, since it seems to have a regional pattern that is suitable for interpretation (Figure 18.1). Altogether, 46 African countries have complete data across all three variables. Most of the countries missing data lack information about HIV prevalence.

One group of countries (shown in the lightest shade of gray) has residents with the highest rates of access to safe drinking water and improved sanitation. In this group, the cluster centroid for safe drinking water is 91.8% of the population and the cluster centroid for access to improved sanitation is 77.1% (see Table 18.7). HIV prevalence is about average for the continent at 5.3%. Most of the countries in this group are on the periphery of the African continent. Key countries include: South Africa and Botswana in Southern Africa; and Egypt, Tunisia, Algeria, and Morocco in Northern (Mediterranean) Africa.

A second group of countries (medium shade of gray) is most notable for having a high HIV prevalence rate (above 10% on average). A number of countries in south central and central Africa are members of this group: (Zimbabwe [Rhodesia], Zambia, Malawi, Burundi, Rwanda, and Uganda). In addition, several countries along the West African coast are in this cluster: Senegal, Cameroon, Gabon, Angola, and Namibia. These countries generally have average levels of access to safe drinking water and improved sanitation.

The third (and largest) group of countries (darkest shade of gray) is located in a wide swath across Central Africa, from the Atlantic Ocean to Indian Ocean coast. Generally, HIV prevalence rates are low (average of 3% of the resident population), but access rates to safe drinking water and good sanitation are poor (even by African standards).

Why do these key factors related to life expectancy vary in the way they do across the African continent? More specifically, why do we see the non-random clustering pattern that is so evident in Figure 18.1? Obviously, the spatial processes underlying these spatial patterns are very complex. Multiple interrelated factors are influencing each of these three life expectancy predictors. You might want to speculate further on additional predictor variables (maybe an income- or education-related measure, for example) that could be included in a "follow-up" exploratory or investigative geographic analysis.

TABLE 18.7

Cluster Analysis of African Countries: Based on Three Predictors of Life Expectancy—GoodWater%, GoodSanit%, and HIV%

Cluster Centroids:

Variable	Cluster 1 (9 countries)	Cluster 2 (13 countries)	Cluster 3 (24 countries)	Grand Centroid (46 countries)
GoodWater%	91.7778	73.2308	59.3333	69.6087
GoodSanit%	77.1111	46.3077	20.1250	38.6739
HIV%	5.3444	10.1000	3.0292	5.4804

Distance between Cluster Centroids	
Cluster 1 and Cluster 2	36.2692
Cluster 1 and Cluster 3	65.6157
Cluster 2 and Cluster 3	30.4741

Example: Grouping States using Basic Health Indicators

Without a doubt, the most frequently examined topic in this book has been the spatial pattern of obesity and associated distributions of other health-related variables such as diabetes and physical inactivity across the United States. Another instructive way to explore obesity and its related variables is to examine the degree to which states cluster spatially when considering all three health-related variables simultaneously. At the county level, for example, we have constructed two crosstab maps and identified distinctive regional patterns of "very healthy" and "very unhealthy" counties, with definitions based on combined levels of obesity, diabetes, and physical inactivity (see Figure 18.2). Counties highlighted on these maps are either in the bottom 10% (or top 10%) of all counties in the United States as measured on all three of these variables.

Using the same three variables (levels of obesity, diabetes, and physical inactivity), we can also examine the "geography of health" pattern at the state level by applying a multivariate cluster analysis. Should we expect the cluster analysis to produce a strong regional patterning of "healthy" and

(continued)

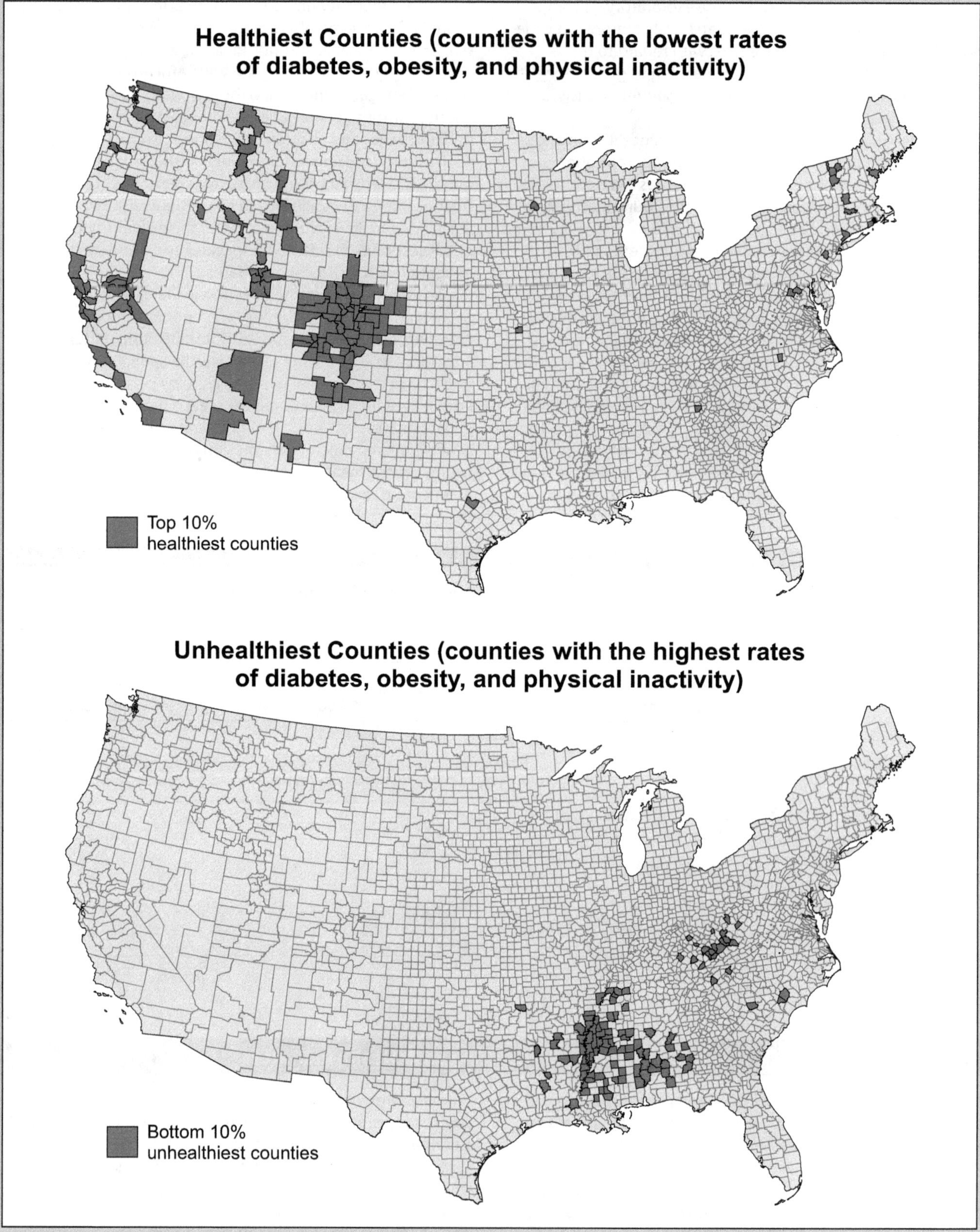

FIGURE 18.2
County Crosstab Map: Obesity, Diabetes, and Physical Inactivity
Source: Centers for Disease Control and Prevention (CDC)

"unhealthy" states? To what extent will the state-level pattern of "health" derived from cluster analysis replicate the county-level pattern of "health" derived from combining the top 10% or bottom 10% of counties on these same three variables? How close or distant are the cluster centroids of the selected model, and what additional information is provided by comparing centroid coordinates?

The multivariate cluster analysis process itself is effectively illustrated with a dendrogram (Figure 18.3) and a summary table (Table 18.8). In this analysis, an "average linkage" method and a "squared-Euclidean" distance measure are used. "Average linkage" defines the distance between two clusters as the mean distance between an observation in one cluster and an observation in the other cluster. "Squared-Euclidean" is simply the square of Euclidean distance. Therefore, distances that are large under Euclidean distance will be considerably larger under the squared-Euclidean measure. Many sources recommend that when using an average linkage method, you should also use a squared-distance measure.

Certain states combine relatively early in the clustering process while others, of course, do not. For example, in this three-dimensional space (obesity, diabetes, and physical inactivity), the states of Iowa and Nebraska are close together; they are more similar in values of health indices and group or cluster at a distance of about 0.3 units in three-dimensional space (see the bottom center of Figure 18.3). Other states do not join a cluster until later in the process. For instance, Georgia does not join a group until a distance of about 1.3 units in

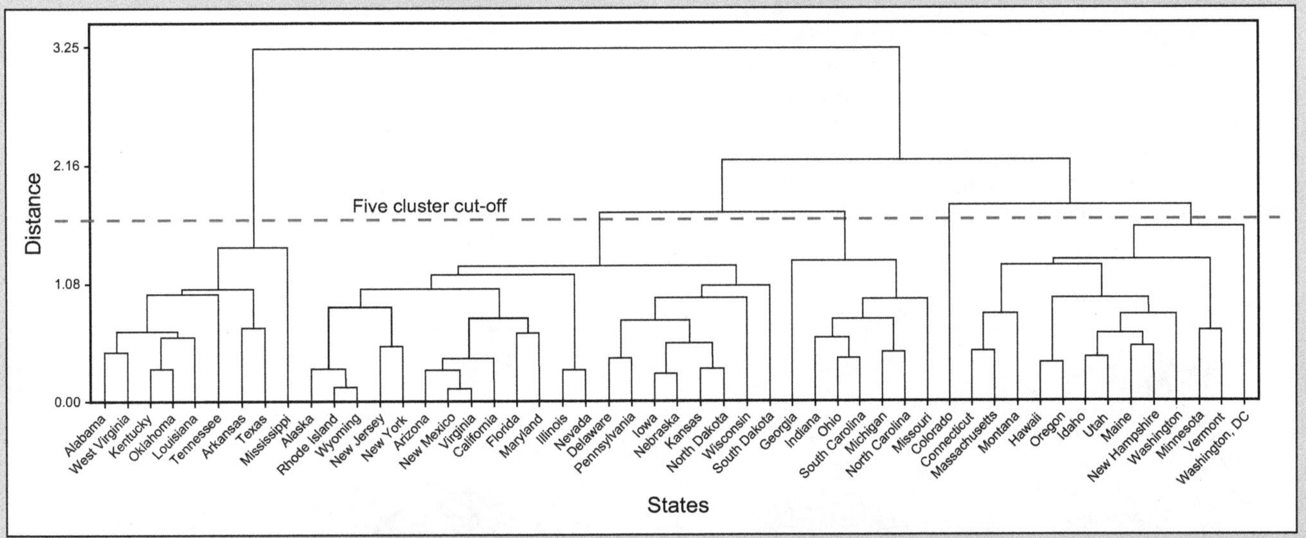

FIGURE 18.3
Dendrogram of State "Health": Based on Cluster Analysis of Levels of Obesity, Diabetes, and Physical Inactivity

TABLE 18.8

Cluster Analysis of U.S. States: Based on Three Predictors—Obesity%, Diabetes%, and PhysicallyInactive%, 2008

Cluster Centroids:

Variable	Cluster 1 (less healthy)	Cluster 2 (average health)	Cluster 3 (very healthy)	Cluster 4 (more healthy than most)	Cluster 5 (less healthy than most)	Grand Centroid
Obesity%	1.411	−0.141	−2.321	−0.953	0.710	0.000
Diabetes%	1.539	−0.326	−1.679	−0.827	0.776	0.000
PhysicallyInactive%	1.493	0.060	−1.605	−1.170	0.302	0.000

Distance between Cluster Centroids	
Cluster 1 and Cluster 2	2.818
Cluster 1 and Cluster 3	5.821
Cluster 1 and Cluster4	4.275
Cluster 1 and Cluster 5	1.579
Cluster 2 and Cluster 3	3.059

Distance between Cluster Centroids	
Cluster 2 and Cluster4	1.556
Cluster 2 and Cluster5	1.413
Cluster 3 and Cluster4	1.670
Cluster 3 and Cluster 5	4.342
Cluster 4 and Cluster 5	2.738

(continued)

three-dimensional space (see Georgia near the center of Figure 18.3). Recall that distance in the three-dimensional space reflects the degree of similarity between states defined by the three health indicator. Shorter distances imply more similarity.

After the entire clustering process is completed, the question becomes: "Where to cut the dendrogram?" We explored cutting the dendrogram at various places (this is a subjective decision) and settled on a five-cluster cut-off (Figure 18.3). This choice was selected because the map of this five-cluster option shows a strong regional patterning and also highlights the Colorado outlier (Figure 18.4). Notice how the dendrogram cut-off line divides the states into clusters that can then be directly placed on the map.

The "least healthy" region is a swath of states ranging from Texas and Oklahoma in the southwest, through the states of the lower Mississippi River valley, then northeast into the Appalachian states of Tennessee, Kentucky, and West Virginia (Cluster 1 in Table 18.8). We know this is the least healthy region because the cluster centroid numbers associated with Cluster 1 are all large positive numbers (indicating high percentages of obesity, diabetes, and physical inactivity). At the other extreme, proudly isolated as a statistical outlier, is the "very healthy" state of Colorado (Cluster 3 in the table).

All of the "more healthy than most" states are in the northern portion of the country, ranging from the Pacific Northwest (Washington, Oregon, Idaho), through Minnesota, then jumping to northern New England (Massachusetts, Vermont, New Hampshire, Maine) (Cluster 4 in Table 18.8). Most of the "less healthy than most" states are along the south Atlantic coast (Georgia, South Carolina, North Carolina) or in the traditionally industrial Midwest (Ohio, Michigan, Indiana) (Cluster 5). The "large middle group" of average health states is widely dispersed nationwide (Cluster 2).

We encourage you to explore these "health" patterns further. What other variables do you hypothesize might be spatially associated with these map patterns? Are some of the descriptive statements and inferential hypotheses mentioned in the obesity example of Chapter 1 still relevant? Once again we see that conducting statistical analyses leads to further research questions—and further statistical analysis!

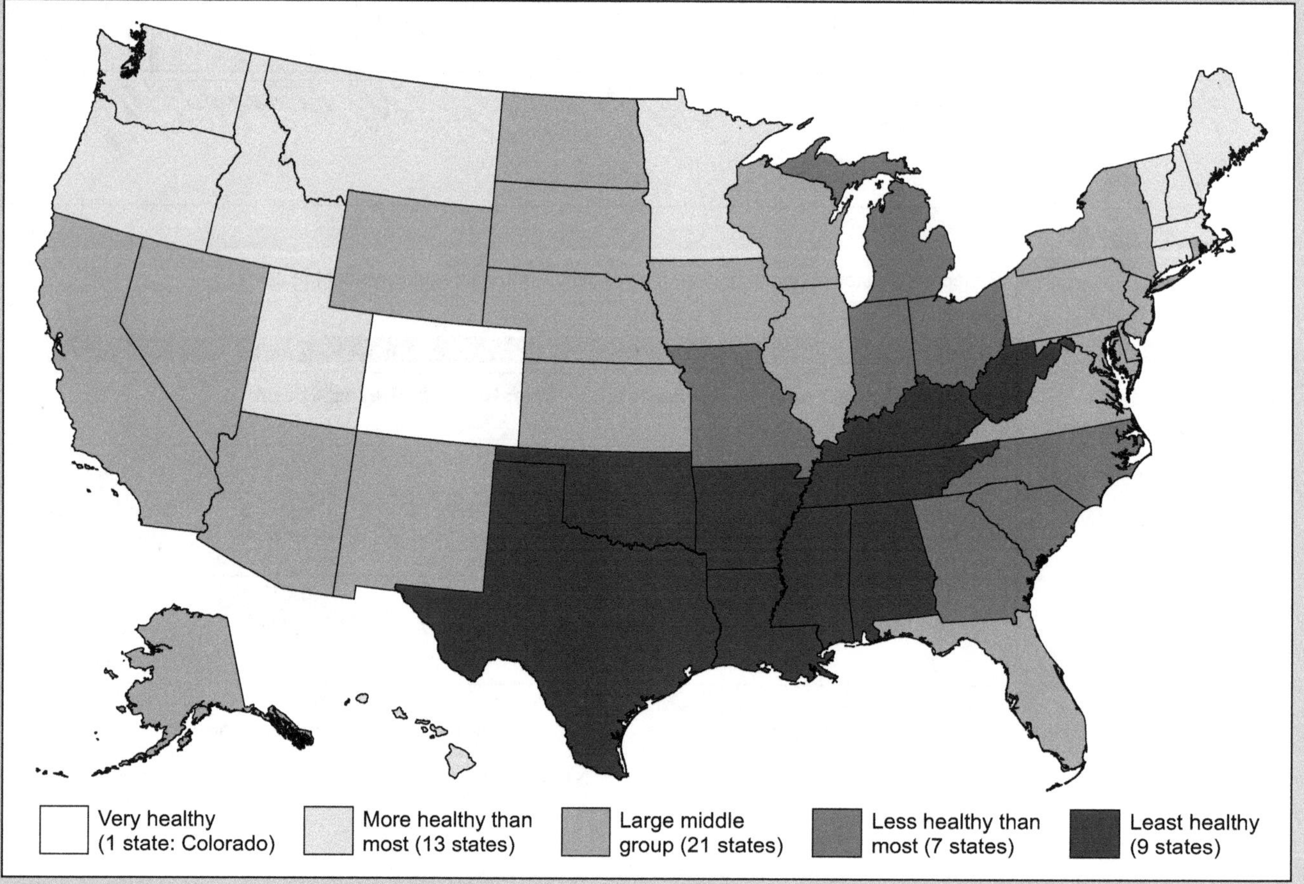

FIGURE 18.4
Cluster Analysis Map of the United States: Based on Three Measures of "Health"—Levels of Obesity, Diabetes, and Physical Inactivity

Example: Grouping Alabama Counties using Basic Health Indicators

We have seen that many of the very unhealthy counties in the United States are in Alabama (Figure 18.2). In fact, Alabama has 17 of the 103 counties meeting the criteria of being in the highest 10% (that is, the least healthy 10%) of all U.S. counties with regard to levels of obesity, diabetes, and physical inactivity. Furthermore, Alabama is one of the nine least healthy states as determined by cluster analysis at the state level (Figure 18.4). To what extent will we find a strong regional patterning of "healthy" and "unhealthy" counties within Alabama itself?

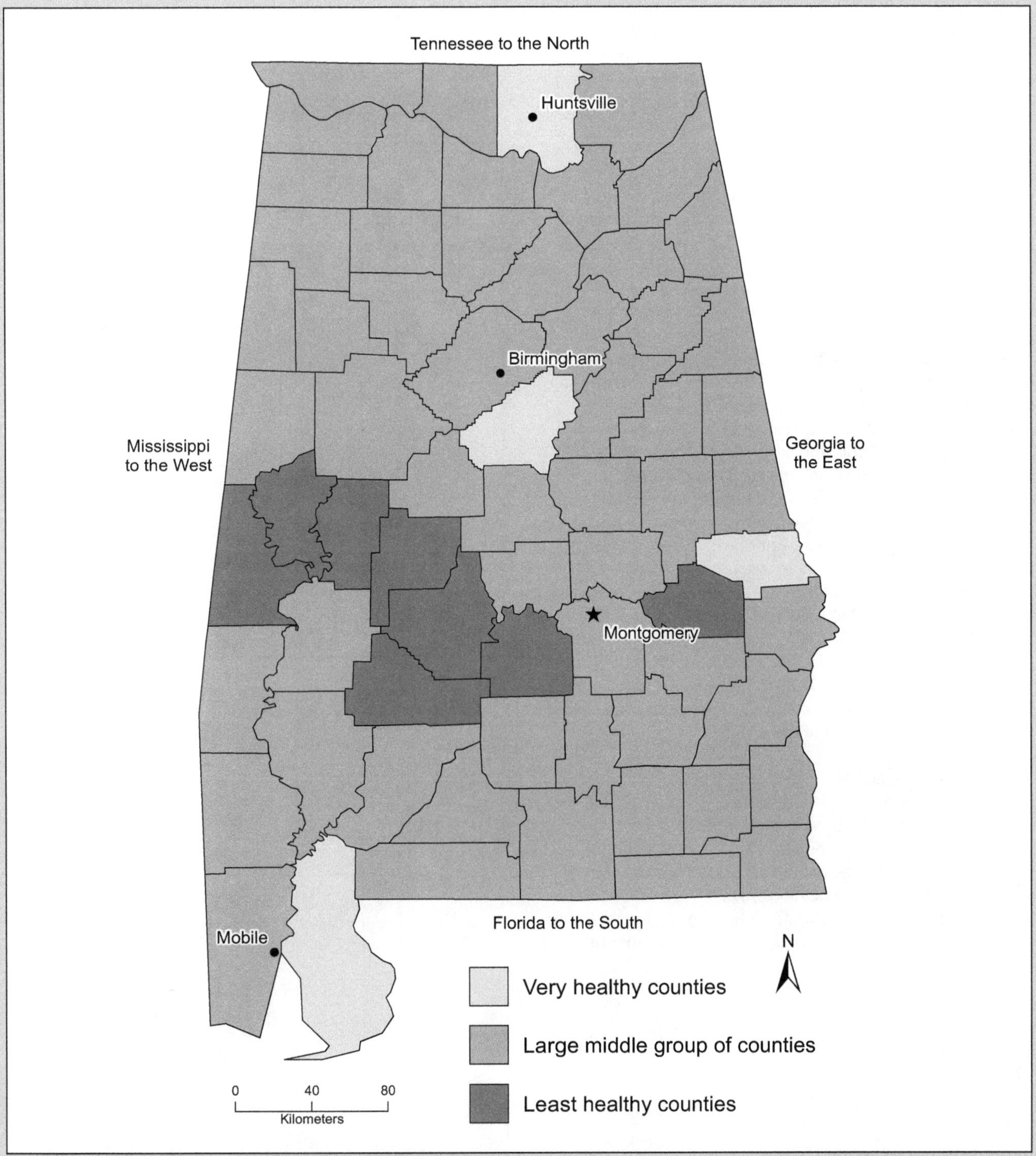

FIGURE 18.5
Cluster Analysis Map of Alabama Counties: Based on Three Measures of "Health"—Levels of Obesity, Diabetes, and Physical Inactivity

(continued)

Using multivariate cluster analysis, we selected the three-cluster solution as best, primarily because the clearest and most interpretable spatial pattern is associated with this option (Figure 18.5).

First, notice that all of the "least healthy" counties are located in a belt running west to east across the central part of the state. Knowledgeable historians and geographers might recognize that these counties are located along the old "cotton belt," with very fertile soils that were most conducive to the development of plantation-style cotton growing in the antebellum South. When this region was initially settled, large plantations were the rule and the overwhelming majority of the population were black slaves.

Even today, these counties have a predominantly black population and are also still quite rural and poor. Of course, the correlations between black, rural, and poor are relatively high in many parts of the South, and it has been widely documented that these segments of our society are relatively unhealthy. As always, keep in mind that "correlation is not causation."

At the other end of the spectrum, several "very healthy" counties form a cluster, even though they are widely dispersed from a locational perspective. These counties have a more suburban character than most other counties in Alabama. Baldwin County is adjacent to Mobile and also has considerable new development along the Gulf Coast. Shelby County is relatively close to Birmingham and Huntsville is located in Madison County. In east-central Alabama, Lee County's very healthy classification is probably attributable in large part to the City of Auburn and Auburn University. A large percentage of Lee County population is young, well-educated, and more likely to not be obese, not have diabetes, and be physically active.

Figure 18.6 shows all of the cluster centroid values as well as a three-dimensional plot of their respective locations. If you carefully examine Figures 18.5 and 18.6, you can see that the cluster centroid of the four "very healthy" counties (Cluster 2) has the lowest levels on all three variables (obesity %, diabetes %, and physical inactivity %). As expected, the "least healthy" set of 8 counties in Figure 18.5 are together as Cluster 3 in Figure 18.6.

In these cluster analysis examples, we relate "poor health" to high levels of obesity, diabetes, and physical inactivity. We also suggest that "good health" is somehow associated with suburban living and proximity to a larger city. In addition, we suggest even more linkages between "good health," younger age and higher level of education. All of these relationships are weakly supported but still strongly suggestive. As geographers, we must always be careful not to attach simple cause-effect labels to variables that happen to be spatially correlated or locations that happen to cluster together.

Nevertheless, as a result of these statistical analyses we are now more knowledgeable about factors associated with poor health. Clearly, with regard to health and obesity (or almost any other variable that has an interesting spatial pattern) it is worthwhile to apply the scientific research process and use of appropriate statistical techniques toward solving problems related to that variable.

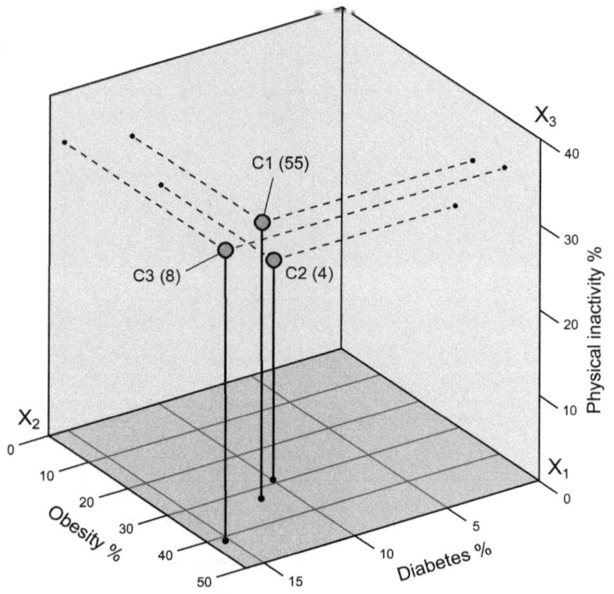

Cluster Centroid Values:

	(55) Cluster 1 (middle group)	(4) Cluster 2 (very healthy)	(8) Cluster 3 (least healthy)	(67) Grand Centroid
Obesity %	33.0	28.7	41.1	33.7
Diabetes %	11.5	9.9	15.2	11.9
Physical inactivity %	32.3	25.6	34.0	32.1

FIGURE 18.6
Centroid Plots of Alabama County Clusters: Based on Three Measures of "Health"—Levels of Obesity, Diabetes, and Physical Inactivity

Key Terms

cluster analysis, *281*
coefficient of multiple determination, *275*
leverage, *278*
multicollinearity, 274
multiple regression, *273*
standardized regression coefficients, *274*
stepwise regression, *277*

Major Goals and Objectives

If you have mastered the material in this chapter, you should now be able to:

1. Explain the basic purposes of multiple regression.
2. Define multicollinearity and explain why it can be a problem in regression modeling.
3. Explain why it is important to determine the relative importance of each predictor (independent) variable.
4. Explain what is measured by the coefficient of multiple determination.
5. Describe the characteristics of a "good" multiple regression model.
6. Explain the basic purposes of cluster analysis.
7. Define the term "cluster centroid."

References and Additional Reading

Draper, N. R. and H. Smith. *Applied Regression Analysis.* 3rd ed. New York: Wiley, 1998.

Everitt, B. S. S. Landau, et.al. *Cluster Analysis.* 5th ed. New York: Wiley, 2011.

Griffith, D. A. and C. G. Amrhein. *Statistical Analysis for Geographers.* Englewood Cliffs, NJ: Prentice Hall, 1997.

Johnston, R. J. *Multivariate Statistical Analysis in Geography: A Primer on the General Linear Model.* London: Longman, 1987.

Montgomery, D. C., E. A. Peck and G. G. Vining. *Introduction to Linear Regression Analysis.* 6th ed. New York: Wiley, 2021.

Rogerson, P. A. *Statistical Methods for Geography: A Student's Guide.* 5th ed. Thousand Oaks, CA: Sage Publications, Inc., 2020

If going further with the geographic examples mentioned in this chapter, sources of data mentioned in earlier chapters remain useful, including: the Centers for Disease Control and Prevention for obesity and related health factors such as diabetes level and level of physical inactivity www.cdc.gov; the World Health Organization and their World Health Statistics publications www.who.int/en; the World Bank for life expectancy and many related world development indicators www.worldbank.org. To learn more about cluster analysis, relevant terminology, and different linkage methods, a good strategy is to examine various statistical software packages that include clustering techniques.

19

Multifactor Analysis

Two-Way Analysis of Variance

In Chapter 11 we explored the concept of ANOVA to compare three or more independent sample means for significant differences, typically referred to as one-way ANOVA. Using one-way ANOVA, we limited our study to a single dependent and independent variable. We can extend this concept by using two-way ANOVA to include other categorical variables to determine how two or more independent variables (or factors) affect the dependent variable under study. Two-way ANOVA has similarities to one-way ANOVA, but introducing another factor allows us to explore new ways that multiple factors can play a part in the differences in a response variable. For example, rather than just comparing differences in obesity using the single categorical variable of census region, we can include race as a second categorical variable. In this way, we not only test whether significant differences exist among census regions, or if there are significant differences among race, but also assess whether race in conjunction with census region plays a part in the level of obesity. Similarly, an agronomist might compare yield results based on the type of fertilizer used *and* the planting density. Using one-way ANOVA limits us to exploring if there are differences in yield based solely on the type of fertilizer used, or solely on the planting density. However, some fertilizers might perform better with a tightly spaced planting strategy, while other fertilizers might perform worse in those conditions. Two-way ANOVA allows us to assess the impact that the two categorical variables of fertilizer type and planting density have on one another, and is termed **interaction**.

Similar to one-way ANOVA, two-way ANOVA assumes that the data follow a normal distribution, and that the groups have similar standard deviations. Also, like one-way ANOVA, we use the F statistic to test for significance.

In one-way ANOVA, the population was split into groups using a single categorical variable (e.g., fertilizer type), termed a **factor**. Two-way ANOVA has two factors (e.g., fertilizer type, planting density), and each factor contains several levels (fertilizer A, fertilizer B; highly spaced planting density; medium-spaced planting-density, tightly spaced planting density). In two-way ANOVA, we are interested in the effects of both factors, and can create a two-way design in a single experiment.

The advantages of using two-way ANOVA as opposed to one-way ANOVA are being more efficient, having more control over the analysis, and the ability to measure the interaction of multiple factors. Including both factors in a single analysis, two-way ANOVA is more efficient by carrying out two separate ANOVA tests at once: one for Factor A, one for Factor B. Also, we can control the variability in one factor by including an additional independent variable. For example, by including a category for planting density, we can determine the effects of the different fertilizers among the different planting densities, and minimize the residual variation of the fertilizer yield by controlling for the planting density. This is similar to how we minimized variation in Chapter 8 by using a stratified design when a second factor was thought to play a role in a dependent variable. Finally, we can explore the interaction of the independent variables with one another, which most likely reflects how the factors work in real life. Interaction represents the combined effect of each factor on the dependent variable. When an interaction exists, the impact of one factor is dependent upon the level of the other factor, allowing us to evaluate the effects of one factor on another. One-way ANOVA only looks at a single factor, and thus, can miss important interactions with other factors. In fact, when an interaction is present, the main effects can be incomplete or misleading.

There are some disadvantages in using two-way ANOVA. As the number of factors and levels increase (e.g., adding additional fertilizer types, additional categories of planting density), the number of combinations can become quite large, increasing the size of the experiment. Data collection at that point may become difficult in order to achieve the appropriate number of unique combinations between the factors. Also, as more combinations occur, the presence of many interactions may be difficult to interpret. Another potential drawback of a two-way ANOVA is that for specific mathematical reasons (not elaborated here) we must require the same number of observations at each level. This is called a **balanced design**. For many, requiring the same number of observations is too restrictive when dealing with real-world geographic problems. Often, this requires the removal of perfectly good sample observations in one factor in order to have the same number of observations in another factor. For example, perhaps our agronomist had 20 observations with minimal spacing, but only 9 observations with highly dense spacing. In this case, they would have to remove 11 of the observations with minimal spacing to match the 9 observations with the dense spacing. While there are techniques for handling unbalanced design with different numbers of observations, they are more challenging to conduct, and therefore are not illustrated in our examples. Therefore, conducting a two-way ANOVA often requires a trade-off between an ideal design vs. the greater efficiency offered by combining multiple classifications into a single test.

We should also note that two-way ANOVA should not be confused with Multivariate ANOVA (MANOVA). MANOVA is an ANOVA experiment with more than one dependent variable. For instance, in our agronomy example, we may not only be interested in the yield for a crop, but also its resistance to disease. In this case, we now have two dependent variables (yield and disease resistance) and two categorical variables of fertilizer type and planting density. Although MANOVA is a useful tool for geographers, it is a more advanced use of ANOVA and is not addressed in our text.

19.1 Two-Way ANOVA

The simplest two-way ANOVA experiment contains only two factors, and the experimental design breaks the factors into separate cells in a matrix. Structured in this fashion, we refer to a two-way ANOVA as a **factorial experiment**. A factorial experiment is one where all levels of each factor are combined with all levels of the other factor. In general, Factor A will have I levels, and Factor B will have J levels, and are referred to as an $I \times J$ ANOVA. The general structure of an $I \times J$ factorial is shown in Table 19.1.

In a factorial experiment, every level of Factor A appears in combination with every level of Factor B, allowing for $I \times J$ combinations. The general form of a two-way ANOVA result is shown in Table 19.2.

Data may be structured to show multiple levels for each factor. For example, Tables 19.3 and 19.4 show a 2 × 2 factorial layout, allowing four unique comparisons, while Table 19.6 shows a 4 × 3 factorial layout with 12 unique combinations.

Factorial experiments measure the **main effect**, **simple effect**, and **interaction** of the individual treatments. The main effect compares the levels of a single factor, averaging over all the levels of another factor-think

TABLE 19.1

An illustration of a Two-Factor A x B Layout

	Factor B				
Factor A	1	2	...	J	*Average*
1	μ_{11}	μ_{12}	...	μ_{1J}	μ_{A1}
2	μ_{21}	μ_{22}	...	μ_{2J}	μ_{A2}
...	...	...	...	...	...
I	μ_{I1}	μ_{I2}	...	μ_{IJ}	μ_{AI}
Average	μ_{B1}	μ_{B2}	...	μ_{BJ}	

TABLE 19.2

Two-Way ANOVA Table Results

Source	SS	df	MS	*F*	*p*-value
Main Effect A (with *a* levels)	SS_A	$a - 1$	SS / df_a	MS_A / MS_E	—
Main Effect B (with *b* levels)	SS_B	$b - 1$	SS_B / df_b	MS_B / MS_E	—
Interaction Effect	SS_{AB}	A*B/(a − 1)(b − 1)	SS_{AB} / df_{ab}	MS_{AB} / MS_E	—
Error (Within)	SSE	ab(n − 1)	SSE / df_E		
Total		$SSTo$			

of this as a one-way ANOVA for a single factor. The simple effect compares one factor at a fixed level to the other factor, and the interaction measures the impact one factor has on the result of another factor over all levels. Interaction among factors is always an important concept to explore when you have two or more factors acting on one another. Figure 19.1 uses a line graph to illustrate the theoretical results of a two-factor experiment with and without the presence of interaction effects. You'll notice that for experiments where there is no presence of interaction, the lines do not have to be parallel, but only exhibit similar changes in one factor over different levels in another factor.

While one-way ANOVA had a single null hypothesis, the addition of a second factor for two-way ANOVA creates three null hypotheses. The three null hypotheses measure the main effect of A (H_0: $\mu_{A1} = \mu_{A2} = ... = \mu_{AJ}$), the main effect of B (H_0: $\mu_{B1} = \mu_{B2} = ... = \mu_{BJ}$), and the interaction between A and B (H_0: there is no interaction between Factors A and B). The alternate hypotheses indicate whether we can reject either of the three null hypotheses. As shown in Table 19.2, each hypothesis includes an *F* statistic and a *p*-value, allowing us to test the significance of the main effects (A, B) and the interaction (A*B).

When interpreting the ANOVA table, it is often useful to first explore the interaction effect. If the interaction between the factors is significant then there is little meaning as to whether the main effects are significant or not. The reason an interaction negates the significance of the main effects is because an interaction assumes the response to changes in one factor depends on the level of the other factor, and the mean for a level of one factor averaged over all levels of the other factor has little meaning. For example, crop yield for one fertilizer may decrease only when planted with a higher density, while the other fertilizer is not affected by the planting density. Should an interaction exist, it is best to summarize the results of the experiment using a pairwise table of means of each A and B combination. For this reason, in a two-way ANOVA, the interaction effects should be tested first.

If the interaction is not significant then we can assume that the main effects operate independent of one another. In other words, the difference between the levels in Factor A are similar at each level of Factor B. We can simply analyze the main effects for each factor separately, treating our experiment as two separate one-way ANOVA tests.

Data for a simple example of a two-factor experiment in Table 19.3 and Figure 19.2 show the average obesity when considering both race and census region. To simplify the concepts, we limit our data set to a 2 × 2 factorial experiment, evaluating obesity between the White and Black population as one factor, and the northeast and south census region as a second factor. We will expand this experiment later into a 4 × 3 factorial experiment to include additional census regions and races.

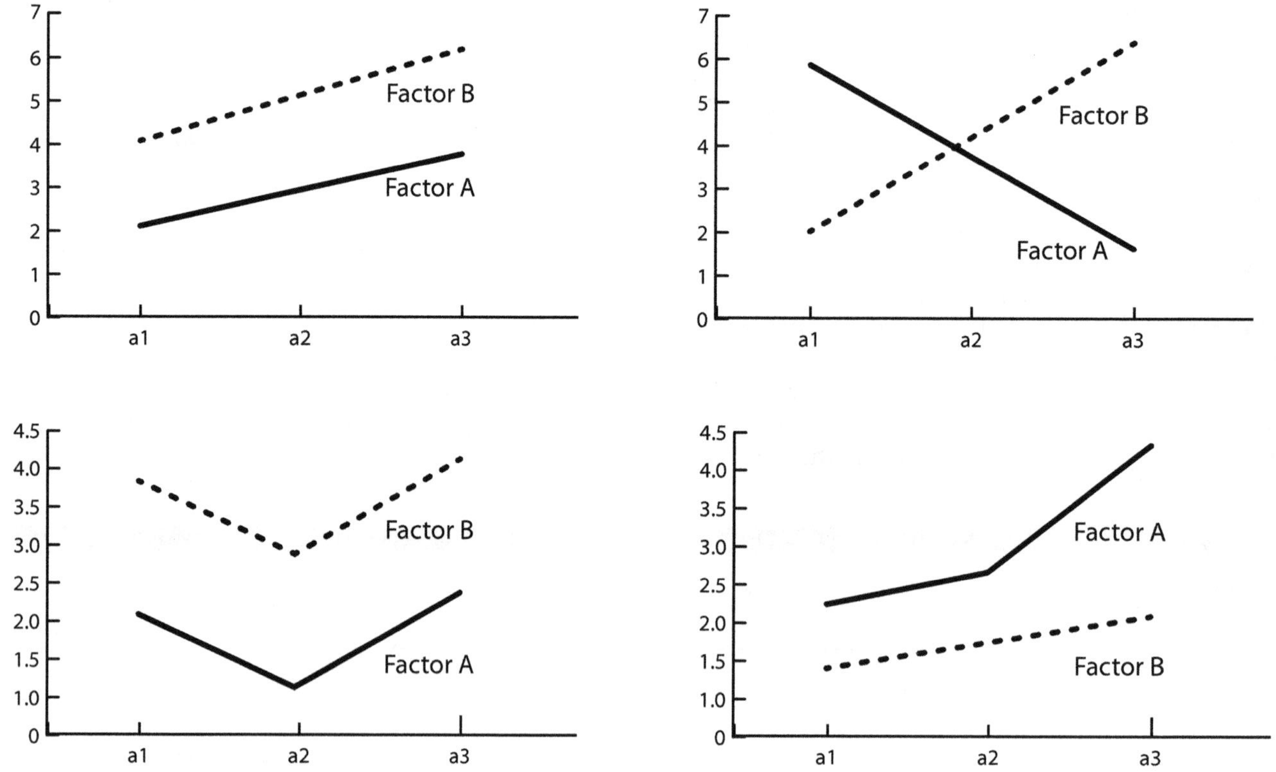

FIGURE 19.1
Illustration of No Interaction (Left) and Interaction (Right) Among Two Factors.

In this example we created a balanced design using eight observations for each census region, and the matrix shows the average obesity value and the raw data related to the combination of factors. For example, the states in the northeast have an average obesity rate for the White population of 28.37%. Note that one of the samples in the South region includes Washington, DC, which has a very low White obesity rate (11.2%). The columns for *Average* show the average obesity value for each factor. For instance, the average obesity rate for the entire northeast is 35.91%. The averages are termed **marginal means**, as they occupy the margins of the table and represent the main effects. In this example, Black obesity is larger than White obesity, and obesity in the South is larger than in the Northeast. Examining the individual cells, we see that the percent obese in the Black population is larger than the percent obese in the White population both in the Northeast and the South. In fact, the average difference of 15.38% between the White and Black population is somewhat alarming. However, the difference in obesity rates between the northeast and south is rather small, under 1% (–0.16, –0.76).

By examining the simple effects, we see that in all cases, whether in the Northeast or the South, the obesity rate for the Black population is higher than the White population. Finally, as we noted, the difference in obesity rates between the census regions for each race is under 1%, meaning the difference in obesity rates for each group are not very different between census regions. In other words, obesity rates across census regions are fairly similar.

This leads us to examine the interaction between Black and White obesity in the Northeast and the South. If interaction is present, we can be confident that the main effects provide an incomplete understanding of the data. Figure 19.2 presents the marginal difference of obesity rates as a line graph. We observe that the relative differences between the Black and White populations are consistent across the two census regions. That is, the lines are generally parallel, and have similar values whether we look at the Northeast or the South. Therefore, the census region has no effect on the rate of obesity for the Black and White population, and thus little to no interaction. In this case, the main effects alone are adequate to explain the differences in obesity rates. Nonetheless, any analysis should proceed with caution as other explanatory variables (e.g., income level, education level) may help play and important part in the differences.

In contrast to the Black and White obesity rates in the northeast and south, we can explore the White and Hispanic[1] obesity rates for the same census regions. In Table 19.4, we notice the Hispanic population has slightly higher obesity rates than the White population. We also observe that the south is more obese than the northeast. However, while the Difference columns and rows in Table 19.3 for the Black and White obesity rates were somewhat similar, the Difference columns for White and Hispanic obesity rates in Table 19.4 are less similar. The differences in obesity rates for each factor are further shown in Figure 19.3. The relative difference of obesity rates between White and Hispanic populations are quite different depending upon the census region. In other words, while the average obesity rate for Whites is consistent between the northeast and south, the Hispanic obesity rates are less consistent based on the census region. On average, Hispanics in the South have a larger obesity rate than Hispanics in the North, and the South appears to have a greater impact on Hispanic obesity than it does on White obesity. This is illustrated in Figure 19.3, as the lines for White and Hispanic obesity rates are not parallel to one another. In this case, there may be an effect of census region on obesity rates, producing a potential

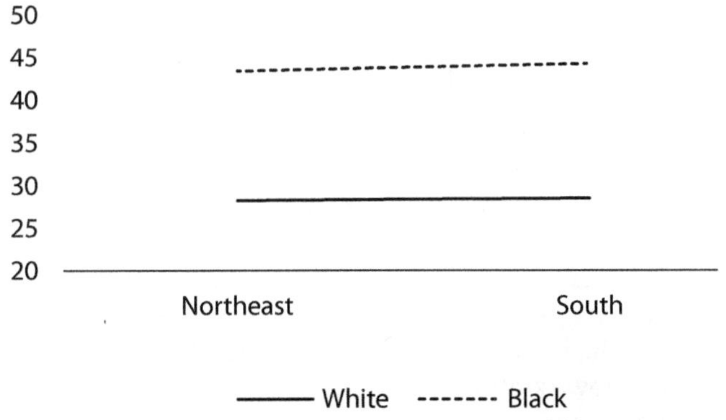

FIGURE 19.2
State-Level White and Black Obesity Rate, Grouped by Census Region, 2020, Showing No Interaction.

TABLE 19.3

State-Level White and Black Average Obesity Rate, Grouped by Census Region, 2020

Census Region	White	Black	Average	Difference
Northeast	28.37 (26.6, 31.1, 25.3, 30.7, 26.6, 31.3, 28.3, 27.1)	43.46 (43.4, 49.8, 35.0, 44.4, 36.7, 44.9, 41.1, 52.4)	35.91	−15.08
South	28.53 (34.3, 36.1, 33.3, 11.2, 27.4, 30.7, 18.8, 36.5)	44.22 (46.2, 48.5, 46.3, 41.1, 38.9, 42.6, 42.4, 47.8)	36.38	−15.68
Average	28.45	43.84	36.15	−15.38
Difference	−0.16	−0.76		

Source: Center for Disease Control

TABLE 19.4

State-Level White and Hispanic Obesity Rate, Grouped by Census Region, 2020

Census Region	White	Hispanic	Average	Difference
Northeast	28.37	29.88	29.13	*−1.51*
South	28.53	32.57	30.55	*−4.03*
Average	28.45	31.23	29.84	*−2.7*
Difference	*−0.16*	*−2.68*	*−1.42*	

Source: Center for Disease Control, 2020.

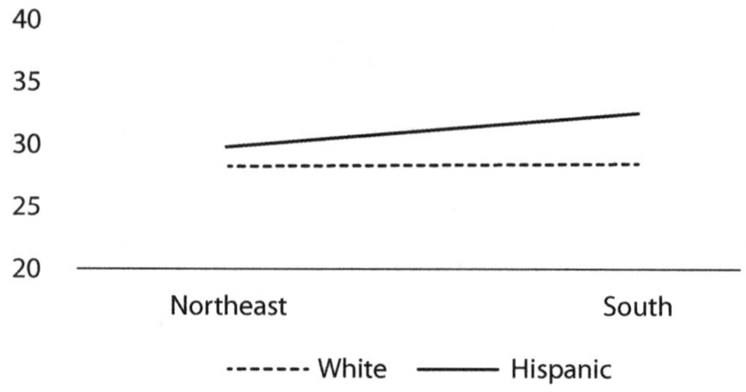

FIGURE 19.3
White and Hispanic Obesity Rate, Grouped by Census Region, 2020, Showing Potential Interaction.

Two-Way Analysis of Variance

Primary Objective: Examine the influence of different categorical independent variables on one dependent variable

Requirements and Assumptions:

1. The populations from which the samples are drawn are normally distributed
2. The observations for within and between groups are independent
3. Equal variances among populations (homoscedastic)
4. Data are measured at the interval or ratio scale
5. The groups contain the same sample size

Hypotheses:

H_0 : There is no difference in the means of Factor A (this is similar to a one-way ANOVA for Factor A):

$$\mu_1 = \mu_2 = \ldots = \mu_k$$

H_0 : There is no difference in the means of Factor B (this is similar to a one-way ANOVA for Factor B):

$$\mu_1 = \mu_2 = \ldots = \mu_k$$

H_0 : There is no interaction between Factors A and B

H_A : There is a difference in the means of Factor A:

$$\mu_{A1} \neq \mu_{A2} \neq \ldots \cdot \mu_{Ak}$$

There is a difference in the means of Factor B, at least one mean differs
There is interaction between Factor A and B

interaction among the factors. Therefore, we must be careful in interpreting the main effect of race, as census region may have an important contribution to explaining any differences. Of course, we have only made descriptive observations at this point. Performing a two-way ANOVA test will allow us to assess the statistical significance of the main effects and any potential interaction.

We now continue exploring our data using a 4 × 3 factorial experiment to calculate the two-factor ANOVA results for the Black, White, and Hispanic obesity rates for all four census regions. Similar to our treatment of multiple regression, we avoid any demonstration of hand calculation, and rely on the results generated using free and commercial software packages.

19.2 Example Applications in Geography

Example: Two-way ANOVA for Differences in State-Level Obesity Rates by Census Region Showing No Interaction

In Chapters 1 and 11 we explored the spatial pattern of obesity across the United States to illustrate the geographic research process, and the utility of one-way ANOVA. Viewing obesity values on the state-level map, we hypothesized that there were differences in obesity among the four different census regions. We tested this hypothesis using ANOVA and confirmed that there were in fact differences in obesity values among the census regions. However, we might consider that race may also play a part in the level of obesity in a particular census region. We could perform a second ANOVA test on obesity and race (e.g., White, Black, Hispanic, Asian), but this would not give us any insight into whether there is any interaction among race and region when assessing obesity. Instead, we will conduct a two-way ANOVA to not only assess whether statistical differences existed in race and region, but also understand any interaction that may exist between the categorical variables. The results of our analysis are shown in Tables 19.5, 19.6, and 19.7.

The *p*-values in Table 19.7 show that the interaction is not statistically significant ($p = 0.21$), meaning we may accept the null hypothesis and state that neither factor has an impact on the other. The factors are independent of one another. Even though no significant interaction is present, we can still examine the line graphs in Figure 19.3 to illustrate the relationship among the data values.

The lack of interaction and evidence of main effects are seen in Figure 19.4. Each line in the graph clearly shows separation of obesity values based on the census region and also the race. While we saw what appeared to be potential interaction among the White and Hispanic populations in the northeast and south in Figure 19.4, our analysis shows that the levels were not statistically significant ($p = 0.21$). As a side note, we ran a two-way ANOVA for the data in Table 19.3, and the interaction effect had a p-value of 0.88. Therefore, while

TABLE 19.5

Worktable for Two-way Analysis of Variance (ANOVA): Differences in State-Level Obesity Rates by Census Region

H_0	1. Factor A: $\mu_1 = \mu_2 = \mu_3 = \mu_4$ 2. Factor B: $\mu_1 = \mu_2 = \mu_3$ 3. No interaction in Factor A and B
H_A	1. Factor A: at least one mean differs 2. Factor B: at least one mean differs 3. Interaction in Factor A and B

TABLE 19.6

State-Level Obesity Rate for White, Black, and Hispanic, Grouped by Census Region, 2020

Census Region	White	Black	Hispanic	Average
Northeast	28.38	43.46	29.89	33.91
South	28.54	44.23	32.58	35.11
Midwest	32.83	44.36	38.34	38.51
West	26.75	44.68	34.29	35.24
Average	29.12	44.18	33.77	

Source: Center for Disease Control, 2020.

TABLE 19.7

Two-Way ANOVA Table Results or Obesity Comparing State-Level Black, White, and Hispanic Population with the Northeast, South, Midwest, and West Census Regions

Source	SS	df	MS	*F*	*p*-value
Main Effect: Census Region	279.73	3	93.24	4.28	0.01
Main Effect: Race	3805.47	2	1902.73	87.33	0.00
Interaction: Census Region * Race	189.59	6	31.60	1.45	0.21
Within (Error)	1830.17	84	21.79		
Total	6104.96	95			

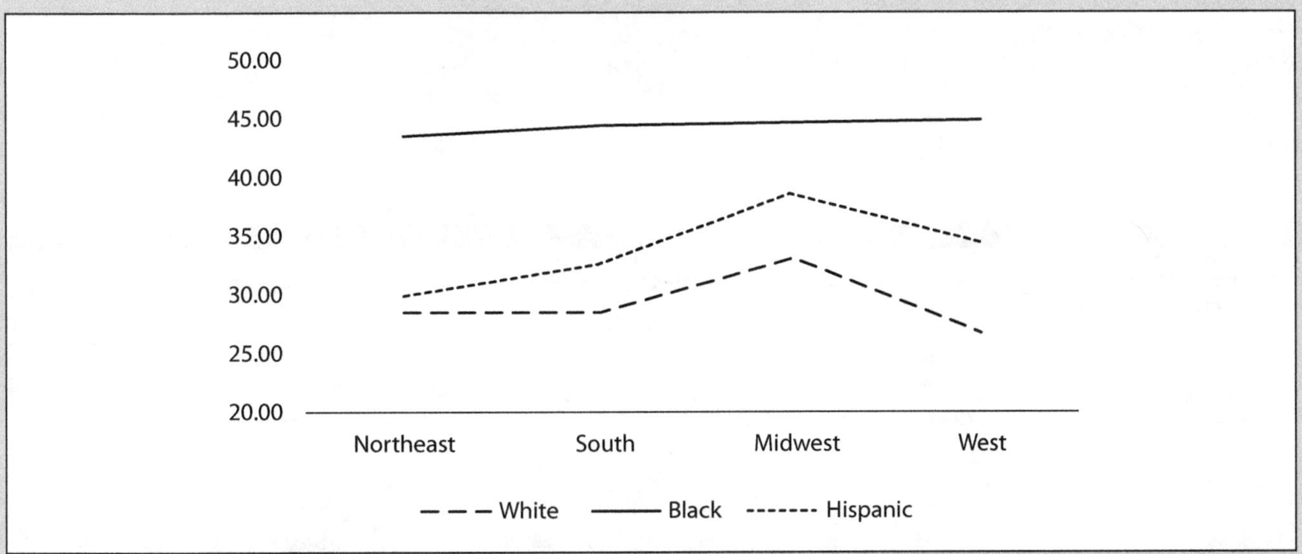

FIGURE 19.4
State-Level White, Black, and Hispanic Obesity Rate, Grouped by Census Regions.

(continued)

there is some interaction in our data, it is not enough to be statistically significant.

With no statistical significance for interaction, we can now examine the *p*-values for the main effects, which are both statistically significant ($p = 0.00$, $p = 0.01$). Therefore, we may reject the null hypotheses for both main effects. The obesity rates for race do not come from the same underlying population, and the obesity rates among the census regions do not come from the same underlying populations. These are the same results we would have seen if we had conducted two one-way ANOVA tests. In this case we have taken advantage of the efficiency offered in a two-way ANOVA test by successfully performing one-way ANOVA tests for race, and for region. Similar to our one-way ANOVA discussion in Chapter 11, should we desire, we can perform pairwise comparisons of the main effects to determine which regions and races were the reason for differences. In fact, many software products perform pairwise tests of the main effects automatically, grouping them by similar values as shown in Table 19.8.

We see that for the main effect of race, all three groups are statistically different.

TABLE 19.8

Grouping Information State-Level White and Hispanic Obesity Rate, Grouped by Census Region, 2020

Race	N	Mean*	Grouping
Black	32	44.18	A
Hispanic	32	33.70	
White	32	29.12	

*Means that do not share a letter are significantly different.

Example: Two-way Analysis of Variance (ANOVA)—Differences by Census Region and Race in State-Level Median Income—Showing Interaction

A balanced, two-way ANOVA was conducted to assess the role that race and geographic location play in income. Data were collected from the US Census Bureau on state-level median income values for Asian, White, Black, and Hispanic populations. We associated each state with a census region, allowing us to form groups based on census region. Tables 19.9, 19.10, and 19.11 show the results of the ANOVA.

The results show statistically significant interaction between race and census region ($p = 0.00$). Therefore, even though the main effects of region and race are statistically significant, we are less concerned with them as the interaction effect indicates that they are not independent of one another. We can see this effect in Figure 19.5, especially among the Hispanic and Black populations. While Asian, Black, and White incomes all decrease in the Midwest, Hispanic incomes actually increase. Also, Black and Asian income increases in the South compared to the Midwest, while White and Hispanic incomes decrease.

With so many comparisons to consider, we can once again return to the grouping information shown in Table 19.12. The table shows groupings for a 95% confidence interval for the interaction of race and census region-means that do not share the same letter are statistically different. One can clearly see that the highest separable incomes are all in the Asian popula-

TABLE 19.9

Worktable for Analysis of Variance (ANOVA): Differences in State-Level Median Income Values for Asian, White, Black, and Hispanic Populations, Grouped by Census Region, 2009–2017

H_0	1. Factor A: $\mu_1 = \mu_2 = \mu_3 = \mu_4$ 2. Factor B: $\mu_1 = \mu_2 = \mu_3 = \mu_4$ 3. No interaction in Factor A and B
H_A	1. Factor A: at least one mean differs 2. Factor B: at least one mean differs 3. Interaction in Factor A and B

TABLE 19.10

Average Median Income Values for Asian, White, Black, and Hispanic Populations, Grouped by Census Region, 2009–2017

Census Region	Asian	Black	White	Hispanic	*Average*
Northeast	$79,746	$42,600	$68,477	$43,667	*$58,622*
South	$71,545	$32,337	$58,106	$45,041	*$51,757*
Midwest	$76,127	$37,293	$56,671	$43,362	*$53,363*
West	$80,203	$44,969	$63,445	$48,063	*$59,170*
Average	$79,746	$42,600	$68,477	$43,667	*$58,622*

Source: US Census Bureau, 2020.

tion. However, not all Asian median incomes are in the same group, as the median incomes for Asians are separated into another group due to the impact of lower incomes in the Midwest. Similarly, we see the median income for the Black population spread over multiple groups.

For whatever reason, incomes for races are not created equal when considering the effect of the census region. What dynamics are in place that cause the Hispanic population to see a relative increase in median incomes in the Midwest compared to other races, or for the White population to have such dramatically lower incomes in the South compared to the other groups? In this case, answering our question regarding interaction has opened up a plethora of other questions we can explore to better understand why these differences exist.

TABLE 19.11

Two-Way ANOVA Income for Asian, White, Black, and Hispanic Populations, Grouped by Census Region, 2020

Source of Variation	SS	df	MS	*F*	*p*-value
Main Effect: Census Region	1,415,105,674	3	471,701,891	67.84	0
Main Effect: Race	31,380,904,017	3	10,460,301,339	1504.42	0
Interaction: Census Region * Race	722,065,579	9	80,229,509	11.54	0
Within (Error)	889,989,738	128	6,953,045		
Total	34,408,065,008	143			

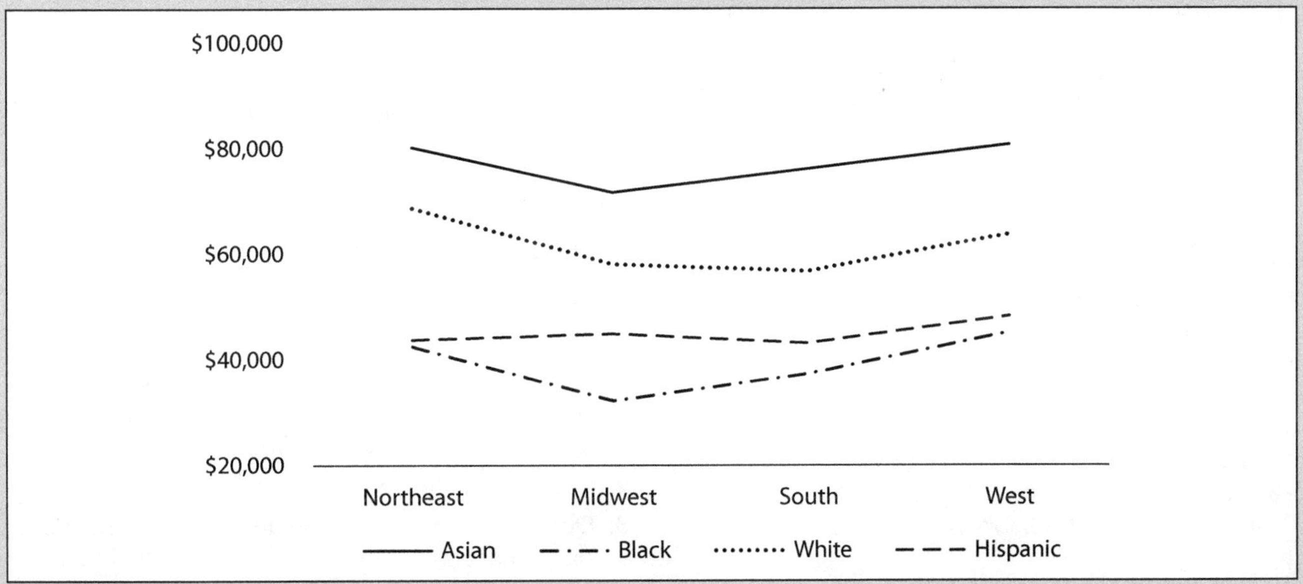

FIGURE 19.5
State-Level Median Income for Asian, Black, White, and Hispanic Populations Grouped by Census Region.

(continued)

TABLE 19.12

Pairwise Comparisons for Median Income, 2009–2017 Census Region*Census Race Grouping

Census Region*Census Race	N	Mean[1]	Grouping							
West*Asian	9	$80,203	A							
Northeast*Asian	9	$79,746	A							
South*Asian	9	$76,127	A							
Midwest*Asian	9	$71,545		B						
Northeast*White	9	$68,476		B						
West*White	9	$63,444			C					
Midwest*White	9	$59,285			C	D				
South*White	9	$56,671				D				
West*Hispanic	9	$48,063					E			
Midwest*Hispanic	9	$45,040					E	F		
West*Black	9	$44,969					E	F		
Northeast*Hispanic	9	$43,666						F		
South*Hispanic	9	$43,361						F		
Northeast*Black	9	$42,599						F		
South*Black	9	$37,293							G	
Midwest*Black	9	$32,336								H

[1] Means that do not share a letter are significantly different.

Key Terms

balanced design, *291*
factor, *290*
factorial experiment, *291*
interaction, *290*
main effects, *291*
marginal means, *293*
two-way analysis of variance (ANOVA) or *F*-test (parametric), *296*

Notes

[1] The term *Hispanic* is actually an ethnicity, and not a race. When comparing groups, we have used the Census Bureau designation of "White, Non-Hispanic" and "Black, Non-Hispanic." For brevity, we have used the term race in our examples without the Non-Hispanic qualifier. For further discussion, please visit www.census.gov/topics/population/hispanic-origin/about/comparing-race-and-hispanic-origin.html

Major Goals and Objectives

If you have mastered the material in this chapter, you should now be able to:

1. Explain why a two-factor ANOVA test is more efficient than a single factor ANOVA test, and what some of the drawbacks are to a two-factor ANOVA test.
2. Explain what a factorial experiment is.
3. Define and understand the differences in main effects, simple effects, and interaction in individual treatments.
4. Be able to make observations about main, simple, and interaction effects in the results of a two-factor ANOVA test.

References and Additional Reading

Cannon, A. R, et al. *STAT2: Modeling with Regression and ANOVA*. 2nd ed. New York: W. H. Freeman, 2019.
Petersen, R. G. *Design and Analysis of Experiments*. New York: Marcel Dekker, 1985.

20

The Basics of Binary Logistic Regression

20.1 The Basics of Logistic Regression

Having previously explored simple and multiple linear regression methods where we modeled the relationship between a quantitative dependent variable and one or more explanatory variables, we now discuss logistic regression, often termed the logistic model or logit model that analyzes the relationship between one or more independent variables and a *binary* dependent variable. Although our focus is on binary logistic regression, there are actually three types of logistic regression models, each with their own goals and objectives: binary logistic regression, ordinal logistic regression, and multinomial logistic regression.

Binary logistic regression is used to predict the relationship between a dependent variable (Y) having only two outcomes, and independent variable(s) (X) that can be a combination of continuous, ordinal, or categorical data types. Examples of a binary response variable would include the results of a COVID-19 patient's outcome to hospitalization (i.e., survives, dies), a homeowner's loan payment (i.e., default, nondefault), or species identification within an ecological habitat (i.e., presence, absence). We describe these outcomes as a "success" or "failure," ordinarily represented as a 1 or 0 respectively.

Ordinal logistic regression is used when the dependent variable (Y) is *ordinal*. That is, the dependent variable has a meaningful order with more than two categories. Examples of ordinal variables might include the response to an opinion poll (e.g., strongly disagree, disagree, agree, strongly agree), grain size on sand samples (e.g., small, medium, large), or soil quality (e.g., good, average, poor).

Multinomial logistic regression is used when a nominal dependent variable has two or more levels (i.e., two or more outcomes). Examples of a dependent variable with two or more levels would include primary household energy use (e.g., oil, natural gas, solar, hydroelectric), or commuter preference (e.g., car, bus, train).

In Chapter 5 we explored the use of binomial probabilities. For binary logistic regression we are similarly interested in the proportion of successes, $p = P(\text{success})$, but now the probability, p, is dependent upon at least one explanatory variable, x. For example, we might want to determine if an individual is categorized as obese (y = 1) or not obese (y = 0). In this case, p is the probability that a person is obese, and the explanatory variables might include (a) level of education, (b) ethnicity, and (c) hours of moderate exercise each week. Unlike our multiple linear regression example, the explanatory variables can be nominal, ordinal, or ratio. For example, the explanatory variable *level of education* is ordinal (e.g., 1 = high school, 2 = college, 3 = post bachelorette), *ethnicity* is nominal (e.g., White, Asian, Black), and *hours of moderate exercise each week* is ratio (e.g., number of hours).

The goal of logistic regression is to calculate the probability of an event occurring vs. the probability of the event not occurring by explaining the impact of the independent variables in terms of odds. Logistic regression prefers to treat the outcome as odds, rather than proportions. Although often thought of as the same thing, probability and odds are different. Probability represents the ratio between the number of positive (or negative) events to the total number of events. Odds are the ratio between probabilities: the probability of an event and the complement of the probability. The odds of an event are calculated as the ratio of the probability that an event occurs to the probability that the event does not occur.

For example, imagine a geographer studying hazards randomly inspects 250 homes that do not have protective shutters for damage after a Category 3 hurricane, and finds 175 homes are damaged. The proportion, $\hat{p}$, of damaged homes is

$$\hat{p} = \frac{175}{250} = .70$$

Therefore, if is the probability of one outcome, then $1 - \hat{p}$ is the probability of the complement of that outcome making the odds equal to

$$\text{odds} = \frac{\hat{p}}{1 - \hat{p}}$$

In our damaged homes example, the odds of a home being damaged after a Category 3 hurricane is

$$\text{odds} = \frac{\hat{p}}{1 - \hat{p}} \ldots = \frac{.70}{.30} = 2.33$$

When referring to odds, we often round to the nearest integer, meaning that the odds of a building sustaining damage after this particular hurricane is approximately 2 to 1. Similarly, the odds of a building not sustaining damage would be 1 to 2.

All of this discussion leads us to the concept of the odds ratio (OR), which measures the odds relative to different events. If we have two events (A, B), we can measure the odds of A occurring relative to B as:

$$\text{odds ratio } \{A, B\} = \frac{\text{odds } \{A\}}{\text{odds } \{B\}} = \frac{\frac{\hat{p}A}{1 - \hat{p}A}}{\frac{\hat{p}B}{1 - \hat{p}B}}$$

The OR represents the odds that an outcome (e.g., building damage) occurs given a particular event (e.g., hurricane category), compared to the odds that the outcome occurs in the absence of the event. Similar to linear regression, logistic regression includes a regression coefficient, b_1 that defines the odds of the outcome based on a unit increase of the independent variable. Recall from our regression example in Chapter 17 that an increase in one mile from Lake Erie resulted in a 7.11-inch increase in lake effect snowfall. However, unlike our linear regression example, we are not measuring absolute changes in a variable (i.e., snowfall in inches), but rather the relative odds of the event occurring. An OR = 1 indicates that the event does not affect the odds of an outcome, while an OR > 1 indicates that the particular event is associated with higher odds for the outcome. Finally, OR < 1 indicates the event is associated with lower odds for the outcome. For example, if we assess the odds ratio of lung cancer based on whether a patient is a smoker (1) or a nonsmoker (0) as 12.8, the odds for lung cancer is 12.8 times more likely for smokers than nonsmokers.

With this knowledge of OR, we can return to the purpose of logistic regression, that is, to fit a curve (rather than a straight line) for a binary response variable as $y = f(x)$. When y is a discreet binary variable (0,1) and x is a numeric value, logistic regression fits the curve to define the relationship of x and y. Due to the response variable being binary, the function forms an S-shaped, or Sigmoid, curve, shown in Figure 20.1

We observe from Figure 20.1 that the response variable on the y-axis only has two possible outcomes (0,1), while the independent variable on the x-axis can take on a continuum of values, in our case 0-8. Both an S-curve and a linear regression line are shown in the figure. Obviously, given the binary aspect of the response variable, the S-curve provides a better fit. Recall from our discussion of linear regression, that the dependent variable Y is modeled as it relates to the explanatory variable X using the equation

$$Y = \alpha + \beta x$$

This relationship does create some problems because in linear regression Y can take on any value, whereas logistic regression constrains the value of p between 0 and 1. Therefore, it is possible when using the linear regression formula to produce results that fall outside 0 and 1 when

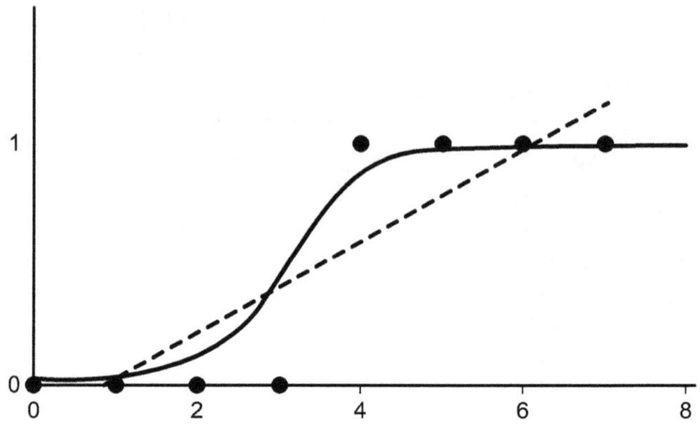

FIGURE 20.1
Linear Regression Line and a Hypothetical S-Curve Fit to Binary Results shows That the S-Curve Provides a Better Fit of the Data.

we have extreme values for x. The solution is to transform the odds using a natural log, making the odds a linear function as

$$\text{logit}(y) = \ln(\text{odds}) = \ln\left(\frac{p}{1-p}\right) = \alpha + \beta x$$

referred to as log odds. Similar to linear regression, y is the response variable, β is the coefficient for the independent variable, and x is the observed value. Having established the relationship between p and x, Figure 20.2 illustrates three possible curves for different values of α and β_1. Therefore, depending upon the relationship between the binary results, we can modify the curve to fit virtually any combination of α and β. Similar to the transition of linear regression to multiple regression, logistic regression may include more than one independent variable as

$$\text{logit}(y) = \ln(\text{odds}) = \ln\left(\frac{p}{1-p}\right) = \alpha + \beta_1 x_1 + \beta_2 x_2 \ldots \beta_n x_n$$

The results are the log odds, which are not intuitive to most people. Therefore, we need to reverse the log using the exponential (exp), sometimes referred to as the antilog, in order to return the log odds to their original probabilities. The formula to return the log odds to their original probabilities is

$$P = \frac{\exp(Y')}{1 + \exp(Y')}$$

where: $Y' = \alpha + \beta x$

In addition to being able to measure binary or ordinal dependent and independent variables, logistic regression does not require many of the assumptions used in linear regression such as a linear relationship between the dependent and independent variables, a normal distribution of errors, homoscedasticity, and ordinal or interval/ratio measurement levels of the independent variables. Logistic regression is able to model nonlinear relationships between the dependent and independent variables. However, logistic regression does follow similar assumptions as linear regression such as using only variables that are statistically meaningful, independence among the observations, and the avoidance of multicollinearity among the independent variables. Logistic regression also has some of its own assumptions that must be met, such as requiring a discrete dependent variable, and a recognition that it typically requires larger sample sizes.

Output from logistic regression includes a number of statistical summaries. However, as an introduction to the topic we will limit our discussion to a summary of the response information, the regression equation that can be used for prediction, coefficients of the independent variables along with their standard errors, p-values, model fit, and odds-ratio for any continuous and categorical predictors.

We will explore four examples of logistic regression including a simple binary logistic regression with one independent variable expanded to a second independent variable, an example using multiple independent variables, an example comparing logistic regression multiple models, and an ordinal logistic regression model.

Logistic Regression Analysis

Primary Objective: Measure the relationship between a categorical dependent variable and one or more independent variables (usually continuous) by plotting the dependent variable's probability scores.

Requirements and Assumptions:

1. **The dependent variable is binary or dichotomous**—binary logistic regression has two possible outcomes (0,1) while ordinal logistic regression can take on more ordinal categories.
2. **Limited multicollinearity between the independent variables**—the independent variables should be independent of each other, meaning there should not be a high level of correlation among the variables.
3. **The independent variables should be linearly related to the log odds**—residual errors are independent of one another in the underlying regression model (no autocorrelation).

Hypotheses:

$H_0: \beta_1 = \beta_2 = \ldots = \beta_k = 0$

H_A: (at least 1 coefficient is not zero)

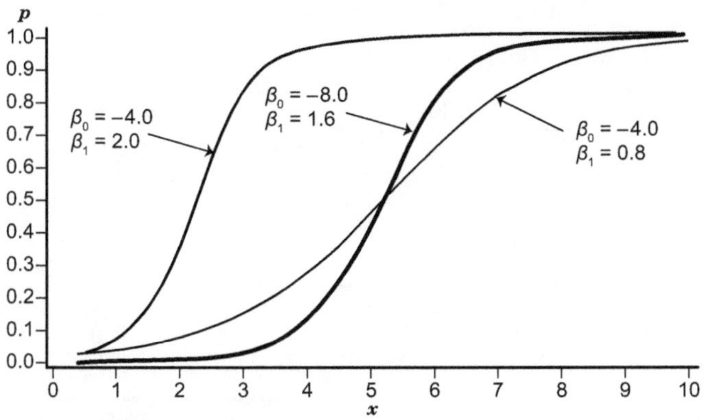

FIGURE 20.2
Hypothetical S-Curves for Different Values of β_0 and β_1.

20.2 A Basic Example

Example: Predicting the Putting Success of an Amateur Golfer

As a simple nongeography example, one of the authors decided to test his skill at putting on a mildly sloped golf green. In theory, a golfer would expect to make more putts the closer they are to the hole. Using logistic regression, we hope to quantify the relationship that distance plays on the success of a putt. The binary dependent variable "success" is coded as 1 for a successful putt, and 0 for a missed putt. The independent variable was coded as 3', 5', 10', 15', and 20', representing the distance of the putt. The results are shown in Table 20.1

From Figure 20.3, one can clearly see that the success in putting decreases as the distance to the hole increases. We also observe the following from the logistic regression output:

- The odds ratio of .76 indicates that every unit increase in putting distance results in a decrease of .24 in putting success. In other words, the golfer would only make 76% of their putts as the unit increases.
- The Sigmoid curve clearly shows the decrease in putting success as the distance increases.

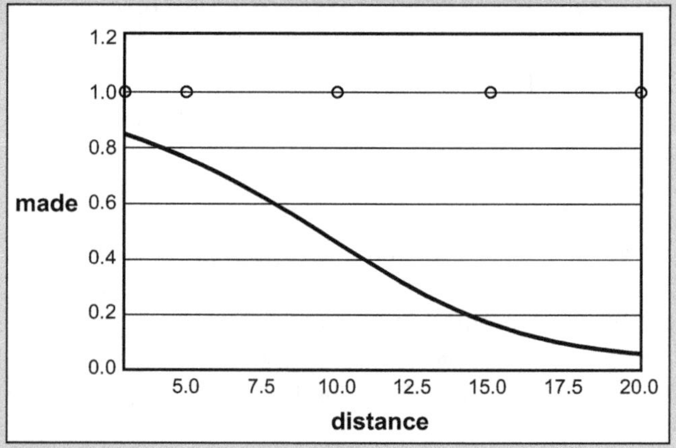

FIGURE 20.3
Sigmoid ("S") Curve for Evaluating Whether the Binary Variable of Whether Golf Putt Was Made (1) or Missed (0) at Different Distances Shows Decreasing Success at Longer Distances.

TABLE 20.1

Results of Binary Logistic Regression of Putting from 3', 5', 10', 15', and 20'

Results of Putting Success from 3', 5', 10', 15', and 20'

Distance	Made	Missed	Percentage Made
3'	17	3	85%
5'	15	5	75%
10'	10	10	50%
15'	3	22	12%
20'	2	23	8%
TOTAL	47	63	

Dependent binary (response) variable is putting success.
Independent (predictor) variables are the distance from the hole.
The regression equation is:

$$\text{Probability of making a putt (1)} = \frac{\exp(Y')}{1+\exp(Y')}$$

where $Y' = 2.175 - .2271$ (distance)

Variable	Coefficient	SE (coefficient)	Odds Ratio	*p*-value
Constant	2.1750	0.48	—	0.000
Distance	−0.2271	0.04	0.797	0.000

Analysis of Variance

Source	DF	Wald Test Chi-Sq	*p*-value
Regression	1	26.23	0.000
Distance	1	26.23	0.000

Model Summary

Deviance R^2	Deviance R^2(adj)	AIC
31.38%	30.72%	107.03

- Unfortunately, the model only explains about 31% of the variability of putting success, likely due to the sample size, and the fact that an amateur golfer is ordinarily wildly inconsistent.

We can use the model's formula to make predictions for the probability of making a putt at any given distance within our distance domain (3'–20'). For example, the probability of making a 12' putt would be 60%, calculated as:

$$Y' = 2.175 - .2271(12') = .41$$

$$.60 = \frac{\exp(.41)}{1+\exp(.41)}$$

Building upon our example, it is generally agreed that putting downhill is more difficult than putting uphill. A common strategy in golf is to land your ball on the green downhill from the hole, allowing you to putt uphill. Now the model is expanded to include putts in the downhill direction, and those observations were added to the original sample in order to assess the impact that putting direction and distance have on putting success, hopefully creating a better model.

Both distance and direction are statistically significant ($p = <0.05$). With data included for the direction of the putt, we can also assess the impact direction has on putting success. The positive coefficient for uphill putts (0.684) indicates that the probability for success increases when putting uphill. Also, the odds ratio for the categorical predictor, *direction*, indicates that the amateur golfer is almost two times as likely to make a putt when putting uphill vs. downhill (1.98). Having this understanding of uphill vs. downhill putts is useful as part of a strategy for approach shots to the green. With such a large odds ratio, the golfer is better served by making sure their approach shots are downhill from the hole.

With this general working knowledge of logistic regression, we will expand our use of logistic regression to concepts related to human and physical geography.

TABLE 20.2

Predicting the success of putting from 3', 5', 10', 15', and 20', and under different direction (uphill/downhill)

Dependent binary (response) variable is putting success.
Independent (predictor) variables are the distance from the hole as a continuous variable, and direction (uphill/downhill) as a categorical predictor.
The regression equation is:

$$\text{Probability of making a putt (1)} = \frac{\exp(Y')}{1+\exp(Y')}$$

where $Y' = 1.44 - .222$ (distance) for downhill putts
$Y' = 2.126 - .222$ (distance) for uphill putts

Variable	Coefficient	SE (coefficient)	Odds Ratio	*p*-value
Constant (downhill)	1.44	0.358	—	0.000
Distance	−0.227	0.032	0.80	0.000
Direction (uphill)	0.684	0.347	—	0.049

Odds Ratio for Categorical Predictors

Level A	Level B	Odds Ratio
Uphill	Downhill	1.98

Analysis of Variance

Source	DF	Wald Test Chi-Sq	*p*-value
Regression	2	46.87	0.000
Distance	1	45.92	0.000
Direction	1	3.88	

Model Summary

Deviance R-Sq	Deviance R-Sq (adj)	AIC
27.15%	26.44%	210

20.3 Multiple Variables

Example: Predicting the Probability of Mortgage Default Based on Financial Factors

For this example, we return to our fictitious town of Middletown. Somewhat concerned over the number of households with negative equity, the town was interested in predicting the probability that a household would default on their homeowner loan based on certain financial factors. To assess the situation, 80 households with negative equity in their mortgage were randomly surveyed to determine if they had defaulted on their mortgage. Planners also collected information on the number of years the homeowners were employed at their current job, current income, level of education, debt to income ratio, and credit card debt. The results of the analysis are shown in Table 20.3. Once again, the dependent variable is binary (default/not default), and the independent variables are logical choices to predict loan default.

The regression showed that employment, debt/income ratio, and credit card debt were statistically significant. Income level and level of education were not statistically significant, and discarded from the model. Middletown officials chose to leave credit card debt in as an explanatory variable, even though it had a p-value = 0.121. With positive coefficients, debt/income ratio and credit card debt showed that increases in these variables increased the probability of mortgage default. The negative coefficient for employment indicates that the longer a homeowner was in their current job, the less likely they were to default on their mortgage. Nonetheless, having an adjusted R^2 of only 22.25% shows that the model does not explain much of the variability in the dependent variable and thus, town officials might want to explore other variables that may have a greater impact on mortgage default.

TABLE 20.3

Predicting the Probability of Home Mortgage Default

Dependent binary (response) mortgage default
Independent (predictor) variables are the number of years in their current job, level of education (categorical), income, debt/income ratio, and credit card debt.
The regression equation is:

$$\text{Probability of loan default (1)} = \frac{\exp(Y')}{1+\exp(Y')}$$

where $Y' = -2.229 - 0.1433$ (year employed) + 0.1372 (debt to income ratio) + 0.306 (credit card debt)

Variable	Coefficient	SE (coefficient)	Odds Ratio	p-value
Constant	−2.229	0.777	—	0.004
Employment	−0.1433	0.0722	.8665	0.047
Debt/income ratio	0.1372	0.0527	1.147	0.009
Credit card debt ratio	0.306	0.197	1.357	0.121

Analysis of Variance

Source	DF	Wald Test Chi-Sq	p-value
Regression	1	13.9	0.003
Employment	1	3.94	0.047
Debt/income ratio	1	6.79	0.009
Credit card debt ratio	1	2.41	0.121

Model Summary

Deviance R-Sq	Deviance R-Sq (adj)	AIC	AICc	BIC	Area Under ROC Curve
31.38%	22.25%	77.83	107.14	87.65	0.8016

20.4 Comparing Two Models

Example: Salamander Presence or Absence Based on Environmental Variables

Sometimes it is valuable to compare the results of different logistic regression models to determine which one provides a better prediction. As an example, researchers performed a study on the abundance estimation and effects of habitat on endangered Eastern tiger salamanders (*Ambystoma tigrinum*) in Maryland and Delaware. The researchers examined 17 ponds for the presence/absence of salamander egg masses and trappings. The researchers also collected data on the pond area (m^2), depth of the pond (cm), water temperature (C°), dissolved oxygen DO (%/L), salinity, and pH. Data collection is expensive, so before expanding their research to a larger area, the researchers might be interested in observing whether a lower cost method for data collection identifying forest coverage and road density from aerial imagery would achieve similar results as the more labor-intensive process of collecting and analyzing water samples. To test these hypotheses, the researchers also used aerial imagery and GIS tools to identify the % of forested area within 300m, 600m, and 900m from a pond, along with the linear miles of road within the same distances. Using data collected by the researchers, we designed two logistic regression experiments to predict the probability of presence/absence of tiger salamanders, and present the results in Table 20.4 and Table 20.5. The first experiment utilized water samples collected at the site, and the second example examined the environmental variables related to forest cover and road density.

In Model 1, only pH was statistically significant. This does make sense as pH is a measure of how acidic or basic water is. The range for pH is from 0 to 14, with 7 being neutral; pHs of less than 7 indicate acidity, while a pH greater than 7 indicates a base. Proper pH is critical for the health of aquatic-based species, and most aquatic life prefer a pH range between 6.5–9.0. We also found that as pH increased by 1 unit, the odds of finding a tiger salamander increased almost 15 times (14.98). One of the reasons for this is that the ponds were mostly acidic or neutral. The sampled ponds did not include high pH values within the domain of values. Model 2 did not have a statistically significant independent variable, but we include the percent of forest within 300m for illustration ($p = -0.140$).

Comparing the two models, neither are all that successful. Model 1 explains 35% of the variability in the predicted outcome, while Model 2 only explains 19.48%. But, having two models also allows us to examine the Akaike Information Criterion (AIC). AIC is a method for evaluating how well a model fits the data used to generate it, and is used to compare different models. As discussed in Chapter 18, the best-fitting model is the one that explains the greatest amount of variation using the fewest possible independent variables, and the lower the

TABLE 20.4

Model 1: Predicting the Presence/Absence of Tiger Salamander Based on Water Quality Variables

Dependent binary (response) presence/absence of Tiger Salamander
Independent (predictor) variables are the pond area (m^2), depth of the pond (cm), water temperature (C°), dissolved oxygen DO (%/L), salinity, and pH.
The regression equation is:

$$\text{Probability of presence (1)} = \frac{\exp(Y')}{1+\exp(Y')}$$

where $Y' = -13.33 + 2.68$ (pH)

Variable	Value	Count
Species Present	1	13
	0	4
TOTAL		17

Variable	Coefficient	SE (coefficient)	Odds Ratio	*p*-value
Constant	−13.33	8.51	—	0.000
pH	2.68	1.62	14.58	0.098

Analysis of Variance

Source	DF	Wald Test Chi-Sq	*p*-value
Regression	1	2.74	0.098
pH	1	2.74	0.098

Deviance R-Sq	Deviance R-Sq (adj)	AIC	AICc
35.54%	30.15%	15.96	16.82

(continued)

AIC, the better the model fit. We see that the AIC value for Model 1 is lower than Model 2 (15.96 vs. 18.94), indicating that Model 1 does a better job at fitting the data. Therefore, while the % of forest coverage may be easier to collect using aerial imagery, it would appear that performing the hard work of field investigation of the pH in the ponds would yield better predictive results.

TABLE 20.5

Model 2: Predicting the Presence/Absence of Tiger Salamander Based on Land Parameters

Dependent binary (response) presence/absence of Tiger Salamander
Independent (predictor) variables are the % forest cover within 300m, % forest cover within 900m, linear meters of roads within 300m, linear meters of roads within 900m, and distance to a building structure in meters.
The regression equation is:

$$\text{Probability of presence (1)} = \frac{\exp(Y')}{1+\exp(Y')}$$

where $Y' = 10.93 + -0.11$ (% forest within 300 m)

Variable	Coefficient	SE (coefficient)	Odds Ratio	*p*-value
Constant	10.93	6.89	—	0.113
% forest (300m)	−0.11	0.07	0.89	0.140

Analysis of Variance

Source	DF	Wald Test Chi-Sq	*p*-value
Regression	1	2.18	0.140
pH	1	2.18	0.140

Deviance R-Sq	Deviance R-Sq (adj)	AIC	AICc
19.48%	14.09%	18.94	19.79

20.5 Ordinal Logistic Regression

Example: Building Damage After Superstorm Sandy

Oftentimes a dependent variable can take on more than two outcomes with a meaningful order. In those cases, we can use ordinal logistic regression. As an example, coastal geographers studied the effectiveness of sand dunes along New Jersey's coast in reducing land damage as a result of Superstorm Sandy. Damage was identified as low, moderate, and high. The researchers believed that dune barriers act like soft sea walls that, even when flattened or breached, can provide protection. They were interested in how effective sand dunes are in preventing the land behind the beaches from the storm surge, and which characteristics of dunes are the most effective.

With the dependent variable reflected as an ordinal value of low, moderate, and high damage, this study was suitable for ordinal logistic regression. The independent predictor variables we included in our model were:

- Dune height: The highest point of the dune relative to the ground in U.S. survey feet.
- Dune width: The wideness of the dune in U.S. survey feet, measured in the middle of each block, which usually represents the average width of a dune.
- Type of structure: The type of the first back-beach structure after dunes, 1 if it is a building, 0 if it is a boardwalk.
- Proximity to structure: The average distance between dunes and back-beach buildings in U.S. survey feet, 0 means the house is on the dune.
- Height of the structure: The average height of the back-beach structure (i.e., boardwalk or building) in U.S. survey feet.
- 60 locations were evaluated, and the ordinary logistic regression results are shown in Table 20.6.

Ordinal regression creates multiple prediction equations. For a dependent variable with k categories, $k - 1$ equations are created, each with a different intercept but sharing the same coefficients for the independent variables. The intercepts represent the points at which the dependent variable might be predicted into a higher category. For example, Intercept 1 (low) represents the coefficient for the formula for low damage, while Intercept 2 (moderate, low) represents the coefficient for moderate and low damage.

Interpreting the results for ordinal regression can be complex, so for our purposes we will simply focus on a few pertinent observations:

- The negative coefficients associated with dune width and dune height indicate that for every 1' increase in dune width, the corresponding odds for the damage class is reduced almost 7% (1 – odds ratio). In other words, as the dune gets wider, the odds for damage is reduced.

- Similarly, increasing the dune height by 1' reduces the odds for the damage class by 37% (1 – odds ratio). In other words, higher dunes appear to reduce the amount of damage.
- The positive coefficient for structure height indicates that for each 1' increase in building height, the odds in damage classification increases by 26%. In other words, as the building height increases, the chance for damage also increases.

The results presented in the ordinal logistic regression may allow coastal geographers a quantitative way to inform policy regarding the use of dunes in protection from hurricanes.

TABLE 20.6

Predicting the Probability of Building Damage Severity Using Dune Characteristics

Dependent ordinal (response) level of damage (low=2, moderate=3, severe=4).
Independent (predictor) variables are the dune width, dune height, proximity to the beach, type of structure, and structure height.

Damage Level	Value	Count
Low	2	21
Moderate	3	17
High	4	22
TOTAL		60

Variable	Coefficient	SE (coefficient)	Odds Ratio	*p*-value
Intercept 1 (low)	−14.254	3.905	—	0.000
Intercept 2 (moderate, low)	−10.940	3.577	—	0.002
Dune Width	−0.069	0.021	0.933	0.001
Dune Height	−0.439	0.193	0.644	0.023
Structure Height	0.234	0.064	1.264	0.001

Key Terms

The Akaike Information Criterion (AIC), *305*
Binary, ordinal, and multinomial logistic regression, *299*
Log odds, *301*
Odds ratio, *300*
S-Curve, *300*

Major Goals and Objectives

If you have mastered the material in this chapter, you should now be able to:

1. Explain the basic purposes of logistic regression, including binary, ordinal, and multinomial.
2. Explain the purpose of The Akaike Information Criterion (AIC), and how it is used to evaluate separate binary logistic regression models.
3. Define odds ratio and explain how it is used.
4. Make predictions based on the results of multiple regression and binary logistic regression.

References and Additional Reading

Anderson, C. J. and Leslie Rutkowski. "Multinomial Logistic Regression," in *Best Practices in Quantitative Methods*, edited by J. Osborne, pp. 390–409. Thousand Oaks, CA: Sage Publications, 2008.

Hosmer, D. W., Jr., S. Lemeshow, and R. X. Sturdivant. *Applied Logistic Regression*. 3rd ed. New York: John Wiley & Sons, 2013.

Sperandei, S. "Understanding logistic regression analysis." *Biochemia Medica*, Vol. 1 (2014): 12–18.

Ozbas, B. and M. Greenberg. *Correlation between Hurricane Sandy Damage along the New Jersey Coast with Land Use, Dunes and Other Local Attributes*. No. CAIT-UTC-019. Rutgers University. Center for Advanced Infrastructure and Transportation, 2013.

Park, H. "An introduction to logistic regression: From basic concepts to interpretation with particular attention to nursing domain." *Journal of Korean Academy of Nursing*, Vol. 43 (2013): 154–64.

Small, H. "Population size estimation and the effects of habitat on state-endangered Eastern Tiger Salamanders (*Ambystoma tigrinum*) in MD and DE." Unpublished Master's Thesis, Salisbury University.

21

Estimation with Spatial Data

This chapter explores several methods for multivariate spatial estimation, referred to as spatial interpolation. Spatial estimation is similar to traditional statistical estimation in that we seek to calculate a best point estimate along with a corresponding confidence interval. However, rather than a single estimate for an entire set of data, spatial estimation calculates a continuous value over an entire geographic space for the unsampled location(s). Also, spatial estimation is multivariate in the sense that in order to estimate an unknown value for a location, we are using the values measured at sample locations, along with their *X*, *Y* coordinates. Therefore, the estimated value for a specific location is a function of *X*, *Y*, and *Z* of the sample locations (in our discussion, *Z* does not have to represent elevation, but rather any continuous variable). For example, we may have a set of soil sample locations with measurements for Potassium (K) and Phosphorus (P) and want to use those observed values to estimate K and P at locations where we have not made measurements. In essence, this is similar to the probability maps discussed in Chapter 5 where we created isolines over a geographic space. In some cases, spatial interpolation creates a grid (or array) of locations over a geographic space and performs estimation on each location. In geographic information science, we often refer to arrayed or gridded data as a raster data structure. To illustrate the statistical concept of spatial estimation, we will explore two common methods of spatial interpolation: Inverse Distance Weighting (IDW) and Ordinary Kriging.

Figure 21.1 presents the original data from Figure 3.13, showing the Potassium (K) levels for 72 soil samples collected in space. To illustrate the calculations for spatial estimation presented in this chapter, we can consider removing a single point in the upper right portion of the map which originally had an observed K value of 40.82 ppm and identify 11 nearby locations surrounding the missing point to be considered to inform our estimation of K at the point where we have removed the observation. The map in Figure 21.1 shows that K increases from north to south, and also from west to east, indicating a spatial trend in the data.

21.1 Estimation Revisited with Spatial Considerations

Before moving into our spatial interpolation example, it is useful to illustrate the logical progression from traditional estimation to spatial estimation. Using all 71 ($n = 71$) locations, we can calculate the mean and 90% confidence interval for our unsampled location as 69.22 ppm +/− 6.37 ppm.

$$\bar{x} = \sum_{i=1}^{n} \frac{x_i}{n} = \frac{4{,}984}{71} = 69.22 \qquad \textbf{(21.1)}$$

$$s = \sum_{i=1}^{n} \sqrt{\frac{(x_i - \bar{x})^2}{n-1}} = 32.50 \qquad \textbf{(21.2)}$$

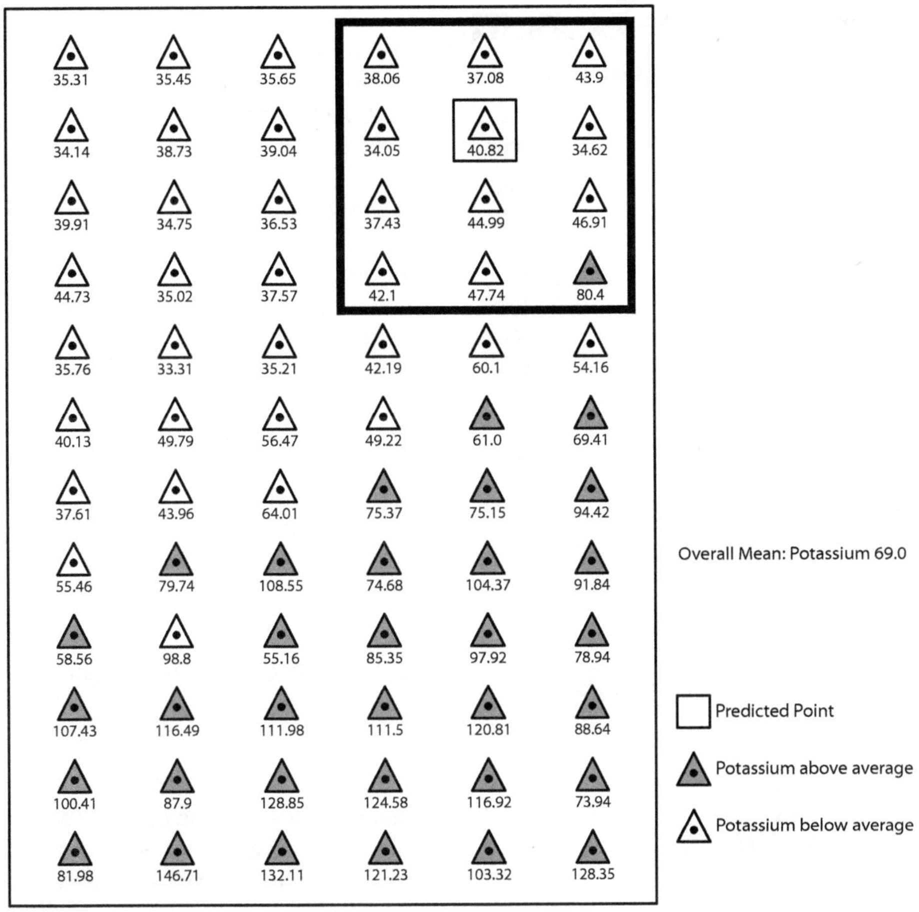

FIGURE 21.1
Potassium (K) Values for the Experimental Plot Presented in Chapter 3 Shows the Location of Soil Samples and corresponding K values. (Modified from Lembo et al., 2006.)

$$\sigma = \bar{x} \pm 1.65\sqrt{\frac{s^2}{n-1}} \tag{21.3}$$

$$= \sqrt{\frac{32.50^2}{70}} = 6.37$$

Therefore, we would state that we are 90% confident that a point drawn from random from this farm field would have a K value between 62.77 and 75.68. Recall however that the actual value of the unsampled location was 40.82.

Clearly, our estimated result is wrong on many levels. The global estimate is an overall or general estimate, but it is probably not the best one we could achieve: intuitively we expect the unknown sample to have a K value that is similar to the surrounding locations (i.e. there is a spatial trend to our data). Our calculated point estimate of 69.22 is nowhere near the values of the sampled locations shown in the upper right portion of the map. Second, the confidence interval does not even contain the true value (40.82). We should not be surprised by this, because our intention of calculating the BPE and CI in this example is not to measure a value chosen at random, but rather, to measure the value at a specific location. As such, the traditional estimation approach does not take the geographic locations into account at all. We expect a certain degree of spatial autocorrelation among the values, and anticipate our estimated value being close to the values of the nearby locations.

We can use the concept of spatial autocorrelation to assist us in making better estimates. Specifically, by evaluating the 11 sample locations surrounding our target unsampled location, we hope to obtain a sample that is more reflective of the geographic area. Table 21.1 shows the X, Y, and K values and distance of the 11 nearest points to the unsampled location.

$$\bar{x} = \sum_{i=1}^{11} \frac{x_i}{n} = \frac{487.3}{11} = 44.30 \tag{21.4}$$

TABLE 21.1

Eight Soil Samples in the Upper Right Portion of Figure 21.1

Location	X	Y	K	Distance
1	181	412	38.06	41.923
2	181	382	34.05	29.593
3	181	352	37.43	42.912
4	182	322	42.1	67.163
5	211	321	47.74	61.086
6	211	351	44.99	31.106
7	210	411	37.08	28.886
8	240	412	43.9	41.542
9	240	382	34.62	29.573
10	241	351	46.91	43.482
11	240	321	80.4	68.225

$$s = \Sigma_{i=1}^{11}\sqrt{\frac{(x_i - \bar{x})^2}{11-1}} = 12.90 \quad (21.5)$$

$$CI = \bar{x} \pm 1.65\sqrt{\frac{s^2}{n-1}}$$

$$= 44.30 \pm 1.65\sqrt{\frac{12.90^2}{10}} \quad (21.6)$$

$$= 44.30 \pm 1.65(4.08) = 5.33$$

Calculating the mean, standard deviation, and the 90% confidence interval using the 11 locations ($df = 10$) results in estimates of 44.30, 12.9, and 5.33 respectively. We may then state that we are 90% confident that the true value of K is between 38.97 and 49.63. Clearly, using only the sample locations in the vicinity of our unsampled location has advantages over using the entire data set, as the result appears more reflective of the values in the geographic area. Also, the confidence interval is much smaller, yielding a result that is more accurate and precise. Nonetheless, the confidence interval is still rather large, which is primarily due to a small sample size of only 11 observations being considered to define the spatial trend in the data. As you recall, sample size has a great influence on the confidence interval. If our sample size was 71, the confidence interval for the same standard deviation would have been reduced to 1.5. This is an important point to remember when we discuss the concept of Kriging, as we can take advantage of a local area information content, but at the same time maintain a high number of sample points to reduce the confidence interval.

Although we have obtained a better prediction when using the 11 nearest observations compared to when calculating an average for the entire data set, the result is still somewhat unsatisfying. Again, using this simple *neighbor* approach we are not actually calculating an estimate for a specific location, but are simply calculating an average for all 11 locations in the geographic region. To take the concept of a spatial relationship further, we notice in Figure 21.1, some locations are nearer than other locations to our unsampled location. This concept of nearness is further realized by examining the Distance column in Table 21.1. Clearly, some locations are closer in distance than other locations. In fact, locations 4, 5, and 11 are twice as far from our unsampled location than locations 1, 2, and 9. Therefore, we should expect locations 1, 2, and 9 to exert greater influence on the unsampled location than locations 4, 5, and 11. The relationship of nearness and its influence on a prediction is the basis for inverse distance weighting and Kriging.

21.2 Inverse Distance Weighting (IDW)

Given our expectation that nearby locations have more influence over the unsampled location, we can explore the concept of inverse distance weighting (IDW). IDW is a deterministic method of multivariate interpolation that estimates values at locations by computing the weighted average of sample locations based on their distance to the unsampled location. IDW is basically a weighted average, with the weights being a function of distance.

Similar to our discussion of weighted mean in Chapter 4, the closer a sample location is from the unsampled location, the greater influence (i.e., weight) it has on the computation. IDW is considered a deterministic interpolation method because the predicted values are based on the surrounding locations and a specific mathematical formula to create a final surface.

IDW assumes that each sampled location has an influence on the unsampled location based on the distance between them. This influence decreases the further away the sampled point is from the unsampled point, so IDW uses the inverse distance (1/d) in its calculations. The inverse distance (1/d) is generally elevated to a power. Usual values for the power are between 1 and 3. If the power = 0, IDW will return the global average of the dataset as its prediction for each unsampled location if the search radius includes all points. If the search radius includes some points and excludes others, IDW will return a local average.

IDW is really an average that includes weights for each sampled location, and that the weights are based on the inverse of the distance to the unsampled location. The constraints of using a weighted average prevent IDW from making estimates above or below the minimum and maximum sample values. IDW is considered an *unbiased estimator*, meaning the weights used to predict an unknown value must add up to 1. IDW is also known as an *exact interpolator* in that when $d = 0$, the predicted value is exactly the value of the sample location. The effect of an exact interpolator is shown in Figure 21.2, illustrating both a profile and overhead view that shows an interpolated line passing directly through the locations used to construct the interpolation.

IDW does offer a certain level of flexibility in terms of the number of sample locations to include and the relative influence a location has on the overall prediction. These variables are referred to as the *search radius* and *power*.

The search radius allows selection of a specific number of locations to include in the calculation. In Figure 21.1, we have highlighted the upper-right portion of the study area to utilize the 11 closest locations to our location of interest. Nothing prevents us from including all 71 locations in the study area, or some other number of locations. The number of locations to include in the estimation depends upon the relative importance of nearby and distant locations. For example, while points further away may have some influence, there may come a point where the influence is so limited, or nonexistent, that including those locations is detrimental to the prediction. As an unbiased estimator, every unsampled location will receive a weight, and all weights must add up to 1. If distant locations have no real-world influence on our location of interest (there is no spatial autocorrelation), it will still receive a weight (albeit small), essentially stealing some power away from the more nearby locations. Unfortunately, IDW does not provide any way for us to know how many locations to include in our estimate. For that, we will need to refer to the variogram, presented later in this chapter.

We can also alter our *power* in the inverse distance calculation as $1/d^p$, where p represents a power function. If $p = 1$, then d remains unchanged from its original distance value, which is considered the default case for IDW. When $p = 2$, then d represents the inverse-distance squared. Increasing the power causes the weights of the locations to decrease rapidly and exhibit a much lower weight (i.e., $1/100^1 = 0.01; 1/100^2 = 0.001$). In this way, we can further penalize or reward a location based on its distance from our unsampled location. Increasing the power an excessive amount may create an artificial bond between the nearby locations, resulting in unappealing *bullseyes* around our sample locations when isolines are introduced. Figure 21.3 illustrates the influence a higher power function introduces on driving a value towards a sample location. Unfortunately, like the search radius, IDW does not really provide us with assistance in determining which power function to utilize. In most cases, a trial-and-error approach are performed. However, when we introduce the variogram later in this chapter, we can have some level of understanding regarding how rapid the changes occur over distance.

IDW has some obvious benefits. As a deterministic method, there are few decisions to make when implementing the solution. Its implementation is simple and straightforward, and therefore fast to compute. However, as an exact interpolator, IDW does not allow flexibility to model potential sampling error or variability at a sample location. Also, IDW assumes that influence on an unsampled location is equal in all directions (*omnidirectional*) and does not emphasize the influence of any particular direction. In many spatial examples, such as wind direction on chemical dispersion, directionality is an important factor.

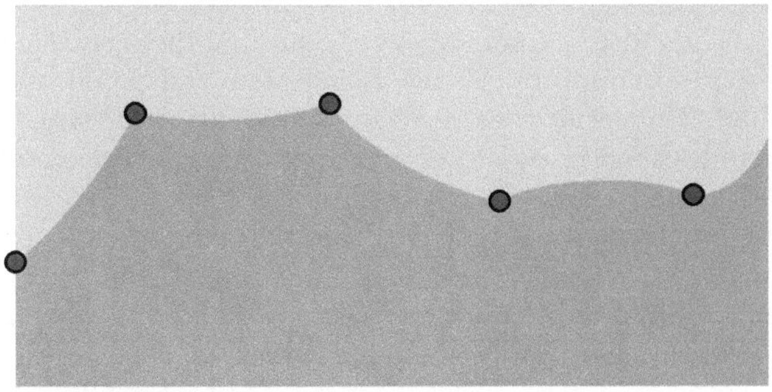

FIGURE 21.2
The Effects of IDW Being an Exact Interpolator Illustrates That Estimations Will Pass Directly Through the Sample Points Used to Form the Estimation.

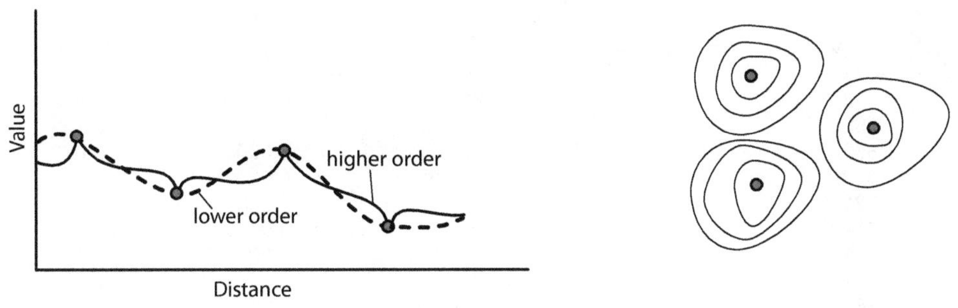

FIGURE 21.3
Effect of Adding a Higher Power Function to IDW Causes an Interpolated Line to Abruptly Pass Through Sample Locations, Creating Bullseyes in a Map View.

The general formula for IDW is

$$T_{unk} = \sum_{i=1}^{n} z_i \left(\frac{\frac{1}{d_i^p}}{\sum_{i=1}^{n} \left(\frac{1}{d_i^p} \right)} \right) \quad (21.7)$$

where: T_{unk} = the estimate for the unsampled value

z_i = the value of sample location i

d_i^p = distance, raised to the power p, of sample location i to the unsampled location

It is often stated that IDW does not allow a user to obtain error estimates for a prediction. Thus, we do not see GIS software provide error estimates as part of their IDW calculations. However, from a mathematical standpoint, we can estimate error based on the algebraic manipulation of weights and covariance among the sample locations. Although the derivation is quite lengthy, Clark and Harper (2000) show the variance of the estimate σ^2 reduces algebraically to:

$$\sigma^2 = \sum_{i=1}^{m} \sum_{i=j}^{m} w_i w_j cov(Z_i, Z_j) \quad (21.8)$$

where: σ^2 is the variance of the prediction

w is the weight of the location giving influence to the unknown value

$ov(Z)$ is the covariance of the sample values

Returning to our example regarding estimation of K at the assumed unsampled location on the map in Figure 21.1 and based on the calculations from Table 21.2 and Equation 21.9, we obtain a best point estimate of 40.5 ppm for K concentration.

$$T_{unk} = \sum_{i=1}^{n} z_i \left(\frac{\frac{1}{d_i^p}}{\sum_{i=1}^{n} \left(\frac{1}{d_i^p} \right)} \right) = 40.5 \quad (21.9)$$

The confidence interval is calculated as:

$$\sigma^2 = \sum_{i=1}^{m} \sum_{i=j}^{m} w_i w_j cov(Z_i, Z_j) = 140.71$$

where

$$cov(Z_i, Z_j) = 150.30 \quad (21.10)$$

Applying the confidence interval formula we calculate the 90% confidence interval as

$$\begin{aligned} CI &= \bar{x} \pm t \sqrt{\frac{\sigma^2}{n}} \\ &= 40.5 \pm 1.37 \sqrt{\frac{140.71}{11}} \quad (21.11) \\ &= 40.45 \pm 4.89 \end{aligned}$$

Therefore, we are 90% confident that the true value of our unsampled location is between and 35.60 and 45.39. This result is more accurate than using the simple average of the 11 points, as it takes nearness into account and actually calculates an estimate for a specific geographic location. Also, the standard error is slightly smaller, providing a more precise estimate. But again, our standard error is somewhat hindered by only having a sample size of 11 to infer spatial trend.

TABLE 21.2

Inverse Distance Weighting Calculation for 11 Soil Samples in the Upper Right Portion of Figure 21.1

ID	X	Y	Potassium (K)	Distance (d)	$\frac{1}{d^2}$	$\frac{\frac{1}{d^2}}{\Sigma \frac{1}{d^2}}$ (weight)	K * weight
1	181	412	38.06	41.923	0.0006	0.076	2.910
2	181	382	34.05	29.593	0.0011	0.153	5.224
3	181	352	37.43	42.912	0.0005	0.073	2.731
4	182	322	42.1	67.163	0.0002	0.030	1.254
5	211	321	47.74	61.086	0.0003	0.036	1.719
6	211	351	44.99	31.106	0.0010	0.139	6.248
7	210	411	37.08	28.886	0.0012	0.161	5.971
8	240	412	43.9	41.542	0.0006	0.078	3.418
9	240	382	34.62	29.573	0.0011	0.154	5.319
10	241	351	46.91	43.482	0.0005	0.071	3.334
11	240	321	80.4	68.225	0.0002	0.029	2.321
			SUM	485.5	0.0074	1.0	40.45

21.3 Ordinary Kriging

Kriging is a geostatistical technique used to predict the values of unsampled locations in space using information from sampled locations. Similar to IDW, Kriging uses weights of neighboring locations. However, while IDW calculates an estimate of an unsampled location based only on the distance to the sampled locations, Kriging expands the concept of spatial autocorrelation to leverage the spatial arrangement among all the sampled locations. Therefore, Kriging uses the spatial arrangement of sample locations and their relationship to one another to make a statistical inference about the unsampled location. In other words, even if two sample locations are not near the unsampled location, the relationship they have with one another can help inform the relationship the unsampled location may have with its neighbors. Thus, the overall structure of the spatial pattern in the data can inform interpolation in addition to the proximity of measurements to unsampled locations. Further, Kriging not only results in a continuous surface of estimated values, but also generates an estimate of uncertainty around each prediction. This latter aspect is a considerable strength as it gives the user confidence in the prediction itself.

As a more advanced geostatistical concept, Kriging offers a number of mathematical adaptations to address specific characteristics of the underlying data including Simple Kriging, Ordinary Kriging, Block Kriging, Co-Kriging, and Indicator Kriging, among others. We will limit our discussion of Kriging to the concept of Ordinary Kriging, which is often viewed as the most common method of Kriging used in spatial analysis. For Ordinary Kriging, we assume that the means and variance of the sample locations are constant across the geographic area.

Obtaining a continuous surface for Kriging is a two-step process. In the first step, we create an *experimental variogram* based on the relationships (spatial and autocorrelation) of the sample locations, and fit a model (known as the variogram model) to the experimental variogram points. The second step is to use the variogram to calculate the weights used to predict values for unsampled locations.

As previously discussed in Chapter 13, a semivariogram is a graphic model that quantifies spatial autocorrelation between sample locations as a function of the distance between one sample location and another. The nugget, sill, and range in the semivariogram not only graphically illustrate the concept of spatial autocorrelation that nearby locations are more similar than distant locations, but also provide quantitative values on the nature of the spatial autocorrelation, allowing us to calculate the weights through a least-squares calculation. The theoretical best-fit line used to model the experimental variogram can only take a few general forms, some of which include a linear fit, general linear fit, spherical fit, exponential, and power. In most geographic examples, the spherical model is used, as it appears to best exhibit the characteristics of spatial autocorrelation (nearby locations are more similar in value than distant locations, and at some distance there is no longer any relationship among the locations).

To revisit our discussion of the variogram in Chapter 13, all pairs of sample points are compared for both their distance away from one another (h) and the squared difference in their measured values. Distances are grouped into categories (e.g., 0–10m, 11–20m, etc.) referred to as lags, and the average squared differences (γ for each lag distance are calculated, allowing us to plot a graph of points for h on the X axis and γ on the Y axis. The graph communicates that for all locations h units apart, they have an average squared difference in their values of γ. The purpose of fitting a line through the pairs of points is to form a simple mathematical relationship between h and γ and use it to predict values of the variable for any distance (not just the distances between sampled locations).

In order to fit a line through the points in a theoretical semivariogram, a combination of functions are used, depending upon the characteristics of the experimental semivariogram. The nugget in the formula is represented as C_0, representing a combination of sampling error and nearby heterogeneity (diversity) as the distance between samples approaches 0. The easiest function to fit the data is the linear model (a) which is essentially a straight line with a single slope term (p) along with a nugget (C_0), as shown in Figure 21.4. The spherical model (b), the most frequently used in geographic analysis, uses three functions to fit a curve from the nugget to the sill, and then fits a straight line at the range. The generalized linear model (c) uses two formulas with two terms representing the slope (p) and a power function (a). We will use the general linear model in a simple example to illustrate the mathematical procedure to make predictions. Similar to the regression line we discussed in Chapter 17, the curve (or line in the case of a linear relationship) in a theoretical semivariogram establishes a simplified relationship among sample locations that relate the differences in sample values to the distances they are from one another. As an example, if $\gamma = 6$ at $h = 300$ m, we can then assume if locations that are 300 m apart anywhere throughout the study area are always six units different in their measured value, then perhaps any unsampled location might be approximately six units different from any sample locations that are 300 m away. Creating a function that quantifies the difference in sample values to the distances those samples are from one another allow us to use many combinations of sample locations to establish a prediction for the unsampled locations based on the unsampled location's distance from the sample locations.

So far our discussions have focused on the distance between sample locations assuming that the variation of a variable is the same in all directions. In many geographic situations, there is often directional influence on a spatial variable. For example, winds out of the northeast may have a directional influence on sulfur dioxide levels downwind from a pollution point source. The ability to account for this level of directionality is known as *anisotropy.* While our implementation of variogram modeling discussed here ignores a directional influence, variogram models can be rather sophisticated and include a directional influence by creating multiple variograms that model specific directions along with distances.

Unlike IDW, where the weight depended only on the distance to the unsampled location, the weights for Kriging are based on both the distance between the sample locations and the unsampled location and also on the overall spatial arrangement of the sample locations with one another. In Ordinary Kriging, the weight uses the theoretical semivariogram model (see Figure 21.4) to quantify the sample locations, the distance to the unsampled location(s), and the spatial relationships among the sample locations and the unsampled location(s). To illustrate this rather complex mathematical relationship, we will return to our estimation of K concentration at the unsampled location in Figure 21.1 and work through an over simplified example using only three locations to estimate the unsampled location. We should note, of course, that our use of three locations is only for illustration and not appropriate in practice.

An Example Using Three Sample Locations

Table 21.3 provides the X, Y, K, and distance to our unsampled location for each of the three sample locations, and Figure 21.5 shows the sample placement along with the distances among all samples and the unsampled location.

Figure 21.6 shows the experimental semivariogram for all 71 sample locations presented in Figure 21.1, broken into eight lags of 50 m in distance, and fit with a generalized linear model.

We can consult the semivariogram in Figure 21.6 to determine the semivariance $\gamma(h)$ for the lag intervals. In our case, we only have three points with distances of

Requirements for Ordinary Kriging

1. The data needs to be sampled from a variable that is continuous in space.
2. The values for the sample locations are spatially autocorrelated.
3. Stationarity—the correlation of location values is based on the distance between them, not on their location in the study area. In other words, variation among the sample location values are similar throughout the study area. If two locations are 10 m apart, the differences in their values should be similar, no matter where those samples are located in the study area.

TABLE 21.3

Example of three locations

Location	X	Y	K	Distance
T_{unk}	210	382	?	—
1	210	411	37.08	28.88
2	240	412	43.9	41.52
3	240	382	34.62	29.57

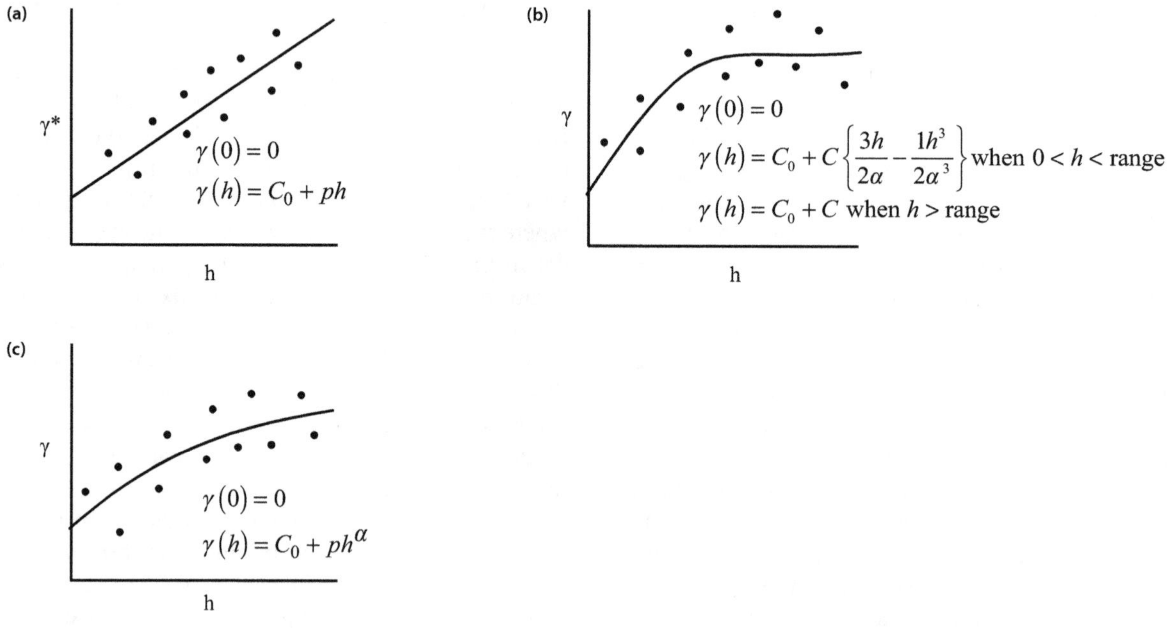

FIGURE 21.4

Three Different Models of the Theoretical Semivariogram Along With Their Formulas That Are Used to Fit the Experimental Semivariogram: (a) Linear (b) Spherical, and (c) Generalized-Linear.

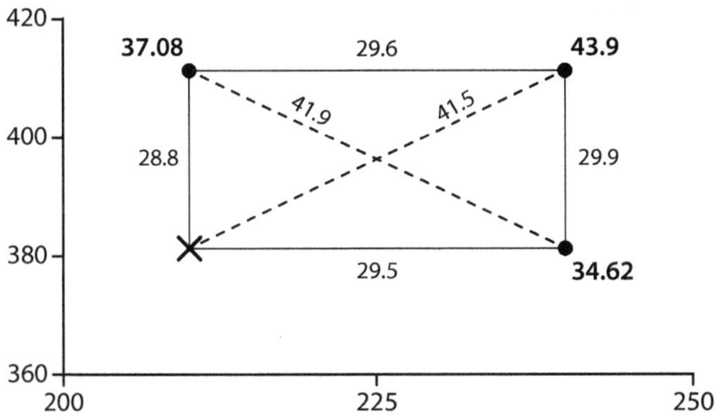

FIGURE 21.5
Three Location Examples to Illustrate Kriging Solution Shows the Sample Points, Unsampled Location, and Distances Among All the Locations.

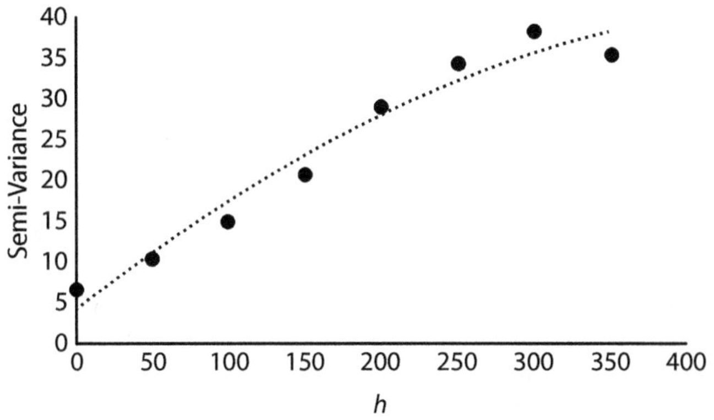

FIGURE 21.6
Variogram of Potassium Values for 71 Sample Locations Shows a Range of Approximately 280 and a Sill of 30.

28.8, 41.5, and 29.5. Therefore, to simplify matters, we will use the semivariogram distances for 30 and 60 to obtain the $\gamma(h)$ for those lags.

Using the model formula for a generalized-linear semivariogram, we will plug in our values for $\gamma(h)$ and h:

$$\begin{aligned}\gamma(h) &= C_0 + ph^a \\ 9.5 &= C_0 + p*30^a \\ 12.4 &= C_0 + p*60^a\end{aligned} \quad \textbf{(21.12)}$$

To simplify our solution to include two unknown variables (p, a) and two equations, we removed the nugget value (C_0), essentially considering it equal to 0, allowing for easier calculation of a and p.

$$\begin{aligned}a &= 0.14 \\ p &= 5.91\end{aligned}$$

We ignore the complex derivation of solving the two nonlinear equations and leave it to the reader to utilize an online calculator. With the function for γ now established, we can present the linear equations necessary to solve for our unknown weights. We have four unknown variables (w_1, w_2, w_3, and λ) and with the knowledge that all the weights must add up to 1, we can create the following four equations to solve for our four unknown variables

$$\begin{aligned}w_1\gamma(g_1,g_1)+w_2\gamma(g_1,g_2)+w_3\gamma(g_1,g_3)+\lambda &= \gamma(g_1,T) \\ w_1\gamma(g_2,g_1)+w_2\gamma(g_2,g_2)+w_3\gamma(g_2,g_3)+\lambda &= \gamma(g_2,T) \\ w_1\gamma(g_3,g_1)+w_2\gamma(g_3,g_2)+w_3\gamma(g_3,g_3)+\lambda &= \gamma(g_3,T) \\ w_1+w_2+w_3 &= 1\end{aligned} \quad \textbf{(21.13)}$$

substituting the distances between the pairs of sample locations (g_i,g_j) and the distance of the sample location to the unsampled location (g_i, T), gives us the following formula

$$\begin{aligned}w_1\gamma(0)+w_2\gamma(29.59)+w_3\gamma(41.92)+\lambda &= \gamma(28.88) \\ w_1\gamma(29.593)+w_2\gamma(0)+w_3\gamma(29.99)+\lambda &= \gamma(41.52) \\ w_1\gamma(41.92)+w_2\gamma(29.99)+w_3\gamma(0)+\lambda &= \gamma(29.57) \\ w_1+w_2+w_3 &= 1\end{aligned} \quad \textbf{(21.14)}$$

We can then use the distances from each sample location to the unsampled location in place of h to determine

$$\begin{aligned}\gamma(h) &= 1.27(h)^{0.53} \\ \gamma(g_1,T) &= 1.27(28.88)^{0.53} = 9.46 \\ \gamma(g_3,T) &= 1.27(41.52)^{0.53} = 9.95 \\ \gamma(g_3,T) &= 1.27(29.57)^{0.53} = 9.49\end{aligned} \quad \textbf{(21.15)}$$

the distance between each sample as

$$\begin{aligned}\gamma(g_1,g_2) &= 1.27(29.59)^{0.53} = 9.49 \\ \gamma(g_1,g_3) &= 1.27(41.92)^{0.53} = 9.97 \\ \gamma(g_2,g_3) &= 1.27(29.91)^{0.53} = 9.51\end{aligned} \quad \textbf{(21.16)}$$

TABLE 21.4

Semivariance for distances represented by three sample locations

(γ)	(h)
9.5	30
12.4	60

and insert

$$\begin{aligned}0w_1+9.49w_2+9.97w_3+\lambda&=9.46\\9.49w_1+0w_2+9.51w_3+\lambda&=9.95\\9.97w_1+9.51w_2+0\ \ +\lambda&=7.49\\w_1+w_2+w_3&=1\end{aligned}$$

using algebraic manipulation, we solve for our unknown weights (w) and λ. If you are uncomfortable solving simultaneous equations, there are numerous websites that allow you to enter the example equations and solve for the unknown variables w_1, w_2, w_3 and λ. In practice, however, the solution is completed using computer programs that leverage matrix algebra. The result of the simultaneous solution for our three-sample example is

$$w_1=0.357;\ w_2=0.28;\ w_3=0.354;\ \lambda=3.19 \quad \textbf{(21.17)}$$

Notice that the weights (w) add up to 1, making this an unbiased estimator. With values for our weights known, we can now solve for our sampled value

$$\begin{gathered}T=w_1g_1+w_2g_2+w_3g_3\\T=0.357(37.08)+0.288(43.90)+0.354(34.62)\\T=38.17\end{gathered} \quad \textbf{(21.18)}$$

and, we can calculate the σ^2 as

$$\begin{gathered}=w_1\gamma(g_1,T)+w_1\gamma(g_1,T)+w_1\gamma(g_1,T)+\lambda-\gamma(T,T)\\=0.357(9.46)+0.288(9.95)+0.354(9.49)+3.19(1)-0\\=12.80\end{gathered} \quad \textbf{(21.19)}$$

$$CI=\bar{x}\pm1.37\sqrt{\frac{12.80}{3}}=2.83$$

Therefore, we are 90% confident that the true value of our unsampled location is between 35.33 and 41.00. Again, the result is not terribly accurate as we only used three locations to model the complex processes in Kriging. Nonetheless, the small example serves as a good illustration of the mathematical principles behind Kriging, and how the weights are formed. We simply decreased the size of our sample in order to better illustrate the mathematical procedure for solving the problem.

21.4 Geographic Examples

Example: Potassium Values for Entire Farm Field

Inverse Distance Weighting Solution

Having worked through the examples of IDW and Kriging using a small subset of the original data, we now expand the example using all 71 locations to estimate the unsampled locations throughout the entire farm field being considered. The field is represented as a raster data set with 138,301 cells (251 columns × 551 rows). The calculations we performed earlier are now expanded to all 138,301 cells, making it obvious that spatial interpolation on a raster data set must be performed using a computer. To perform the estimation, we used ArcGIS Pro developed by Environmental Systems Research Institute (Esri), the leading commercial GIS. Using brute-force trial and error, we can experiment with different parameters for the power and search radius. Figure 21.7 illustrates the IDW solution using a power = 1 and a search radius to include all 71 locations; power = 2 search radius to include all 71 locations; power = 2 search radius to include 24 locations; and power = 4 search radius to include 24 locations.

Applying the different parameters yields very interesting results, consistent with the theoretical concepts we discussed earlier. In Figure 21.7a, using all 71 sample locations and a power = 1 provides a smooth map with a single overall trend. The smoothness is a result of using all 71 sample locations to estimate each cell. As we stated, the weights in the example must add up to 1, and by including every sampled location, we steal some power away from nearby locations. Also, a power = 1 does not dramatically reduce the influence of locations further away or dramatically increase the influence of nearby locations.

Figure 21.7b also uses all 71 locations, but changes the power from 1 to 2. The influence of increasing the power is evident in the creation of a few bullseyes around the sample locations, and an increased sinuosity of the pattern. Figure 21.7c maintains a power of 2, but reduces the search radius to include only the closest 24 sample locations. Here again, we observe greater sinuosity and additional bullseyes around the sample locations as fewer samples are contributing to the overall weight, allowing the nearby samples to have a greater influence on the prediction. Figure 21.7d increases the power to 4, which is ordinarily not advisable, but included here to illustrate the impact an excessive power has on the overall predictions. Here we have a very unpleasant looking map as the power function rapidly drives prediction values closer to the sample locations, creating many bullseyes.

Each of the four results (a-d) are simply a response to the formulas used to create them. None of these results are

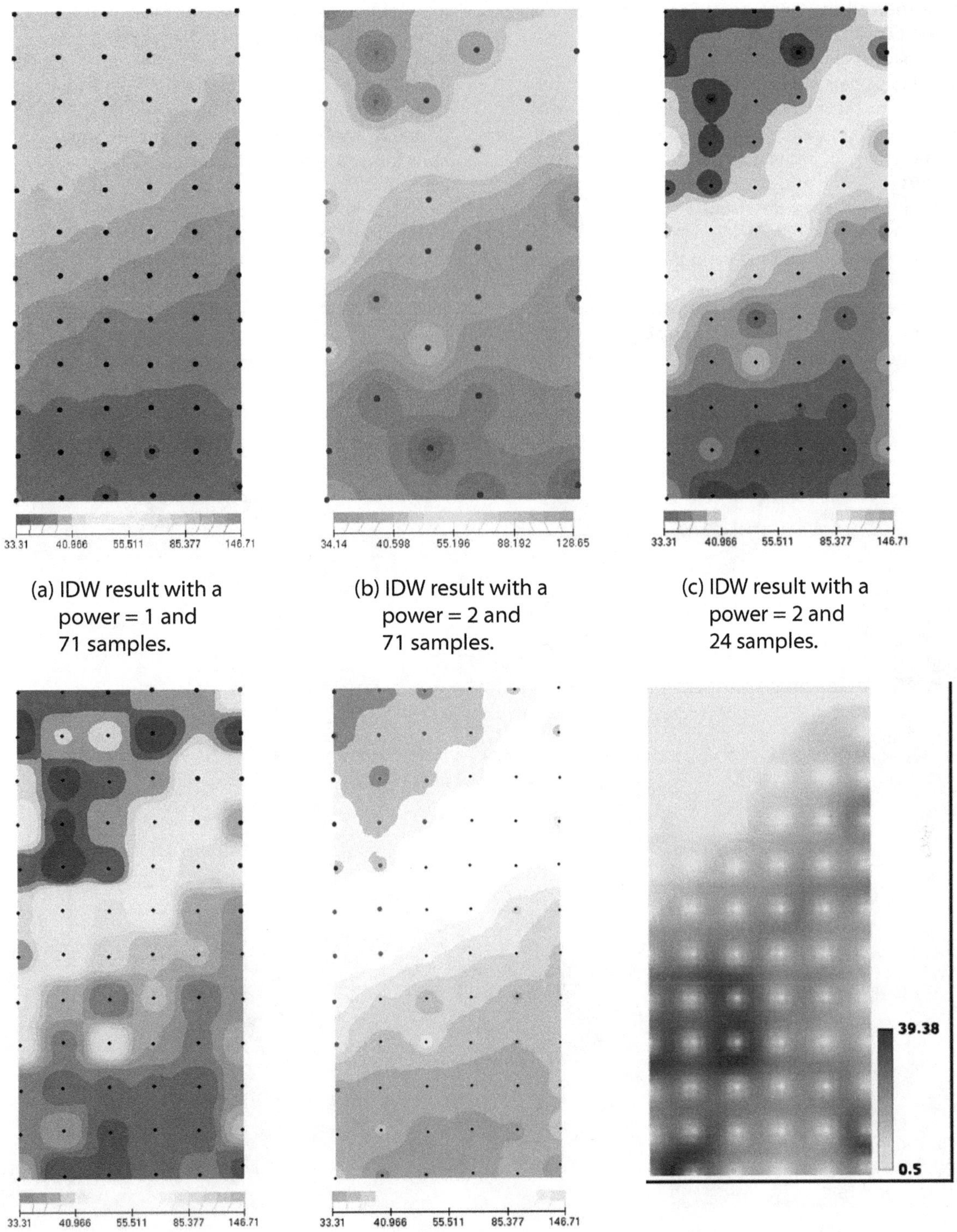

(a) IDW result with a power = 1 and 71 samples.

(b) IDW result with a power = 2 and 71 samples.

(c) IDW result with a power = 2 and 24 samples.

(d) IDW result with a power = 4 and 24 samples.

(e) IDW result with a power = 1.5 and 60 samples.

(f) Standard errors for IDW solution.

FIGURE 21.7

Results of IDW Interpolation Under Different Power and Search Radius Values Illustrates the Differences Each Parameter Has on the Overall Prediction.

(continued)

wrong, although we could easily argue that some are more appropriate than others. This is not unlike our discussion in Chapter 2 on classification schemes. Understanding the purpose and impact of our choices in the power and search radius can help us better understand the most important parameters to choose, and not get bogged down in a trial-and-error approach for selecting the parameters.

Fortunately, we do not have to blindly apply random combinations of power and search radius. As previously stated, the variogram provides useful information on the distance at which autocorrelation no longer has an effect on the sample values. Although the variogram is often associated with Kriging, an astute quantitative geographer will recognize its value and leverage the information in the variogram to make an informed decision about the IDW parameters. Including locations beyond the range has no benefit, as they have no distance-based impact on the location values. Also, the slope of the variogram provides a glimpse into the rate of change in sample values as distance increases. Figure 21.6 shows the variogram for our 71 sample locations. The range (*h*) is just under 300 m, meaning most points within that distance are spatially autocorrelated. The slope of the line is not terribly steep either. Therefore, we are probably justified in using a larger number of sample locations (reflective of 300 m apart), and keeping our power relatively low. Figure 21.7e uses a distance of 300 (around 60 locations) and a power of 1.5 based on the information we obtain from the variogram. And finally, Figure 21.7f shows the standard errors of the IDW solution. We note that the map of standard errors required us to automate Equation 21.8 ourselves as GIS software vendors have ignored the generation of standard error calculation in their IDW solutions. You will notice that the standard errors are the smallest nearest the location of the sample locations.

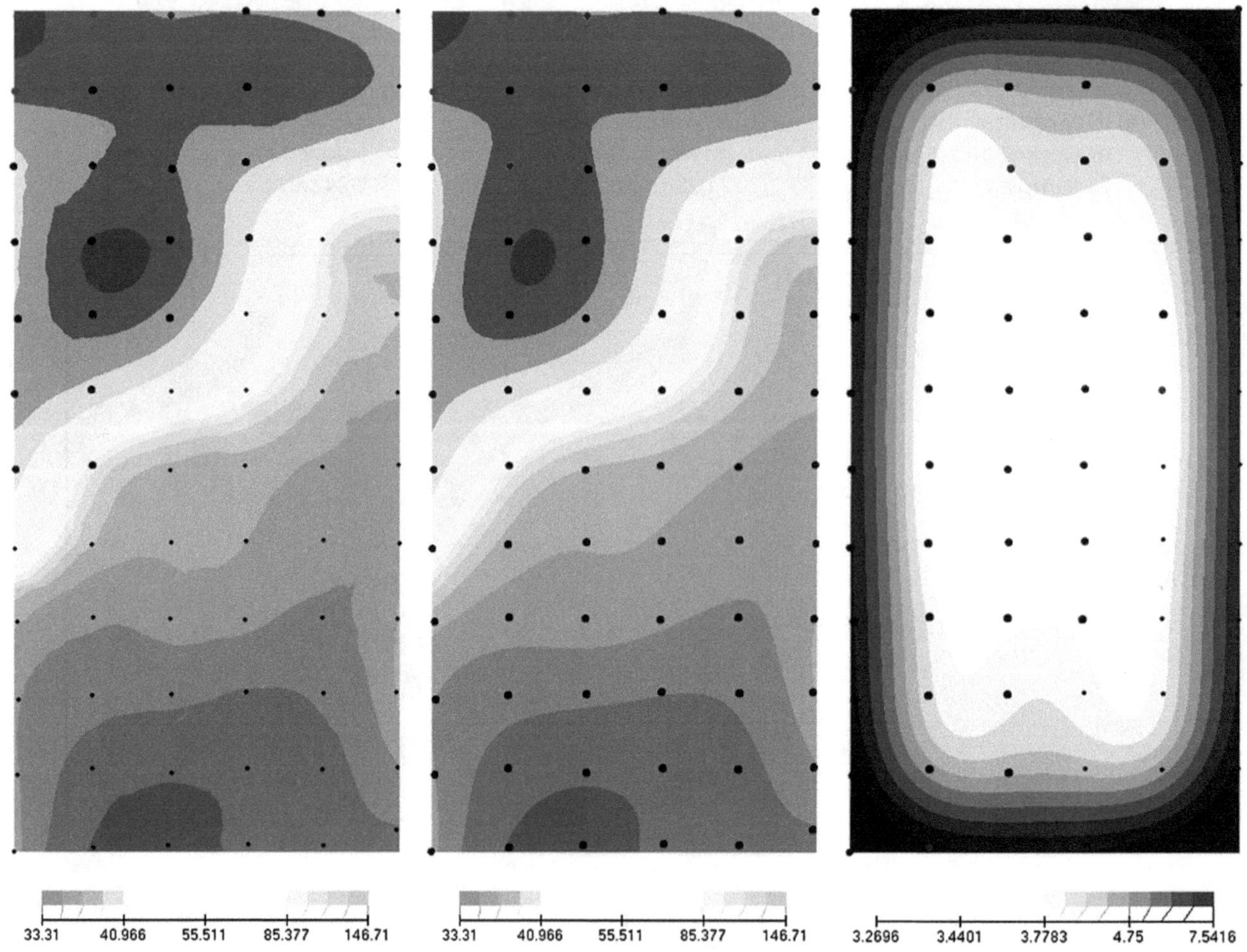

(a) 24 samples (b) Samples = 71 (c) Standard error

FIGURE 21.8
Results of Kriging interpolation and Standard Error.

Ordinary Kriging Solution

Using the variogram, we can also apply the concept of Kriging to our Potassium levels in the farm field. Once again, we can reference the variogram in Figure 21.6 to obtain estimates for the nugget, range, and sill, and then enter those values into the Kriging solution module in ArcGIS Pro to generate a surface. Figure 21.8 illustrates the result of Kriging when we consider 24 and 71 sample locations. The results are more aesthetically pleasing than the IDW results and the readily available GIS software provides us with standard errors for our predictions. Our use of B&W printing makes it a little difficult to determine the range of values in our Figure 21.8c. Nonetheless, it is fairly obvious that we have higher confidence in the predictions for points in the center of the map than along the edges.

Example: Last Spring Frost

We return to our example of the last spring frost, introduced in Chapter 1, and analyzed in Chapter 17 using linear regression. Our goal is to create a continuous surface of estimated last spring frost (LSF) dates over the entire southeastern United States. To perform the analysis, we had to project the sample locations from a spherical latitude/longitude coordinate system to a planar Universal Transverse Mercator system. We notice from the variogram in Figure 21.9 that sample locations are no longer spatially autocorrelated beyond 400 miles (range). We also observe a nugget value of four days, indicating that for sample locations very near one another, there is a small level of variability due to sampling error. Using these values with the ArcGIS Pro software, creates the interpolated surface shown in Figure 21.10. We can use the resulting surface in conjunction with other geographic layers as part of a specific GIS analysis process. In addition, we have a corresponding surface of standard errors, allowing us to estimate the confidence interval for each estimated cell in the raster data set.

Figure 21.10 illustrates the impact of latitude on the LSF value, and also the general trend of later LSF values as we move over the Appalachian Mountains. The rise in elevation clearly makes *winter hold on a little longer*, resulting in later dates for the last spring frost. We also observe earlier LSF values closer to the Atlantic Ocean, an indication of the moderating effect of the ocean on temperatures. Figure 21.11 is a map showing

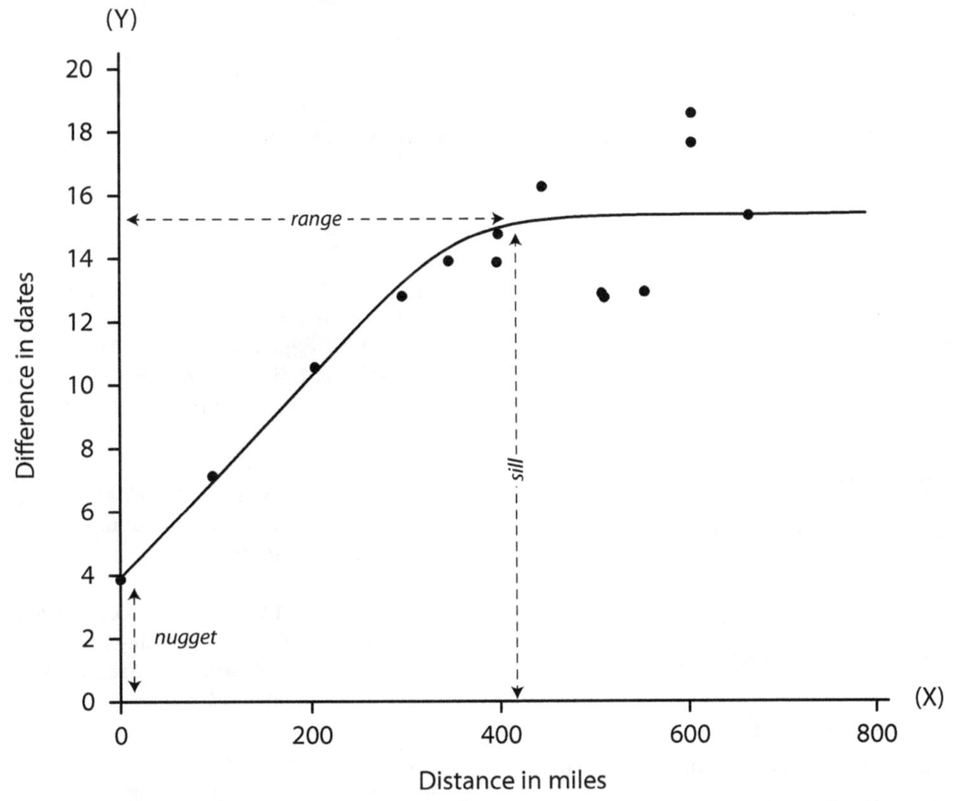

FIGURE 21.9
Variogram for the Last Spring Frost (LSF) Data.

(continued)

standard errors of the prediction locations, giving us an idea of where greater uncertainty exists in the Kriging solution.

Even though we have attempted to simplify the presentation of spatial estimation, this has been one of the more mathematically rigorous chapters in the book. We have only briefly touched on the use of interpolation methods to estimate continuous values at unsampled locations over a geographic area. Entire textbooks and university courses are devoted to the study of interpolation and estimation of spatial data. In fact, it would not be uncommon to see a university offer multiple courses on Kriging alone. Even so, our treatment of the subject should allow the reader to observe how spatial estimation methods are similar to traditional statistical estimation principles, with certain considerations for spatial autocorrelation factored in. Also, when attempting to use interpolation techniques offered in GIS software, the reader will be better prepared to make the most appropriate decisions to achieve satisfactory results.

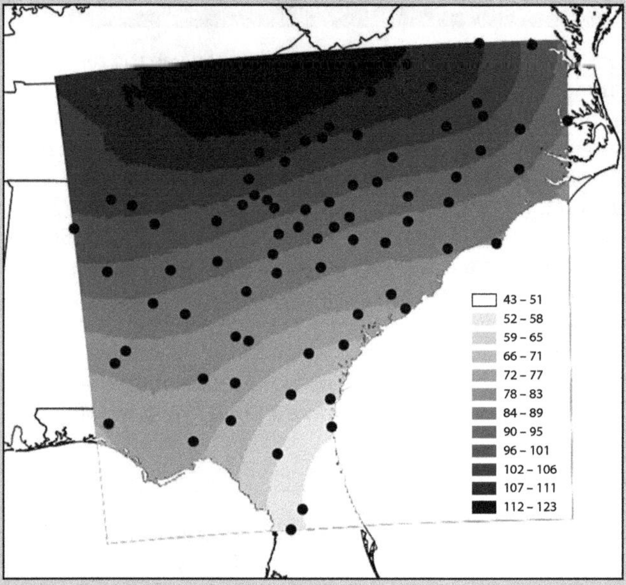

FIGURE 21.10
Results of Kriging Interpolation for the Last Spring Frost Example.

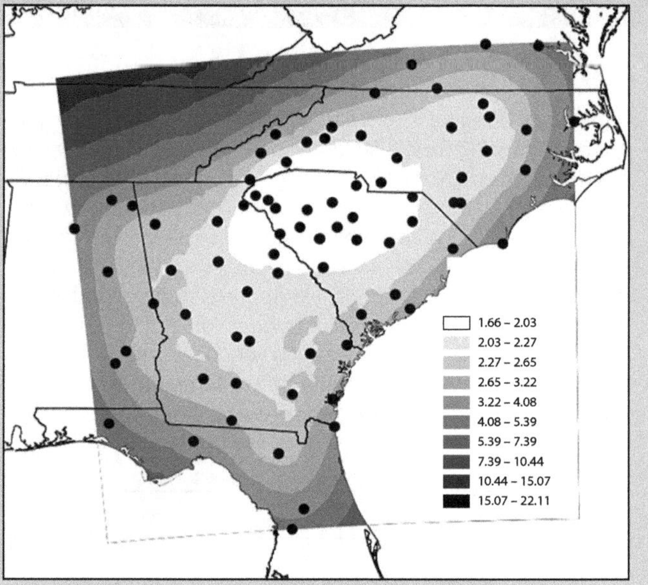

FIGURE 21.11
Standard Errors for the Last Spring Frost Example.

Key Terms

estimation, *308*
interpolation, *308*
inverse distance, *310*
variogram, *313*

Major Goals and Objectives

If you have mastered the material in this chapter, you should now be able to:

1. Understand the purposes of spatial interpolation.
2. Understand the principle of the variogram and its influence on estimation.
3. Explain the basic objectives of Kriging.
4. Explain the basic objectives of inverse distance weighting.
5. Define and understand the differences between inverse distance weighting and Kriging.

References and Additional Reading

Clark, I., and W. V. Harper. *Practical Geostatistics*. Columbus, OH: Ecosse North America LLC, 2000.

Bolstad, P. *GIS Fundamentals: A First Text on Geographic Information Systems*. 6th ed. Ann Arbor, MI: XanEdu, 2019.

Mitchell, A. and L. S. Griffin. *The Esri Guide to GIS Analysis. Vol. 2: Spatial Measurements and Statistics*. 2nd ed. Redlands, CA: Esri Press, 2021.

Babak, O. and C. V. Deutsch. (2009). "Statistical approach to inverse distance interpolation." *Stochastic Environmental Research and Risk Assessment: Research Journal*, Vol. 23 (2009): 543.

O'Sullivan, D. and D. J. Unwin. *Geographic Information Analysis*. 2nd ed. New York: John Wiley and Sons, 2010.

Cressie, N. A. C. *Statistics for Spatial Data*. Revised ed. New York: John Wiley and Sons, 2015.

Lembo, A., M. Lew, M. Laba, and P. Baveyye. "Use of spatial SQL to assess the practical significance of the modifiable areal unit problem." *Computer and Geosciences*, Vol. 2 (2006): 270–274.

McKillup S. and M. D. Dyar. *Geostatistics Explained: An Introductory Guide for Earth Scientists*. Cambridge: Cambridge University Press, 2010.

Part 8

Epilogue

22

Problem Solving and Policy Determination in Practical Geographic Situations

22.1 Geographic Problem Solving and Policy Situations

One of the primary goals of this textbook is to provide you with an introduction to the broad spectrum of statistical techniques that geographers use to approach and solve spatial problems. It is neither possible nor desirable to discuss all quantitative procedures available to geographers in a single introductory text. For example, we do not present intermediate and advanced multivariate techniques such as principal components factor analysis, canonical correlation, and discriminate analysis. Nor do we provide any specifics regarding more specialized procedures in geographic information sciences (GIS), such as Geary's *C*, the general *G*-statistic, geographically weighted regression, and nearest neighbor hierarchical clustering. If you are interested in learning more about these additional techniques, we encourage you to consult some of the more-advanced references listed at the end of various chapters.

As suggested in the Preface, there are two important general skills you should master to have a basic competency in using statistical techniques to help solve practical geographic problems and conduct geographic research. First, whenever you encounter a geographic problem that grabs or captures your interest, you should be able to develop specific descriptive statements and inferential hypotheses related to that problem. Then you should be able to select the appropriate statistical technique(s) and graphic procedure(s) that allow you to best construct these descriptive statements and test the inferential hypotheses. We might summarize this first general skill as "asking the proper questions and selecting the appropriate statistical approaches."

Second, when presented with the results or computer-produced output of any statistical analysis, you should be able to interpret those results and outputs properly. The best way to learn this second general skill is to practice, practice, and practice. Unfortunately, no magical shortcut can provide you with this expertise. If you can explain the results contained in the output from a statistical software package in plain, simple English (and in the context of the specific variables used in your research problem), then you have successfully reached your goal.

The remainder of this chapter presents a series of geographic situations or scenarios and their solutions. With each scenario, you will be given enough information to make an informed decision regarding which statistical method or graphic technique to use. In some cases, you may be able to interpret the situation in more than one acceptable way, thereby allowing alternative correct answers. If you select an answer that is different from that listed in Section 22.2, you should be able to explain and defend the logic of your alternative solution.

Perhaps the most difficult decisions involve scenarios that require use of an inferential statistic. Multiple steps may be required to identify the single best statistical test. Recall from Chapter 9 (Table 9.7) that the selection of an appropriate inferential test may require consideration of such dimensions as the following:

1. type of question (questions about differences, questions about similarities, questions about explicitly spatial relationships);

2. level of measurement [nominal (categorical), ordinal, interval/ratio];
3. parameter (mean, proportion, etc.);
4. number of samples (one, two, three or more); and
5. relationship of samples (independent, dependent).

To help you choose the most appropriate inferential technique, the body of Table 9.7 is replicated (Table 22.1), with a filled-in framework that includes every inferential technique presented in Chapters 9–18. It can be used as a decision-making guide for many of the geographic situations that follow. Remember that with many of these scenarios, enough information is provided to direct you to a single best answer, but other problems in the list may have more than one reasonable solution. Do not view the list of solutions as definitive; if you are innovative, you may identify valid alternatives that we have not listed that logically apply to the situation.

TABLE 22.1

Organizational Structure for Selection of Appropriate Inferential Test

	Questions about differences			
Level of measurement	**One sample**	**Dependent sample (matched-pairs)**	**Two samples**	**Three or more (k) samples**
Nominal scale (categorical variables)	Chi-square goodness-of-fit (χ^2)			
Ordinal scale	Chi-square goodness-of-fit (χ^2) Kolmogorov-Smirnov goodness-of-fit	Wilcoxon matched-pairs signed-ranks	Wilcoxon (Mann-Whitney) rank sum	Kruskal-Wallis analysis of variance by ranks
Interval/ratio scale	One-sample difference of means One-sample difference of proportions test	Matched-pairs (*t*-test)	Two-sample difference of means Two-sample difference of proportions	Analysis of variance (ANOVA)

	Questions about similarities		
	Questions about strength and direction of relationship between two variables	**Questions about form or nature of relationship between one dependent and one independent variable**	**Questions about form or nature of relationship between one dependent variable and two or more independent variables**
Nominal scale (categorical variables)	Contingency analysis		
Ordinal scale	Contingency analysis Spearman rank correlation coefficient		
Interval/ratio scale	Pearson correlation coefficient	Simple linear regression	Multiple regression

	Questions about explicitly spatial relationships
Nominal scale (categorical variables)	Nearest neighbor (point data) Quadrat analysis (point data) Join count (area data)
Ordinal scale	Moran's *I* (not covered in text)
Interval/ratio scale	Moran's *I* (area data)

1. Wanda Wolf, the head of scientific research at Redrock National Park, wants to determine the attitudes of park visitors regarding various Park activities and services. She decides to give a survey questionnaire to every 10th car entering the Park. What type of sample is Wanda planning to take?
2. A cartographer is studying the map interpretation abilities of students at a university. A random sample of 12 seniors and 16 freshmen (28 students in all) answer a set of questions related to the map, and the 28 scores are placed in a single, overall rank order. How would the cartographer test the hypothesis that seniors have better interpretive abilities?
3. A Manitoba farmer is concerned about the threat of hailstorms and wants to know whether he should purchase insurance. He has a 35-year record of hailstorm frequency and knows that anywhere from 0 to 4 storms have hit his farm in the past, with a mean frequency of 1.4 hailstorms per year. How can he calculate the likelihood of a hailstorm hitting his farm next year?
4. An economic geographer wants to examine the strength of relationship between population size and the number of retail establishments for a set of Australian cities. What technique should be applied?
5. An oceanographer wants to analyze the pattern of manganese nodules in a portion of the North Atlantic to determine if the pattern is more clustered than random. What technique should she use?
6. A small village in northern Ontario has a 112-year record showing the number of days in May with measurable precipitation (greater than 0.10 centimeters). Some years are totally dry, with no days of measurable precipitation, but during one very wet year, rainfall occurred 16 days of the month. The mean number of days with measurable rainfall is 3.2. What method should be used to calculate the likelihood of 5 or more days of rainfall for this upcoming May?
7. A soil scientist is interested in observing the spatial structure of pH values at different locations and their similarities with nearby locations. What kind of graph might he make to determine the distance at which the pH values are no longer similar?
8. As part of a recreation planning survey at a park, questions are asked of all persons visiting the park on a particular day. What type of sample is built into this sample design?
9. A regional planner obtains random samples of home costs from four Michigan counties. Suppose the home costs appear normal in each county and that prices seem to vary about the same amount from county to county. What test should the planner use to learn if the differences in home cost by county are significant?
10. A geographer wants to see if monthly rents for student housing vary according to the distance of the rental property from campus. She suspects the closer an apartment unit to campus, the higher the rent. How can she test whether this distance decay effect is present?
11. Officials in Kamloops, British Columbia, know from a recent survey that 3.2% of their residents are "senior elderly," that is, age 80 and older. What will they be calculating if they try to estimate the likelihood that the true population proportion of senior elderly is between 3.0% and 3.4%?
12. A work-study student assigned to the administrative office at a university is trying to find out what activities a sample of students want in the soon-to-be-constructed student union. The only stipulation is that any student has an equal chance of being included in the sample. What type of sample should the work-study student take?
13. The Barranca Board of Education wants some information about local residents' attitudes toward spending more money on public education. Plans are to take a random sample of 285 voters, learn their political affiliation (Democrat, Republican, Independent) and their attitude about increasing public education expenditures (strongly agree, agree, neither agree nor disagree, disagree, strongly disagree). What technique should be used to determine if there is a relationship between political party and attitude toward greater public education funding?
14. A social services planner in Chicago, Illinois, is concerned about people's access to a set of service centers scattered around the metropolitan area and does not want anyone living too far from any of the centers. How could she test whether the pattern of these centers is more dispersed than random?
15. From a recent nationwide study it is known that the typical American watches 25 hours of television per week (with a population standard deviation of 5.6 hours). Suppose 50 residents of New Orleans, Louisiana, are randomly selected and their viewing hours monitored. What technique can be used to determine whether the viewing habits of New Orleans residents differ from the national mean?
16. A geographer is looking at two maps of Nashville, Tennessee. One map has a point indicating the location of every pawnshop, while the other map has a point representing every grocery store. What descriptive spatial measure should the geographer use to compare their levels of spatial dispersion?
17. A demographer has recorded the number of children per family in a sample of 22 households in the Indian state of Kerala. She suspects from a quickly

sketched histogram that this sample has been taken from an extremely skewed population. What inferential test can she use to test this suspicion?

18. Geographic consultant Pierre Portage is studying visitor activity patterns in the Quebec park system. Managers want to know if park attendance figures differ from June and July, using the same set of parks in each case. What technique should Pierre advise them to apply?
19. A geographer has two isoline maps of Pakistan: one showing mean annual precipitation and the other showing population density. Suppose the same random sample of points is selected from each map. Assume that both variables are from normally distributed populations. What technique will determine the degree to which the two map patterns are associated?
20. A physical geographer is preparing to take soil samples from an area that is 20% marshland. She suspects that environmental degradation is most severe in the marshland portion of the study area, so she wants to ensure that area is proportionally oversampled. What type of spatial point sampling should she apply?
21. Environmentalists are investigating the spatial extent of thermal plume (higher water temperature) around a water discharge pipe at Calvert Cliffs nuclear power plant near Chesapeake Bay. Suppose personnel want to estimate surface water temperatures at various distances from the discharge point. What statistical procedure would be most appropriate?
22. A choropleth map of states in Mexico uses the following legend categories: 11.37 to 16.26; 16.27 to 21.36; 21.37 to 26.36; 26.37 to 30.14. What method of classification is being used?
23. An economic geographer is studying the spatial patterns of income in Melbourne, Australia, and it is known that the mean family income is $38,346. A sample of 15 Melbourne households reveals an average of $40,487. How should the geographer test to see if the sample of 15 families is representative of the entire population of Melbourne?
24. Suppose you have collected data on infant mortality rate and number of hospitals per 100,000 people from 50 less-developed countries. You want to look at a graphic relationship between these two set of numbers. What graphic will you construct?
25. A climatologist is looking at modified climographs from 10 weather stations and is focusing on the plots of mean monthly precipitation for each of the stations. Some graphs have a very high spike (showing that almost all the annual precipitation falls in just a couple of months of the year), whereas other graphs indicate precipitation falls rather evenly throughout the year. What descriptive statistic should be used to measure and quantify these visual observations?
26. Steve "Smokey" Jones is a ranger in Plywood National Forest, and he has access to wildfire records over a 65-year period. Some years have no fires, but in one very dry year 16 fires burned fifty acres or more. He has discovered the mean annual number of such extensive fires is 1.8. How can Smokey calculate the odds of three or more wildfires in Plywood next year?
27. You would like to make a county-level choropleth map showing only the counties that have all three of these undesirable characteristics: very high infant mortality rate (worst 10% of all counties), very high suicide rate (worst 10% of all counties), and very high high-school dropout rate (worst 10% of all counties). What kind of map would you construct?
28. A geography student has been studying the relationship of distance from Lake Ontario and the amount of "lake effect snow." Some friends of hers live in this study area and have a cabin 10 miles from the lake. What technique should she apply to estimate how much snowfall they received last winter?
29. London, England, has a long, normally distributed record of annual precipitation. Someone wants to know the probability that more than 800 millimeters of precipitation will fall next year. As a step in this process, what statistic needs to be calculated?
30. A fluvial geomorphologist is studying surface runoff levels in two different environments: (1) a natural setting with no modification by human activity; and (2) an urban setting, highly modified through development. To test for significant differences in runoff, spatial random samples are taken from both settings. Suppose the runoff data are placed into a single overall rank order from high to low within the two sampled areas. What test should be used to see whether the urban runoff exceeds that in the natural setting?
31. A sample of residents is surveyed to determine the number of miles traveled per week not related to work. A year later, following a large increase in gasoline prices, the same individuals are sampled again to record possible changes in discretionary travel behavior. How could you test whether the change in gasoline price is related to decrease automobile use?
32. From a representative sample of Canadian voters, how would a political geographer determine if the percentage of voters favoring placement of a "strict quota" on annual immigration volume differs significantly by gender?
33. A cultural geographer wants to examine the strength of possible relationship between ethnic/racial group identification and type of restaurant most frequented. Suppose 300 people are surveyed,

and assigned to 1 of 5 ethnic/racial groups and 1 of 10 restaurant types. How would you test to determine if restaurant preferences differ significantly by ethnic/racial group?

34. A researcher wants to depict graphically the mean average snowfall amounts that are 10% likely to be exceeded across western and central Canada (British Columbia, Alberta, Saskatchewan, and Manitoba). The graphic is based on data from 85 weather stations in this region. What will the researcher be depicting graphically?
35. A regional planner is studying income inequalities across a set of counties in northern Appalachia. Median household income by county is known for both 2000 and 2010. How can the planner descriptively measure the relative dispersion of income among counties in 2000, and compare that measure with the 2010 result?
36. Suppose you have the complete historical record of major earthquakes (say 6.0 or higher on the Richter scale) for that portion of the San Andreas Fault on the upper San Francisco Peninsula. Let's suppose the confirmed record indicates the date of each of 22 such incidents over the last 3,000 years. What technique could you use to predict the date of the next major earthquake?
37. An economic geographer is consulting for Flor-Mart (the giant store chain), helping them better determine the demographic profile of their customers. She has an alphabetical master list of all credit card holders and wants every 20th cardholder to complete a detail questionnaire. The five-percent sample results, with careful follow-up of non-respondents, will be very useful to FM management. What type of sampling is she suggesting?
38. Political geographer Gerry Mander has a metropolitan area map showing the 78 election districts. For each district, Gerry knows whether the majority of registered voters is Republican or Democrat. What technique should he use to learn if political party affiliation is more clustered than random?
39. Mean annual precipitation totals vary considerably across the State of California. Suppose a climatologist wants to examine the spatial structure of these precipitation totals in the state and the amount of similarity for nearby locations. What kind of graph could you draw to show the distance at which mean annual precipitation totals are now longer similar to one another?
40. The U.S. Department of Housing and Urban Development is attempting to reduce crime rates in various public housing projects. Data are collected from a sample of projects for a number of variables thought to be associated with crime, including: resident age, income, employment status, and participation in support programs. What test can be used to determine the relative influence of these various factors on crime?
41. Geoscientist Gary Gabbro is studying erosion rates of several soil types in a county. He eventually wants to use ANOVA to evaluate differences in erosion rates by soil type. He has taken systematic spatial samples from each, but is not sure if any of the underlying populations from which these samples have been taken is normally distributed (an ANOVA assumption). What test should he run before conducting an ANOVA?
42. Chinese agricultural officials are conducting a study of crop damage resulting from insect infestations in the lower Liao River Valley. It is hypothesized that one factor associated with the extent of insect damage is the amount of fertilizer used by different farmers. What test would be used to learn the strength of relationship between fertilizer use levels and the amount of insect damage?
43. An urban geographer wants to analyze a map pattern depicting the number of existing home sales by census block group in Detroit, Michigan. He suspects some blockbusting is occurring and wants to determine the extent of home sales clustering ("hot spots") in certain portions of the metro area. How should he proceed?
44. A random sample exit poll of Jefferson County voters resulted in the following opinions regarding a commercial-retail blue law that would prohibit businesses from being open on Sunday. In Drumlintown, 48% were in favor of such a law, while in the rest of the County, 54% were in favor. How would you test the hypothesis that support for the blue law is higher outside Drumlintown than in the city?
45. A geography student has a table listing all of the states, followed by a zero if the state has not passed legislation making English the official language, or followed by a one if it has. If a map is made using this information, what technique can she use to determine if the spatial pattern making English the official language is random?
46. A medical geographer wants to predict the number of physicians living in a city that has a population of 10,000 to 12,000, based on the median family income of the city. Suppose this relationship has been tested for other cities of this size, but not for this particular place. What statistical technique is appropriate?
47. A seismologist in New Zealand has been monitoring earthquakes in the country for years. Many years no "noticeable" quakes (above 3 on the Richter scale) occurred, but one year there were six such quakes. How will the seismologist calculate the

probability of more than one earthquake of this magnitude next year?

48. Suppose an urban planner in San Francisco, California, wants to determine if residents in rental units differ from residents in owner-occupied units with regard to the percentage favoring rent control legislation. If random samples of residents are taken from both types of residences, what statistical test should be used?

49. Transportation planners in Minneapolis, Minnesota, are concerned about the effect of weather on the ridership levels of public transit. The hypothesis is that days with "bad" weather conditions (rain, snow) tend to attract more riders than "good" days (dry, sunny). Given daily ridership data and knowing the overall population variability in these rates, how could the hypothesized difference in ridership levels be statistically validated?

50. A researcher working with a 50-year climatological record of Atlanta, Georgia, knows the number of days each year with temperatures above 90°F. When she constructs a histogram showing the distribution of these 50 values, it doesn't look at all symmetric. A few years had many more "very hot days" than the others. What descriptive statistic can she apply to measure this lack of symmetry?

51. Suppose you have applied the natural breaks (single-linkage) method of classification to a set of province-level lung cancer rate data. You would like to examine the step-by-step classification process graphically. What graphic will you examine?

52. Experts handling snow-making equipment for Tamarac Lodge and Ski Resort suspect a relationship between overnight wind direction and the need to make snow during the late winter/early spring season. Using sample data on these variables from the last 10 years, how could they test this relationship?

53. In a list of 50 test scores, the mean score is 74.1 out of 100. What statistic should be calculated to determine how much the typical score varies from this mean?

54. From a random sample of trees in a national forest, a biogeographer wants to identify the pattern of diseased trees to see if they are randomly distributed of clustered. This will provide guidance concerning the most appropriate type of treatment (widespread aerial spraying or concentrated treatment of specific areas from ground level). What test should be used?

55. Jill Doakes, currently with the U.S. Bureau of the Census, wants to estimate the 2040 location of the geographic center of population. Given county-level population estimates for 2040, what descriptive statistic should she use?

56. The cost of a "market basket" of groceries is obtained from random samples of grocery stores in Boston, Atlanta, Minneapolis, and San Diego. The total cost of the same list of food items (same national brands and sizes) is recorded for each store in every city surveyed. What test should be applied to learn if the market basket costs between cities are different?

57. A region contains three major rock types: limestone, calcareous marl, and sandstone. Suppose similar-sized areas are sampled in each rock type, and the number of springs in each is tabulated. What test will determine if the density of springs varies by rock type?

58. A Japanese city has five major shopping centers. A random sample of 25 shoppers at each center is asked how much money they spent shopping in the center that day. What statistical test could be used to see if the typical shopper at the different centers spends differing amounts of money? Assume the pattern of expenditures at each center is very positively skewed.

59. Elmwood, Oklahoma, has two restaurants. Suppose a random sample of patrons at each restaurant is asked the distance they have traveled to eat (the distribution of distances is severely non-normal). What test is appropriate to discover if patrons of the two restaurants travel different distances?

60. A geographer plans to make a choropleth map of Texas counties, showing median family income. However, he wants to use a method of classification that will highlight the unusual counties (outliers), thereby allowing map viewers to quickly spot "extreme" counties. What method of classification should he use?

61. A climatologist is conducting a study of "lake effect snow" along the shores of Lake Ontario. She has collected annual snowfall data from numerous monitoring sites at varying distances from the Lake. What statistical technique should be used to measure the strength of relationship between annual snowfall and distance from the lake?

62. The housing department at Everbrown University has just completed a survey of 86 students, asking the distance of their off-campus housing unit from campus and the monthly rental cost of their unit. Given this information, what technique should housing personnel use to predict the monthly rent for a student living 1.5 miles from campus?

63. Let us suppose that there have been 12 confirmed tornado touch-downs in Ellis County, Kansas, since 1900. What statistical technique would you recommend to predict the date of the next tornado touch-down in the County?

64. An earthquake in eastern Turkey resulted in the following number of deaths and injuries in 11 villages located at increasing distances from the epicenter: 520, 410, 320, 310, 50, 170, 210, 140, 100, 50, and 20.

That is, 520 deaths and injuries occurred in the village closest to the epicenter, 410 occurred in the next closest village, and so on. If no further information is available, how can you measure the strength of relationship between proximity to the epicenter and number of casualties and injuries?

65. A random sample of residents whose homes are on the floodplain of the Mississippi River is surveyed regarding their attitudes toward various flood management practices. After a flood has occurred, the same set of homeowners is resurveyed about these issues. What test should be used to determine if attitudes toward flood management policies have changed?

66. A 2012–2013 survey of winter guests at Tamarac Lodge estimates that 58% of the respondents did not ski during their visit. A consultant hypothesizes that an even larger percentage will not ski this upcoming winter. When the data becomes available, how will she test this hypothesis?

67. In previous years, the overall mean travel time for a boat trip down the Colorado River through Grand Canyon National Park was 144 hours. Because of River congestion and overuse, park personnel hypothesize the trip will take significantly more time this year. After the data becomes available, how can you determine if boat trips now take longer?

68. A glaciologist studying temperature variation in Greenland has proposed that temperatures are significantly warmer today than 300 years ago. Using 100 test sites around the subcontinent, he has measured the average temperature over the last year and estimated the average temperature at those sites 300 years ago from the chemical composition of ice bores. How should he test for warming?

69. A political geographer is interested in knowing if the strength of support for a Mississippi Republican candidate for the Senate in the last election was spatially autocorrelated. The candidate received 65% of the vote statewide. Each Mississippi county is identified as either above or below the statewide percentage and this pattern is plotted on a map. What statistical technique should be used?

70. A geographer believes that medical service facilities in a major urban area are clustered near high accessibility locations (such as primary commercial routes and highway interchanges). If each sampled medical service facility is plotted as a point on an area map, what technique could be applied to analyze this pattern?

71. A cartographer is studying the relative effectiveness of different map variations. The same map is produced in two distinct ways: five-color and five shades of gray. The same series of questions about the map are asked of 15 people using each of the two map variations, and the score for each participant is recorded. How can the cartographer test the hypothesis that higher average scores are received when the five-color version of the map is used?

72. An environmental planner has been collecting sulfur dioxide level readings from a sample of monitoring sites at different distances from a coal-burning utility power plant in the Ohio River Valley. What statistical technique can she use to predict the sulfur dioxide level at an unmonitored site 12 miles downwind from the power plant?

73. Texas is known for taking capital punishment seriously; more executions occur here than in any other state. However, a geographer is curious about whether attitudes toward capital punishment vary from one part of the state to another. As one segment of his study, random survey samples of residents are taken from El Paso (west Texas) and Beaumont (east Texas). In each city, he has estimated the percentage of respondents favoring the death penalty. What technique should be used to see if these attitudes differ statistically?

74. A hurricane has recently moved up the Atlantic coast of the United States, causing damage in major coastal cities. Suppose the amount of damage (in millions of dollars) is as follows: (the cities are listed in location from south to north; the southernmost city is listed first and the northernmost city is listed last) 786, 451, 507, 410, 202, 51, 171, 35, 27, 58. How can these data be evaluated to learn if location northward or southward is significantly related to the amount of damage?

75. The agricultural extension agent in an Oregon county is concerned about the adverse effects of hailstorms on farms in her area. Hailstorm activity has been recorded over the last 55 years. Some years, no hailstorms occur anywhere in the county. However, in one extreme year, nine separate hailstorms occurred. How can she calculate the probability of the county experiencing exactly two hailstorms next year?

76. The Vermont Tourist Bureau wants to target its television advertising for the upcoming winter season. The Bureau director feels that visitors come equally from Boston, New York, and metropolitan regions in southern Canada. Many resort owners, however, believe that Boston is the clear leader for tourists. Using data samples from the previous season of the seven largest resorts in the state, how could this question be resolved?

77. The chair of a local environmental action group hypothesizes that a relationship exists between the volume of recycled material left along the curb for pickup and various homeowner characteristics, including: age, income, education, and a measure of

homeowner community involvement. What method should be used to predict the likelihood of a particular homeowner participating in the recycling program?

78. Suppose you have a set of data listing the total number of murders from automatic high-capacity guns per thousand people by state since the year 2000. You would like to display these data graphically, showing the location of each state on this distribution and also showing the interquartile range. What graphic will you construct?

79. Suppose a demographer for the Ontario provincial government is looking at both the number of migrants moving into economic development areas and the number of migrants leaving the same area. How can she test whether the number of in- and out-migrants is statistically different?

80. A tornado-spotter wants to analyze the initial "touch-down" point pattern for tornadoes that initially touched down in Arkansas. The overall pattern appears somewhat clustered, but there is some disagreement. What inferential technique can be used to determine whether the pattern is more clustered than random?

81. Personnel at an Ontario Provincial Park want to locate a new campground on a riverfront site that will not be flooded too frequently. Budget managers advise a location that will be flooded no more frequently than once a decade. On one possible site, it has flooded 8 times in the last century. What procedure should be used to determine the probability that this site will not be flooded too frequently in the next 30 years?

82. An urban geographer is part of a team planning to collect data from city residents. One action that has been suggested is to survey every home in selected city blocks (to save time and money). What sampling strategy would this be?

83. Hydrologists want to analyze the volume of material carried by two rivers (expressed in grams of solid material per liter of water). One river drains a predominantly agricultural area, while the other drains a predominantly forested area. To test for differences in volume of material carried, 12 random water samples are taken from both rivers, one sample a week through the summer months. What test should be used?

84. An economic development planner is interested in evaluating changing patterns of poverty in Appalachia. She has a series of county-level choropleth maps showing each county as either above regional median family income (shaded) or below regional median family income (unshaded). How can she test whether poverty is becoming more spatially concentrated over time?

85. The community of Sliding Stone, Pennsylvania, has five clearly identifiable neighborhoods. A student at the local university thinks lot sizes differ by neighborhood, so he selects a random sample of 10 lots from each neighborhood. It is assumed that lot sizes vary within each neighborhood by differing amounts. What statistical method should he use to verify his hypothesis?

86. A newspaper reporter for a small regional paper wants to determine if the local Labour Party candidate for Parliament has a similar level of support among voters in two adjacent towns. The proportion of a random sample of voters in each town favoring the candidate is recorded. How can the reported discover if the support level for the candidate differs between the two towns?

87. Suppose there are two choropleth maps of Texas counties, one showing median family income in one of five ordinal categories, and another showing the percentage of labor force in profession occupation classified into four ordinal categories. What technique should be used to learn if the two choropleth map patterns differ?

88. A private high school is situated in the center of a five-neighborhood region. Each neighborhood has about the same high school age population. A random selection of 100 students at the school resulted in the following counts (22, 24, 13, 24, and 17). That is, 22 students from neighborhood 1, 24 from neighborhood 2, and so on). What technique can be used to discover if student enrollments vary by neighborhood?

89. An urban geographer is examining citizens' attitudes toward growth in a rapidly expanding suburban area. The research design is set up to survey a random sample of 230 people at two different times. First, attitudes toward growth will be measured before construction of a major new regional shopping mall. Later, the same people will be resurveyed after the mall has been open a year. How can the geographer test whether respondents' attitudes toward growth have changed?

90. Researchers continue to explore the various factors that might be related to the homicide rate in a major city. Data are available by census tract for many variables, including age, income, education, and unemployment rate. What test could be used to estimate the homicide area in a census tract not included in the original study?

91. Suppose you are studying historical changes in housing characteristics in your hometown. You want to see if house size (square feet of living area) and the age of the home are associated. What method will you use to see if a relationship exists?

92. In another study concerned about crop damage caused by insect infestations in the Liao River Valley of China, insecticide use levels seem to be critical. Officials want to predict the amount of crop damage expected on a farm not included in the original study, solely based on the level of insecticide use. What technique should be used?

93. A recreation geographer is testing the effectiveness of an attendance prediction model (number of visitors from a set of cities to a set of regional parks). It is hypothesized that attendance will be directly related to city population size and inversely related to distance from the city to the park. What statistical technique could be used to test the validity of this model?

94. Suppose you have data from the Centers for Disease Control (CDC) listing HIV rates for all counties in the United States. You want to construct a choropleth map based on these data, and your goal is have about the same number of counties allocated to each of the five categories you plan to have on the map. What classification method are you going to use?

95. The World Health Organization (WHO) has collected data on tuberculosis infection rates from a sample of neighborhoods in Lagos, Nigeria. If these neighborhoods are classified as either "more polluted" or "less polluted" based on levels of industrial activity, how can WHO researchers test whether tuberculosis infection rates differ by pollution level of the neighborhood?

96. A cartography student has proposed a research paper project. She wants to contrast the map interpretation skills of male versus female students. Suppose map interpretation test scores from 18 males and 20 females are placed in an overall rank order, from highest score to lowest score. What technique should she use to learn if male and female map interpretation skills differ?

97. On the wall of the fire department headquarters in Chicago, Illinois, a large map shows the location of fire stations and recent large fires. Someone is using a ruler to measure the straight-line distance from a fire station to a fire. A visiting geographer says, "No, that is not the true distance!" What measure of distance should be used?

98. Health officials are concerned about the recent diffusion of influenza in Moscow, Russia. If the number of cases is known for each local administrative unit for several consecutive weeks, the morbidity rate for influenza could be displayed on a series of choropleth maps (one map for each week), and the degree of clustering, randomness, or dispersion could be analyzed for each map pattern. What technique would be used?

99. The SAT scores of 50 Middlebury students, 41 Salisbury students, and 83 Boysenbury students are randomly selected for analysis, and an overall ranking of scores is created. What statistical test should be used to learn if SAT scores between these three schools are different?

100. After completing a local pollution abatement program, Florence Floss wants to know if mean pollution levels from a sample of 10 sites are significantly lower than national Environmental Protection Agency standards. The national mean EPA standard and variance around that national standard are known from existing reports. What statistical procedure should Ms. Floss follow?

101. In Banff National Park, Alberta, some observers think the number of black bear sightings has increased dramatically. Park headquarters has data listing the number of sightings at each campground that existed 20 years ago. How could a resource manager test this historical data against a newly collected set of sightings from the same campgrounds to determine if any increase has occurred?

102. You suspect that teenage pregnancy rates vary considerably from state to state. After collecting the state-level teenage pregnancy data, you want to classify the information and display it on a choropleth (area pattern) map. However, you want to classify the data so that variation within each category is minimized while variation between categories is maximized (as much as possible). What simple classification do you propose?

103. Wolves are now being reintroduced into areas where they formerly thrived, but have recently been eliminated. Of course, this strategy is controversial, usually opposed by area farmers and ranchers. Suppose personnel from the Department of the Interior plan to release 24 wolves from a single departure point in Yellowstone National Park. Each wolf is tagged and released individual at one hour intervals. You have the data showing the movement pattern of each wolf. How can you best summarize these data graphically to show the general direction of movement?

104. City planners want to determine the attitudes of residents concerning possible construction of a new public swimming pool. They want their proposed 5-percent sample to be representative of the entire community, and they decide to use number of children per household as an indicator variable. If the mean number of children per household is known community-wide, how should planners test to see if the 5-percent sample mean is representative of the community-wide mean?

105. Coral Reef is conducting a study of hazard perception in a community along the Susquehanna River. A

questionnaire Coral has written lists three possible management strategies: strengthen the levee, re-channel a portion of the water from upstream, and relocate selected structures. The study protocol contains three distinct groups: people living on the 25-year flood zone, people living within the 100-year flood zone, and people living on higher ground. How can Ms. Reef test whether the preferred management strategy is related to the respondent's location?

106. A planner in Tampa, Florida, has assisted in the design of a newly planned community project, which has on-site health care facilities. He hypothesizes that residents in the planned community are older than the metropolitan-wide mean. If the age statistics for all Tampa residents is known, what statistical technique should be applied to learn if planned community residents are older?

107. The United Nations Population Fund wants to evaluate the success of a family planning program in selected small villages in Somalia. Previous studies have shown that a 38% rate of effective contraceptive use after two years is a suitable international standard for success. How should one test whether the Somali program has been effective?

108. In Pennsylvania counties, unemployment rate data are available for 2011 and 2012. How can Terrance Tundra (a planner that works with Commonwealth labor statistics) measure the relative variability of 2013 county unemployment rates, to compare descriptively with the previous two years?

109. A climatologist has measured sulfur dioxide levels in the lower atmosphere repeatedly at seven different weather stations along a line of latitude across central Canada for an entire year. How can he test for differences in sulfur dioxide levels between stations?

110. A geographer with the World Bank is studying global economic development trends and has information regarding the percentage of labor force in "telecommunications" jobs for a sample of countries. He hypothesizes that countries in "upper," "middle," and "lower" income categories have different percentages of their labor force in this job sector. A potential problem is that some sample distributions appear to be from highly skewed populations. What statistical test should be used?

111. A geographer is studying total fertility rates in African countries and wants to correlate these data with other variables using Pearson's correlation coefficient. How should she test sample total fertility rate data from each African country to determine if it is from a normally distributed population and for subsequent input into the Pearson's correlation test?

112. A spatial analyst for the Center for Climatological Research is examining multiple point pattern maps (one map pattern per year) showing the distribution of hailstorms across a five-county region in northeast Colorado. How can these data be analyzed to compare directly the amount of relative dispersion on each of the maps?

113. Complete historical climate records for Fairbanks, Alaska, indicated that 18.6% of the days have clear weather, with no appreciable cloudiness. Suppose a sample of 50 days in the last year is selected at random, and 26% of these days had clear weather. What test would you use to determine if the sample is typical or representative of the overall climate record?

114. In 2010, the Ontario Tourism Bureau compiled a "Top 10" list from a Province-wide sample survey, asking residents to rank cities in the Province by "livability." The Bureau wants to again ask a sample of Ontario residents to provide a current list. What statistical test should be used to measure the degree to which the two Top 10 lists are related?

115. Experts studying inflation rates for people throughout Europe hypothesize that people living in countries that currently belong to the European Union (EU) are enjoying lower inflation rates than people living in non-EU countries. Large random samples of individuals in EU and non-EU countries are taken, and the mean inflation rate is calculated for each group. What technique is used to test for significant difference?

116. The Chinook City Council feels that the turnover rate of land parcels is not equivalent across all types of land use. The Council decides to tabulate type of land use (residential, commercial-retail, industrial) against the timing of the last sale of that land parcel (within the last year, 1 to 5 years ago, 6 to 10 years ago, and more than 10 years ago). If they randomly select 200 land parcels from the City records, how should they determine if a relationship exists between land use and date of last sale?

117. A planner wants to examine the pattern of housing quality in a lower-income suburb of Birmingham, England. Suppose a random sample of housing units is taken from each of Birmingham's local election districts, and each district is then categorized as having a higher or lower number of substandard units than the metro-wide mean, and a binary map is drawn. The planner suspects that the substandard units are mostly concentrated near the city center. How can this hypothesis be tested?

118. Joe Doakes is working on a new "Atlas of Australia" and has been asked to construct an isoline map showing the likelihood that more than 20 inches of precipitation will occur at various weather stations across the country. What should Joe create to complete this task?

119. A geographer wants to learn if mean soybean yields per acre on three soil types are significantly different. From a sample of farms on each soil type, he collects the soybean yield values. It is assumed that the populations from which the samples are taken is normally distributed. What statistical text should he use?

120. Marcia Monadnock is the recreation activities planner at a ski resort in the Catskill Mountains. She thinks that weather conditions in New York City on the Fridays before a potential ski weekend influence the number of people who will actually ski that weekend. Specifically, she hypothesizes that if it is a "warm" Friday there will be fewer skiers, while if it is a "cold" Friday there will be more skiers. Using visitation data from a sample of winter weekends over the last decade, how can she determine if her hypothesis is correct?

121. Religious affiliation (Protestant, Catholic, Jewish, etc.) and voting preference (Democrat, Republican, and Independent) data are collected from a sample of 250 people in a key New York City election precinct. How can the hypothesis be tested that religious affiliation is related to voting preference?

122. Transportation planners in metropolitan Eskerville know that the mean commuting distance metrowide is 5.4 miles (each way). A detailed study is made of 25 randomly selected commuters, and their mean commuting distance is 7.1 miles. How should planners test whether the sample of commuters is typical of all commuters in Eskerville?

123. A World Health Organization researcher in Kenya is studying the distribution of three medical problems: "river blindness," HIV infection, and an infectious form of "sleeping sickness." Random samples are collected from 100 people on a river floodplain and 125 people on an adjacent highlands area. How can the researcher determine whether the incidence of these medical problems is related to the type of physical location in which they live?

124. Fred Foehn lives 500 meters from an interstate highway. From his front yard, Fred has been measuring noise levels at noon on a systematic sample of days over the last year. At the same time, he also records the primary wind direction (north, south, east, or west). Fred hypothesizes that noise level is related to wind direction. How can he test this hypothesis?

125. Along a segment of bayfront beach, environmental researcher Donna Dune is studying selected measures of soil chemistry. She thinks that the human impacts on soil chemistry will be most severe in the tidal wetland portion of the beach, which encompasses about 15% of the total study area. Donna therefore decides to "oversample" in the tidal wetland area, taking almost 50% of her soil samples from that area. What type of spatial sampling design is she advocating?

126. You would like to summarize graphically the point distribution and variation of UFO (unidentified flying object) sightings in New Mexico since the year 2000. What graphic summaries would it be logical to display?

127. A vocational-technical community college is proposed in a community. A comprehensive citywide poll conducted two years ago indicated that 58% of all community residents were in favor of financing construction of this facility. Since then, economic difficulties may have altered residents' views on this issue, so a new sample survey has been proposed. How can community leaders learn if the view about financing this project has remained unchanged?

128. A corn farmer in central Nebraska has discovered that at least three inches of precipitation are needed during a critical three week period in the growing season to avoid irrigation. If he has to irrigate, he will lose money that year. Historic data indicate that in 15 of the last 18 years precipitation has exceeded three inches during this time. How can the farmer determine if it is profitable to plant corn?

129. A health services provider is surveying attitudes regarding "freedom of choice" for women in continuing pregnancy across subareas of Cleveland, Ohio. She wants to predict the level of support for freedom of choice in a critical precinct of the City. Fortunately, she has information from previous precinct-level models (but not this particular precinct) demonstrating that several factors influence these attitudes, including: median age, percent Catholic, median household income, and median school years completed. What statistical procedure will she utilize?

130. The historical record at a site along the lower Brahmaputra River in Bangladesh indicates that major flooding occurs at least once during the rainy season 4 years out of 10 (40% of the time). How would you estimate the probability that major flooding will occur at least once in 6 of the next 10 years?

131. Georgianna Peach, a medical services provider in Savanna, Georgia, is concerned about attendance at a health care clinic in town. She wants to develop and test a model that will predict clinic use level as a function of population size within one mile, number of high school educated people within one mile, and number of households below poverty level within one mile. What statistical technique is she proposing?

132. Dilbert Doakes is a consultant for the Los Angeles Fire Department, and he has been asked to look at the spacing of fire stations throughout the city. Of particular concern are structures that may be too far

from the nearest fire station to be protected adequately. What statistical technique should he use to help address this concern?

133. Suppose the mean age of New Zealand's population is 34.4 years. Raymond Yachtclub (the new planning director of Wellington) wants to determine if Wellington has a typical age profile, with a mean age similar to the national figure. If the ages of 80 Wellington residents are collected, what technique will he use?

134. According to a recent nationwide study, only 54% of all Americans can locate Mexico on a world map (this is a true fact!). From a sample of 60 students in a college geography course, suppose 63% are able to locate Mexico. What test should be used to learn if the sample of students is typical?

135. From samples of less-developed counties and more-developed countries, suppose data are collected on mortality rates caused by heart disease and other heart-related problems. What technique should be used to determine if there is a difference between LDC and MDC death rates as a result of these causes?

136. Coralie Ollis is a geographer doing some consulting for a major car insurance company. She has pulled a random sample of 375 accident claims from company files, and her hypothesis is that the number of insurance claims varies significantly by season of the year (spring, summer, fall, and winter). How will she test this hypothesis statistically?

137. The mayor of Concord, Kentucky, received 60.8% of the vote in his successful election to that office. He is running for re-election, and a survey of 110 expected voters is planned. What technique will tell his campaign manager if the current level of support differs significantly from the first election?

138. A climatologist has 85 years of annual snowfall data for Flagstaff, Arizona. What inferential statistical procedure can be applied to learn if the distribution of annual snowfalls is normal?

139. Suppose an economic geographer is given a choropleth map of North Carolina counties, with each county classified in a binary fashion as either above or below the statewide mean regarding the percentage of population below the poverty level. She hypothesizes that the far western (Appalachian) counties will emerge as a poorer-than-average cluster. What technique should she use?

140. A staff cartographer is assigned to design a world map showing birth rates by country. The study director wants the following categories to be used: less than 9.9, 10.0–19.9, 20.0–29.9, 30.0–39.9, and 40.0 and greater. What method of classification is to be applied in this map design?

141. A transportation engineer records noise levels at two adjacent locations, both 200 yards from a major highway. The first location has an intervening noise barrier about 10 yards from the highway, while the second location does not have any intervening noise barrier. Based on random samples (taken at the same times at both locations), how can the engineer test whether there is a significant difference in noise levels at the two locations?

142. A political geographer is interested in determining the degree of spatial autocorrelation for illiteracy rates in a large U.S. city. Using school district boundary lines, he wants to know if school districts having high illiteracy rates are located near other school districts of high illiteracy. What kind of statistical test should he perform?

143. The U.S. Geological Survey wants to map the likelihood that a magnitude 5.0 or greater earthquake (Richter scale) will occur at various locations across southern California sometime in the next decade. What technique should be applied?

144. Urban geographers have long thought that home lot size is larger the greater the distance from the city center. What technique would you apply to confirm this hypothesis and predict the lot size of a home six miles from the city center?

145. A demographer is studying population growth rates within Nigeria to see if this variable is related spatially to religious affiliation. The demographer has collected a systematic spatial sample from 250 households across the country. From each household he has recorded their religious affiliation and the number of children. How should he now proceed statistically?

146. How would a geomorphologist test the relationship between type of bedrock material and level of surface soil acidity for a sample of sites in a river valley? That is, do different types of bedrock material support soils having different levels of acidity?

147. Legislators in Aspen, Colorado want to learn if year-round residents differ from seasonal residents with regard to the percentage favoring a law to strengthen the existing noise control ordinance. If random samples are taken from both types of residents, what test should be used?

148. From two small villages in the Scottish Highlands, a researcher has collected random samples of villager height. How can the researcher determine if residents of one village are significantly taller (or shorter) than residents of the other village?

149. Winter tourism levels may be increasing in the Canadian Rockies. A survey conducted in 2005 calculated the mean seasonal revenue (due to tourism) for 45 communities in this region. What statis-

tical test would you use to learn if seasonal revenue from last year has increased since 2005 for these same communities?

150. A community in Indiana has six neighborhood parks. The local recreation planner takes a random sample of visitors to each park, asking them the distance from their home to the park. (The sample data is assumed to be from non-normal populations). What test can be used to determine if park visitors travel different distances to the six neighborhood parks?

151. A university student in Toronto, Ontario, is conducting research about attitudes toward a newly proposed rent control initiative. He knows that about 58% of Toronto residents live in rental units, and he wants to ensure that about 58% of the respondents to his survey live in rental units. What type of sampling design should he utilize?

152. A university student in Atlanta, Georgia, wants to write a course research paper on the distribution of poverty in the metro area. She collects poverty rate data by census tract, and wants to determine if tracts with high poverty levels are located near other tracts of high poverty. What kind of test will she perform for her research paper?

22.2 Answers to Geographic Problems and Policy Situations

1. systematic sample
2. two-sample difference test–Wilcoxon/Mann-Whitney
3. Poisson probability
4. Pearson's correlation coefficient
5. quadrat analysis
6. Poisson probability
7. variogram
8. cluster sample
9. analysis of variance
10. Pearson's correlation coefficient
11. confidence interval
12. random sample
13. contingency analysis
14. nearest neighbor analysis
15. one-sample difference of means–Z
16. standard distance
17. Kolmogorov-Smirnov goodness-of-fit, normal
18. matched-pairs t
19. Pearson's correlation coefficient
20. stratified disproportional
21. simple linear regression
22. equal intervals based on range
23. one-sample difference of means–t
24. scatterplot
25. kurtosis
26. Poisson probability
27. contingency (crosstab) map, conditional coverage
28. simple linear regression
29. standard score
30. two-sample difference test–Wilcoxon/Mann-Whitney
31. matched-pairs t
32. two-sample difference of proportions
33. contingency analysis
34. probability map
35. coefficient of variation
36. geometric probability
37. systematic sample
38. join count analysis
39. variogram
40. multiple regression
41. Kolmogorov-Smirnov goodness-of-fit (normal)
42. Pearson's correlation coefficient
43. Moran's I–local
44. two-sample difference of proportions
45. join count analysis
46. simple linear regression
47. Poisson probability
48. two-sample difference of proportions
49. two-sample difference of means–Z
50. skew
51. dendrogram
52. chi-square or Pearson's correlation coefficient
53. standard deviation
54. quadrat analysis
55. weighted mean center
56. analysis of variance
57. chi-square goodness-of-fit–uniform
58. Kruskal-Wallis analysis of variance
59. two-sample difference test–Wilcoxon/Mann-Whitney
60. natural breaks—either single linkage or Jenks
61. Pearson's correlation coefficient
62. simple linear regression
63. geometric probability
64. Spearman's rank correlation coefficient
65. matched-pairs t
66. two-sample difference of proportions
67. one-sample difference of means–or t
68. matched-pairs t
69. join count analysis
70. quadrat analysis
71. two-sample difference of means–t
72. simple linear regression

73. two-sample difference of proportions
74. Spearman's rank correlation coefficient
75. Poisson probability
76. contingency analysis
77. multiple regression
78. boxplot
79. matched-pairs *t*
80. quadrat analysis
81. binomial probability
82. cluster sampling
83. two-sample difference of means
84. join count analysis
85. Kruskal-Wallis analysis of variance
86. two-sample difference of proportions
87. Spearman's rank correlation coefficient
88. chi-square goodness-of-fit uniform
89. matched-pairs *t*
90. multiple regression
91. Pearson's correlation coefficient
92. simple linear regression
93. chi-square goodness-of-fit proportional
94. quantile breaks
95. two-sample difference of proportion
96. two-sample difference test–Wilcoxon/Mann-Whitney
97. Manhattan or network
98. Moran's *I*
99. Kruskal-Wallis analysis of variance
100. one-sample difference of means
101. matched-pairs *t*
102. Jenks' method of natural breaks
103. linear directional mean
104. one-sample difference of means–*Z* or *t*
105. contingency analysis
106. one-sample difference of means–*Z* or *t*
107. one-sample difference of proportions
108. coefficient of variation
109. analysis of variance
110. Kruskal-Wallis analysis of variance
111. Kolmogorov-Smirnov goodness-of-fit, normal
112. coefficient of variation
113. one-sample difference of proportions
114. Spearman's correlation coefficient
115. two-sample difference of means
116. contingency analysis
117. join count analysis
118. probability map
119. analysis of variance
120. two-sample difference of means–*t*
121. contingency analysis
122. one-sample difference of means–*t*
123. contingency analysis
124. analysis of variance
125. disproportional stratified point sample
126. mean center and standard deviational ellipse
127. one-sample difference of proportion
128. binomial probability
129. multiple regression
130. binomial probability
131. multiple regression
132. nearest neighbor analysis
133. one-sample difference of means–*Z*
134. one-sample difference of proportions
135. two-sample difference of means–*Z* or *t*
136. chi-square goodness-of-fit uniform
137. one-sample difference of proportions
138. Kolmogorov-Smirnov goodness-of-fit, normal
139. join count analysis
140. equal interval not based on range
141. two-sample difference of means
142. Moran's *I*
143. probability map
144. simple linear regression
145. contingency analysis
146. chi-square goodness-of-fit uniform
147. two-sample difference of proportions
148. two-sample difference of means
149. matched-pairs *t*
150. Kruskal-Wallis analysis of variance
151. stratified proportional sample
152. Moran's *I*

Appendix

Statistical Tables

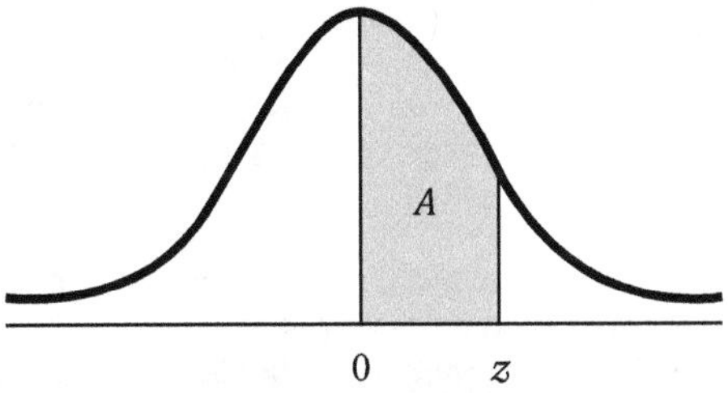

TABLE A

The Normal Table

Note: To get *A* for a given value of *Z*, insert a decimal point before the four digits. For example, $Z = 1.43$ gives $A = 0.4236$.

z	0.00	0.01	0.02	0.03	0.04	0.05	0.06	0.07	0.08	0.09
0.0	0000	0040	0080	0120	0160	0199	0239	0279	0319	0359
0.1	0398	0438	0478	0517	0557	0596	0636	0675	0714	0753
0.2	0793	0832	0871	0910	0948	0987	1026	1064	1103	1141
0.3	1179	1217	1255	1293	1331	1368	1406	1443	1480	1517
0.4	1554	1591	1628	1664	1700	1736	1772	1808	1844	1879
0.5	1915	1950	1985	2019	2054	2088	2123	2157	2190	2224
0.6	2257	2291	2324	2357	2389	2422	2454	2486	2517	2549
0.7	2580	2611	2642	2673	2704	2734	2764	2794	2823	2852
0.8	2881	2910	2939	2967	2995	3023	3051	3078	3106	3133
0.9	3159	3186	3212	3238	3264	3289	3315	3340	3365	3389
1.0	3413	3438	3461	3485	3508	3531	3554	3577	3599	3621
1.1	3643	3665	3686	3708	3729	3749	3770	3790	3810	3830
1.2	3849	3869	3888	3907	3925	3944	3962	3980	3997	4015
1.3	4032	4049	4066	4082	4099	4115	4131	4147	4162	4177
1.4	4192	4207	4222	4236	4251	4265	4279	4292	4306	4319
1.5	4332	4345	4357	4370	4382	4394	4406	4418	4429	4441
1.6	4452	4463	4474	4484	4495	4505	4515	4525	4535	4545
1.7	4554	4564	4573	4582	4591	4599	4608	4616	4625	4633
1.8	4641	4649	4656	4664	4671	4678	4686	4692	4699	4706
1.9	4713	4719	4726	4732	4738	4744	4750	4756	4761	4767
2.0	4772	4778	4783	4788	4793	4798	4803	4808	4812	4817
2.1	4821	4826	4830	4834	4838	4842	4846	4850	4854	4857
2.2	4861	4864	4868	4871	4875	4878	4881	4884	4887	4890
2.3	4893	4896	4898	4901	4904	4906	4909	4911	4913	4916
2.4	4918	4920	4922	4925	4927	4929	4931	4932	4934	4936
2.5	4938	4940	4941	4943	4945	4946	4948	4949	4951	4952
2.6	4953	4955	4956	4957	4959	4960	4961	4962	4963	4964
2.7	4965	4966	4967	4968	4969	4970	4971	4972	4973	4974
2.8	4974	4975	4976	4977	4977	4978	4979	4979	4980	4981
2.9	4981	4982	4982	4983	4984	4984	4985	4985	4986	4986
3.0	4987	4987	4987	4988	4988	4989	4989	4989	4990	4990
3.1	4990	4991	4991	4991	4992	4992	4992	4992	4993	4993
3.2	4993	4993	4994	4994	4994	4994	4994	4995	4995	4995
3.3	4995	4995	4996	4996	4996	4996	4996	4996	4996	4997
3.4	4997	4997	4997	4997	4997	4997	4997	4997	4998	4998
3.5	4998	4998	4998	4998	4998	4998	4998	4998	4998	4998

The area, *A*, stays at 0.4998 until $Z = 3.62$. From $Z = 3.63$ to 3.90 $A = 0.4999$. For $Z > 3.90$, $A = 0.5000$, to four decimal places.
From Leon F. Marzillier, *Elementary Statistics* ©1990 by Wm. C. Brown Publishers.

TABLE B

Table of Random Numbers

31871	60770	59235	41702	89372	28600	30013	18266	65044	61045
87134	32839	17850	37359	27221	92409	94778	17902	09467	86757
06728	16314	81076	42172	46446	09226	96262	77674	70205	98137
95646	67486	05167	07819	79918	83949	45605	18915	79458	54009
44085	87246	47378	98338	40368	02240	72593	52823	79002	88190
83967	84810	51612	81501	10440	48553	67919	73678	83149	47096
49990	02051	64575	70323	07863	59220	01746	94213	82977	42384
65332	16488	04433	37990	93517	18395	72848	97025	38894	46611
42309	04063	55291	72165	96921	53350	34173	39908	11634	87145
84715	41808	12085	72525	91171	09779	07223	75577	20934	92047
63919	83977	72416	55450	47642	01013	17560	54189	73523	33681
97595	78300	93502	25847	19520	16896	69282	16917	04194	25797
17116	42649	89252	61052	78332	15102	47707	28369	60400	15908
34037	84573	49914	59688	18584	53498	94905	14914	23261	58133
08813	14453	70437	49093	69880	99944	40482	04254	62842	68089
67115	41050	65453	04510	35518	88843	15801	86163	49913	46849
14596	62802	33009	74095	34549	76634	64270	67491	83941	55154
70258	26948	60863	47666	58512	91404	97357	85710	03414	56591
83369	81179	32429	34781	00006	65951	40254	71102	60416	43296
83811	49358	75171	34768	70070	76550	14252	97378	79500	97123
14924	71607	74638	01939	77044	18277	68229	09310	63258	85064
60102	56587	29842	12031	00794	90638	21862	72154	19880	80895
33393	30109	42005	47977	26453	15333	45390	89862	70351	36953
92592	78232	19328	29645	69836	91169	95180	15046	45679	94500
27421	73356	53897	26916	52015	26854	42833	64257	49423	39440
26528	22550	36692	25262	61419	53986	73898	80237	71387	32532
07664	10752	95021	17030	76784	86861	12780	44379	31261	18424
37954	72029	29624	09119	13444	22645	78345	79876	37582	75549
66495	11333	81101	69328	84838	76395	35997	07259	66254	47451
72506	28524	39595	49356	92733	42951	47774	75462	64409	69116
09713	70270	28077	15634	36525	91204	48443	50561	92394	60636
51852	70782	93498	44669	79647	06321	04020	00111	24737	05521
31460	22222	18801	00675	57562	97923	45974	75158	94918	40144
14328	05024	04333	04135	53143	79207	85863	04962	89549	63308
84002	98073	52998	05749	45538	26164	68672	97486	32341	99419
89541	28345	22887	79269	55620	68269	88765	72464	11586	52211
50502	39890	81465	00449	09931	12667	30278	63963	84192	25266
30862	61996	73216	12554	01200	63234	41277	20477	71899	05347
36735	58841	35287	51112	47322	81354	51080	72771	53653	42108
11561	81204	68175	93037	47967	74085	05905	86471	47671	18456

From Leon F. Marzillier, *Elementary Statistics* ©1990 by Wm. C. Brown Publishers.

TABLE C

Student's *t* Distribution

	Degrees of freedom															
t	1	2	3	4	5	6	7	8	9	10	11	12	13	14	15	16
0.1	0317	0353	0367	0374	0379	0382	0384	0386	0387	0388	0389	0390	0391	0391	0392	0392
0.2	0628	0700	0729	0744	0753	0760	0764	0768	0770	0773	0774	0776	0777	0778	0779	0780
0.3	0928	1038	1081	1104	1119	1129	1136	1141	1145	1148	1151	1153	1155	1157	1159	1160
0.4	1211	1361	1420	1452	1472	1485	1495	1502	1508	1512	1516	1519	1522	1524	1526	1528
0.5	1476	1667	1743	1783	1809	1826	1838	1847	1855	1861	1865	1869	1873	1876	1878	1881
0.6	1720	1953	2046	2096	2127	2148	2163	2174	2183	2191	2197	2202	2206	2210	2213	2215
0.7	1944	2218	2328	2387	2424	2449	2467	2481	2492	2501	2508	2514	2519	2523	2527	2530
0.8	2148	2462	2589	2657	2700	2729	2750	2766	2778	2788	2797	2804	2810	2815	2819	2823
0.9	2333	2684	2828	2905	2953	2986	3010	3028	3042	3054	3063	3071	3078	3083	3088	3093
1.0	2500	2887	3045	3130	3184	3220	3247	3267	3283	3296	3306	3315	3322	3329	3334	3339
1.1	2651	3070	3242	3335	3393	3433	3461	3483	3501	3514	3526	3535	3544	3551	3557	3562
1.2	2789	3235	3419	3518	3581	3623	3654	3678	3696	3711	3723	3734	3742	3750	3756	3762
1.3	2913	3384	3578	3683	3748	3793	3826	3851	3870	3886	3899	3910	3919	3927	3934	3940
1.4	3026	3518	3720	3829	3898	3945	3979	4005	4025	4041	4055	4066	4075	4084	4091	4097
1.5	3128	3638	3847	3960	4030	4079	4114	4140	4161	4177	4191	4203	4212	4221	4228	4235
1.6	3222	3746	3960	4075	4148	4196	4232	4259	4280	4297	4310	4322	4332	4340	4348	4354
1.7	3307	3844	4062	4178	4251	4300	4335	4362	4383	4400	4414	4426	4435	4444	4451	4458
1.8	3386	3932	4152	4269	4341	4390	4426	4452	4473	4490	4503	4515	4525	4533	4540	4546
1.9	3458	4026	4232	4349	4421	4469	4504	4530	4551	4567	4580	4591	4601	4609	4616	4622
2.0	3524	4082	4303	4419	4490	4538	4572	4597	4617	4633	4646	4657	4666	4674	4680	4686
2.1	3585	4147	4367	4482	4551	4598	4631	4655	4674	4690	4702	4712	4721	4728	4735	4740
2.2	3642	4206	4424	4537	4605	4649	4681	4705	4723	4738	4750	4759	4768	4774	4781	4786
2.3	3695	4259	4475	4585	4651	4694	4725	4748	4765	4779	4790	4799	4807	4813	4819	4824
2.4	3743	4308	4521	4628	4692	4734	4763	4784	4801	4813	4824	4832	4840	4846	4851	4855
2.5	3789	4352	4561	4666	4728	4767	4795	4815	4831	4843	4852	4860	4867	4873	4877	4882
2.6	3831	4392	4598	4700	4759	4797	4823	4842	4856	4868	4877	4884	4890	4895	4900	4903
2.7	3871	4429	4631	4730	4786	4822	4847	4865	4878	4888	4897	4903	4909	4914	4918	4921
2.8	3908	4463	4661	4756	4810	4844	4867	4884	4896	4906	4914	4920	4925	4929	4933	4936
2.9	3943	4494	4687	4779	4831	4863	4885	4901	4912	4921	4928	4933	4938	4942	4945	4948
3.0	3976	4523	4712	4800	4850	4880	4900	4915	4925	4933	4940	4945	4949	4952	4955	4958
3.1	4007	4549	4734	4819	4866	4894	4913	4927	4936	4944	4949	4954	4958			
3.2	4036	4573	4753	4835	4880	4907	4925	4937	4946	4953	4958					
3.3	4063	4596	4771	4850	4893	4918	4934	4946	4954							
3.4	4089	4617	4788	4864	4904	4928	4943	4953								
3.5	4114	4636	4803	4876	4914	4936	4950									
3.6	4138	4654	4816	4886	4922	4943										
3.7	4160	4670	4829	4896	4930	4950										
3.8	4181	4686	4840	4904	4937											
3.9	4201	4701	4850	4912	4943											
4.0	4220	4714	4860	4919	4948											
4.1	4239	4727	4869	4926	4953											
4.2	4256	4739	4877	4932												
4.3	4273	4750	4884	4937												
4.4	4289	4760	4891	4942												
4.5	4304	4770	4898	4946												
4.6	4319	4779	4903	4950												
4.7	4333	4788	4909													
4.8	4346	4796	4914													
4.9	4359	4804	4919													
5.0	4372	4811	4923													
5.8	4456	4858	4949													
6.4	4507	4882														
7.0	4548	4901														
10.0	4683	4951														
12.8	4752															
31.9	4900															
63.7	4950															

TABLE C

(continued)

	Degrees of freedom														
t	**17**	**18**	**19**	**20**	**21**	**22**	**23**	**24**	**25**	**26**	**27**	**28**	**29**	**30**	**∞**
0.1	0392	0393	0393	0393	0394	0394	0394	0394	0394	0394	0395	0395	0395	0395	0398
0.2	0781	0781	0782	0782	0783	0783	0784	0784	0785	0785	0785	0785	0786	0786	0793
0.3	0928	1162	1163	1164	1164	1165	1166	1166	1167	1167	1168	1168	1168	1169	1179
0.4	1529	1531	1532	1533	1534	1535	1536	1537	1537	1538	1538	1539	1540	1540	1554
0.5	1883	1884	1886	1887	1889	1890	1891	1892	1893	1894	1894	1895	1896	1896	1915
0.6	2218	2220	2222	2224	2225	2227	2228	2229	2230	2231	2232	2233	2234	2235	2257
0.7	2533	2536	2538	2540	2542	2544	2545	2547	2548	2549	2550	2551	2552	2553	2580
0.8	2826	2829	2832	2834	2837	2839	2841	2842	2844	2845	2847	2848	2849	2850	2881
0.9	3097	3100	3103	3106	3108	3111	3113	3115	3116	3118	3120	3121	3122	3124	3159
1.0	3343	3347	3351	3354	3357	3359	3361	3364	3366	3367	3369	3371	3372	3373	3413
1.1	3567	3571	3575	3578	3581	3584	3586	3589	3591	3593	3595	3597	3598	3600	3643
1.2	3767	3772	3776	3779	3782	3785	3788	3791	3793	3795	3797	3799	3801	3802	3849
1.3	3945	3950	3954	3958	3962	3965	3968	3970	3973	3975	3977	3979	3981	3982	4032
1.4	4103	4107	4112	4116	4119	4123	4126	4128	4131	4133	4136	4138	4139	4141	4192
1.5	4240	4245	4250	4254	4258	4261	4264	4267	4269	4272	4274	4276	4278	4280	4332
1.6	4360	4365	4370	4374	4377	4381	4384	4387	4389	4392	4394	4396	4398	4400	4452
1.7	4463	4468	4473	4477	4481	4484	4487	4490	4492	4495	4497	4499	4501	4503	4554
1.8	4552	4557	4561	4565	4569	4572	4575	4578	4580	4583	4585	4587	4589	4590	4641
1.9	4627	4632	4636	4640	4644	4647	4650	4652	4655	4657	4659	4661	4663	4665	4713
2.0	4691	4696	4700	4704	4707	4710	4713	4715	4718	4720	4722	4724	4725	4727	4772
2.1	4745	4750	4753	4757	4760	4763	4766	4768	4770	4772	4774	4776	4777	4779	4821
2.2	4790	4794	4798	4801	4804	4807	4809	4812	4814	4816	4817	4819	4820	4822	4861
2.3	4828	4832	4835	4838	4841	4843	4846	4848	4850	4851	4853	4854	4856	4857	4893
2.4	4859	4863	4866	4869	4871	4874	4876	4877	4879	4881	4882	4884	4885	4886	4918
2.5	4885	4888	4891	4894	4896	4898	4900	4902	4903	4905	4906	4907	4908	4909	4938
2.6	4906	4910	4912	4914	4916	4918	4920	4921	4923	4924	4925	4926	4927	4928	4953
2.7	4924	4927	4929	4931	4933	4935	4936	4937	4939	4940	4941	4942	4943	4944	
2.8	4938	4941	4943	4945	4946	4948	4949	4950	4951	4952	4953	4954	4955	4956	
2.9	4950	4952	4954	4956	4957	4958	4960								

Note: The column headed by ∞ is for use when $df > 30$. It coincides with column 1 of table A.

For $t > 2.9$ and $df > 16$, $A > 0.4950$.

All missing entries in the table have $A > 0.4950$. The reason this value of A was chosen is that for $A > 0.495$, p-value < 0.005. Therefore, if the value you want for A is missing, you may assume the significance level is 0.005 or higher.

For $t > 5.0$, selected values for t are given for $df = 1, 2, 3$, to show when significance is reached at various significance levels.

For example, $t = 7.0$, $df = 2$ gives $A = 0.4901$, p-value $= 0.0099$ (< 0.01). Therefore, there is significance at the .01 level. Any value of t less than 7.0 would not be significant at the .01 level.

TABLE D

Values of *t* for Selected Confidence Levels

	Level of confidence									
df	**0.68**	**0.80**	**0.85**	**0.90**	**0.925**	**0.95**	**0.975**	**0.98**	**0.99**	**0.995**
1	1.82	3.08	4.17	6.31	8.45	12.7	25.5	31.8	63.7	12.7
2	1.31	1.89	2.28	2.92	3.44	4.30	6.21	6.97	9.92	14.1
3	1.19	1.64	1.92	2.35	2.68	3.18	4.18	4.54	5.84	7.45
4	1.13	1.53	1.78	2.13	2.39	2.78	3.50	3.75	4.60	5.60
5	1.10	1.48	1.70	2.02	2.24	2.57	3.16	3.36	4.03	4.77
6	1.08	1.44	1.65	1.94	2.15	2.45	2.97	3.14	3.71	4.35
7	1.07	1.41	1.62	1.89	2.09	2.36	2.84	3.00	3.50	4.03
8	1.06	1.40	1.59	1.86	2.05	2.31	2.75	2.90	3.36	3.83
9	1.05	1.38	1.57	1.83	2.01	2.26	2.69	2.82	3.25	3.69
10	1.05	1.37	1.55	1.81	1.99	2.23	2.63	2.76	3.17	3.58
11	1.04	1.36	1.55	1.80	1.97	2.20	2.59	2.72	3.11	3.50
12	1.04	1.36	1.54	1.78	1.95	2.18	2.56	2.68	3.05	3.43
13	1.03	1.35	1.53	1.77	1.94	2.16	2.53	2.65	3.01	3.37
14	1.03	1.35	1.52	1.76	1.92	2.14	2.51	2.62	2.98	3.33
15	1.03	1.34	1.52	1.75	1.91	2.13	2.49	2.60	2.95	3.29
16	1.03	1.34	1.51	1.75	1.90	2.12	2.47	2.58	2.92	3.25
17	1.02	1.33	1.51	1.74	1.90	2.11	2.46	2.57	2.90	3.22
18	1.02	1.33	1.50	1.73	1.89	2.10	2.45	2.55	2.88	3.20
19	1.02	1.33	1.50	1.73	1.88	2.09	2.43	2.54	2.86	3.17
20	1.02	1.33	1.50	1.72	1.88	2.09	2.42	2.53	2.85	3.15
21	1.02	1.32	1.50	1.72	1.87	2.08	2.41	2.52	2.83	3.14
22	1.02	1.32	1.49	1.72	1.87	2.07	2.41	2.51	2.82	3.12
23	1.02	1.32	1.49	1.71	1.86	2.07	2.40	2.50	2.81	3.10
24	1.02	1.32	1.49	1.71	1.86	2.06	2.39	2.49	2.80	3.09
25	1.01	1.32	1.49	1.71	1.86	2.06	2.38	2.49	2.79	3.08
26	1.01	1.31	1.48	1.71	1.85	2.06	2.38	2.48	2.78	3.07
27	1.01	1.31	1.48	1.70	1.85	2.05	2.37	2.47	2.77	3.06
28	1.01	1.31	1.48	1.70	1.85	2.05	2.37	2.47	2.76	3.05
29	1.01	1.31	1.48	1.70	1.85	2.05	2.36	2.46	2.76	3.04
30	1.01	1.31	1.48	1.70	1.84	2.04	2.36	2.46	2.75	3.03
∞	1.00	1.28	1.44	1.64	1.78	1.96	2.24	2.33	2.58	2.81

If you want to construct a 0.95 confidence interval estimate with $df = 15$, use $t = 2.13$. Use the last line when $df > 30$. The values in it coincide with the normal table values.

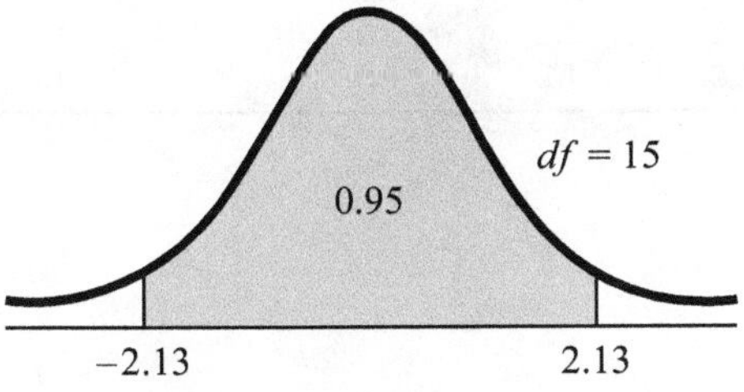

TABLE E

Critical Values of the *F* Distribution

$\alpha = .05$

df_2 \ df_1	1	2	3	4	5	6	7	8	9	10	11	12
1	161	200	216	225	230	234	237	239	241	242	243	244
2	18.5	19.0	19.2	19.2	19.3	19.3	19.4	19.4	19.4	19.4	19.4	19.4
3	10.1	9.6	9.3	9.1	9.0	8.9	8.9	8.8	8.8	8.8	8.8	8.7
4	7.7	6.9	6.6	6.4	6.3	6.2	6.1	6.0	6.0	6.0	5.9	5.9
5	6.6	5.8	5.4	5.2	5.1	5.0	4.9	4.8	4.8	4.7	4.7	4.7
6	6.0	5.1	4.8	4.5	4.4	4.3	4.2	4.2	4.1	4.1	4.0	4.0
7	5.6	4.7	4.4	4.1	4.0	3.9	3.8	3.7	3.7	3.6	3.6	3.6
8	5.3	4.5	4.1	3.8	3.7	3.6	3.5	3.4	3.4	3.4	3.3	3.3
9	5.1	4.3	3.9	3.6	3.5	3.4	3.3	3.2	3.2	3.1	3.1	3.1
10	5.0	4.1	3.7	3.5	3.3	3.2	3.1	3.1	3.0	3.0	2.9	2.9
11	4.8	4.0	3.6	3.4	3.2	3.1	3.0	3.0	2.9	2.8	2.8	2.8
12	4.8	3.9	3.5	3.3	3.1	3.0	2.9	2.8	2.8	2.8	2.7	2.7
13	4.7	3.8	3.4	3.2	3.0	2.9	2.8	2.8	2.7	2.7	2.6	2.6
14	4.6	3.7	3.3	3.1	3.0	2.8	2.8	2.7	2.6	2.6	2.6	2.5
15	4.5	3.7	3.3	3.1	2.9	2.8	2.7	2.6	2.6	2.6	2.5	2.5
16	4.5	3.6	3.2	3.0	2.8	2.7	2.7	2.6	2.5	2.5	2.4	2.4
17	4.4	3.6	3.2	3.0	2.8	2.7	2.6	2.6	2.5	2.4	2.4	2.4
18	4.4	3.6	3.2	2.9	2.8	2.7	2.6	2.5	2.5	2.4	2.4	2.3
19	4.4	3.5	3.1	2.9	2.7	2.6	2.5	2.5	2.4	2.4	2.3	2.3
20	4.4	3.5	3.1	2.9	2.7	2.6	2.5	2.4	2.4	2.4	2.3	2.3
21	4.3	3.5	3.1	2.8	2.7	2.6	2.5	2.4	2.4	2.3	2.3	2.2
22	4.3	3.4	3.0	2.8	2.7	2.6	2.5	2.4	2.3	2.3	2.3	2.2
23	4.3	3.4	3.0	2.8	2.6	2.5	2.4	2.4	2.3	2.3	2.2	2.2
24	4.3	3.4	3.0	2.8	2.6	2.5	2.4	2.4	2.3	2.3	2.2	2.2
25	4.2	3.4	3.0	2.8	2.6	2.5	2.4	2.3	2.3	2.2	2.2	2.2
26	4.2	3.4	3.0	2.7	2.6	2.5	2.4	2.3	2.3	2.2	2.2	2.2
27	4.2	3.4	3.0	2.7	2.6	2.5	2.4	2.3	2.2	2.2	2.2	2.1
28	4.2	3.3	3.0	2.7	2.6	2.4	2.4	2.3	2.2	2.2	2.2	2.1
29	4.2	3.3	2.9	2.7	2.6	2.4	2.4	2.3	2.2	2.2	2.1	2.1
30	4.2	3.3	2.9	2.7	2.5	2.4	2.3	2.3	2.2	2.2	2.1	2.1
40	4.1	3.3	2.9	2.7	2.5	2.4	2.3	2.3	2.2	2.1	2.0	2.0
50	4.0	3.2	2.8	2.6	2.4	2.3	2.2	2.1	2.1	2.0	2.0	2.0
60	4.0	3.2	2.8	2.5	2.4	2.2	2.2	2.1	2.0	2.0	2.0	1.9
70	4.0	3.1	2.7	2.5	2.4	2.2	2.1	2.1	2.0	2.0	1.9	1.9
80	4.0	3.1	2.7	2.5	2.3	2.2	2.1	2.0	2.0	2.0	1.9	1.9
100	3.9	3.1	2.7	2.5	2.3	2.2	2.1	2.0	2.0	1.9	1.9	1.8
120	3.9	3.1	2.7	2.4	2.3	2.2	2.1	2.0	2.0	1.9	1.9	1.8
∞	3.8	3.0	2.6	2.4	2.2	2.1	2.0	1.9	1.9	1.8	1.8	1.8

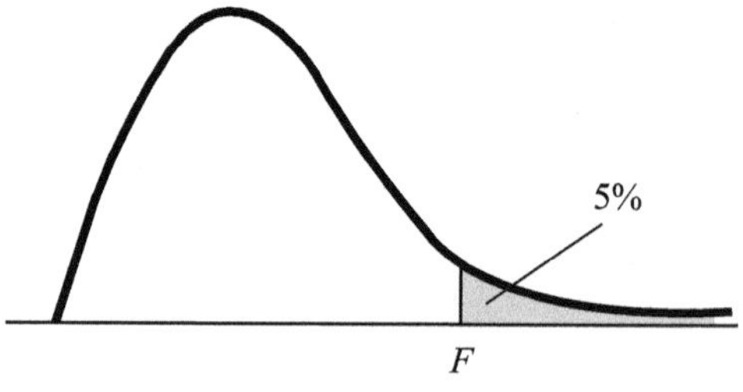

TABLE E

(continued)

α=.05

df_1

df_2	15	20	24	30	40	50	60	75	100	120	∞
1	246	248	249	250	251	252	253	253	253	253	254
2	19.4	19.4	19.4	19.4	19.4	19.5	19.5	19.5	19.5	19.5	19.5
3	8.7	8.7	8.6	8.6	8.6	8.6	8.6	8.6	8.6	8.6	8.5
4	5.9	5.8	5.8	5.8	5.7	5.7	5.7	5.7	5.7	5.7	5.6
5	4.6	4.6	4.5	4.5	4.5	4.4	4.4	4.4	4.4	4.4	4.4
6	3.9	3.9	3.8	3.8	3.8	3.8	3.7	3.7	3.7	3.7	3.7
7	3.5	3.4	3.4	3.4	3.3	3.3	3.3	3.3	3.3	3.3	3.2
8	3.2	3.2	3.1	3.1	3.0	3.0	3.0	3.0	3.0	3.0	2.9
9	3.0	2.9	2.9	2.9	2.8	2.8	2.8	2.8	2.8	2.8	2.7
10	2.8	2.8	2.7	2.7	2.7	2.6	2.6	2.6	2.6	2.6	2.5
11	2.7	2.6	2.6	2.6	2.5	2.5	2.5	2.5	2.4	2.4	2.4
12	2.6	2.5	2.5	2.5	2.4	2.4	2.4	2.4	2.4	2.3	2.3
13	2.5	2.5	2.4	2.4	2.3	2.3	2.3	2.3	2.3	2.2	2.2
14	2.5	2.4	2.4	2.3	2.3	2.2	2.2	2.2	2.2	2.2	2.1
15	2.4	2.3	2.3	2.2	2.2	2.2	2.2	2.2	2.1	2.1	2.1
16	2.4	2.3	2.2	2.2	2.2	2.1	2.1	2.1	2.1	2.1	2.0
17	2.3	2.2	2.2	2.2	2.1	2.1	2.1	2.0	2.0	2.0	2.0
18	2.3	2.2	2.2	2.1	2.1	2.0	2.0	2.0	2.0	2.0	1.9
19	2.2	2.2	2.1	2.1	2.0	2.0	2.0	2.0	1.9	1.9	1.9
20	2.2	2.1	2.1	2.0	2.0	2.0	2.0	1.9	1.9	1.9	1.8
21	2.2	2.1	2.1	2.0	2.0	1.9	1.9	1.9	1.9	1.9	1.8
22	2.2	2.1	2.1	2.0	2.0	1.9	1.9	1.9	1.8	1.8	1.8
23	2.1	2.0	2.0	2.0	1.9	1.9	1.9	1.8	1.8	1.8	1.8
24	2.1	2.0	2.0	1.9	1.9	1.9	1.8	1.8	1.8	1.8	1.7
25	2.1	2.0	2.0	1.9	1.9	1.8	1.8	1.8	1.8	1.8	1.7
26	2.1	2.0	2.0	1.9	1.8	1.8	1.8	1.8	1.8	1.8	1.7
27	2.1	2.0	1.9	1.9	1.8	1.8	1.8	1.8	1.7	1.7	1.7
28	2.0	2.0	1.9	1.9	1.8	1.8	1.8	1.8	1.7	1.7	1.6
29	2.0	1.9	1.9	1.8	1.8	1.8	1.8	1.7	1.7	1.7	1.6
30	2.0	1.9	1.9	1.8	1.8	1.8	1.7	1.7	1.7	1.7	1.6
40	1.9	1.8	1.8	1.7	1.7	1.7	1.6	1.6	1.6	1.6	1.5
50	1.9	1.8	1.7	1.7	1.6	1.6	1.6	1.6	1.5	1.5	1.4
60	1.8	1.8	1.7	1.6	1.6	1.6	1.5	I.S	1.5	1.5	1.4
70	1.8	1.7	1.7	1.6	1.6	1.5	1.5	1.5	1.4	1.4	1.4
80	1.8	1.7	1.6	1.6	1.5	1.5	1.5	1.4	1.4	1.4	1.3
100	1.8	1.7	1.6	1.6	1.5	1.5	1.4	1.4	1.4	1.4	1.3
120	1.8	1.7	1.6	1.6	1.5	1.5	1.4	1.4	1.4	1.3	1.2
∞	1.7	1.6	1.5	1.5	1.4	1.4	1.3	1.3	1.2	1.2	1.0

From Leon F. Marzillier, *Elementary Statistics* ©1990 by Wm. C. Brown Publishers.

TABLE E

(continued)

$\alpha = 0.01$

df_2 \ df_1	1	2	3	4	5	6	7	8	9	10	11	12
1	4052	4999	5403	5625	5764	5859	5928	5981	6022	6056	6082	6106
2	98.5	99.0	99.2	99.2	99.3	99.3	99.3	99.4	99.4	99.4	99.4	99.4
3	34.1	30.8	29.5	28.7	28.2	27.9	27.7	27.5	27.3	27.2	27.1	27.0
4	21.2	18.0	16.7	16.0	15.5	15.2	15.0	14.8	14.7	14.5	14.4	14.4
5	16.3	13.3	12.1	11.4	11.0	10.7	10.4	10.3	10.2	10.0	10.0	9.9
6	13.7	10.9	9.8	9.2	8.8	8.5	8.3	8.1	8.0	7.9	7.8	7.7
7	12.2	9.6	8.4	7.8	7.5	7.2	7.0	6.8	6.7	6.6	6.5	6.5
8	11.3	8.6	7.6	7.0	6.6	6.4	6.2	6.0	5.9	5.8	5.7	5.7
9	10.6	8.0	7.0	6.4	6.1	5.8	5.6	5.5	5.4	5.3	5.2	5.1
10	10.0	7.6	6.6	6.0	5.6	5.4	5.2	5.1	5.0	4.8	4.8	4.7
11	9.6	7.2	6.2	5.7	5.3	5.1	4.9	4.7	4.6	4.5	4.5	4.4
12	9.3	6.9	6.0	5.4	5.1	4.8	4.6	4.5	4.4	4.3	4.2	4.2
13	9.1	6.7	5.7	5.2	4.9	4.6	4.4	4.3	4.2	4.1	4.0	4.0
14	8.9	6.5	5.6	5.0	4.7	4.5	4.3	4.1	4.0	3.9	3.9	3.8
15	8.7	6.4	5.4	4.9	4.6	4.3	4.1	4.0	3.9	3.8	3.7	3.7
16	8.5	6.2	5.3	4.8	4.4	4.2	4.0	3.9	3.8	3.7	3.6	3.6
17	8.4	6.1	5.2	4.7	4.3	4.1	3.9	3.8	3.7	3.6	3.5	3.4
18	8.3	6.0	5.1	4.6	4.2	4.0	3.8	3.7	3.6	3.5	3.4	3.4
19	8.2	5.9	5.0	4.5	4.2	3.9	3.8	3.6	3.5	3.4	3.4	3.3
20	8.1	5.8	4.9	4.4	4.1	3.9	3.7	3.6	3.4	3.4	3.3	3.2
21	8.0	5.8	4.9	4.4	4.0	3.8	3.6	3.5	3.4	3.3	3.2	3.2
22	7.9	5.7	4.8	4.3	4.0	3.8	3.6	3.4	3.4	3.3	3.2	3.1
23	7.9	5.7	4.8	4.3	3.9	3.7	3.5	3.4	3.3	3.2	3.1	3.1
24	7.8	5.6	4.7	4.2	3.9	3.7	3.5	3.4	3.2	3.2	3.1	3.0
25	7.8	5.6	4.7	4.2	3.9	3.6	3.5	3.3	3.2	3.1	3.0	3.0
26	7.7	5.5	4.6	4.1	3.8	3.6	3.4	3.3	3.2	3.1	3.0	3.0
27	7.7	5.5	4.6	4.1	3.8	3.6	3.4	3.3	3.1	3.1	3.0	2.9
28	7.6	5.4	4.6	4.1	3.8	3.5	3.4	3.2	3.1	3.0	3.0	2.9
29	7.6	5.4	4.5	4.0	3.7	3.5	3.3	3.2	3.1	3.0	2.9	2.9
30	7.6	5.4	4.5	4.0	3.7	3.5	3.3	3.2	3.1	3.0	2.9	2.8
40	7.3	5.2	4.3	3.8	3.5	3.3	3.1	3.0	2.9	2.8	2.7	2.7
50	7.2	5.1	4.2	3.7	3.4	3.2	3.0	2.9	2.8	2.7	2.6	2.6
60	7.1	5.0	4.1	3.6	3.3	3.1	3.0	2.8	2.7	2.6	2.6	2.5
70	7.0	4.9	4.1	3.6	3.3	3.1	2.9	2.8	2.7	2.6	2.5	2.4
80	7.0	4.9	4.0	3.6	3.2	3.0	2.9	2.7	2.6	2.6	2.5	2.4
100	6.9	4.8	4.0	3.5	3.2	3.0	2.8	2.7	2.6	2.5	2.4	2.4
120	6.8	4.8	4.0	3.5	3.2	3.0	2.8	2.7	2.6	2.5	2.4	2.3
∞	6.6	4.6	3.8	3.3	3.0	2.8	2.6	2.5	2.4	2.3	2.2	2.2

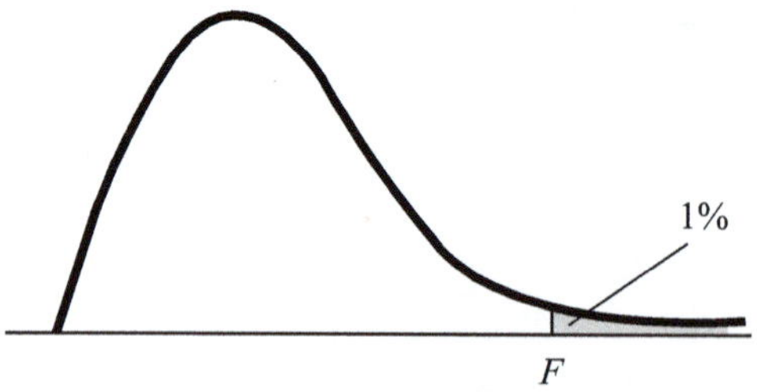

From Leon F. Marzillier, *Elementary Statistics* ©1990 by Wm. C. Brown Publishers.

TABLE E

(continued)

$\alpha = 0.01$

df_2 \ df_1	15	20	24	30	40	50	60	75	100	120	∞
1	6157	6209	6235	6261	6287	6302	6313	6323	6334	6339	6366
2	99.4	99.4	99.5	99.5	99.5	99.5	99.5	99.5	99.5	99.5	99.5
3	26.9	26.7	26.6	26.5	26.4	26.3	26.2	26.3	26.2	26.2	26.1
4	14.2	14.0	13.9	13.8	13.8	13.7	13.6	13.6	13.6	13.6	13.5
5	9.7	9.6	9.5	9.4	9.3	9.2	9.2	9.2	9.1	9.1	9.0
6	7.6	7.4	7.3	7.2	7.1	7.1	7.1	7.0	7.0	7.0	6.9
7	6.3	6.2	6.1	6.0	5.9	5.8	5.8	5.8	5.8	5.7	5.6
8	5.5	5.4	5.3	5.2	5.1	5.1	5.0	5.0	5.0	5.0	4.9
9	5.0	4.8	4.7	4.6	4.6	4.5	4.4	4.4	4.4	4.4	4.3
10	4.6	4.4	4.3	4.2	4.2	4.1	4.1	4.0	4.0	4.0	3.9
11	4.2	4.1	4.0	3.9	3.9	3.8	3.8	3.7	3.7	3.7	3.6
12	4.0	3.9	3.8	3.7	3.6	3.6	3.5	3.5	3.5	3.4	3.4
13	3.8	3.7	3.6	3.5	3.4	3.4	3.3	3.3	3.3	3.2	3.2
14	3.7	3.5	3.4	3.4	3.3	3.2	3.2	3.1	3.1	3.1	3.0
15	3.5	3.4	3.3	3.2	3.1	3.1	3.0	3.0	3.0	3.0	2.9
16	3.4	3.3	3.2	3.1	3.0	3.0	2.9	2.9	2.9	2.8	2.8
17	3.3	3.2	3.1	3.0	2.9	2.9	2.8	2.8	2.8	2.8	2.7
18	3.2	3.1	3.0	2.9	2.8	2.8	2.8	2.7	2.7	2.7	2.6
19	3.2	3.0	2.9	2.8	2.8	2.7	2.7	2.6	2.6	2.6	2.5
20	3.1	2.9	2.9	2.8	2.7	2.6	2.6	2.6	2.5	2.5	2.4
21	3.0	2.9	2.8	2.7	2.6	2.6	2.6	2.5	2.5	2.5	2.4
22	3.0	2.8	2.8	2.7	2.6	2.5	2.5	2.5	2.4	2.4	2.3
23	2.9	2.8	2.7	2.6	2.5	2.5	2.4	2.4	2.4	2.4	2.3
24	2.9	2.7	2.7	2.6	2.5	2.4	2.4	2.4	2.3	2.3	2.2
25	2.8	2.7	2.6	2.5	2.4	2.4	2.4	2.3	2.3	2.3	2.2
26	2.8	2.7	2.6	2.5	2.4	2.4	2.3	2.3	2.2	2.2	2.1
27	2.8	2.6	2.6	2.5	2.4	2.3	2.3	2.2	2.2	2.2	2.1
28	2.8	2.6	2.5	2.4	2.4	2.3	2.3	2.2	2.2	2.2	2.1
29	2.7	2.6	2.5	2.4	2.3	2.3	2.2	2.2	2.2	2.1	2.0
30	2.7	2.6	2.5	2.4	2.3	2.2	2.2	2.2	2.1	2.1	2.0
40	2.5	2.4	2.3	2.2	2.1	2.0	2.0	2.0	1.9	1.9	1.8
50	2.4	2.3	2.2	2.1	2.0	1.9	1.9	1.9	1.8	1.8	1.7
60	2.4	2.2	2.1	2.0	1.9	1.9	1.8	1.8	1.7	1.7	1.6
70	2.3	2.2	2.1	2.0	1.9	1.8	1.8	1.7	1.7	1.7	1.5
80	2.3	2.1	2.0	1.9	1.8	1.8	1.8	1.7	1.6	1.6	1.5
100	2.2	2.1	2.0	1.9	1.8	1.7	1.7	1.6	1.6	1.5	1.4
120	2.2	2.0	2.0	1.9	1.8	1.7	1.6	1.6	1.5	1.5	1.4
∞	2.0	1.5	1.8	1.7	1.6	1.5	1.5	1.4	1.4	1.3	1.0

TABLE F

***p*-values for χ^2**

χ^2	Degrees of freedom 2	3	4
3.2	2019		
3.3	1920		
3.4	1827		
3.5	1738		
3.6	1653		
3.7	1572		
3.8	1496		
3.9	1423		
4.0	1353		
4.1	1287		
4.2	1225		
4.3	1165		
4.4	1108		
4.5	1054		
4.6	1003	2035	
4.7	0954	1951	
4.8	0907	1870	
4.9	0863	1793	
5.0	0821	1718	
5.1	0781	1646	
5.2	0743	1577	
5.3	0707	1511	
5.4	0672	1447	
5.5	0639	1386	
5.6	0608	1328	
5.7	0578	1272	
5.8	0550	1218	
5.9	0523	1166	2067
6.0	0498	1116	1991
6.1	0474	1068	1918
6.2	0450	1023	1847
6.3	0429	0979	1778
6.4	0408	0937	1712
6.5	0388	0897	1648
6.6	0302	0858	1586
6.7	0351	0821	1526
6.8	0334	0786	1468
6.9	0317	0752	1413
7.0	0302	0719	1359
7.1	0287	0688	1307

χ^2	Degrees of freedom 2	3	4	5	6	7
7.2	0273	0658	1257	2062		
7.3	0260	0629	1209	1993		
7.4	0247	0602	1162	1926		
7.5	0235	0576	1117	1860		
7.6	0224	0550	1074	1797		
7.7	0213	0526	1032	1736		
7.8	0202	0503	0992	1676		
7.9	0193	0481	0953	1618		
8.0	0183	0460	0916	1562		
8.1	0174	0440	0880	1508		
8.2	0166	0421	0845	1456		
8.3	0158	0402	0812	1405		
8.4	0150	0384	0780	1355		
8.5	0143	0367	0749	1307	2037	
8.6	0136	0351	0719	1261	1974	
8.7	0129	0336	0691	1216	1912	
8.8	0123	0321	0663	1173	1851	
8.9	0117	0307	0636	1131	1793	
9.0	0111	0293	0611	1091	1736	
9.1	0106	0280	0586	1051	1680	
9.2	0101	0267	0563	1013	1626	
9.3	0096	0256	0540	0977	1574	
9.4	0091	0244	0518	0941	1523	
9.5	0087	0233	0497	0907	1473	
9.6	0082	0223	0477	0874	1425	
9.7	0078	0213	0456	0842	1379	
9.8	0074	0203	0439	0811	1333	2002
9.9	0071	0194	0421	0781	1289	1943
10	0067	0186	0404	0752	1247	1886

All missing entries on this page give *p*-values > 0.20.
See the next page for *p*-values corresponding to larger values of χ^2.

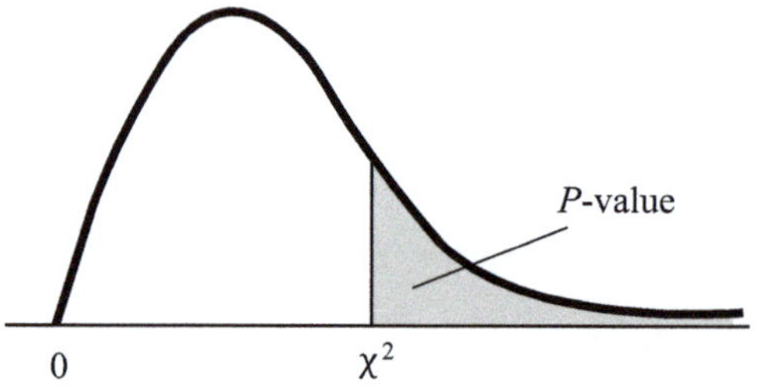

TABLE F

(continued)

	Degrees of freedom								
χ^2	**2**	**3**	**4**	**5**	**6**	**7**	**8**	**9**	**10**
11	0041	0117	0266	0514	0884	1386	2017		
12		0074	0174	0348	0620	1006	1512	2133	
13		0046	0113	0234	0430	0721	1118	1626	2237
14			0073	0156	0296	0512	0818	1223	1730
15			0047	0104	0203	0360	0591	0909	1321
16				0068	0138	0251	0424	0669	0996
17				0045	0093	0174	0301	0487	0744
18					0062	0120	0212	0352	0550
19					0042	0082	0149	0252	0403
20						0056	0103	0179	0293
21						0038	0071	0127	0211
22							0049	0089	0151
23								0062	0107
24								0043	0076
25									0053
26									0037

The missing entries in the top right-hand corner give p-value > 0.20. All other missing entries give p-value < 0.005.

Index